A SECOND COURSE IN
STOCHASTIC PROCESSES

A SECOND COURSE IN STOCHASTIC PROCESSES

SAMUEL KARLIN
STANFORD UNIVERSITY

HOWARD M. TAYLOR
CORNELL UNIVERSITY

Academic Press
An Imprint of Elsevier
San Diego New York Boston
London Sydney Tokyo Toronto

Academic Press
An Imprint of Elsevier
525 B Street, Suite 1900, San Diego, California 92101-4495, USA
http://www.academicpress.com

Academic Press
Harcourt Place, 32 Jamestown Road, London NW1 7BY, UK
http://www.academicpress.com

Library of Congress Cataloging in Publication Data

Karlin, Samuel, Date
    A second course in stochastic processes

    Continues A first course in stochastic processes.
    Includes bibliographical references and index.
    1. Stochastic processes. 1. Taylor, Howard M., joint author. II. Title.
QA274.K372   519.2   80-533

ISBN-13: 978-0-12-398650-4   ISBN-10: 0-12-398650-8

Printed and bound in the United Kingdom

Transferred to Digital Printing, 2011

CONTENTS

Chapter 12

SUMS OF INDEPENDENT RANDOM VARIABLES AS A MARKOV CHAIN

Chapter 13

ORDER STATISTICS, POISSON PROCESSES, AND APPLICATIONS

Chapter 14

CONTINUOUS TIME MARKOV CHAINS

Chapter 15

DIFFUSION PROCESSES

Chapter 16

COMPOUNDING STOCHASTIC PROCESSES

Chapter 17

FLUCTUATION THEORY OF PARTIAL SUMS OF INDEPENDENT IDENTICALLY DISTRIBUTED RANDOM VARIABLES

Chapter 18

QUEUEING PROCESSES

PREFACE

This *Second Course* continues the development of the theory and applications of stochastic processes as promised in the preface of *A First Course*. We emphasize a careful treatment of basic structures in stochastic processes in symbiosis with the analysis of natural classes of stochastic processes arising from the biological, physical, and social sciences.

Apart from expanding on the topics treated in the first edition of this work but not incorporated in *A First Course,* this volume presents an extensive introductory account of the fundamental concepts and methodology of diffusion processes and the closely allied theory of stochastic differential equations and stochastic integrals. A multitude of physical, engineering, biological, social, and managerial phenomena are either well approximated or reasonably modeled by diffusion processes; and modern approaches to diffusion processes and stochastic differential equations provide new perspectives and techniques impinging on many subfields of pure and applied mathematics, among them partial differential equations, dynamical systems, optimal control problems, statistical decision procedures, operations research, studies of economic systems, population genetics, and ecology models.

A new chapter discusses the elegant and far-reaching distributional formulas now available for a wide variety of functionals (e.g., first-passage time, maximum, order statistics, occupation time) of the process of sums of independent random variables. The identities, formulas, and results in this chapter have important applications in queueing and renewal theory, for statistical decision procedures, and elsewhere.

The logical dependence of the chapters in *A Second Course* is shown by the diagram below (consult also the preface to *A First Course* on the relationships of Chapters 1–9).

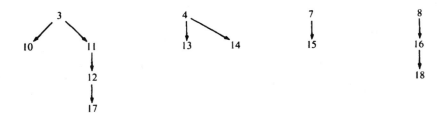

The book can be coupled to *A First Course* in several ways, depending on the background and interests of the students. The discussion of Markov chains in *A First Course* can be supplemented with parts of the more advanced Chapters 10–12, and 14. The material on fluctuation theory of sums of independent random variables (Chapters 12 and 17), perhaps supplemented by some parts of the chapter on queueing processes (Chapter 18), may be attractive and useful to students of operations research and statistics. Chapter 16, on compounding stochastic processes, is designed as an enticing introduction to a hierarchy of relevant models, including models of multiple species population growth, of migration and demographic structures, of point processes, and compositions of Poisson processes (Lévy processes).

We strongly recommend devoting a semester to diffusion processes (Chapter 15). The dependence relationships of the sections of Chapter 15 are diagrammed below. Section 1 provides a generalized description of various characterizations of diffusion. The examples of Section 2, which need not be absorbed in their totality, are intended to hint at the rich diversity of natural models of diffusion processes; the emphasis on biological examples reflects the authors' personal interests, but diffusion models abound in other sciences as well. Sections 3–5 point up the utility and tractability of diffusion process analysis. Section 6 takes up the boundary classification of one-dimensional diffusion processes; Section 7, on the same topic, is more technical. Section 8 provides constructions of diffusions with different types of boundary behavior. Sections 9 and 10 treat a number of topics motivated by problems of population genetics and statistics. The formal (general) theory of Markov processes with emphasis on applications to diffusions is elaborated in Sections 11 and 12. Section 13 exhibits the spectral representations for several classical diffusion models, which are of some interest because of their connections with classical special functions. The key concepts, a host of examples, and some methods of stochastic differential equations and stochastic integrals are introduced in Sections 14–16.

Chapter 15

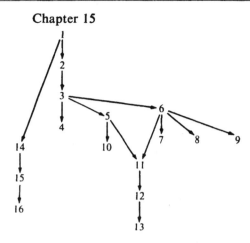

As noted in the earlier prefaces, we have drawn freely on the thriving literature of applied and theoretical stochastic processes without citing specific articles. A few representative books are listed at the end of each chapter and may be consulted profitably as a guide to more advanced material.

We express our gratitude to Stanford University, the Weizmann Institute of Science in Israel, and Cornell University for providing a rich intellectual environment and facilities indispensable for the writing of this text. The first author is grateful for the continuing grant support provided by the National Science Foundation and the National Institutes of Health that permitted an unencumbered concentration on a number of the concepts of this book and on its various drafts. We are also happy to acknowledge our indebtedness to many colleagues who have offered constructive criticisms, among them Professor M. Taqqu of Cornell, Dr. S. Tavaré of the University of Utah, Professors D. Iglehart and M. Harrison of Stanford, and Professor J. Kingman of Oxford. Finally, we thank our students P. Glynn, E. Cameron, J. Raper, R. Smith, L. Tierney, and P. Williams for their assistance in checking the problems, and for their helpful reactions to early versions of Chapter 15.

PREFACE TO *A FIRST COURSE*

The purposes, level, and style of this new edition conform to the tenets set forth in the original preface. We continue with our tack of developing simultaneously theory and applications, intertwined so that they refurbish and elucidate each other.

We have made three main kinds of changes. First, we have enlarged on the topics treated in the first edition. Second, we have added many exercises and problems at the end of each chapter. Third, and most important, we have supplied, in new chapters, broad introductory discussions of several classes of stochastic processes not dealt with in the first edition, notably martingales, renewal and fluctuation phenomena associated with random sums, stationary stochastic processes, and diffusion theory.

Martingale concepts and methodology have provided a far-reaching apparatus vital to the analysis of all kinds of functionals of stochastic processes. In particular, martingale constructions serve decisively in the investigation of stochastic models of diffusion type. Renewal phenomena are almost equally important in the engineering and managerial sciences especially with reference to examples in reliability, queueing, and inventory systems. We discuss renewal theory systematically in an extended chapter. Another new chapter explores the theory of stationary processes and its applications to certain classes of engineering and econometric problems. Still other new chapters develop the structure and use of diffusion processes for describing certain biological and physical systems and fluctuation properties of sums of independent random variables useful in the analyses of queueing systems and other facets of operations research.

The logical dependence of chapters is shown by the diagram below. Section 1 of Chapter 1 can be reviewed without worrying about details. Only Sections 5 and 7 of Chapter 7 depend on Chapter 6. Only Section 9 of Chapter 9 depends on Chapter 5.

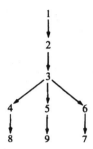

An easy one-semester course adapted to the junior-senior level could consist of Chapter 1, Sections 2 and 3 preceded by a cursory review of Section 1, Chapter 2 in its entirety, Chapter 3 excluding Sections 5 and/or 6, and Chapter 4 excluding Sections 3, 7, and 8. The content of the last part of the course is left to the discretion of the lecturer. An option of material from the early sections of any or all of Chapters 5–9 would be suitable.

The problems at the end of each chapter are divided into two groups: the first, more or less elementary; the second, more difficult and subtle.

The scope of the book is quite extensive, and on this account, it has been divided into two volumes. We view the first volume as embracing the main categories of stochastic processes underlying the theory and most relevant for applications. In *A Second Course* we introduce additional topics and applications and delve more deeply into some of the issues of *A First Course.* We have organized the edition to attract a wide spectrum of readers, including theorists and practitioners of stochastic analysis pertaining to the mathematical, engineering, physical, biological, social, and managerial sciences.

The second volume of this work, *A Second Course in Stochastic Processes,* will include the following chapters: (10) Algebraic Methods in Markov Chains; (11) Ratio Theorems of Transition Probabilities and Applications; (12) Sums of Independent Random Variables as a Markov Chain; (13) Order Statistics, Poisson Processes, and Applications; (14) Continuous Time Markov Chains; (15) Diffusion Processes; (16) Compounding Stochastic Processes; (17) Fluctuation Theory of Partial Sums of Independent Identically Distributed Random Variables; (18) Queueing Processes.

As noted in the first preface, we have drawn freely on the thriving literature of applied and theoretical stochastic processes. A few representative references are included at the end of each chapter; these may be profitably consulted for more advanced material.

We express our gratitude to the Weizmann Institute of Science, Stanford University, and Cornell University for providing a rich intellectual environment and facilities indispensable for the writing of this text. The first author is grateful for the continuing grant support provided by the Office of Naval Research that permitted an unencumbered concentration on a number of the concepts and drafts of this book. We are also happy to acknowledge our indebtedness to many colleagues who have offered a variety of constructive criticisms. Among others, these include Professors P. Brockwell of La Trobe, J. Kingman of Oxford, D. Iglehart and S. Ghurye of Stanford, and K. Itô and S. Stidham, Jr. of Cornell. We also thank our students M. Nedzela and C. Macken for their assistance in checking the problems and help in reading proofs.

SAMUEL KARLIN
HOWARD M. TAYLOR

PREFACE TO FIRST EDITION

Stochastic processes concern sequences of events governed by probabilistic laws. Many applications of stochastic processes occur in physics, engineering, biology, medicine, psychology, and other disciplines, as well as in other branches of mathematical analysis. The purpose of this book is to provide an introduction to the many specialized treatises on stochastic processes. Specifically, I have endeavored to achieve three objectives: (1) to present a systematic introductory account of several principal areas in stochastic processes, (2) to attract and interest students of pure mathematics in the rich diversity of applications of stochastic processes, and (3) to make the student who is more concerned with application aware of the relevance and importance of the mathematical subleties underlying stochastic processes.

The examples in this book are drawn mainly from biology and engineering but there is an emphasis on stochastic structures that are of mathematical interest or of importance in more than one discipline. A number of concepts and problems that are currently prominent in probability research are discussed and illustrated.

Since it is not possible to discuss all aspects of this field in an elementary text, some important topics have been omitted, notably stationary stochastic processes and martingales. Nor is the book intended in any sense as an authoritative work in the areas it does cover. On the contrary, its primary aim is simply to bridge the gap between an elementary probability course and the many excellent advanced works on stochastic processes.

Readers of this book are assumed to be familiar with the elementary theory of probability as presented in the first half of Feller's classic *Introduction to*

Probability Theory and Its Applications. In Section 1, Chapter 1 of my book the necessary background material is presented and the terminology and notation of the book established. Discussions in small print can be skipped on first reading. Excercises are provided at the close of each chapter to help illuminate and expand on the theory.

This book can serve for either a one-semester or a two-semester course, depending on the extent of coverage desired.

In writing this book, I have drawn on the vast literature on stochastic processes. Each chapter ends with citations of books that may profitably be consulted for further information, including in many cases bibliographical listings.

I am grateful to Stanford University and to the U.S. Office of Naval Research for providing facilities, intellectual stimulation, and financial support for the writing of this text. Among my academic colleagues I am grateful to Professor K. L. Chung and Professor J. McGregor of Stanford for their constant encouragement and helpful comments; to Professor J. Lamperti of Dartmouth, Professor J. Kiefer of Cornell, and Professor P. Ney of Wisconsin for offering a variety of constructive criticisms; to Dr. A. Feinstein for his detailed checking of substantial sections of the manuscript, and to my students P. Milch, B. Singer, M. Feldman, and B. Krishnamoorthi for their helpful suggestions and their assistance in organizing the exercises. Finally, I am indebted to Gail Lemmond and Rosemarie Stampfel for their superb technical typing and all-around administrative care.

SAMUEL KARLIN

CONTENTS OF *A FIRST COURSE*

Chapter 10

ALGEBRAIC METHODS IN MARKOV CHAINS

Many important results concerning Markov chains can be obtained by using either purely algebraic methods or a combination of probabilistic and algebraic techniques. We will develop a number of these techniques in the present chapter. In order not to disrupt the continuity of presentation, we present here only a brief summary of some basic facts of matrix theory needed immediately. A fairly complete discussion of these results is given in the Appendix to *A First Course*.

1: Preliminaries

Of fundamental importance in considerations of Markov chains is the computation of the n-step transition probabilities. (Special methods are developed in Sections 4–6 applicable in the case where the Markov chain is a random walk.) To this end, we develop the necessary machinery involving the theory of eigenvalues and eigenvectors.[†]

(a) Spectral Representation

Let $\mathbf{A}$ be an $n \times n$ matrix. A nonzero n-dimensional vector $\mathbf{x}$ which satisfies the relation $\mathbf{Ax} = \lambda\mathbf{x}$ for some number λ is called a right eigenvector of $\mathbf{A}$, with corresponding eigenvalue λ. If $\mathbf{xA} = \lambda\mathbf{x}$, we call $\mathbf{x}$ a left eigenvector of $\mathbf{A}$. If there exists a complete linearly independent family $\mathbf{x}^{(1)}, \ldots, \mathbf{x}^{(n)}$ of right (or, alternatively, left) eigenvectors of $\mathbf{A}$, then there exists a linearly independent family $\boldsymbol{\phi}^{(1)}, \ldots, \boldsymbol{\phi}^{(n)}$ of right eigenvectors of $\mathbf{A}$ and a linearly independent family $\boldsymbol{\psi}^{(1)}, \ldots, \boldsymbol{\psi}^{(n)}$ of left eigenvectors of $\mathbf{A}$ which are biorthogonal. This means that

$$(\boldsymbol{\phi}^{(i)}, \boldsymbol{\psi}^{(j)})^{\ddagger} \equiv \sum_{k=1}^{n} \varphi_{ik}\overline{\psi}_{jk} = \delta_{ij} = \begin{cases} 0 & \text{if } i \neq j, \\ 1 & \text{if } i = j, \end{cases}$$

[†] The reader unfamiliar with the basic theory of eigenvalues and eigenvectors of matrices should consult the Appendix of *A First Course in Stochastic Processes* at this point.

[‡] $(\mathbf{a}, \mathbf{b})$ denotes the inner product of the vectors $\mathbf{a}$ and $\mathbf{b}$.

where $\boldsymbol{\phi}^{(i)} = (\varphi_{i1}, \ldots, \varphi_{in})$, $\boldsymbol{\psi}^{(j)} = (\psi_{j1}, \ldots, \psi_{jn})$, and $\bar{\psi}_{jh}$ denotes the complex conjugate of ψ_{jh}. In this case, the matrix $\mathbf{A}$ is said to be *diagonalizable*. Let

$$
\boldsymbol{\Phi} = \begin{Vmatrix} \varphi_{11} & \cdots & \varphi_{n1} \\ \vdots & & \vdots \\ \varphi_{1n} & \cdots & \varphi_{nn} \end{Vmatrix}, \qquad \boldsymbol{\Psi} = \begin{Vmatrix} \bar{\psi}_{11} & \cdots & \bar{\psi}_{1n} \\ \vdots & & \vdots \\ \bar{\psi}_{n1} & \cdots & \bar{\psi}_{nn} \end{Vmatrix},
$$

$$
\boldsymbol{\Lambda} = \begin{Vmatrix} \lambda_1 & 0 & \cdots & 0 \\ 0 & \lambda_2 & \cdots & 0 \\ \vdots & \vdots & & \vdots \\ 0 & 0 & \cdots & \lambda_n \end{Vmatrix},
$$

where $\lambda_1, \ldots, \lambda_n$ are the (not necessarily distinct) eigenvalues associated with the eigenvectors $\boldsymbol{\phi}^{(1)}, \ldots, \boldsymbol{\phi}^{(n)}$. (Notice that we have not labeled the element of $\boldsymbol{\Phi}$ in the usual order.) Then $\mathbf{A}$ possesses a spectral representation as a product of three special matrices:

$$
\mathbf{A} = \boldsymbol{\Phi}\boldsymbol{\Lambda}\boldsymbol{\Psi}.
$$

Using the relation $(\boldsymbol{\phi}^{(i)}, \boldsymbol{\psi}^{(j)}) = \delta_{ij}$, we can verify by direct calculation that $\boldsymbol{\Psi}\boldsymbol{\Phi} = \boldsymbol{\Phi}\boldsymbol{\Psi} = \mathbf{I}$ ($\mathbf{I}$ = the identity matrix). Then $\mathbf{A}^2 = \boldsymbol{\Phi}\boldsymbol{\Lambda}\boldsymbol{\Psi}\boldsymbol{\Phi}\boldsymbol{\Lambda}\boldsymbol{\Psi} = \boldsymbol{\Phi}\boldsymbol{\Lambda}^2\boldsymbol{\Psi}$ and generally

$$
\mathbf{A}^m = \boldsymbol{\Phi}\boldsymbol{\Lambda}^m\boldsymbol{\Psi}, \tag{1.1}
$$

where obviously

$$
\boldsymbol{\Lambda}^m = \begin{Vmatrix} \lambda_1^m & 0 & \cdots & 0 \\ 0 & \lambda_2^m & \cdots & 0 \\ \vdots & \vdots & \ddots & \vdots \\ 0 & 0 & \cdots & \lambda_n^m \end{Vmatrix}.
$$

When $\mathbf{A}$ is a Markov matrix, formula (1.1) provides a convenient representation of the mth-step transition probability matrix. Its effective use requires determining a complete set of left and right eigenvectors.

(b) *Positive Matrices*

Let $\mathbf{A}$ be a real matrix which has at least one positive element and no negative elements; we write $\mathbf{A} > \mathbf{0}$ and call $\mathbf{A}$ positive. If every element of $\mathbf{A}$ is positive, we write $\mathbf{A} \gg \mathbf{0}$ and call $\mathbf{A}$ strictly positive. The following results are known.

To each $\mathbf{A} > \mathbf{0}$ there corresponds a number $r(\mathbf{A}) \geq 0$, the spectral radius of $\mathbf{A}$, which is zero if and only if $\mathbf{A}^m = \mathbf{0}$ for some integer $m > 0$. In any case there are positive vectors $\mathbf{f}, \mathbf{x} > \mathbf{0}$, such that $\mathbf{A}\mathbf{x} = r(\mathbf{A})\mathbf{x}$, $\mathbf{f}\mathbf{A} = r(\mathbf{A})\mathbf{f}$. If λ is any eigenvalue of $\mathbf{A}$, then $|\lambda| \leq r(\mathbf{A})$; if $|\lambda| = r(\mathbf{A})$, then $\eta = \lambda/r(\mathbf{A})$ is a root of unity, i.e., $\eta^k = 1$ for some integer k, and $\eta^m r(\mathbf{A})$ is an eigenvalue of $\mathbf{A}$ for $m = 1, 2, \ldots$. Finally, suppose that $\mathbf{A}^m \gg \mathbf{0}$ for some $m > 0$; then $\mathbf{x}$ and $\mathbf{f}$ are strictly positive vectors and uniquely determined up to a constant factor. Moreover $|\lambda| < r(\mathbf{A})$ if λ is an eigenvalue of $\mathbf{A}$ different from $r(\mathbf{A})$.

2: Relations of Eigenvalues and Recurrence Classes

The foregoing results have immediate application to the study of finite-state Markov chains.

Let $\mathbf{P} = \|P_{ij}\|$, $i, j = 1, \ldots, n$, be a matrix of transition probabilities. Evidently $\mathbf{P} > \mathbf{0}$. Let $\mathbf{x}$ be any vector satisfying $\sum_{i=1}^{n} x_i = 1$. Then

$$\mathbf{xP} = \left(\sum_{i=1}^{n} x_i P_{i1}, \sum_{i=1}^{n} x_i P_{i2}, \ldots, \sum_{i=1}^{n} x_i P_{in} \right).$$

Now

$$\sum_{j=1}^{n} \left(\sum_{i=1}^{n} x_i P_{ij} \right) = \sum_{i=1}^{n} x_i \sum_{j=1}^{n} P_{ij} = \sum_{i=1}^{n} x_i = 1. \tag{2.1}$$

We claim that $\mathbf{xP} \geq \lambda\mathbf{x}$ cannot hold with $\mathbf{x} > \mathbf{0}$ for any value of $\lambda > 1$, so that $r(\mathbf{P}) \leq 1$. In fact, summing the components of both sides in $\mathbf{xP} \geq \lambda\mathbf{x}$ as in (2.1) yields $\sum_{i=1}^{n} x_i \geq \lambda \sum_{i=1}^{n} x_i$. Since $\sum_{i=1}^{n} x_i > 0$, we can cancel this factor, which implies that $\lambda \leq 1$.

On the other hand, the vector $(1, \ldots, 1)$ is immediately seen to be a right eigenvector of $\mathbf{P}$ with eigenvalue 1; thus $r(\mathbf{P}) = 1$.

The property that 1 is an eigenvalue with a corresponding positive left eigenvector for any finite Markov matrix can also be deduced from Theorem 1.3 of Chapter 3.[†] We know that in a finite state Markov chain at least one state (and therefore at least one class) is positive recurrent. Relabeling the states if necessary, we may assume that the states $i = 1, \ldots, s$ form a positive recurrent class. Therefore $P_{ij} = 0$ for any pair i, j for which $i \in \{1, \ldots, s\}$ and $j \in \{s + 1, \ldots, n\}$. Thus, $\mathbf{P}$ has the form

$$\mathbf{P} = \left\| \begin{matrix} \mathbf{P}_1 & \mathbf{0} \\ \mathbf{B} & \mathbf{C} \end{matrix} \right\| \tag{2.2}$$

and $\mathbf{P}_1$ forms an $s \times s$ Markov matrix. Now the basic limit theorem of Markov chains (see Theorem 1.3 of Chapter 3) asserts the existence of $\pi_1, \ldots, \pi_s$ such that $\pi_i > 0$,

$$\sum_{i=1}^{s} \pi_i P_{ij} = \pi_j, \qquad j = 1, \ldots, s,$$

and

$$\sum_{i=1}^{s} \pi_i = 1.$$

† Chapters 1–9 are in *A First Course in Stochastic Processes* (Second Edition, 1975).

Let $\mathbf{x}^0 = (\pi_1, \ldots, \pi_s, 0, \ldots, 0)$; we may verify at once because of the special structure of $\mathbf{P}$ as displayed in (2.2) that $\mathbf{x}^0\mathbf{P} = \mathbf{x}^0$. A slightly more detailed analysis yields the following:

Theorem 2.1. *If* $\mathbf{P}$ *is a finite Markov matrix, then the multiplicity of the eigenvalue* 1 *is equal to the number of recurrent classes associated with* $\mathbf{P}$.

Proof. We have seen above that if C_1 is a recurrent class of states, then there exists a left eigenvector $\mathbf{x}^{(1)} > 0$ for the eigenvalue 1 such that $x_i^{(1)} = 0$ if $i \notin C_1$. Similarly, with each recurrent class $C_2, C_3, \ldots$ there is associated a positive eigenvector $\mathbf{x}^{(2)}, \mathbf{x}^{(3)}, \ldots$, with eigenvalue 1 such that $x_i^{(h)} = 0$ if $i \notin C_h$. Since distinct classes are disjoint, it is clear that $\mathbf{x}^{(1)}, \mathbf{x}^{(2)}, \ldots$ are linearly independent vectors, and so the multiplicity of the eigenvalue 1 is at least the number of distinct recurrent classes. To prove the reverse inequality suppose that $\mathbf{x}\mathbf{P} = \mathbf{x}$. Then $\mathbf{x}\mathbf{P}^m = \mathbf{x}$ for $m = 1, 2, \ldots$, i.e.,

$$\sum_{i=1}^{n} x_i P_{ij}^m = x_j, \qquad j = 1, \ldots, n, \quad m = 1, 2, \ldots.$$

But if j is a transient state, we know that $\lim_{m \to \infty} P_{ij}^{(m)} = 0$ for all i. It follows that $x_j = 0$ for every transient state j, and so we can write

$$\sum_{h=1}^{r} \sum_{i \in C_h} x_i P_{ij} = x_j, \qquad j \in \bigcup_{h=1}^{r} C_h,$$

where $C_1, \ldots, C_r$ are the recurrent classes. Also $P_{ij} = 0$ if i and j are in distinct recurrent classes; therefore, we have

$$\sum_{i \in C_h} x_i P_{ij} = x_j \qquad \text{for} \quad j \in C_h, \quad h = 1, \ldots, r.$$

If $x_i \neq 0$ for some $i \in C_h$, then by Theorem 1.3 of Chapter 3 there exists a constant a_h such that

$$x_i = a_h x_i^{(h)}, \qquad i \in C_h.$$

Thus

$$\mathbf{x} = \sum_{h=1}^{r} a_h \mathbf{x}^{(h)},$$

from which we see that the $\mathbf{x}^{(h)}$ form a basis for the manifold of left eigenvectors with eigenvalue 1. ∎

PROBABILISTIC INTERPRETATION OF EIGENVALUES AND EIGENVECTORS

Let us now consider the manifold of right eigenvectors of $\mathbf{P}$ with eigenvalue 1. It turns out that there is a basis for this manifold which has a very simple probabilistic interpretation. In fact, if $C_1, \ldots, C_r$ are the recurrent classes associated with $\mathbf{P}$, we define $p_i^{(h)}$ to be the probability that starting from i the state of the system will eventually lie in C_h, i.e.,

$$p_i^{(h)} = \Pr\{X_n \in C_h \text{ for some } n = 1, 2, \ldots \mid X_0 = i\}.$$

Clearly

$$p_i^{(h)} = \begin{cases} 1 & \text{for} \quad i \in C_h, \\ 0 & \text{for} \quad i \in C_j, \quad j \neq h, \end{cases} \tag{2.3}$$

since it is not possible to leave a recurrent class. If we define $\mathbf{p}^{(h)} = (p_1^{(h)}, \ldots, p_n^{(h)})$, $h = 1, \ldots, r$, the preceding equations show at once that the vectors $\mathbf{p}^{(1)}, \ldots, \mathbf{p}^{(r)}$ are linearly independent. Furthermore, the $p_i^{(h)}$ satisfy the equations

$$p_i^{(h)} = \sum_{j=1}^{n} P_{ij} p_j^{(h)}, \qquad i = 1, \ldots, n, \quad h = 1, \ldots, r,$$

[Eq. (3.4) of Chapter 3], which shows that $\mathbf{p}^{(1)}, \ldots, \mathbf{p}^{(r)}$ are right eigenvectors of $\mathbf{P}$ with eigenvalue 1. As the $\mathbf{p}^{(i)}$ are linearly independent and their number r is the multiplicity of the eigenvalue 1, they form a basis for the right eigenmanifold corresponding to the eigenvalue 1. Finally, we observe by direct evaluation with the aid of (2.3) that

$$(\mathbf{x}^{(i)}, \mathbf{p}^{(j)}) = \begin{cases} 1 & \text{if} \quad i = j, \\ 0 & \text{if} \quad i \neq j, \end{cases}$$

since the only nonzero components of $\mathbf{x}^{(i)}$ are those whose indices are in C_i, and their sum is just 1.

Let us assume now that $\mathbf{P}$ has a spectral representation, and that the eigenvalues $\lambda_1, \lambda_2, \ldots, \lambda_n$ are labeled so that $1 = \lambda_1 = \cdots = \lambda_r \geq |\lambda_{r+1}| \geq |\lambda_{r+2}| \geq \cdots$ and $\lambda_{r+1} \neq 1$. Then we can take $\boldsymbol{\phi}^{(1)} = \mathbf{p}^{(1)}, \ldots, \boldsymbol{\phi}^{(r)} = \mathbf{p}^{(r)}$ and $\boldsymbol{\psi}^{(1)} = \mathbf{x}^{(1)}, \ldots, \boldsymbol{\psi}^{(r)} = \mathbf{x}^{(r)}$ (see Appendix to *A First Course*). From

$$\mathbf{P}^m = \boldsymbol{\Phi}\boldsymbol{\Lambda}^m\boldsymbol{\Psi}$$

we obtain

$$P_{ij}^m = \sum_{h=1}^{n} \varphi_{hi} \lambda_h^m \overline{\psi}_{hj} = \varphi_{1i}\overline{\psi}_{1j} + \cdots + \varphi_{ri}\overline{\psi}_{rj} + \sum_{h=r+1}^{n} \varphi_{hi} \lambda_h^m \overline{\psi}_{hj}.$$

Suppose that $\mathbf{P}$ has no eigenvalue, different from 1, whose modulus equals 1; then $|\lambda_h| < 1$, $h = r + 1, \ldots, n$, and as $m \to \infty$,

$$\sum_{h=r+1}^{n} \varphi_{hi} \lambda_h^m \overline{\psi}_{hj} \to 0$$

and the rate of convergence is of the order at least $|\lambda_{r+1}|^m$. We shall see shortly that $|\lambda_h| < 1$, $h = r + 1, \ldots, n$, if and only if $\mathbf{P}$ has no periodic recurrent classes (Theorem 3.1 below). Assuming that $\mathbf{P}$ has no periodic recurrent classes and recalling that $x_j^{(h)} = \psi_{hj}$, $h = 1, \ldots, r$, $j = 1, \ldots, n$, is different from zero if and only if $j \in C_h$, we see that

$$\varphi_{1i}\bar{\psi}_{1j} = \varphi_{2i}\bar{\psi}_{2j} = \cdots = \varphi_{ri}\bar{\psi}_{rj} = 0 \qquad \text{for} \quad j \text{ transient.}$$

Thus, if j is transient, $P_{ij}^m = \sum_{h=r+1}^n \varphi_{hi}\lambda_h^m\bar{\psi}_{hj}$ and this tends to zero at the rate $|\lambda_{r+1}|^m$ as $m \to \infty$. Now if $i, j \in C_h$, then among the first r terms in the expression for P_{ij}^m the only nonvanishing one is $\varphi_{hi}\bar{\psi}_{hj}$; but $\varphi_{hi} = 1$ (recall that $\varphi_{hi} = p_i^{(h)}$) and $\psi_{hj} = \pi_j = \lim_{m \to \infty} P_{ij}^m$. We see generally for all states j that $\pi_j - P_{ij}^m$ goes to zero at least as fast as $|\lambda_{r+1}|^m$ as $m \to \infty$.

Now let us assume that in addition to $|\lambda_{r+1}| < 1$ we have the special situation that $|\lambda_{r+2}| < |\lambda_{r+1}|$. Let, as usual, T denote the set of all transient states, $i, j \in T$; we wish to find the following limit:

$$\lim_{m \to \infty} \Pr\{X_m = j \,|\, X_0 = i, X_m \in T\},$$

i.e., the limiting value ($m \to \infty$) of the probability that starting from state i the process is in the transient state j, given that at time m, X_m is in a transient state. We have

$$\Pr\{X_m = j \,|\, X_0 = i, X_m \in T\} = \frac{P_{ij}^m}{\sum_{j \in T} P_{ij}^m}.$$

As we have seen before, for j transient $P_{ij}^m = \sum_{h=r+1}^n \varphi_{hi}\lambda_h^m\bar{\psi}_{hj}$. Since $|\lambda_{r+1}| > |\lambda_{r+2}|$, we readily find that

$$\lim_{m \to \infty} \frac{P_{ij}^m}{\sum_{j \in T} P_{ij}^m} = \frac{\varphi_{r+1,i}\bar{\psi}_{r+1,j}}{\sum_{j \in T} \varphi_{r+1,i}\bar{\psi}_{r+1,j}} = \frac{\bar{\psi}_{r+1,j}}{\sum_{j \in T} \bar{\psi}_{r+1,j}},$$

assuming that the denominator does not vanish. If the denominator does vanish, we have to examine the terms in $\sum_{h=r+1}^n \varphi_{hi}\lambda_h^m\bar{\psi}_{hj}$ containing λ_{r+2} and other eigenvalues whose modulus equals $|\lambda_{r+2}|$, and so forth.

3: Periodic Classes

We wish to give a more complete description of the structure of a periodic chain. The simplest class with period d is clearly one in which there are d states $1, \ldots, d$ and

$$P_{12} = P_{23} = \cdots = P_{d-1,d} = P_{d1} = 1, \qquad \mathbf{P} = \begin{Vmatrix} 0 & 1 & 0 & \cdots & 0 \\ 0 & 0 & 1 & \cdots & 0 \\ \vdots & & & & \\ 1 & 0 & 0 & \cdots & 0 \end{Vmatrix}.$$

A less trivial example may be formed by replacing the individual states $1, \ldots, d$ by disjoint families $C_1, \ldots, C_d$ of states and defining the P_{ij} in such a way that

$P_{ij} \neq 0$ only if $i \in C_1, j \in C_2$, or $i \in C_2, j \in C_3, \ldots$, or $i \in C_d, j \in C_1$. The matrix $\mathbf{P}$ then takes the form

$$
\mathbf{P} = \begin{Vmatrix} 0 & \mathbf{P}_1 & 0 & \cdots \\ 0 & 0 & \mathbf{P}_2 & \cdots \\ \vdots & \vdots & \vdots & \\ \mathbf{P}_d & 0 & 0 & \cdots \end{Vmatrix}.
$$

At the same time, we can define P_{ij} so that every two states communicate. We will now prove that every periodic class is of this form. Let d be the period of the class W and assume that the states are labeled by $1, 2, \ldots, M$. Let C_1 consist of the states of W which can be reached from state 1 in some multiple of d transitions; i.e., $j \in C_1$ if and only if $P_{1j}^{nd} > 0$ for some integer $n > 0$. For each $r = 1, \ldots, d - 1$, we define C_{r+1} to consist of those states which can be reached from state 1 in r plus some multiple of d transitions; i.e., $j \in C_{r+1}$ if and only if $P_{1j}^{nd+r} > 0$ for some integer $n \geq 0$.

First we show that if $j \in C_1$ then $P_{j1}^h > 0$ implies $h = md$ for some $m > 0$. In fact, since $j \in C_1$ implies that $P_{1j}^{nd} > 0$ for some $n > 0$, it follows that $P_{jj}^{nd+h} \geq P_{j1}^h P_{1j}^{nd} > 0$ (cf. Chapter 2, Theorem 3.1), and so by the definition of period $nd + h$ must be divisible by d; hence h is. Next we show that if $i \in C_1, j \in C_{r+1}$ then $P_{ij}^h > 0$ implies that $h = nd + r$ for some $n \geq 0$. In fact, let $P_{ji}^s > 0$ for some $s > 0$; $P_{i1}^{qd} > 0$ for some $q > 0$; and $P_{1j}^{md+r} > 0$ for some $m \geq 0$. Thus, if $w = s + dq + md + r$ then $P_{11}^w \geq P_{1j}^{md+r} P_{ji}^s P_{i1}^{qd} > 0$, and w is a multiple of d; therefore $s + r$ also is a multiple of d. But $P_{jj}^{h+s} \geq P_{ji}^s P_{ij}^h > 0$, so that $h + s$ is divisible by d. Combining these two results, we infer that $h - r$ is divisible by d, and therefore, $h = nd + r$ for some $n \geq 0$.

We leave it to the reader to verify that the above results imply that $C_1, \ldots, C_d$ are disjoint and nonempty, that $\bigcup_{i=1}^d C_i = W$, and that $i \in C_r$ requires $P_{ij} = 0$ for every $j \notin C_{r+1}$ where $C_{d+1} = C_1$.

Having thus analyzed the matrix of a periodic class, we can now demonstrate an earlier assertion concerning the occurrence of eigenvalues of modulus 1 of a Markov transition matrix.

Theorem 3.1. *If $\mathbf{P}$ is the transition matrix of a finite irreducible periodic Markov chain with period d, then the dth roots of unity are eigenvalues of $\mathbf{P}$, each with multiplicity 1, and there are no other eigenvalues of modulus 1.*

Proof. Let $D_1, \ldots, D_d$ be the "moving classes" of the process as established above, i.e., $i \in D_r$ implies $P_{ij} = 0$ for every $j \notin D_{r+1}$. It is no loss of generality to assume that $D_1 = \{1, \ldots, n_1\}$, $D_2 = \{n_1 + 1, \ldots, n_1 + n_2\}, \ldots, D_d = \{M - n_d + 1, \ldots, M\}$. From the definition of the moving class it follows that

$$
\mathbf{P}^d = \begin{Vmatrix} \mathbf{A}_1 & 0 & \cdots & 0 \\ 0 & \mathbf{A}_2 & \cdots & 0 \\ \vdots & \vdots & \ddots & \\ 0 & 0 & \cdots & \mathbf{A}_d \end{Vmatrix},
$$

where $\mathbf{A}_i$ is an $n_i \times n_i$ Markov matrix. Furthermore, for each i, $\mathbf{A}_i^m \gg \mathbf{0}$ for some integer $m > 0$ (see Problem 5, Chapter 2). Thus, $\mathbf{A}_i$ has a strictly positive left eigenvector $\boldsymbol{\mu}^{(i)}$, with eigenvalue 1, of algebraic multiplicity 1. Owing to the form of $\mathbf{P}^d$, it is clear that, by adjoining an appropriate number of zeros on one or both sides of each $\boldsymbol{\mu}^{(i)}$, we determine linearly independent vectors $\mathbf{x}^{(1)}, \ldots, \mathbf{x}^{(d)}$ such that

$$\mathbf{x}^{(i)} = \mathbf{x}^{(i)}\mathbf{P}^d, \qquad i = 1, \ldots, d.$$

Let us consider the vectors $\mathbf{y}^{(1)} = \mathbf{x}^{(1)}$, $\mathbf{y}^{(2)} = \mathbf{x}^{(1)}\mathbf{P}, \ldots, \mathbf{y}^{(d)} = \mathbf{x}^{(1)}\mathbf{P}^{d-1}$. Since the only nonzero components of $\mathbf{x}^{(1)}$ are those with indices $1, 2, \ldots, n_1$, and observing that $P_{ij}^{(h)}$ may differ from zero only if the moving class in which i lies agrees with that of j after precisely h steps, we see that the only nonzero components of $\mathbf{y}^{(i)}$ are those with indices $(n_1 + \cdots + n_{i-1} + 1, \ldots, n_1 + \cdots + n_i)$. This implies that the vectors $\mathbf{y}^{(i)}$ ($i = 1, \ldots, d$) are linearly independent. Furthermore

$$\mathbf{y}^{(i)}\mathbf{P}^d = \mathbf{x}^{(1)}\mathbf{P}^{i-1}\mathbf{P}^d = \mathbf{x}^{(1)}\mathbf{P}^d\mathbf{P}^{i-1} = \mathbf{x}^{(1)}\mathbf{P}^{i-1} = \mathbf{y}^{(i)}.$$

It follows that if we restrict attention to the n_i-dimensional linear space obtained by considering only those components of $\mathbf{y}^{(i)}$ whose indices lie in D_i, we obtain a left eigenvector with eigenvalue 1 for $\mathbf{A}_i$.

Because the eigenvalue 1 has simple multiplicity for $\mathbf{A}_i$, it follows that each $\mathbf{y}^{(i)}$ is a constant multiple of $\mathbf{x}^{(i)}$. Actually, if we normalize each $\mathbf{x}^{(1)}, \ldots, \mathbf{x}^{(d)}$ by the condition $\sum_{i=1}^{n} x_i^{(h)} = 1$, $h = 1, \ldots, d$, then, in fact, $\mathbf{y}^{(h)} = \mathbf{x}^{(h)}$, $h = 1, \ldots, d$. Accordingly, we may write $\mathbf{x}^{(2)} = \mathbf{x}^{(1)}\mathbf{P}$, $\mathbf{x}^{(3)} = \mathbf{x}^{(2)}\mathbf{P}, \ldots, \mathbf{x}^{(1)} = \mathbf{x}^{(d)}\mathbf{P}$.

Let $\omega = e^{2\pi i/d}$. Combining the above equations in the indicated manner, we obtain

$$(\mathbf{x}^{(1)} + \mathbf{x}^{(2)} + \mathbf{x}^{(3)} + \cdots + \mathbf{x}^{(d)})\mathbf{P}$$
$$= \mathbf{x}^{(1)} + \mathbf{x}^{(2)} + \cdots + \mathbf{x}^{(d)},$$

$$(\mathbf{x}^{(1)} + \omega\mathbf{x}^{(2)} + \omega^2\mathbf{x}^{(3)} + \cdots + \omega^{d-1}\mathbf{x}^{(d)})\mathbf{P}$$
$$= \omega^{-1}(\mathbf{x}^{(1)} + \omega\mathbf{x}^{(2)} + \cdots + \omega^{d-1}\mathbf{x}^{(d)}),$$

$$(\mathbf{x}^{(1)} + \omega^2\mathbf{x}^{(2)} + \omega^4\mathbf{x}^{(3)} + \cdots + \omega^{2(d-1)}\mathbf{x}^{(d)})\mathbf{P}$$
$$= \omega^{-2}(\mathbf{x}^{(1)} + \omega^2\mathbf{x}^{(2)} + \cdots + \omega^{2(d-1)}\mathbf{x}^{(d)}),$$

$$\vdots$$

$$(\mathbf{x}^{(1)} + \omega^{(d-1)}\mathbf{x}^{(2)} + \omega^{2(d-1)}\mathbf{x}^{(3)} + \cdots + \omega^{(d-1)^2}\mathbf{x}^{(d)})\mathbf{P}$$
$$= \omega^{-(d-1)}(\mathbf{x}^{(1)} + \omega^{(d-1)}\mathbf{x}^{(2)} + \cdots + \omega^{(d-1)^2}\mathbf{x}^{(d)}).$$

The linear independence of the $\mathbf{x}^{(i)}$ ensures that none of the vectors appearing above are zero. These relations exhibit the property that the dth roots of unity are all eigenvalues of $\mathbf{P}$.

Suppose next that $\mathbf{xP} = \lambda\mathbf{x}$ for some nonzero $\mathbf{x}$. Then $\mathbf{xP}^d = \lambda^d\mathbf{x}$. Looking at the contracted vectors $\mathbf{z}^{(1)} = (x_1, \ldots, x_{n_1})$, $\mathbf{z}^{(2)} = (x_{n_1+1}, \ldots, x_{n_1+n_2})$, $\ldots$, $\mathbf{z}^{(d)} = (x_{M-n_d+1}, \ldots, x_M)$, we see that

$$\mathbf{z}^{(i)}\mathbf{A}_i = \lambda^d\mathbf{z}^{(i)}, \qquad i = 1, \ldots, d.$$

Since at least one of the $\mathbf{z}^{(i)}$ is nonzero, and for each $\mathbf{A}_i$ there is an m such that $\mathbf{A}_i^m \gg 0$, either $\lambda^d = 1$ or $|\lambda^d| < 1$. If $\lambda^d = 1$, then there are constants $c_1, \ldots, c_d$ such that

$$\mathbf{z}^{(i)} = c_i\mathbf{x}^{(i)}, \qquad i = 1, \ldots, d,$$

and so we see that $\mathbf{x} = c_1\mathbf{x}^{(1)} + \cdots + c_d\mathbf{x}^{(d)}$.

Now

$$\lambda\mathbf{x} = \mathbf{xP} = c_1\mathbf{x}^{(2)} + c_2\mathbf{x}^{(3)} + \cdots + c_d\mathbf{x}^{(1)}$$

or

$$\lambda c_1\mathbf{x}^{(1)} + \cdots + \lambda c_d\mathbf{x}^{(d)} = c_d\mathbf{x}^{(1)} + c_1\mathbf{x}^{(2)} + \cdots + c_{d-1}\mathbf{x}^{(d)}.$$

Since the $\mathbf{x}^{(i)}$ are linearly independent, we have

$$\lambda c_1 = c_d, \qquad \lambda c_2 = c_1, \qquad \ldots, \qquad \lambda c_d = c_{d-1},$$

or

$$c_{d-1} = \lambda c_d = (\lambda^{-1})^{d-1}c_d, \qquad c_{d-2} = \lambda^2 c_d = (\lambda^{-1})^{d-2}c_d, \qquad \ldots,$$

$$c_1 = \lambda^{d-1}c_d = \lambda^{-1}c_d$$

since $\lambda^d = 1$, and this means that $\mathbf{x}$ is plainly a constant multiple of one of the eigenvectors of $\mathbf{P}$ already constructed. ■

The case of an arbitrary Markov matrix $\mathbf{P}$ follows easily from the preceding theorem. We have

Theorem 3.2. *If $\mathbf{P}$ is a finite Markov matrix, then any eigenvalue of $\mathbf{P}$ of modulus 1 is a root of unity. The dth roots of unity are eigenvalues of $\mathbf{P}$ if and only if $\mathbf{P}$ has a recurrent class with period d. The multiplicity of each dth root of unity is just the number of recurrent classes of period d.*

The proof is essentially identical with that of Theorem 3.1. Since $\lambda\mathbf{x} = \mathbf{xP}$ implies

$$\lambda^m\mathbf{x} = \mathbf{xP}^m$$

or

$$\lambda^m x_j = \sum_{i=1}^{n} x_i P_{ij}^m,$$

then, letting $m \to \infty$, we see that $x_j = 0$ if j is transient. We may therefore restrict attention to the recurrent states, and the theorem immediately reduces to the case considered in the previous theorem.

4: Special Computational Methods in Markov Chains

Let **P** be the transition probability matrix of a random walk on the nonnegative integers with probability $\frac{1}{2}$ of going to each of the two neighboring states from state k ($k \geq 1$) and with a reflecting barrier at the origin; that is,

$$\mathbf{P} = \left\| \begin{matrix} 0 & 1 & 0 & 0 & \cdots \\ \frac{1}{2} & 0 & \frac{1}{2} & 0 & \cdots \\ 0 & \frac{1}{2} & 0 & \frac{1}{2} & \cdots \\ \vdots & \vdots & \vdots & \vdots & \end{matrix} \right\|.$$

To obtain the probability of reaching state l from state k in n steps we could multiply matrix **P** by itself n times and seek out the element P_{kl}^n in the kth row and lth column of the matrix $\mathbf{P}^n$. This method, however, is very cumbersome and lengthy in practice.

A second approach is to attempt to generalize the method of eigenvalues and eigenvectors as developed in Section 2. In the case of infinite matrices this cannot always be done. However, for matrices of the same form as **P** above or, more generally, transition probability matrices corresponding to random walks, there is available an infinite analog to the representation formula (1.1).

We proceed to obtain P_{kl}^n in a manner which will illustrate a general method applicable to arbitrary random walks.

Adding the two trigonometric identities

$$\cos(\alpha \pm \beta) = \cos \alpha \cos \beta \mp \sin \alpha \sin \beta$$

leads to the identity

$$\cos \alpha \cos \beta = \tfrac{1}{2} \cos(\alpha + \beta) + \tfrac{1}{2} \cos(\alpha - \beta). \tag{*}$$

Let $\alpha = \theta$ and $\beta = k\theta$ ($k = 1, 2, \ldots$). We get

$$\cos \theta \cos k\theta = \tfrac{1}{2} \cos(k + 1)\theta + \tfrac{1}{2} \cos(k - 1)\theta. \tag{4.1}$$

Since the elements in the kth row of matrix **P** are

$$P_{k,0} = P_{k,1} = 0, \ldots, \quad P_{k,k-2} = 0, \quad P_{k,k-1} = \tfrac{1}{2}, \quad P_{k,k} = 0,$$

$$P_{k,k+1} = \tfrac{1}{2}, \quad P_{k,k+2} = 0, \ldots, \quad k = 2, 3, \ldots,$$

$$P_{1,0} = \tfrac{1}{2}, \quad P_{1,1} = 0, \quad P_{1,2} = \tfrac{1}{2}, \quad P_{1,3} = 0, \ldots,$$

$$P_{0,0} = 0, \quad P_{0,1} = 1, \quad P_{0,2} = 0, \ldots,$$

Equation (4.1) can be written as

$$\cos\theta\cos k\theta = \sum_{r=0}^{\infty} P_{kr}\cos r\theta, \qquad k = 0, 1, \ldots. \tag{4.2}$$

Next, multiply this equation by $\cos\theta$. Then

$$\cos^2\theta\cos k\theta = \sum_{r=0}^{\infty} P_{kr}\cos\theta\cos r\theta. \tag{4.3}$$

But by (4.2) $\cos\theta\cos r\theta$ can be expressed as

$$\cos\theta\cos r\theta = \sum_{s=0}^{\infty} P_{rs}\cos s\theta.$$

Substituting this in (4.3), we have

$$\cos^2\theta\cos k\theta = \sum_{r=0}^{\infty} \left(P_{kr}\sum_{s=0}^{\infty} P_{rs}\cos s\theta \right)$$

$$= \sum_{s=0}^{\infty} \left(\cos s\theta\sum_{r=0}^{\infty} P_{kr}P_{rs} \right)$$

$$= \sum_{s=0}^{\infty} P_{ks}^2\cos s\theta.$$

Although it simplifies the notation to take all summations from 0 to ∞, note that all but a finite number of terms are equal to zero.

After $n - 1$ iterations of the procedure of multiplying by $\cos\theta$ and then interchanging the order of summation we finally obtain

$$\cos^n\theta\cos k\theta = \sum_{r=0}^{\infty} P_{kr}^n\cos r\theta. \tag{4.4}$$

Multiply both sides of this equation by $\cos s\theta$ and integrate with respect to θ from 0 to 2π:

$$\int_0^{2\pi} \cos^n\theta\cos k\theta\cos s\theta\, d\theta = \int_0^{2\pi} \sum_{r=0}^{\infty} P_{kr}^n\cos r\theta\cos s\theta\, d\theta$$

$$= \sum_{r=0}^{\infty} P_{kr}^n\int_0^{2\pi} \cos r\theta\cos s\theta\, d\theta. \tag{4.5}$$

Using the identity (*) with $\alpha = r\theta$ and $\beta = s\theta$, it is simple to show that

$$\int_0^{2\pi} \cos r\theta\cos s\theta\, d\theta = \begin{cases} 0 & \text{for} \quad r \neq s, \\ \pi & \text{for} \quad r = s \geq 1, \\ 2\pi & \text{for} \quad r = s = 0. \end{cases} \tag{4.6}$$

From (4.6) and (4.5) we immediately obtain

$$
P_{ks}^n = \begin{cases} \dfrac{1}{\pi} \displaystyle\int_0^{2\pi} \cos^n \theta \cos k\theta \cos s\theta \, d\theta, & s \neq 0, \\[3mm] \dfrac{1}{2\pi} \displaystyle\int_0^{2\pi} \cos^n \theta \cos k\theta \, d\theta, & s = 0. \end{cases} \tag{4.7}
$$

This integral can be computed without difficulty for given n, k, and s.

The general method, of which the preceding technique constitutes a particularly simple example, is the following.

Suppose we are given a random walk process on the nonnegative integers whose matrix $\mathbf{P}$ of one-step transition probabilities is given by

$$
\mathbf{P} = \begin{Vmatrix} r_0 & p_0 & 0 & 0 & \cdots \\ q_1 & r_1 & p_1 & 0 & \cdots \\ 0 & q_2 & r_2 & p_2 & \cdots \\ \vdots & \vdots & \vdots & \vdots & \end{Vmatrix}, \tag{4.8}
$$

where $q_n + r_n + p_n = 1$, $q_n > 0$, $p_n > 0$, $r_n \geq 0$, for $n = 1, 2, \ldots$, and $r_0 + p_0 = 1$, $p_0 > 0$, $r_0 \geq 0$. (Note for future reference, however, that none of the following general results are dependent upon the conditions $q_n + r_n + p_n = 1$, $n = 1, 2, \ldots$, and $r_0 + p_0 = 1$.) Let us consider the following system of equations:

$$
x Q_k(x) = q_k Q_{k-1}(x) + r_k Q_k(x) + p_k Q_{k+1}(x), \qquad k = 1, 2, \ldots, \tag{4.9}
$$

with the "initial" specifications $Q_0(x) \equiv 1$ and $Q_1(x) = (x - r_0)/p_0$. Since $p_n > 0$ for all $n = 0, 1, 2, \ldots$, it is clear that $Q_n(x)$, $n \geq 2$, are determined recursively from (4.9) and that $Q_n(x)$ is a polynomial in x of exact degree n. Now a theorem whose proof is beyond the scope of this book asserts that there exists a function $\sigma(x)$ on the interval $[-1, 1]$ that is nondecreasing and not identically constant, such that

$$
\int_{-1}^{1} Q_k(x) Q_s(x) \, d\sigma(x) \begin{cases} = 0 & \text{if } k \neq s, \\ > 0 & \text{if } k = s, \end{cases} \quad k = 0, 1, 2, \ldots. \tag{4.10}
$$

We express the property (4.10) by the statement that the functions $Q_k(x)$, $k = 0, 1, \ldots$, are "orthogonal polynomials with respect to the distribution $\sigma(x)$ over the interval $[-1, 1]$." The function $\sigma(x)$ is unique up to an additive constant. This general theorem enables us to derive an explicit expression for the P_{ks}^n. In fact, Eq. (4.9), in view of the prescriptions for $Q_0(x)$ and $Q_1(x)$, may be written

$$
x Q_k(x) = \sum_{r=0}^{\infty} P_{kr} Q_r(x), \qquad k = 0, 1, \ldots. \tag{4.11}
$$

Multiplying both sides by x and substituting (4.11) into the right side of the resulting equation, we have

$$x^2 Q_k(x) = \sum_{r=0}^{\infty} P_{kr} \sum_{s=0}^{\infty} P_{rs} Q_s(x) = \sum_{s=0}^{\infty} P_{ks}^2 Q_s(x), \qquad k = 0, 1, \ldots . \quad (4.12)$$

Proceeding in this fashion, we obtain

$$x^n Q_k(x) = \sum_{r=0}^{\infty} P_{kr}^n Q_r(x), \qquad k = 0, 1, \ldots, \quad n = 1, 2, \ldots . \quad (4.13)$$

Multiplying both sides by $Q_s(x)$ and integrating over $[-1, 1]$ with respect to $d\sigma(x)$, we find, by virtue of the orthogonality relations (4.10), that

$$\int_{-1}^{1} x^n Q_k(x) Q_s(x)\, d\sigma(x) = \sum_{r=0}^{\infty} P_{kr}^n \int_{-1}^{1} Q_r(x) Q_s(x)\, d\sigma(x)$$

$$= P_{ks}^n \int_{-1}^{1} Q_s^2(x)\, d\sigma(x).$$

Thus

$$P_{ks}^n = \frac{\int_{-1}^{1} x^n Q_k(x) Q_s(x)\, d\sigma(x)}{\int_{-1}^{1} Q_s^2(x)\, d\sigma(x)}, \quad (4.14)$$

which is the desired formula.

As remarked earlier, it should be observed that this procedure bears a similarity to the diagonalization method of Section 1. In fact, equations (4.9) assert simply that for each value of x the infinite vector $(Q_0(x), Q_1(x), \ldots)$ is a formal eigenvector of the matrix (4.8) for the eigenvalue x. Since there is a continuum of eigenvalues, it is reasonable to expect that a discrete sum analogous to that obtained in Sections 1 and 2 for P_{ij}^n is in general impossible. Actually, it turns out that the appropriate generalization of the spectral representation (1.1) is (4.14). The underlying mathematical phenomenon involves the existence of a "continuous spectrum" in addition to the (possibly empty) discrete spectrum, which is generally the case when one is dealing with infinite matrices. The precise mathematical elaboration of these ideas is beyond the level of this book. We will, nevertheless, illustrate this theory with the discussion of some additional examples.

It might appear that the method, elegant though it may be in theory, is of little value in practice. To find P_{ks}^n, it is necessary to determine the polynomials $\{Q_k(x)\}_{k=0}^{\infty}$ and further to obtain the distribution $\sigma(x)$, concerning which we have so far asserted nothing more than its existence. The actual situation, however, is far better than this pessimistic evaluation. First of all, a great deal is known about general orthogonal polynomials, from which one can deduce important theoretical results concerning the behavior of the P_{ks}^n, and in particular, ratios of them, as $n \to \infty$.

Second, a random walk which arises in a concrete problem will in all likelihood have a transition probability matrix which is far more regular than the general form (4.8). For example, one might have $p_0 = p_1 = p_2 = \cdots$, $q_1 = q_2 = q_3 = \cdots$, or else $p_n = p_{n+1} = \cdots$, $q_n = q_{n+1} = \cdots$ for some integer n. In these cases, as well as others, it can be shown that the polynomials $Q_n(x)$ are combinations of various classical polynomial systems which have been studied extensively.

5: Examples

(a) *The Symmetric Random Walk with Reflecting Barrier*

In order to bring the computations of Section 4 into the general form involving orthogonal polynomials, we have only to put

$$Q_k(x) = \cos k \, (\arccos x), \qquad k = 0, 1, \ldots.$$

The $Q_k(x)$ are orthogonal over the interval $[-1, 1]$ with respect to the distribution $d\sigma(x) = \rho(x) \, dx$, where $\rho(x) = (1/\pi)(1 - x^2)^{-1/2}$, since

$$\int_{-1}^{1} Q_k(x) Q_l(x) \rho(x) \, dx = C \int_{0}^{\pi} \cos k\theta \cos l\theta \, d\theta = 0 \qquad \text{if} \quad k \neq l,$$

as the change of variable $x = \cos \theta$ shows.

(b) *Another Random Walk with Reflecting Barrier*

As a further example, consider the random walk on the nonnegative integers whose transition probability matrix is

$$\mathbf{P} = \begin{Vmatrix} 0 & 1 & 0 & 0 & 0 & \cdots \\ q & 0 & p & 0 & 0 & \cdots \\ 0 & q & 0 & p & 0 & \cdots \\ \vdots & \vdots & \vdots & \vdots & \vdots & \end{Vmatrix}$$

with $q, p > 0$, $q + p = 1$.

Multiplying both sides of relation (4.1) by $2\sqrt{pq}(\sqrt{q/p})^k$ we obtain

$$2\sqrt{pq} \cos \theta (\sqrt{q/p})^k \cos k\theta = p(\sqrt{q/p})^{k+1} \cos(k+1)\theta$$
$$+ q(\sqrt{q/p})^{k-1} \cos(k-1)\theta, \qquad k = 1, 2, \ldots$$

$$(5.1)$$

Thus the polynomials

$$Q_k(x) = (\sqrt{q/p})^k \cos k\theta, \quad 2\sqrt{pq} \cos \theta = x, \qquad k = 0, 1, 2, \ldots$$

satisfy the system of equations (4.9) corresponding to the matrix **P** above except for $k = 0$. Here $Q_0(x) \equiv 1$ and $Q_1(x) = x/2p$, whereas we want the initial conditions $Q_0(x) \equiv 1$, $Q_1(x) = x$.

To remedy this problem, we start with the identity

$$\cos \theta \sin(k + 1)\theta = \tfrac{1}{2} \sin k\theta + \tfrac{1}{2} \sin(k + 2)\theta, \qquad k = 0, 1, 2, \dots . \quad (5.2)$$

Multiplying both sides by $2\sqrt{pq}(\sqrt{q/p})^k$ and dividing by $\sin \theta$, we convert (5.2) into

$$2\sqrt{pq}(\cos \theta)(\sqrt{q/p})^k \frac{\sin(k + 1)\theta}{\sin \theta}$$

$$= q(\sqrt{q/p})^{k-1} \frac{\sin k\theta}{\sin \theta} + p(\sqrt{q/p})^{k+1} \frac{\sin(k + 2)\theta}{\sin \theta}, \qquad k = 1, 2, \dots .$$

Let

$$Z_k(\theta) = (\sqrt{q/p})^k \frac{\sin(k + 1)\theta}{\sin \theta}, \qquad k = 0, 1, \dots .$$

Then

$$Z_0(\theta) \equiv 1 \quad \text{and} \quad Z_1(\theta) = \sqrt{q/p} \frac{\sin 2\theta}{\sin \theta},$$

while

$$2\sqrt{pq}(\cos \theta)Z_k(\theta) = qZ_{k-1}(\theta) + pZ_{k+1}(\theta), \qquad k = 1, 2, \dots .$$

Let

$$R_k(x) = Z_k(\theta); \qquad x = 2\sqrt{pq} \cos \theta.$$

Note that $R_0(x) = 1$ and $R_1(x) = x/p$, while

$$xR_k(x) = qR_{k-1}(x) + pR_{k+1}(x), \qquad k = 1, 2, \dots .$$

It also follows that $R_k(x)$ is a polynomial of degree k. Finally, let $P_k(x) = (2p - 1)R_k(x) + (2 - 2p)Q_k(x)$, $k = 0, 1, \dots .$ Then $P_0(x) \equiv 1$, $P_1(x) = x$, and moreover

$$xP_k(x) = qP_{k-1}(x) + pP_{k+1}(x), \qquad k = 1, 2, \dots .$$

since both $R_k(x)$ and $Q_k(x)$ satisfy the same relations. Thus the $P_k(x)$ are the polynomials corresponding to the transition matrix **P**.

The detailed procedure for obtaining the distribution $\sigma(x)$ with respect to which the $P_k(x)$ are orthogonal on $[-1, 1]$ is beyond the scope of this book. Therefore, we will simply present it, leaving to the reader the verification that it enjoys the desired properties. If $p \geq \frac{1}{2}$, then $\sigma(x)$ is constant outside the interval $[-\sqrt{4pq}, \sqrt{4pq}]$, and in that interval

$$d\sigma(x) = \frac{C\sqrt{4pq - x^2}}{1 - x^2} \, dx.$$

If $p > \frac{1}{2}$, then $\sigma(x)$ has, in addition to the "density" specified above for the interval $[-\sqrt{4pq}, \sqrt{4pq}]$, two "jumps" of magnitude $\frac{1}{2}(1 - 2p)/q$ at the points -1 and $+1$.

The constant C serves as a normalizing factor to guarantee that $\int_{-1}^{1} d\sigma(x) = 1$.

(c) Random Walk with Absorbing Barrier

Next we discuss the problem of random walks on the integers $-1, 0, 1, 2, 3, \ldots$ with probability $\frac{1}{2}$ of going from a state $k \geq 0$ to each of its two neighboring states and with an absorbing barrier at state -1; that is, the transition probability matrix is

$$\mathbf{P} = \begin{array}{c} \\ \\ \begin{array}{c} -1 \\ 0 \\ 1 \\ \vdots \end{array} \end{array} \begin{array}{c} \text{states:} \\ \begin{array}{cccc} -1 & 0 & 1 & 2 \quad \cdots \end{array} \\ \left\| \begin{array}{cccc} 1 & 0 & 0 & 0 \quad \cdots \\ \frac{1}{2} & 0 & \frac{1}{2} & 0 \quad \cdots \\ 0 & \frac{1}{2} & 0 & \frac{1}{2} \quad \cdots \\ \vdots & \vdots & \vdots & \vdots \end{array} \right\| \end{array}.$$

Although $\mathbf{P}$ is not of the form considered in the general method, we can still follow a procedure analogous to that used in Section 4.

The key to our analysis is the identity

$$\cos \theta \sin(k + 1)\theta = \tfrac{1}{2} \sin k\theta + \tfrac{1}{2} \sin(k + 2)\theta, \qquad k = 0, 1, 2, \ldots . \quad (5.3)$$

Since the kth row of $\mathbf{P}$ consists of elements

$$P_{k, -1} = 0, \quad P_{k, 0} = 0, \ldots, \quad P_{k, k-1} = \tfrac{1}{2}, \quad P_{k, k} = 0, \quad P_{k, k+1} = \tfrac{1}{2},$$

$$P_{k, k+2} = 0, \ldots, \qquad k = 1, 2, \ldots,$$

$$P_{0, -1} = \tfrac{1}{2}, \quad P_{0, 0} = 0, \quad P_{0, 1} = \tfrac{1}{2}, \quad P_{0, 2} = 0, \ldots,$$

the relation can be written, for $k = 0, 1, \ldots,$ as

$$\cos \theta \sin(k + 1)\theta = \sum_{r = -1}^{\infty} P_{kr} \sin(r + 1)\theta.$$

Multiplying this by $\cos \theta$ and substituting

$$\cos \theta \sin(r + 1)\theta = \sum_{s = -1}^{\infty} P_{rs} \sin(s + 1)\theta$$

into the right side of the resulting equation yields

$$\cos^2 \theta \sin(k + 1)\theta = \sum_{r=-1}^{\infty} P_{kr} \sum_{s=-1}^{\infty} P_{rs} \sin(s + 1)\theta$$

$$= \sum_{s=-1}^{\infty} \sin(s + 1)\theta \sum_{r=-1}^{\infty} P_{kr} P_{rs}$$

$$= \sum_{s=-1}^{\infty} P_{ks}^2 \sin(s + 1)\theta.$$

Repeating this procedure $n - 1$ times leads to

$$\cos^n \theta \sin(k + 1)\theta = \sum_{r=-1}^{\infty} P_{kr}^n \sin(r + 1)\theta.$$

Now multiply both sides of this equation by $\sin(s + 1)\theta$ and integrate with respect to θ over $[0, 2\pi]$:

$$\int_0^{2\pi} \cos^n \theta \sin(k + 1)\theta \sin(s + 1)\theta \, d\theta$$

$$= \int_0^{2\pi} \sum_{r=-1}^{\infty} P_{kr}^n \sin(r + 1)\theta \sin(s + 1)\theta \, d\theta$$

$$= \sum_{r=-1}^{\infty} P_{kr}^n \int_0^{2\pi} \sin(r + 1)\theta \sin(s + 1)\theta \, d\theta \qquad (s = 0, 1, \dots). \quad (5.4)$$

Using elementary trigonometric identities, it is easily shown that

$$\int_0^{2\pi} \sin(r + 1)\theta \sin(s + 1)\theta \, d\theta = \begin{cases} 0 & \text{if} \quad r \ne s \\ \pi & \text{if} \quad r = s \end{cases} \quad (r, s = 0, 1, 2, \dots). \quad (5.5)$$

It follows from (5.4) and (5.5) that we can express the n-step transition probabilities by the formula

$$P_{ks}^n = \frac{1}{\pi} \int_0^{2\pi} \cos^n \theta \sin(k + 1)\theta \sin(s + 1)\theta \, d\theta,$$

$$k, s = 0, 1, 2, \dots, \quad n = 0, 1, \dots. \quad (5.6)$$

This is just the general method as applied to the matrix

$$\mathbf{P}' = \begin{Vmatrix} 0 & \frac{1}{2} & 0 & 0 & \cdots \\ \frac{1}{2} & 0 & \frac{1}{2} & 0 & \cdots \\ 0 & \frac{1}{2} & 0 & \frac{1}{2} & \cdots \\ \vdots & \vdots & \vdots & \vdots & \end{Vmatrix}$$

obtained by deleting the first row and column from

$$\mathbf{P} = \begin{Vmatrix} 1 & 0 & 0 & 0 & \cdots \\ \frac{1}{2} & 0 & \frac{1}{2} & 0 & \cdots \\ 0 & \frac{1}{2} & 0 & \frac{1}{2} & \cdots \\ \vdots & \vdots & \vdots & \vdots & \end{Vmatrix}.$$

The validity of its use in computing the transition probabilities P_{ks}^n for k, $s = 0, 1, \ldots$ is based on the observation that in computing P_{ks}^n it is not necessary to consider any path which leads to state -1, since such a path cannot leave this state. The orthogonal polynomials are

$$Q_k(x) = \frac{\sin(k + 1)\theta}{\sin \theta}, \qquad k = 0, 1, 2, \ldots,$$

where $x = \cos \theta$, and it is easy to check that the functions $Q_k(x)$ are orthogonal with respect to $d\sigma(x) = (2\pi)^{-1}(1 - x^2)^{+1/2} \, dx$ over $[-1, 1]$.

As an application of the preceding result let us compute the probability that starting from state k absorption into state -1 occurs exactly at the nth transition. Absorption into state -1 can obviously occur at the nth step only if the process is in state 0 at the $(n - 1)$th step and then absorption occurs at the next step. But the probability of being in state 0 at the $(n - 1)$th step, having started from state k, is, by (5.6),

$$P_{k0}^{n-1} = \frac{1}{\pi} \int_0^{2\pi} \cos^{n-1} \theta \sin(k + 1)\theta \sin \theta \, d\theta,$$

while $P_{0, -1} = \frac{1}{2}$. Hence the probability of absorption into state -1 at time n starting from state k is

$$A_k^n = \frac{1}{2\pi} \int_0^{2\pi} \cos^{n-1} \theta \sin(k + 1)\theta \sin \theta \, d\theta. \tag{5.7}$$

6: Applications to Coin Tossing

The random walks discussed above are related to coin-tossing problems. Suppose that two gamblers agree to carry out a series of coin tossings with a fair coin. They agree that each time the coin shows heads gambler I wins one unit from gambler II; otherwise he loses one unit to gambler II. Let

$$X_i = \begin{cases} +1 & \text{if gambler I wins,} \\ -1 & \text{if gambler I loses,} \end{cases}$$

at the ith toss of the coin. Then $\Pr\{X_i = +1\} = \Pr\{X_i = -1\} = \frac{1}{2}$ and $S_n = \sum_{i=1}^n X_i$ $(n \geq 1)$ is the net gain of gambler I after n tosses of the coin. Further, set $S_0 = 0$. One of the simplest questions that can be asked concerning this contest is the following: *What is the probability that after n tosses of the coin gambler I's net gain will be zero?* Clearly gambler I's net gain cannot be zero if n is odd; hence we consider only the case $n = 2m$. Now if I's net gain is zero in $2m$ trials, evidently he won in m trials and lost in m trials. The desired probability is clearly

$$\mu_{2m} = 2^{-2m} \binom{2m}{m}, \qquad m = 1, 2, \ldots .$$

Next, what is the probability that the net gain of gambler I will equal zero for the first time after $n = 2m$ tosses of the coin? Clearly S_n describes a symmetric random walk process on all the integers. Therefore our question can be formulated this way: *What is the probability $f''_{0,0}$ of a first return to zero occurring at the nth step?* Starting from zero the first step can be $+1$ or -1 with probability $\frac{1}{2}$ for either choice. Because of the obvious symmetry about the origin the probability of first return to zero from $+1$ must equal that from -1 and so our question will be answered if we find the probability of first passage to zero from state $+1$ in $2m - 1$ steps. But this must equal the probability of first passage to state -1 starting from state 0 in $2m - 1$ steps, because of the homogeneous nature of the process. This equals the probability A_0^{2m-1} of absorption into state -1 in $2m - 1$ steps starting from state 0 in a random walk process on the integers $\{-1, 0, 1, 2, \ldots\}$ with absorbing barrier at state -1, which is given by formula (5.7) with $k = 0$ and $n = 2m - 1$:

$$A_0^{2m-1} = \frac{1}{2\pi} \int_0^{2\pi} \cos^{2m-2} \theta \sin^2 \theta \, d\theta = \frac{1}{\pi} \int_0^{\pi} \cos^{2m-2} \theta \sin^2 \theta \, d\theta.$$

To evaluate the latter integral make the substitution $x = \cos \theta$. Then

$$A_0^{2m-1} = \frac{1}{\pi} \int_{-1}^{1} x^{2m-2}(1 - x^2)^{1/2} \, dx$$

$$= \frac{2}{\pi} \frac{1}{2} \int_0^1 (x^2)^{m-3/2}(1 - x^2)^{1/2} 2x \, dx.$$

Next substitute $t = x^2$. Then

$$A_0^{2m-1} = \frac{1}{\pi} \int_0^1 t^{m-3/2}(1 - t)^{1/2} \, dt = \frac{1}{\pi} B(m - \tfrac{1}{2}, \tfrac{3}{2}), \tag{6.1}$$

where

$$B(\alpha, \beta) = \int_0^1 t^{\alpha-1}(1 - t)^{\beta-1} \, dt$$

is the beta function, which can be expressed in terms of the gamma function:

$$B(\alpha, \beta) = \frac{\Gamma(\alpha)\Gamma(\beta)}{\Gamma(\alpha + \beta)}.$$

Then

$$A_0^{2m-1} = \frac{1}{\pi} \frac{\Gamma(m - \frac{1}{2})\Gamma(\frac{3}{2})}{\Gamma(m + 1)} = \frac{1}{\pi} \frac{(m - \frac{3}{2})(m - \frac{5}{2}) \cdots \frac{3}{2} \cdot \frac{1}{2}\Gamma(\frac{1}{2})\frac{1}{2}\Gamma(\frac{1}{2})}{m(m - 1) \cdots 2 \cdot 1}$$

by well-known properties of the gamma function. Since $\Gamma(\frac{1}{2}) = \sqrt{\pi}$ we find that the probability that the total gain of gambler I will equal zero for the first time after $2m$ tosses of the coin is given by

$$f_{0,0}^{2m} = \begin{cases} \frac{1}{2} & \text{for} \quad m = 1, \\ \dfrac{(2m - 3)(2m - 5) \cdots 3 \cdot 1}{2^m \cdot m!} & \text{for} \quad m \geq 2. \end{cases} \tag{6.2}$$

A straightforward computation yields the following interesting result:

$$f_{0,0}^{2m} = \mu_{2m-2} - \mu_{2m}, \qquad m = 1, 2, \ldots,$$

where we define $\mu_0 = 1$.

$$\sum_{k=m+1}^{\infty} f_{0,0}^{2k} = \sum_{k=m+1}^{\infty} (\mu_{2k-2} - \mu_{2k}) = \mu_{2m} - \lim_{n \to \infty} \mu_{2n}$$

$$= \mu_{2m} - \lim_{n \to \infty} 2^{-2n} \binom{2n}{n} = \mu_{2m}. \tag{6.3}$$

Interpreting the two ends of (6.3), we have

$$\mu_{2m} = \Pr\{S_{2m} = 0\} = \Pr\{S_1 \neq 0, S_2 \neq 0, \ldots, S_{2m} \neq 0\}. \tag{6.4}$$

Next we want to answer the question, *What is the probability that gambler I will have a net gain zero for the kth time after $2m$ tosses of the coin ($k > 1$)?* In terms of the random walk described by the process S_n this question can be formulated in the following way: What is the probability that the kth return to the origin will occur at the $2m$th step? Again our first step from state zero can be taken to state $+1$ and we can similarly assume that after each of the first $k - 1$ returns to the origin we always make our next step to state $+1$. Now, since our random walk is homogeneous, we can freely interchange intermediate steps without affecting the probability of reaching one state from another. This way we can "take initially" our first step "to the right" (to state $+1$), and all steps to state $+1$ after each of the first $k - 1$ returns to the origin, thus reaching state k in the first k steps. Then the required probability will be the probability of reaching state zero for the first time in $2m - k$ steps from state k or reaching state -1 for the first time in $2m - k$ steps from state $k - 1$. The latter probability, however, is that of absorption into state -1 at the $(2m - k)$th step starting from state $k - 1$, in the random walk process on the integers $\{-1, 0, 1, 2, \ldots\}$ with absorbing barrier at state -1, and is given by formula (5.7) with $n = 2m - k$ and $k - 1$ in place of k:

$$A_{k-1}^{2m-k} = \frac{1}{2\pi} \int_0^{2\pi} \cos^{2m-k-1} \theta \sin k\theta \sin \theta \, d\theta. \tag{6.5}$$

The value of this integral is

$$A_{k-1}^{2m-k} = \frac{1}{2^{2m-k}} \binom{2m-k}{m} \frac{k}{2m-k}. \tag{6.6}$$

(For a validation of (6.6) the reader may consult Problems 6 and 7.)

As a further question, we consider the sequence $S_1, S_2, \ldots, S_{2m}$ *and ask for the probability that exactly k members of this sequence will equal zero.* This will be the probability $z_{k,2m}$ that in $2m$ tosses of the coin gambler I's net gain will be zero exactly k times.

By (6.4) $z_{0,2m} = \mu_{2m}$. Now we evaluate $z_{1,2m}$. Let B_r be the event that among $S_1, \ldots, S_{2m}$ only S_{2r} vanishes. Then for $r < m$

$$\begin{aligned}
B_r &= \{S_1 \neq 0, \ldots, S_{2r-1} \neq 0, S_{2r} = 0, S_{2r+1} \neq 0, \ldots, S_{2m} \neq 0\} \\
&= \{S_1 \neq 0, \ldots, S_{2r-1} \neq 0, S_{2r} = 0\} \\
&\cap \{S_{2r+1} - S_{2r} \neq 0, \ldots, S_{2m} - S_{2r} \neq 0\}.
\end{aligned}$$

Now clearly the events $\{S_1 \neq 0, \ldots, S_{2r-1} \neq 0, S_{2r} = 0\}$ and $\{S_{2r+1} - S_{2r} \neq 0, \ldots, S_{2m} - S_{2r} \neq 0\}$ are independent, and the probability of the latter is just μ_{2m-2r}. Thus

$$\Pr\{B_r\} = f_{0,0}^{2r} z_{0,2m-2r}, \qquad 1 \leq r \leq m,$$

where $z_{0,0} = 1$ by definition and so

$$z_{1,2m} = \sum_{r=1}^{m} \Pr\{B_r\} = \sum_{r=1}^{m} f_{0,0}^{2r} z_{0,2m-2r}. \tag{6.7}$$

But μ_{2m-2r} is the probability of gambler I's net gain being zero after $2m - 2r$ coin tossings and, as pointed out before, is equal to $\Pr\{S_{2m} - S_{2r} = 0\}$. But the events $\{S_1 \neq 0, \ldots, S_{2r-1} \neq 0, S_{2r} = 0\}$ and $\{S_{2m} - S_{2r} = 0\}$ are also independent, and so we obtain finally

$$\Pr\{B_r\} = \Pr\{S_1 \neq 0, \ldots, S_{2r-1} \neq 0, S_{2r} = 0, S_{2m} - S_{2r} = 0\}.$$

The events on the right-hand side are mutually exclusive and their union is the event $\{S_{2m} = 0\}$; hence

$$\sum_{r=1}^{m} \Pr\{B_r\} = \Pr\{S_{2m} = 0\} = \mu_{2m}.$$

Thus $z_{1,2m} = \mu_{2m} = z_{0,2m}$ for $m \geq 1$. In the same way we can show that

$$z_{k,2m} = \sum_{r=1}^{m-k+1} f_{0,0}^{2r} z_{k-1,2m-2r}, \qquad k \geq 2, m \geq 1. \tag{6.8}$$

Comparing this with (6.7) and using $z_{1,2m} = z_{0,2m}$ and the property that $f_{0,0}^{2m} = \mu_{2m-2} - \mu_{2m} = z_{0,2m-2} - z_{1,2m}$, we obtain

$$z_{2,2m} = z_{1,2m} - f_{0,0}^{2m} = 2z_{1,2m} - z_{0,2m-2}, \qquad m \geq 1.$$

Substituting this into (6.8), by induction we obtain the recursion relation

$$z_{k, 2m} = 2z_{k-1, 2m} - z_{k-2, 2m-2}.$$ (6.9)

If we write $z_{k, 2m} = 2^{k-2m} a_{k, 2m}$, (6.9) reduces to

$$a_{k-1, 2m} = a_{k, 2m} + a_{k-2, 2m-2}.$$ (6.10)

These recursion relations are satisfied by

$$a_{k, 2m} = \binom{2m - k}{m}.$$ (6.11)

Now $z_{0, 2m}$ and $z_{1, 2m}$ are known, and direct substitution shows that $a_{0, 2m}$ and $a_{1, 2m}$ are given by (6.11). Clearly (6.10) uniquely determines $a_{k, 2m}$ for $k \geq 2$; hence it follows that (6.11) is the correct evaluation of $a_{k, 2m}$, and so

$$z_{k, 2m} = 2^{k-2m} \binom{2m - k}{m}.$$ (6.12)

To answer another question about the sequence $S_1, \ldots, S_{2n}$ we define the concept of a sign change in the sequence. We say that a sign change occurs at time k if $S_k = 0$ and $S_{k-1} S_{k+1} = -1$. Then we ask the question, *What is the probability that there are exactly r sign changes in the sequence $S_1, \ldots, S_{2n}$?* We will have r sign changes in the sequence if among the $S_2, S_4, \ldots, S_{2n-2}$ there are exactly k zeros ($k = r, r + 1, \ldots, n - 1$) and at exactly r out of these k states the process changes direction, i.e., the required probability is

$$C_{r, 2n} = \sum_{k=r}^{n-1} \Pr\{\text{exactly } k \text{ of } S_2, S_4, \ldots, S_{2n-2} \text{ are zero}\}$$

$$\times \Pr\{r \text{ out of } k \text{ times we change direction at zero}\}.$$

But

$$\Pr\{\text{exactly } k \text{ of } S_2, S_4, \ldots, S_{2n-2} \text{ are zero}\}$$

$$= \Pr\{\text{exactly } k \text{ of } S_1, S_2, S_3, S_4, \ldots, S_{2n-2} \text{ are zero}\}$$

$$= 2^{k-2n+2} \binom{2n - k - 2}{n - 1},$$

as is given by formula (6.12). Further, a change in direction at a zero can occur with probability $\frac{1}{2}$. Hence

$$\Pr\{r \text{ out of } k \text{ times we change directions at zero}\}$$

$$= \binom{k}{r} \frac{1}{2^r} \frac{1}{2^{k-r}} = \binom{k}{r} 2^{-k},$$

so

$$C_{r,2n} = \sum_{k=r}^{n-1} 2^{k-2n+2} \binom{2n-k-2}{n-1}\binom{k}{r} 2^{-k}$$

$$= 2^{2-2n} \sum_{j=0}^{n-r-1} \binom{2n-r-j-2}{n-1}\binom{r+j}{r}$$

$$= 2^{2-2n} \sum_{j=0}^{n-r-1} \binom{2n-r-j-2}{n-r-j-1}\binom{r+j}{j}.$$

Elementary Problems

1. Verify the statement of Theorem 2.1 for the Markov matrix

$$\left\| \begin{matrix} p & 1-p \\ q & 1-q \end{matrix} \right\|, \qquad \text{where} \quad 0 \le p \le 1 \quad \text{and} \quad 0 \le q \le 1.$$

For what values of p and q is there exactly one recurrent class in the two-state Markov chain corresponding to this matrix?

2. Let $\mathbf{P}, \mathbf{Q}$ be finite probability transition matrices of order n such that $\mathbf{PQ} = \mathbf{I} = \mathbf{QP}$. Show that $\mathbf{P}$ and $\mathbf{Q}$ are permutation matrices, i.e., matrices with only one nonzero entry in any row or column.

3. Suppose we have a two-state Markov chain $\{X_n, n \ge 0\}$ with states 0 and 1, and $P_{00} = 1 - \alpha$, $P_{01} = \alpha$, $P_{11} = 1 - \beta$, $P_{10} = \beta$ $(0 < \alpha, \beta < 1)$. Let N be the first index $n \ge 1$ such that $X_{n-1} = X_n = 0$, and let $d_0 = E[N|X_0 = 0]$. Prove that

$$d_0 = 1 + \frac{\alpha}{1-\alpha}\frac{1+\beta}{\beta}.$$

Hint: Consider the relationships between d_0 and $d_1 = E[N|X_0 = 1]$.

4. Consider the 3×3 Markov transition matrix

$$\mathbf{P} = \left\| \begin{matrix} \tfrac{1}{2} & \tfrac{1}{2} & 0 \\ \tfrac{1}{2} & 0 & \tfrac{1}{2} \\ 0 & \tfrac{1}{2} & \tfrac{1}{2} \end{matrix} \right\|.$$

Determine the corresponding eigenvalues and right and left eigenvectors and thereby establish the spectral representation $\mathbf{P} = \mathbf{\Phi \Lambda \Psi}$, where

$$\mathbf{\Phi} = \left\| \begin{matrix} 1/\sqrt{3} & 1/\sqrt{2} & 1/\sqrt{6} \\ 1/\sqrt{3} & 0 & -2/\sqrt{6} \\ 1/\sqrt{3} & -1/\sqrt{2} & 1/\sqrt{6} \end{matrix} \right\|, \qquad \mathbf{\Psi} = \left\| \begin{matrix} 1/\sqrt{3} & 1/\sqrt{3} & 1/\sqrt{3} \\ 1/\sqrt{2} & 0 & -1/\sqrt{2} \\ 1/\sqrt{6} & -2/\sqrt{6} & 1/\sqrt{6} \end{matrix} \right\|$$

and

$$\mathbf{\Lambda} = \left\| \begin{matrix} 1 & 0 & 0 \\ 0 & \tfrac{1}{2} & 0 \\ 0 & 0 & -\tfrac{1}{2} \end{matrix} \right\|.$$

5. Let $\{X_n, n \geq 0\}$ be a Markov chain on the states $0, 1, \ldots, N$. Suppose that states $0, \ldots, r - 1$ are transient while states $r, \ldots, N$ are absorbing ($p_{ii} = 1$ for $r \leq i \leq N$). The transition matrix has the form

$$\mathbf{P} = \left\| \begin{matrix} \mathbf{Q} & \mathbf{R} \\ \mathbf{0} & \mathbf{I} \end{matrix} \right\|,$$

where $\mathbf{0}$ is an $(N - r + 1) \times r$ matrix all of whose components are zero, $\mathbf{I}$ is an $(N - r + 1) \times (N - r + 1)$ identity matrix and $q_{ij} = p_{ij}$ for $0 \leq i, j < r$.

(a) Show that the n-step transition matrix can be written as

$$\mathbf{P}^n = \left\| \begin{matrix} \mathbf{Q}^n & (\mathbf{I} + \mathbf{Q} + \cdots + \mathbf{Q}^{n-1})\mathbf{R} \\ \mathbf{0} & \mathbf{I} \end{matrix} \right\|.$$

(b) For transient states i and j, let n_{ij} be the mean total number of visits to state j before absorption, conditioned on $X_0 = i$, and let $\mathbf{N}$ be the $r \times r$ matrix whose elements are n_{ij}. Show that $n_{ij} = \sum_{n=0}^{\infty} p_{ij}^{(n)}$, whence $\mathbf{N} = \mathbf{I} + \mathbf{Q} + \mathbf{Q}^2 + \cdots$. Establish that the matrix $\mathbf{I} - \mathbf{Q}$ is invertible with inverse $\mathbf{N} = (\mathbf{I} - \mathbf{Q})^{-1}$ by directly verifying that $\mathbf{N}(\mathbf{I} - \mathbf{Q}) = \mathbf{I}$.

(c) For transient state i and absorbing state j, let b_{ij} be the probability of absorption in j conditioned on $X_0 = i$. Show that $b_{ij} = \lim_{n \to \infty} p_{ij}^{(n)}$ for $0 \leq i < r$ and $r \leq j \leq N$, whence $\mathbf{B} = \mathbf{NR}$ where $\mathbf{B}$ is the $r \times (N - r + 1)$ matrix having elements b_{ij}.

Remark: $\mathbf{N}$ is called the *fundamental matrix* of the chain.

6. Let $\mathbf{P} = \|p_{ij}\|_{i,j=1}^{N}$ be the transition matrix of an irreducible aperiodic Markov chain $\{X_n, n \geq 0\}$. Let $T_j = \min\{n \geq 1 : X_n = j\}$ be the first hitting time to state j and let $m_{ij} = E[T_j | X_0 = i]$.

(a) Establish the formula (*) $\mathbf{M} = \mathbf{1} + \mathbf{P}(\mathbf{M} - \mathbf{M}_{dg})$ where $\mathbf{1}$ is an $N \times N$ matrix all of whose elements are 1, $\mathbf{M} = \|m_{ij}\|_{i,j=1}^{N}$ and

$$\mathbf{M}_{dg} = \left\| \begin{matrix} m_{11} & 0 & \cdots & 0 \\ 0 & m_{22} & \cdots & 0 \\ \vdots & & \ddots & \vdots \\ 0 & 0 & \cdots & m_{NN} \end{matrix} \right\|.$$

(b) By multiplying Eq. (*) on the left by the stationary distribution π and using $\pi = \pi\mathbf{P}$ show that $\pi\mathbf{M}_{dg} = (1, 1, \ldots, 1)$, whence $\pi_i = 1/m_{ii}$ for $i = 1, \ldots, N$.

7. Determine $P_{ij}^{(n)}$ for the Markov chain with transition probabilities

$$P_{ij} = \begin{cases} \dfrac{1}{(i+1)(i+2)}, & j < i + 1, \\[2mm] \dfrac{(i+1)}{(i+2)} & j = i + 1, \quad i = 1, 2, \ldots, \\[2mm] 0, & j > i + 1. \end{cases}$$

Hint: Use induction.

Answer:

$$P_{ij}^{(n)} = \begin{cases} \dfrac{n}{(i+n)(i+n+1)}, & j < i+n, \\[2ex] \dfrac{i+1}{i+n+1}, & j = i+n, \quad i = 1, 2, \ldots, \\[2ex] 0, & j > i+n. \end{cases}$$

Problems

1. Consider a finite state Markov chain $\{X_n\}$, $n = 0, 1, \ldots$, with two classes, one of which is an absorbing state. For definiteness, let 0 denote the absorbing barrier and let $i = 1$, $2, \ldots, N$ represent the states of the other class. Let the eigenvalues of the transition probability matrix arranged in decreasing order be $\lambda_0 = 1 > |\lambda_2| > |\lambda_3| \geq |\lambda_4| \geq \cdots$. (Note: we are assuming that $1 > |\lambda_2| > |\lambda_3|$.) Let b_j be the limiting probability of being in state j given that absorption into state 0 has not occurred and the initial state is i, i.e., $b_j = \lim_{n \to \infty} \Pr\{X(n) = j \mid X(n) \neq 0, X(0) = i\}$. (See page 6.) Determine the rate of approach to zero of

$$\frac{P_{ij}^n}{1 - P_{i0}^n} - b_j, \quad j = 1, 2, \ldots, N \quad (i \geq 1).$$

2. Consider the finite random walk on the integers $0, 1, 2, \ldots, N$ represented by the matrix of one-step transition probabilities

$$\mathbf{P} = \begin{Vmatrix} 0 & 1 & 0 & 0 & \cdots & & & \\ \frac{1}{2} & 0 & \frac{1}{2} & 0 & \cdots & & & \\ 0 & \frac{1}{2} & 0 & \frac{1}{2} & \cdots & & & \\ & \vdots & \vdots & \vdots & & & & \\ & & & & \cdots & \frac{1}{2} & 0 & \frac{1}{2} \\ & & & & \cdots & 0 & 1 & 0 \end{Vmatrix}.$$

Find the formula for the r-step transition probabilities using the method of orthogonal polynomials.

3. Consider the same random walk as in Problem 2, but now ignore states 0 and N. (States 0 and N act as absorbing barriers.) Then the $(N-1) \times (N-1)$ matrix of the one-step transition probabilities corresponding to the transient states will be

$$\mathbf{P} = \begin{Vmatrix} 0 & \frac{1}{2} & 0 & 0 & \cdots & 0 & 0 \\ \frac{1}{2} & 0 & \frac{1}{2} & 0 & \cdots & 0 & 0 \\ 0 & \frac{1}{2} & 0 & \frac{1}{2} & \cdots & 0 & 0 \\ \vdots & \vdots & \vdots & \vdots & & & \\ 0 & 0 & 0 & 0 & & \frac{1}{2} & 0 \end{Vmatrix}.$$

Find the formula for the r-step transition probabilities, given that absorption has not taken place.

Hint: The relevant orthogonal polynomials are $Q_n(x) = \sin n\theta$, $x = \cos \theta$ where θ varies over the finite set $\theta = k\pi/N$, $k = 0, 1, \ldots, 2N - 1$.

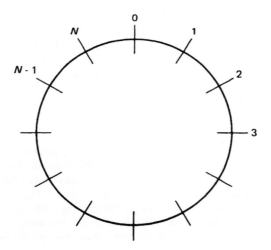

4. Consider the random walk on the circle, i.e., along $N + 1$ points numbered $0, 1, 2, \ldots, N$, placed symmetrically along the circumference of a circle. Let the matrix of first-step transition probabilities be given by

$$\mathbf{P} = \begin{Vmatrix} 0 & \tfrac{1}{2} & 0 & 0 & \cdots & 0 & 0 & \tfrac{1}{2} \\ \tfrac{1}{2} & 0 & \tfrac{1}{2} & 0 & \cdots & 0 & 0 & 0 \\ 0 & \tfrac{1}{2} & 0 & \tfrac{1}{2} & \cdots & 0 & 0 & 0 \\ \vdots & \vdots & \vdots & \vdots & & \vdots & \vdots & \vdots \\ 0 & 0 & 0 & 0 & \cdots & \tfrac{1}{2} & 0 & \tfrac{1}{2} \\ \tfrac{1}{2} & 0 & 0 & 0 & \cdots & 0 & \tfrac{1}{2} & 0 \end{Vmatrix}.$$

Find the formula for the r-step transition probabilities, $\mathbf{P}^r = \| P_{nm}^{(r)} \|$.

5. Consider the random walk on the circle as in Problem 4, but let the matrix of first-step transition probabilities be given by

$$\mathbf{P} = \begin{Vmatrix} 0 & p & 0 & \cdots & 0 & q \\ q & 0 & p & \cdots & 0 & 0 \\ 0 & q & 0 & \cdots & 0 & 0 \\ \vdots & \vdots & \vdots & & \vdots & \vdots \\ 0 & 0 & 0 & \cdots & 0 & p \\ p & 0 & 0 & \cdots & q & 0 \end{Vmatrix}.$$

Find the formula for the r-step transition probabilities.

Hint: Let $Z(\theta) = pe^{i\theta} + qe^{-i\theta}$ in the analysis of Problem 4.

6. Prove

$$\frac{1}{\pi} \int_0^{2\pi} \cos^n \theta \cos k\theta \cos l\theta \, d\theta$$

$$= \begin{cases} \left[\binom{n}{(n + k - l)/2} + \binom{n}{(n + k + l)/2} \right] \frac{1}{2^{n+1}} & \text{if } n + k + l \text{ is even,} \\ 0 & \text{otherwise.} \end{cases}$$

(This is P_{kl}^n of formula (4.7) for $l \neq 0$.)
Evaluate $\lim_{n \to \infty} \sqrt{n} P_{kl}^n$.

Hint: Let k be a nonnegative integer. Prove

$$\frac{1}{2\pi} \int_0^{2\pi} \cos^n \theta \cos k\theta \, d\theta = \begin{cases} 0 & \text{if } n + k \text{ is odd,} \\ 2^{-n} \binom{n}{(n + k)/2} & \text{otherwise.} \end{cases}$$

To this end, use the identity

$$\cos^{n+1} \theta \cos(k - 1)\theta = \cos^n \theta \cos k\theta + \cos^n \theta \sin(k - 1)\theta \sin \theta,$$

integrate by parts, and derive the recursion relation

$$\int \cos^n \theta \cos k\theta \, d\theta = \frac{n - k + 2}{n + 1} \int \cos^{n+1} \theta \cos(k - 1)\theta \, d\theta.$$

7. Evaluate

$$P_{kl}^n = \frac{1}{\pi} \int_0^{2\pi} \cos^n \theta \sin(k + 1)\theta \sin(l + 1)\theta \, d\theta.$$

Hint: Use the result of Problem 6.

8. Let $\mathbf{P} = \|P_{ij}\|$ denote the transition probability matrix of a finite state Markov chain $\{X_n\}_0^\infty$ consisting of three classes $\{0\}$, $\{1, 2, \ldots, N - 1\}$, and $\{N\}$ of which 0 and N are absorbing states and the other class is transient. We introduce the family of matrices

$$\mathbf{P}(\theta) = \|P_{ij} e^{\theta(j - i)}\| = \|P_{ij}(\theta)\| \qquad (\theta \text{ real})$$

and the moment-generating functions $M^{(t)}(\theta|k)$,

$$M^{(t)}(\theta|X_0 = k) = E[\exp(\theta(X_t - X_0))|X_0 = k] = \mathbf{e}_k' \mathbf{P}^t(\theta)\mathbf{e} = \sum_{j=0}^N P_{kj}^t(\theta),$$

where $\mathbf{e}_k'$ indicates the row vector $(0, \ldots, 1, 0, \ldots, 0)$ with a unit in the position k, $\mathbf{e}$ is the $N + 1$ column vector of unit elements, and $\mathbf{P}^t(\theta)$ is the tth power of $\mathbf{P}(\theta)$. Let π_{k0} and $\pi_{kN} = 1 - \pi_{k0}$ $(1 \leq k \leq N - 1)$ denote the probability of ultimate absorption in states 0 and N, respectively, for the Markov chain $\{X_n\}_0^\infty$ with initial state k. Prove that

$$\lim_{t \to \infty} M^{(t)}(\theta|k) = \pi_{k0} e^{-k\theta} + \pi_{kN} e^{(N-k)\theta}.$$

9. Under the same notation as in Problem 8, suppose real a and b exist satisfying

$$M(a|k) \leq 1 \leq M(b|k) \qquad (k = 1, 2, \ldots, N - 1) \qquad (+)$$

[here $M(\theta|k) = M^{(1)}(\theta|k)$]. Prove that

$$\frac{e^{kb} - 1}{e^{Nb} - 1} \leq \pi_{kN} \leq \frac{e^{ka} - 1}{e^{Na} - 1}.$$

Hint: Verify $M^{(t)}(a|k) \leq 1 \leq M^{(t)}(b|k)$ $(k = 1, 2, \ldots, N - 1)$ holds for all t and use the result of the previous problem.

10. Consider a Markov chain on $(0, 1, \ldots, N)$ with transition probability matrix

$$P_{ij} = \binom{N}{j} p_i^j (1 - p_i)^{N-j}, \qquad \text{where} \quad p_i = \frac{(1 + \sigma)i}{N + \sigma i}, \quad 0 < \sigma < 1$$

(see Example G of Section 2, Chapter 2).

Verify that

$$M(\theta|k) = e^{-k\theta}(p_k e^{\theta} + 1 - p_k)^N$$

[see Problem 8 regarding the definition of $M(\theta|k)$].

11. Under the conditions of Problem 10 show that

$$a = \log \frac{1}{1 + \sigma}, \qquad b = \log \frac{1 - \sigma}{1 + \sigma}$$

fulfill the requirements of $(+)$ in Problem 9 and therefore deduce the bounds

$$\frac{[(1 - \sigma)/(1 + \sigma)]^k - 1}{[(1 - \sigma)/(1 + \sigma)]^N - 1} \leq \pi_{kN} \leq \frac{1/(1 + \sigma)^k - 1}{1/(1 + \sigma)^N - 1}.$$

12. *Some Coin-Tossing Relations.* Let $\{X_i\}$, $1 \leq i < \infty$, be independent, identically distributed, random variables such that

$$\Pr\{X_i = +1\} = \Pr\{X_i = -1\} = \tfrac{1}{2}.$$

Let $S_n = \sum_{i=1}^n X_i$ for $1 \leq n < \infty$ and

$$P(m, n) = \Pr\{S_{2j} = 0 \text{ for some } j \text{ satisfying } m \leq j < m + n\}.$$

Prove that $P(m, n) + P(n, m) = 1$ for $m \geq 1, n \geq 1$.

Hint: Recall that $\Pr\{S_{2n} = 0\} = \Pr\{S_1 \neq 0, S_2 \neq 0, \ldots, S_{2n} \neq 0\}$. Now assume the result holds for $m = k$ and arbitrary $n \geq 1$. Then justify the following equalities:

$$\begin{aligned}
1 - P(k + 1, n) &= \Pr\{S_{2j} \neq 0 \text{ for } k \leq j < k + n + 1\} \\
&\quad + \Pr\{S_{2k} = 0 \text{ and } S_{2j} \neq 0 \text{ for } k + 1 \leq j < k + n + 1\} \\
&= \Pr\{S_{2j} \neq 0 \text{ for } k \leq j < k + n + 1\} \\
&\quad + \Pr\{S_{2k} = 0\}\Pr\{S_{2j} \neq 0 \text{ for } 1 \leq j < n + 1\} \\
&= \Pr\{S_{2j} = 0 \text{ for some } j \text{ with } n + 1 \leq j < k + n + 1\} \\
&\quad + \Pr\{S_{2j} \neq 0 \text{ for } 1 \leq j < k + 1\}\Pr\{S_{2n} = 0\} \\
&= \Pr\{S_{2j} = 0 \text{ for some } j \text{ with } n + 1 \leq j < k + n + 1\} \\
&\quad + \Pr\{S_{2n} = 0 \text{ and } S_{2j} \neq 0 \text{ for } n + 1 \leq j < k + n + 1\} \\
&= P(n, k + 1).
\end{aligned}$$

13. Let $\{X_n, n \geq 0\}$ be a Markov chain on the states $0, 1, \ldots, N$. Suppose that states $0, \ldots, r - 1$ are transient while states $r, \ldots, N$ are absorbing ($p_{ii} = 1$ for $r \leq i \leq N$). The transition matrix has the form

$$\mathbf{P} = \begin{Vmatrix} \mathbf{Q} & \mathbf{R} \\ \mathbf{0} & \mathbf{I} \end{Vmatrix},$$

where $\mathbf{0}$ is an $(N - r + 1) \times r$ matrix all of whose components are zero, $\mathbf{I}$ is an $(N - r + 1) \times (N - r + 1)$ identity matrix and $q_{ij} = p_{ij}$ for $0 \leq i, j < r$.

For a transient state j, let T_j be the number of visits to j before absorption and let

$$f_i(\mathbf{s}) = E[s_0^{T_0} s_1^{T_1} \cdots s_{r-1}^{T_{r-1}} | X_0 = i], \qquad |s_j| < 1, \quad j = 0, \ldots, r - 1.$$

(a) Show that

$$\mathbf{f}(\mathbf{s}) = \mathbf{S}(\mathbf{I} - \mathbf{QS})^{-1}(\mathbf{I} - \mathbf{Q})\mathbf{1},$$

where

$$\mathbf{f}(\mathbf{s}) = \begin{pmatrix} f_0(\mathbf{s}) \\ f_1(\mathbf{s}) \\ \vdots \\ f_{r-1}(\mathbf{s}) \end{pmatrix}, \qquad \mathbf{1} = \begin{pmatrix} 1 \\ 1 \\ \vdots \\ 1 \end{pmatrix}, \qquad \text{and} \qquad \mathbf{S} = \begin{pmatrix} s_0 & 0 & \cdots & 0 \\ 0 & s_1 & \cdots & 0 \\ \vdots & \vdots & & \vdots \\ 0 & 0 & \cdots & s_{r-1} \end{pmatrix}.$$

(b) Let $v_{ij} = E[T_j^2 | X_0 = i]$ and let $\mathbf{V} = \|v_{ij}\|$. By differentiating the result of (a), show

$$\mathbf{V} = \mathbf{N}(2\mathbf{N} - \mathbf{I})\mathbf{1} \qquad \text{where} \qquad \mathbf{N} = (\mathbf{I} - \mathbf{Q})^{-1}$$

(see Elementary Problem 5).

(c) Show that the variance of T_j given $X_0 = i$ is $\text{Var}[T_j | X_0 = i] = n_{ij}(2n_{jj} - 1 - n_{ij})$ for $0 \leq i, j < r$.

14. Let $\mathbf{P} = \|p_{ij}\|$ be the transition matrix of a Markov chain $\{X_n, n \geq 0\}$ on the states $0, 1, \ldots, N$. Suppose that states 0 and N are absorbing ($p_{00} = p_{NN} = 1$) while states $1, \ldots, N - 1$ are transient. Let $T_k = \min\{n \geq 0 : X_n = k\}$ be the first passage time to state k for $k = 0$ or N, and then specify the finite-dimensional distributions of a process $\{Y_n, n \geq 0\}$ on states $1, \ldots, N$ using the formula

$$\Pr\{Y_1 = j_1, \ldots, Y_n = j_n | Y_0 = i\} = \frac{\Pr\{X_1 = j_1, \ldots, X_n = j_n \text{ and } T_N < T_0 | X_0 = i\}}{\Pr\{T_N < T_0 | X_0 = i\}}.$$

Informally, we speak of $\{Y_n\}$ as the process $\{X_n\}$ conditioned on eventual absorption in state N.

(a) Show that $\{Y_n\}$ is a Markov chain having transition probabilities $q_{ij} = p_{ij} u_j / u_i$, where $u_i = \Pr\{T_N < T_0 | X_0 = i\}$.

(b) Show that the n-step transition probabilities satisfy $q_{ij}^{(n)} = p_{ij}^{(n)} u_j / u_i$.

(c) If m_{ij} (respectively, n_{ij}) is the mean number of visits to state j of $\{Y_n\}$, (respectively, $\{X_n\}$) before absorption, conditioned on $X_0 = Y_0 = i$, show that $m_{ij} = n_{ij} u_j / u_i$ for $i, j = 1, \ldots, N - 1$.

15. In Section 2 it is indicated that for a *finite* absorbing Markov chain the limits

$$b_j = \lim_{n \to \alpha} \Pr\{X_n = j \,|\, X_n \in T, X_0 = i\}, \qquad j \in T,$$

exist, and that the elements b_j form the left eigenvector of $\mathbf{P}$ corresponding to λ_1, the largest nonunit eigenvalue of $\mathbf{P}$. (Assume T, the set of transient states, is closed and aperiodic.) Normalize so that $\sum_{j \in T} b_j = 1$. Then $\{b_j\}$ is known as the quasi-stationary distribution (or asymptotic conditional distribution) of $\{X_n\}$.

Show that if $\Pr\{X_0 = j\} = b_j$ then $\Pr\{X_n = j \,|\, X_n \in T\} = b_j$ for all $n \geq 1$.

16. *(continuation).* Let $\{c_j\}_{j \in T}$ be the elements in the right eigenvector of $\mathbf{P}$ corresponding to λ_1 and normalized so that $\sum_{j \in T} c_j b_j = 1$. Show that

$$\lim_{m \to \infty} \lim_{k \to \infty} \Pr\{X_m = j \,|\, X_{m+k} \in T\} = b_j c_j.$$

The distribution $\{b_j c_j\}$ is called the *product distribution*.

17. *(continuation).* Let $\{X_n^*\}$ be a new Markov chain whose transition matrix P^* has elements

$$P_{ij}^* = \frac{b_j P_{ji}}{\lambda_1 b_i}.$$

Show that the product distribution is the stationary distribution for this chain.

18. *(continuation).* Let $\{X_n^{\#}\}$ be the Markov chain whose transition matrix $\mathbf{P}^{\#}$ has elements

$$P_{ij}^{\#} = \frac{c_i P_{ij}}{\lambda_1 c_j}.$$

Find the stationary distribution.

19. Suppose the chain of Problem 15 has at least two absorbing states. Form a new Markov chain by conditioning an eventual absorption into a particular fixed state. (i) Show that the product distribution on the transient states is unchanged. (ii) Evaluate the quasi-stationary distribution of the new chain.

NOTES

Chapter 16 of Feller [1] is devoted to a discussion of algebraic methods in Markov chains. Also relevant is the book by Kemeny and Snell [2].

REFERENCES

1. W. Feller, "An Introduction to Probability Theory and Its Applications," Vol. 1, 2nd ed. Wiley, New York, 1957.
2. J. G. Kemeny and J. L. Snell, "Finite Markov Chains." Van Nostrand-Reinhold, New York, 1960.

Chapter 11

RATIO THEOREMS OF TRANSITION PROBABILITIES AND APPLICATIONS

This chapter provides an introduction to some important and expanding areas in probability theory. Section 6, on optimal stopping, requires only the definitions of *regular* and *superregular* from the previous section as a prerequisite.

1: *Taboo Probabilities*

In the null recurrent irreducible Markov chain the average of the transition probabilities $(1/n) \sum_{m=0}^{n-1} P_{ij}^m$ tends to zero as n increases to ∞. Even so, recurrence to any given state occurs with certainty. This says that the relative frequency of visits to any given state approaches zero with the passage of time yet each state is visited infinitely often with probability 1. It is meaningful to determine the number of visits to a given state i relative to the number of visits to a different state j as the number of trials increases to infinity. For this purpose, it is useful to introduce the "taboo transition probabilities"

$$_kP_{ij}^n = \Pr\{X_n = j, X_v \neq k, v = 1, 2, \ldots, n-1 | X_0 = i\} \quad \text{for} \quad k \neq j, n \geq 1,$$

the probability that the process will go from state i to state j in n steps without entering state k in the intervening time. In this context state k is called a taboo state. Similarly, we define for $k \neq j, n \geq 1$,

$$_kf_{ij}^n = \Pr\{X_n = j, X_v \neq j, X_v \neq k, v = 1, 2, \ldots, n-1 | X_0 = i\},$$

the probability that the process will go from state i to state j for the first time at the nth transition without visiting state k in the intervening time. For convenience we define for $k \neq i$

$$_kP_{ij}^0 = \begin{cases} 0 & \text{if} \quad i \neq j, \\ 1 & \text{if} \quad i = j, \end{cases}$$

and

$$_kf_{ij}^0 \equiv 0 \qquad \text{for all} \quad i, j.$$

Now for $i \neq j, n \geq 0$,

$$f_{ij}^n = \sum_{v=0}^n {}_jP_{ii}^v {}_jf_{ij}^{n-v}, \tag{1.1}$$

since the event of first transition from i to j in n steps can be decomposed into n mutually exclusive events of returning to state i in v steps without visiting state j in the intervening time and then entering state j for the first time in $n - v$ steps without returning to state i, $v = 0, 1, \ldots, n - 1$.

The key to the derivation of (1.1) consists of classifying the paths according to the *last* time prior to n at which state i is visited. This should be contrasted with the method underlying the derivation of formula (5.1), Chapter 2, which exploits the idea of the *first* occurrence of a given event.

Generally, relations involving taboo states are verified by considering the first or last occurrence of some event. This duality between first and last plays an important role in many areas of probability theory. This is particularly true in the case of sums of identically and independently distributed random variables where there is a complete equivalence between these concepts.

The identity (1.1) can be expressed by an equivalent generating function equation. This may be deduced by paralleling the method that led from (5.9) to (5.10) of Chapter 2. To this end, we define the generating functions

$$_if_{ij}(s) = \sum_{n=1}^\infty {}_if_{ij}^n s^n,$$

$$_jP_{ii}(s) = \sum_{n=0}^\infty {}_jP_{ii}^n s^n.$$

Then owing to the convolution form of relation (1.1) we obtain for $i \neq j$

$$f_{ij}(s) = {}_jP_{ii}(s)\,{}_if_{ij}(s). \tag{1.2}$$

Now, since

$$\sum_{n=1}^\infty f_{ij}^n \leq 1 \qquad \text{and} \qquad \sum_{n=1}^\infty {}_if_{ij}^n \leq 1,$$

by Abel's Lemma, part (a) (Lemma 5.1 of Chapter 2), we conclude that

$$\lim_{s \to 1-} f_{ij}(s) = \sum_{n=1}^\infty f_{ij}^n \qquad \text{and} \qquad \lim_{s \to 1-} {}_if_{ij}(s) = \sum_{n=1}^\infty {}_if_{ij}^n.$$

If i and j communicate, then there is an $n \geq 1$ with $f_{ij}^n > 0$, and then also by (1.1)

$$_if_{ij}^{n-v} > 0 \qquad \text{for some} \quad v = 0, 1, \ldots, n - 1.$$

Thus,

$$\sum_{n=1}^{\infty} {}_i f_{ij}^n > 0$$

and then

$$\lim_{s \to 1-} {}_j P_{ii}(s) = \frac{\sum_{n=1}^{\infty} f_{ij}^n}{\sum_{n=1}^{\infty} {}_i f_{ij}^n} < \infty.$$

Hence, by Abel's theorem, part (b),

$$_j P_{ii}^* = \sum_{n=0}^{\infty} {}_j P_{ii}^n = \lim_{s \to 1-} {}_j P_{ii}(s) < \infty.$$

2: Ratio Theorems

The proofs of the two theorems of this section require the following lemma:

Lemma 2.1.   Let

$$c_n = \sum_{v=0}^{n} a_{n-v} b_v, \qquad n = 0, 1, 2, \ldots,$$

where $0 \le a_n \le K$ (K a positive constant) and $\sum_{n=0}^{\infty} a_n = \infty$. Then the relation $\lim_{n \to \infty} b_n = b$ with b finite implies

$$\lim_{n \to \infty} \frac{c_n}{\sum_{v=0}^{n} a_v} = b.$$

Proof.   Observe that $\sum_{v=0}^{n} a_{n-v} = \sum_{v=0}^{n} a_v$. Hence,

$$\frac{c_n}{\sum_{v=0}^{n} a_v} - b = \frac{\sum_{v=0}^{n} a_{n-v} b_v}{\sum_{v=0}^{n} a_v} - b = \frac{\sum_{v=0}^{n} a_{n-v}(b_v - b)}{\sum_{v=0}^{n} a_{n-v}}.$$

Since $b_n \to b$, which is finite, there is an $M > 0$ such that $|b_n| < M$ for all $n \ge 0$. Now choose $N = N(\varepsilon)$ such that

$$|b_n - b| < \varepsilon \qquad \text{for all} \quad n \ge N.$$

Then

$$\left| \frac{c_n}{\sum_{v=0}^{n} a_v} - b \right| \le \left| \frac{\sum_{v=0}^{N-1} a_{n-v}(b_v - b)}{\sum_{v=0}^{n} a_v} \right| + \left| \frac{\sum_{v=N}^{n} a_{n-v}(b_v - b)}{\sum_{v=0}^{n} a_v} \right|$$

$$\le 2M \left| \frac{\sum_{v=0}^{N-1} a_{n-v}}{\sum_{v=0}^{n} a_v} \right| + \varepsilon \left| \frac{\sum_{v=N}^{n} a_{n-v}}{\sum_{v=0}^{n} a_v} \right|$$

$$\le \frac{2MNK}{\sum_{v=0}^{n} a_v} + \varepsilon.$$

Letting $n \to \infty$ and then $\varepsilon \downarrow 0$, the lemma is proved. ■

Next we state three relations similar to (1.1) whose proofs are similar to the proof of (1.1).

For $k \neq j$, $i \neq j$, and $n \geq 0$,

$$_kP_{ij}^n = \sum_{v=0}^{n} {_k}f_{ij}^v \, {_k}P_{jj}^{n-v}, \tag{2.1}$$

$$P_{ij}^n = \sum_{v=0}^{n} P_{ii}^v \, {_i}P_{ij}^{n-v}, \tag{2.2}$$

$$P_{ij}^n = \sum_{v=0}^{n} f_{ij}^v P_{jj}^{n-v}. \tag{2.3}$$

Equation (2.3) is the same as (5.9) of Chapter 2.

Theorem 2.1. *Let i and j be arbitrary states and assume that j is recurrent; then*

$$\lim_{m \to \infty} \frac{\sum_{n=0}^{m} P_{ij}^n}{\sum_{n=0}^{m} P_{jj}^n} = f_{ij}^*,$$

where

$$f_{ij}^* = \sum_{n=1}^{\infty} f_{ij}^n. \tag{2.4}$$

Proof. From Eq. (2.3)

$$\sum_{n=0}^{m} P_{ij}^n = \sum_{n=0}^{m} \sum_{v=0}^{n} f_{ij}^v P_{jj}^{n-v} = \sum_{n=0}^{m} \sum_{\mu=0}^{n} f_{ij}^{n-\mu} P_{jj}^\mu = \sum_{n=0}^{m} \sum_{\mu=0}^{\infty} f_{ij}^{n-\mu} P_{jj}^\mu,$$

since $f_{ij}^{n-\mu} = 0$ for $\mu > n$ (by convention).

Since both summations are in fact finite they can be interchanged.

$$\sum_{n=0}^{m} P_{ij}^n = \sum_{\mu=0}^{\infty} P_{jj}^\mu \sum_{n=0}^{m} f_{ij}^{n-\mu} = \sum_{\mu=0}^{\infty} P_{jj}^\mu \sum_{r=0}^{m-\mu} f_{ij}^r.$$

Define

$$F_{ij}^m = \sum_{r=0}^{m} f_{ij}^r \qquad \text{for} \quad m = 0, 1, 2, \ldots$$

and

$$F_{ij}^m = 0 \qquad \text{for} \quad m = -1, -2, \ldots.$$

Then

$$\sum_{n=0}^{m} P_{ij}^n = \sum_{\mu=0}^{\infty} F_{ij}^{m-\mu} P_{jj}^\mu = \sum_{\mu=0}^{m} F_{ij}^{m-\mu} P_{jj}^\mu = \sum_{v=0}^{m} P_{jj}^{m-v} F_{ij}^v.$$

Now we apply Lemma 2.1 with

$$a_v = P_{jj}^v, \qquad b_v = F_{ij}^v, \qquad \text{and} \qquad c_m = \sum_{n=0}^{m} P_{ij}^n.$$

The conditions of the lemma are satisfied, since

$$0 \le P_{jj}^v \le 1 \qquad \text{and} \qquad \sum_{v=0}^{\infty} P_{jj}^v = \infty,$$

as j is recurrent. Since

$$\lim_{m \to \infty} b_m = \lim_{m \to \infty} F_{ij}^m = \sum_{r=0}^{\infty} f_{ij}^r = f_{ij}^*,$$

we conclude by virtue of the lemma that

$$\lim_{m \to \infty} \frac{\sum_{n=0}^{m} P_{ij}^n}{\sum_{n=0}^{m} P_{jj}^n} = f_{ij}^*$$

and the theorem is proved. ■

Rewriting Eq. (2.1) with $k = i \ne j$ gives

$$_iP_{ij}^n = \sum_{v=0}^{n} {_if_{iji}^v} P_{jj}^{n-v} \qquad \text{for} \quad n \ge 0. \tag{2.5}$$

Passing to the corresponding generating functions we have

$$_iP_{ij}(s) = {_if_{ij}(s)} {_iP_{jj}(s)}.$$

Since we have previously shown that $_iP_{jj}^* = \sum_{n=0}^{\infty} {_iP_{jj}^n} < \infty$, provided states i and j communicate, invoking Abel's lemma yields

$$\lim_{s \to 1-} {_iP_{ij}(s)} = \left(\lim_{s \to 1-} {_if_{ij}(s)} \right)\left(\lim_{s \to 1-} {_iP_{jj}(s)} \right) < \infty.$$

Thus, by Abel's lemma, part (b), we have

$$_iP_{ij}^* = \sum_{n=0}^{\infty} {_iP_{ij}^n} = \lim_{s \to 1-} {_iP_{ij}(s)} < \infty. \tag{2.6}$$

Theorem 2.2. *If i and j are in the same recurrent class, then*

$$\lim_{m \to \infty} \frac{\sum_{n=0}^{m} P_{ij}^n}{\sum_{n=0}^{m} P_{ii}^n} = {_iP_{ij}^*}.$$

Remark.   For $i \ne j$ we introduce the random variables

$$U_n = U(i, j, n) = \begin{cases} 1 & \text{if the process starting in state } i \\ & \text{is in state } j \text{ after } n \text{ steps without} \\ & \text{returning to state } i \text{ in the intervening} \\ & \text{time,} \\ 0 & \text{otherwise} \end{cases}$$

Then $E[U_n] = {}_iP^n_{ij}$ and

$$E\left[\sum_{n=1}^{\infty} U_n\right] = \sum_{n=1}^{\infty} {}_iP^n_{ij} = {}_iP^*_{ij}.$$

Thus it follows under the conditions of Theorem 2.2 that ${}_iP^*_{ij}$ is the expected number of visits to state j between successive visits to state i.

Proof.   From Eq. (2.2),

$$\sum_{n=0}^{m} P^n_{ij} = \sum_{n=0}^{m} \sum_{v=0}^{n} P^v_{ii}\,{}_iP^{n-v}_{ij} = \sum_{n=0}^{m} \sum_{v=0}^{\infty} P^v_{ii}\,{}_iP^{n-v}_{ij},$$

since ${}_iP^{n-v}_{ij} = 0$ for $v > n$. Interchanging the summations, we find

$$\sum_{n=0}^{m} P^n_{ij} = \sum_{v=0}^{\infty} P^v_{ii} \sum_{n=0}^{m} {}_iP^{n-v}_{ij} = \sum_{v=0}^{\infty} P^v_{ii}\,{}_i\tilde{P}^{m-v}_{ij},$$

where

$${}_i\tilde{P}^m_{ij} = \sum_{v=0}^{m} {}_iP^v_{ij} \qquad \text{for} \quad m = 0, 1, 2, \ldots$$

and

$${}_i\tilde{P}^m_{ij} = 0 \qquad \text{for} \quad m = -1, -2, \ldots.$$

Then,

$$\sum_{n=0}^{m} P^n_{ij} = \sum_{v=0}^{m} P^v_{ii}\,{}_i\tilde{P}^{m-v}_{ij} = \sum_{v=0}^{m} P^{m-v}_{ii}\,{}_i\tilde{P}^v_{ij}.$$

Now, we can apply Lemma 2.1 with $a_v = P^v_{ii}$, $b_v = {}_i\tilde{P}^v_{ij}$, and $c_m = \sum_{n=0}^{m} P^n_{ij}$, since $0 \le |P^v_{ii}| \le 1$ and $\sum_{v=0}^{\infty} P^v_{ii} = \infty$ as i is recurrent. But

$$\lim_{m\to\infty} b_m = \lim_{m\to\infty} {}_i\tilde{P}^m_{ij} = \sum_{v=0}^{\infty} {}_iP^v_{ij} = {}_iP^*_{ij},$$

and thus it follows by Lemma 2.1 that

$$\lim_{m\to\infty} \frac{\sum_{n=0}^{m} P^n_{ij}}{\sum_{n=0}^{m} P^n_{ii}} = {}_iP^*_{ij}.$$

This proves the theorem.   ■

If i and j communicate, we can write

$$\lim_{m\to\infty} \frac{\sum_{n=0}^{m} P^n_{jj}}{\sum_{n=0}^{m} P^n_{ii}} = \lim_{m\to\infty} \frac{\sum_{n=0}^{m} P^n_{jj} \sum_{n=0}^{m} P^n_{ij}}{\sum_{n=0}^{m} P^n_{ij} \sum_{n=0}^{m} P^n_{ii}}$$

since $\sum_{n=0}^{m} P^n_{ij} > 0$ for m sufficiently large.

Now if both i and j are recurrent and in the same class, then according to our last two theorems the first ratio on the right-hand side tends to $1/f_{ij}^* = 1$, the second ratio approaches $_iP_{ij}^*$, and therefore

$$\lim_{m \to \infty} \frac{\sum_{n=0}^{m} P_{jj}^n}{\sum_{n=0}^{m} P_{ii}^n} = {}_iP_{ij}^*.$$

From (1.1) and (2.5) and provided $i \neq j$, we obtain, respectively,

$$f_{ij}(s) = {}_jP_{ii}(s)\,{}_if_{ij}(s),$$

$$_iP_{ij}(s) = {}_if_{ij}(s)\,{}_iP_{jj}(s).$$

Then from Abel's lemma we have the identities $f_{ij}^* = {}_jP_{ii}^*\,{}_if_{ij}^*$, $_iP_{ij}^* = {}_if_{ij}^*\,{}_iP_{jj}^*$. If i and j belong to the same recurrent class, then $f_{ij}^* = 1$ and, therefore,

$$_iP_{ij}^* = \frac{_iP_{jj}^*}{_jP_{ii}^*}.$$

3: Existence of Generalized Stationary Distributions

In the case of an irreducible positive recurrent class the stationary distribution $\{\pi_i\}_{i=0}^{\infty}$ constitutes a convergent positive solution (i.e., $\sum_{i=0}^{\infty} \pi_i < \infty$) to the system of equations

$$\sum_{i=0}^{\infty} x_i P_{ij} = x_j, \qquad j = 0, 1, 2, \ldots.$$

This is the content of Theorem 1.3, Chapter 3. Our next theorem proves that this property characterizes positive recurrence.

Theorem 3.1. *Assume the Markov chain is irreducible. If the system of equations*

$$\sum_{j=0}^{\infty} x_j P_{ji} = x_i, \qquad i = 0, 1, 2, \ldots, \tag{3.1}$$

has a solution for which

$$\sum_{j=0}^{\infty} |x_j| < \infty \qquad and \qquad x_j \text{ are not all zero,}$$

then the Markov chain is positive recurrent.

Proof. By simple iteration we can obtain for any $n \geq 1$

$$\sum_{j=0}^{\infty} x_j P_{ji}^n = x_i.$$

Let $\tilde{P}_{ji}^m = (1/m) \sum_{n=0}^{m-1} P_{ji}^n$. Then

$$\sum_{j=0}^{\infty} x_j \tilde{P}_{ji}^m = x_i.$$

Now let $m \to \infty$. Since

$$\sum_{j=0}^{\infty} |x_j \tilde{P}_{ji}^m| \le \sum_{j=0}^{\infty} |x_j| < \infty,$$

we can interchange the limit and the summation. Then

$$x_i = \lim_{m \to \infty} \sum_{j=0}^{\infty} x_j \tilde{P}_{ji}^m = \sum_{j=0}^{\infty} x_j \lim_{m \to \infty} \tilde{P}_{ji}^m.$$

But

$$\lim_{m \to \infty} \tilde{P}_{ji}^m = \lim_{m \to \infty} \frac{1}{m} \sum_{n=0}^{m-1} P_{ji}^n = \pi_i \ge 0.$$

Hence,

$$x_i = \pi_i \sum_{j=0}^{\infty} x_j.$$

Since $\sum_{j=0}^{\infty} |x_j| < \infty$ and there is an i such that $x_i \ne 0$, we have shown that for some i

$$\pi_i \ne 0 \quad \text{and therefore} \quad \pi_i > 0, \qquad i = 0, 1, 2, \ldots. \quad \blacksquare$$

A converse to the theorem above is also true, in the following strengthened form involving inequalities.

Theorem 3.2. *If an irreducible Markov chain is positive recurrent and $x_j \ge 0$, $j = 0, 1, 2, \ldots$ are solutions of the set of inequalities*

$$\sum_{j=0}^{\infty} x_j P_{ji} \le x_i, \qquad i = 0, 1, 2, \ldots,$$

then

$$\sum_{j=0}^{\infty} x_j < \infty.$$

Proof. As before we have for any $n \ge 1$

$$\sum_{j=0}^{\infty} x_j P_{ji}^n \le x_i$$

and further, for any $m \geq 1$,

$$\sum_{j=0}^{\infty} x_j \tilde{P}_{ji}^m \leq x_i, \qquad \text{where} \quad \tilde{P}_{ji}^m = \frac{1}{m} \sum_{n=0}^{m-1} P_{ji}^n.$$

Since $x_j \geq 0$, $\tilde{P}_{ji}^m \geq 0$, $\sum_{j=0}^M x_j \tilde{P}_{ji}^m \leq x_i$ for any $M > 0$. If we let $m \to \infty$,

$$\lim_{m \to \infty} \sum_{j=0}^{M} x_j \tilde{P}_{ji}^m = \pi_i \sum_{j=0}^{M} x_j \leq x_i.$$

Since $\pi_i > 0$, the partial sums $\sum_{j=0}^M x_j$ are uniformly bounded for all $M > 0$. Therefore

$$\sum_{j=0}^{\infty} x_j < \infty. \quad \blacksquare$$

According to Theorem 3.1 for irreducible null recurrent processes there cannot exist a convergent solution of (3.1). Nevertheless, there do exist interesting positive solutions, as attested to by the following theorem.

Theorem 3.3.   *For a recurrent irreducible Markov chain the positive sequence given by*

$$v_0 = 1, \qquad v_i = {}_0 P_{0i}^*, \quad i = 1, 2, \ldots,$$

is a solution of the system of equations

$$v_i = \sum_{j=0}^{\infty} v_j P_{ji}, \qquad i = 0, 1, 2, \ldots \tag{3.2}$$

[see (2.6) for the definition of ${}_0 P_{0i}^*$].

Proof.   We have, by the definition of ${}_0 P_{0i}^*$,

$$\sum_{j=0}^{\infty} v_j P_{ji} = \sum_{j=1}^{\infty} {}_0 P_{0j}^* P_{ji} + P_{0i} = P_{0i} + \sum_{j=1}^{\infty} \sum_{n=1}^{\infty} {}_0 P_{0j}^n P_{ji}.$$

The double series on the right-hand side is composed of all nonnegative terms, and hence, the order of summation can be interchanged to yield

$$\sum_{j=0}^{\infty} v_j P_{ji} = P_{0i} + \sum_{n=1}^{\infty} \sum_{j=1}^{\infty} {}_0 P_{0j}^n P_{ji}.$$

Now

$$\sum_{j=1}^{\infty} {}_0 P_{0j}^n P_{ji} = \begin{cases} {}_0 P_{0i}^{n+1} & \text{if } i \neq 0, \\ f_{00}^{n+1} & \text{if } i = 0. \end{cases}$$

Thus, for $i \neq 0$

$$\sum_{j=0}^{\infty} v_j P_{ji} = P_{0i} + \sum_{n=1}^{\infty} {}_0P_{0i}^{n+1} = \sum_{n=0}^{\infty} {}_0P_{0i}^{n+1} = {}_0P_{0i}^* = v_i,$$

since

$$P_{0i} = {}_0P_{0i} \quad \text{for} \quad i \neq 0.$$

For $i = 0$,

$$\sum_{j=0}^{\infty} v_j P_{j0} = P_{00} + \sum_{n=1}^{\infty} f_{00}^{n+1} = \sum_{n=0}^{\infty} f_{00}^{n+1} = f_{00}^* = 1 = v_0,$$

and the verification of (3.2) is hereby complete. ∎

Theorem 3.4. *For a recurrent irreducible Markov chain the system*

$$v_i = \sum_{j=0}^{\infty} v_j P_{ji}, \qquad i = 0, 1, 2, \ldots, \tag{3.3}$$

$$v_0 = 1, \qquad v_i \geq 0, \quad i = 1, 2, \ldots, \tag{3.4}$$

has a unique solution.

Proof. Since by the previous theorem we know that $v_0 = 1$, $v_i = {}_0P_{0i}^*$, is a solution of (3.3) that also satisfies (3.4), we must prove that there is no other solution to (3.3) fulfilling condition (3.4).

Let $\{a_i\}$ be a sequence that satisfies (3.3) and (3.4). Then

$$a_i = \sum_{j=0}^{\infty} a_j P_{ji}, \qquad i = 0, 1, 2, \ldots.$$

Multiplying both sides by P_{ik} and summing over i, we have

$$a_k = \sum_{i=0}^{\infty} a_i P_{ik} = \sum_{i=0}^{\infty} P_{ik} \sum_{j=0}^{\infty} a_j P_{ji}$$

$$= \sum_{j=0}^{\infty} a_j \sum_{i=0}^{\infty} P_{ji} P_{ik} = \sum_{j=0}^{\infty} a_j P_{jk}^2.$$

The interchange of the order of summation is justified since all terms in the above expressions are nonnegative. Similarly, by repetition of this procedure we obtain for any $n \geq 1$

$$a_i = \sum_{j=0}^{\infty} a_j P_{ji}^n, \qquad i = 0, 1, 2, \ldots.$$

Since the Markov chain is irreducible and recurrent, for each i there is an $n \geq 1$ such that $P_{0i}^n > 0$. Then

$$a_i = \sum_{j=0}^{\infty} a_j P_{ji}^n \geq a_0 P_{0i}^n > 0 \qquad \text{as} \quad a_0 > 0.$$

Therefore, $a_i > 0$ for all i.

Next we introduce the quantities

$$Q_{ij} = \frac{a_j}{a_i} P_{ji}. \tag{3.5}$$

Clearly,

$$Q_{ij} \geq 0 \qquad \text{and} \qquad \sum_{j=0}^{\infty} Q_{ij} = \frac{\sum_{j=0}^{\infty} a_j P_{ji}}{a_i} = 1.$$

Thus, we may regard the Q_{ij} as transition probabilities of a Markov chain. The second-order transition probabilities are given by

$$Q_{ij}^2 = \sum_{k=0}^{\infty} Q_{ik} Q_{kj} = \sum_{k=0}^{\infty} \frac{a_k}{a_i} P_{ki} \frac{a_j}{a_k} P_{jk}$$

$$= \frac{a_j}{a_i} \sum_{k=0}^{\infty} P_{jk} P_{ki} = \frac{a_j}{a_i} P_{ji}^2.$$

Similarly, the n-step transition probabilities are given by

$$Q_{ij}^n = \frac{a_j}{a_i} P_{ji}^n \qquad \text{for} \quad n \geq 1.$$

It follows that

$$\sum_{n=0}^{\infty} Q_{ii}^n = \sum_{n=0}^{\infty} P_{ii}^n = \infty,$$

and the Q_{ij} are transition probabilities of a recurrent Markov chain $\mathcal{Q}$. We now apply the ratio theorems. From Theorem 2.1 we can write

$$\lim_{m \to \infty} \frac{\sum_{n=0}^{m} Q_{i0}^n}{\sum_{n=0}^{m} Q_{00}^n} = f_{i0}^*(Q) = 1,$$

where $f_{i0}^*(Q)$ is defined with respect to the Markov chain $\mathcal{Q}$ in the usual way. Its value is 1 since $\mathcal{Q}$ is a recurrent irreducible Markov chain. But we also have by Theorem 2.2

$$\lim_{m \to \infty} \frac{\sum_{n=0}^{m} Q_{i0}^n}{\sum_{n=0}^{m} Q_{00}^n} = \frac{a_0}{a_i} \lim_{m \to \infty} \frac{\sum_{n=0}^{m} P_{0i}^n}{\sum_{n=0}^{m} P_{00}^n} = \frac{a_0}{a_i} {}_0 P_{0i}^*.$$

Since $a_0 = 1$, we have shown that

$$a_i = {}_0P^*_{0i}, \qquad i = 1, 2, \ldots.$$

This proves uniqueness. ■

4: Interpretation of Generalized Stationary Distributions

The Markov chain $\mathscr{Q}$ with transition probability matrix $\|Q_{ij}\|$ associated with $\mathbf{P} = \|P_{ij}\|$ through formula (3.5) in terms of some positive solution of (3.3) is referred to as the *reversed process* of $\mathscr{P}$. In the positive recurrent case where $v_i = c\pi_i$ ($c = $ a constant), Q_{ij} possesses the following interpretation. Assuming that the initial distribution of the state variable is $\{\pi_i\}$, i.e., the Markov chain process $\{X(n); n \geq 0\}$ begins in state k with probability π_k, we calculate the conditional probability that the initial state was j, given the state after one transition is i. By Bayes' rule this is simply

$$Q_{ij} = \Pr\{X(0) = j \,|\, X(1) = i\} = \frac{\Pr\{X(1) = i \,|\, X(0) = j\} \, \Pr\{X(0) = j\}}{\Pr\{X(1) = i\}}. \qquad (4.1)$$

Since the process is stationary,

$$\Pr\{X(1) = i\} = \pi_i, \qquad \Pr\{X(0) = j\} = \pi_j,$$

and (4.1) obviously becomes

$$Q_{ij} = \frac{P_{ji}\pi_j}{\pi_i}. \qquad (4.2)$$

Iteration of (4.2) does indeed reverse the time scale. We easily deduce that if $X(0)$ has the initial stationary distribution $\{\pi_j\}$, then

$$Q^n_{ij} = \Pr\{X(0) = j \,|\, X(n) = i\}.$$

The name "reversed process" assigned to $\|Q_{ij}\|$ is manifestly apt.

The above method of introducing the reversed process applies whenever a positive (not necessarily convergent) solution of (3.3) is available. This device will be of further use in Theorem 5.2 below.

As indicated by Theorem 3.1, the solution of (3.3) is divergent in the null recurrent case and convergent for the positive recurrent case. An interesting interpretation can be given to any solution of (3.3) convergent or not. For the positive recurrent case the values $\{v_i\}_{i=0}^{\infty}$ are proportional to the stationary probabilities of being in the various states. In the general case the values v_i can be interpreted as the stationary expected number of particles in state i under suitable conditions of equilibrium. The precise sense of this equilibrium phenomenon is described in the following theorem.

Theorem 4.1. *Suppose a denumerable number of particles are independently undergoing a Markov process determined by* $\mathbf{P} = \|P_{ij}\|$. *Suppose* $A_i(n)$ *represents the number of particles in state i at time n. If* $A_i(0)$, $i = 1, 2, \ldots$, *are independent Poisson random variables with means* v_i, *where* $\sum_k v_k P_{ki} = v_i$, $v_i > 0$, *then* $A_i(n)$, $i = 1, 2, \ldots$, *are again independent random variables (r.v.'s) with the same distribution as* $A_i(0)$, $i = 1, 2, \ldots$, *respectively.*

Remark. The statement that an infinite number of random variables are independent and each Poisson distributed shall mean that any finite subcollection has this property.

Proof. We proceed by induction.

Let $A_k(n; i)$ be the number of particles in state k at time n which were in state i at time $n - 1$. Define the vectors $\mathbf{A}(n)$ and $\mathbf{A}(n; i)$ by

$$\mathbf{A}(n) = (A_1(n), A_2(n), \ldots); \qquad \mathbf{A}(n; i) = (A_1(n; i), A_2(n; i), \ldots).$$

Then $\mathbf{A}(n) = \sum_i \mathbf{A}(n; i)$. Now, by the induction hypothesis (that $A_i(n - 1)$, $i = 1, 2, \ldots$ are independent r.v.'s) and because of the assumption that the particles act independently, it follows that the vectors $\mathbf{A}(n; i)$, $i = 1, 2, \ldots$, are independent random variables. We shall further show that the components $A_k(n; i)$, $k = 1, 2, \ldots$, of each vector $\mathbf{A}(n; i)$ are independent random variables. It then follows that $A_k(n) = \sum_i A_k(n; i)$, $k = 1, 2, \ldots$, are independent r.v.'s.

Now for any finite number of components and integers $a_1, a_2, \ldots, a_r$, we have

$$\Pr\{A_{k_1}(n; i) = a_1, A_{k_2}(n; i) = a_2, \ldots, A_{k_r}(n; i) = a_r\}$$

$$= \sum_{a=0}^{\infty} \Pr\{A_i(n - 1) = a\}$$

$$\times \Pr\{A_{k_1}(n; i) = a_1, \ldots, A_{k_r}(n; i) = a_r | A_i(n - 1) = a\}. \quad (4.3)$$

By the induction hypothesis the first factor in the sum is equal to

$$\exp(-v_i) \frac{v_i^a}{a!}. \quad (4.4)$$

Because of the conditioning and the fact that the particles act independently, the second factor is a multinomial probability distribution. Thus

$$\Pr\{A_{k_1}(n; i) = a_1, \ldots, A_{k_r}(n; i) = a_r | A_i(n - 1) = a\}$$

$$= \frac{a!}{a_1! a_2! \cdots a_r!} \left(\prod_{v=1}^{r} (P_{ik_v})^{a_v} \right) \frac{(1 - \sum_{v=1}^{r} P_{ik_v})^{a - \sum_{v=1}^{r} a_v}}{(a - \sum_{v=1}^{r} a_v)!}. \quad (4.5)$$

This expression is taken to be zero in the case $\sum_{v=1}^{r} a_v > a$.

Inserting (4.4) and (4.5) into (4.3) gives

$$\Pr\{A_{k_1}(n; i) = a_1, A_{k_2}(n; i) = a_2, \ldots, A_{k_r}(n; i) = a_r\}$$

$$= \sum_{a = \sum_{v=1}^{r} a_v} \prod_{m=1}^{r} \frac{(v_i P_{ik_m})^{a_m}}{a_m!} \exp(-v_i P_{ik_m})$$

$$\times \frac{[(1 - \sum_{m=1}^{r} P_{ik_m})v_i]^{a - \sum_{m=1}^{r} a_m}}{(a - \sum_{m=1}^{r} a_m)!} \exp\left[-v_i\left(1 - \sum_{m=1}^{r} P_{ik_m}\right)\right]$$

$$= \prod_{m=1}^{r} \left[\frac{(v_i P_{ik_m})^{a_m}}{a_m!} \exp(-v_i P_{ik_m})\right] \sum_{a = a^*}^{\infty} \frac{[(1 - \sum_{m=1}^{r} P_{ik_m})v_i]^{a - a^*}}{(a - a^*)!}$$

$$\times \exp\left[-v_i\left(1 - \sum_{m=1}^{r} P_{ik_m}\right)\right]. \tag{4.6}$$

But the sum plainly equals 1 (set $a = a^* + \alpha$, sum over α) since it is just the sum of the Poisson density for the parameter

$$\left(1 - \sum_{m=1}^{r} P_{ik_m}\right)v_i.$$

The resulting factorization in (4.6) shows that $\{A_k(n; i)\}$ are independently Poisson distributed with means $\{v_i P_{ik}\}$, respectively.

Hence, $A_k(n) = \sum_{i=0}^{\infty} A_k(n; i)$ are independent Poisson variables with respective means $\sum_{i=0}^{\infty} v_i P_{ik} = v_k$.

Finally we note that $A_i(0)$, $i = 1, 2, \ldots$, were prescribed to be independent and Poisson distributed with means v_i, which together with the induction argument implies the result. ∎

5: Regular, Superregular, and Subregular Sequences for Markov Chains

Various criteria for determining recurrence, transience, and positive recurrence of a Markov chain were given and applied to some Markov chain queueing models in Theorems 4.1–4.2 of Chapter 3. The conditions set forth there pertained to the nature of the solutions of either of the systems of equations

$$\sum_{i=0}^{\infty} \xi_i P_{ik} = \xi_k, \qquad k = 0, 1, 2, \ldots, \tag{5.1}$$

and

$$\sum_{k=0}^{\infty} P_{ik} \eta_k = \eta_i, \qquad i = 0, 1, 2, \ldots.$$

The modern point of view on this problem is to present these developments as a corollary of the theory of regular, superregular, and subregular sequences. We now elaborate some of the simpler aspects of this elegant theory, which also embodies the elements of potential theory for Markov matrix operators. The classical potential theory appears in treating these same ideas in the case of Brownian motion. This interplay between potential theory and probability has received much attention in recent years and provides insights from one field to the other.

Let **P** be a given transition probability matrix. A nonnegative vector (i.e., sequence) $\mathbf{u} = \{u(j)\}_{j=0}^{\infty}$ is said to be, relative to **P**,

$$\text{right regular if} \quad \sum_j P_{ij}u(j) = u(i) \quad \text{(abbreviated } r\text{-regular)},$$

$$\text{right superregular if} \quad \sum_j P_{ij}u(j) \le u(i) \quad (r\text{-superregular)},$$

$$\text{right subregular if} \quad \sum_j P_{ij}u(j) \ge u(i) \quad (r\text{-subregular)}.$$

A right superregular sequence $\{u(i)\}$ is said to be minimal if $0 \le \xi(i) \le u(i)$ $(i \ge 0)$ and $\xi(i)$ regular implies $\xi(i) = cu(i)$ for all i and for some constant c. A nonnegative vector $\{v(i)\}_{i=0}^{\infty}$ is said to be

$$\text{left regular if} \quad \sum_i v(i)P_{ij} = v(j) \quad (l\text{-regular)},$$

$$\text{left superregular if} \quad \sum_i v(i)P_{ij} \le v(j) \quad (l\text{-superregular)},$$

$$\text{left subregular if} \quad \sum_i v(i)P_{ij} \ge v(j) \quad (l\text{-subregular)}.$$

Our first result is a representation theorem for right regular sequences in terms of minimal regular sequences.

Theorem 5.1. *Let* **u** *be a vector which is r-superregular with respect to* **P**. *Then*

$$a(i) = \lim_{n \to \infty} \sum_j P_{ij}^{(n)}u(j)$$

exists for all i, and **a** *is an r-regular vector with respect to* **P**. *Moreover, if* **b** *is a vector which is r-regular with respect to* **P** *and $b(i) \le u(i)$ for all i, then $b(i) \le a(i)$ for all i. If we write*

$$u(i) = a(i) + c(i), \quad i \ge 0, \tag{5.2}$$

where

$$c(i) = u(i) - a(i),$$

then $c(i)$ is minimal r-superregular.

Proof. We have, using the definition of r-superregular,

$$\sum_j P_{ij}^{(n)}u(j) = \sum_j \sum_k P_{ik}^{(n-1)}P_{kj}u(j)$$

$$= \sum_k P_{ik}^{(n-1)} \sum_j P_{kj}u(j) \le \sum_k P_{ik}^{(n-1)}u(k).$$

In concise notation we can write this as $\mathbf{P}^n\mathbf{u} \leq \mathbf{P}^{n-1}\mathbf{u}$, where inequality between vectors is understood as being componentwise. Thus for each i,

$$u(i) \geq \sum_j P_{ij} u(j) \geq \sum_j P_{ij}^{(2)} u(j) \geq \cdots.$$

Since all the terms are nonnegative, $a(i)$ exists and $a(i) \leq u(i)$ for all i. Now

$$\sum_j P_{ij} \sum_k P_{jk}^{(n)} u(k) = \sum_k P_{ik}^{(n+1)} u(k).$$

If we let $n \to \infty$, the right side converges to $a(i)$, while the left side, if we formally carry out the passage to the limit under the summation, tends to

$$\sum_j P_{ij} a(j);$$

in other words,

$$\sum_j P_{ij} a(j) = a(i).$$

To justify the passage to the limit within the summation, let i be considered as fixed. For $\varepsilon > 0$ there exists $N(\varepsilon)$ such that

$$\sum_{j > N(\varepsilon)} P_{ij} u(j) \leq \varepsilon.$$

Thus

$$\sum_{j > N(\varepsilon)} P_{ij} \sum_k P_{jk}^{(n)} u(k) \leq \sum_{j > N(\varepsilon)} P_{ij} u(j) \leq \varepsilon \qquad \text{for all } n$$

Now

$$\sum_j P_{ij} \sum_k P_{jk}^{(n)} u(k) = \sum_{j \leq N(\varepsilon)} P_{ij} \sum_k P_{jk}^{(n)} u(k) + \sum_{j > N(\varepsilon)} P_{ij} \sum_k P_{jk}^{(n)} u(k).$$

As $n \to \infty$, we have already seen that the left side tends to $a(i)$. Since the first sum over j on the right side is a finite sum, its limit is just

$$\sum_{j \leq N(\varepsilon)} P_{ij} a(j).$$

The second term is at most ε. Thus

$$a(i) = \sum_{j \leq N(\varepsilon)} P_{ij} a(j) + d(i), \qquad \text{where} \quad 0 \leq d(i) \leq \varepsilon,$$

which implies readily that $a(i) = \sum_j P_{ij} a(j)$, and $\mathbf{a} = \{a(i)\}$ is r-regular with respect to P. Finally, suppose that

$$b(i) = \sum_j P_{ij} b(j) \leq u(i) \qquad \text{for all } i.$$

Then by induction

$$b(i) = \sum_j P_{ij}^{(n)} b(j) \leq \sum_j P_{ij}^{(n)} u(j) \qquad \text{for all } n, i.$$

Hence

$$b(i) \leq \lim_{n \to \infty} \sum_j P_{ij}^{(n)} u(j) = a(i) \qquad \text{for all } i.$$

It is trivial to verify that $c(i) = u(i) - a(i)$ is r-superregular. It remains only to establish the minimal character of $c(i)$. Suppose

$$0 \leq \xi(i) \leq c(i), \qquad i \geq 0, \tag{5.3}$$

where $\{\xi(i)\}$ is r-regular. Applying the matrix P to (5.3) n times yields

$$\xi = \mathbf{P}^n \xi \leq \mathbf{P}^n \mathbf{c}.$$

But by the definition of $\mathbf{c}$ we see that $\mathbf{P}^n \mathbf{c} = \mathbf{P}^n \mathbf{u} - \mathbf{P}^n \mathbf{a} = \mathbf{P}^n \mathbf{u} - \mathbf{a}$ tends to the zero vector. This conclusion verifies that $\{c(i)\}$ is minimal r-superregular as claimed. ∎

The representation (5.2) in the case of Brownian motion becomes the classical Riesz representation involving superharmonic functions, harmonic functions, and potential functions. The development of this elegant theory is beyond the level of this book.

For transient Markov chains we can construct r-superregular functions very simply. Recall that in this case $\sum_{n=0}^{\infty} P_{ik}^n < \infty$ for all $i, k \geq 0$.

We claim that for k_0 fixed

$$u_i = \sum_{n=0}^{\infty} P_{ik_0}^n, \qquad i = 0, 1, \ldots, \tag{5.4}$$

is r-superregular. Indeed,

$$\sum_{j=0}^{\infty} P_{ij} u_j = \sum_{n=0}^{\infty} P_{ik_0}^{n+1} = u_i - P_{ik_0}^0 \leq u_i. \tag{5.5}$$

The above construction exhibits an abundance of nonconstant positive r-superregular vectors. The nonconstant character of (5.4) is evident since inequality prevails in (5.5) when $i = k_0$. The situation is quite different for the recurrent case. The next theorem asserts that the only r-superregular sequence is the constant vector. This generalizes the criteria of Theorem 4.1, Chapter 3.

Theorem 5.2. *An irreducible Markov chain with transition matrix* $\mathbf{P}$ *is recurrent if and only if every nonnegative vector* $\mathbf{v}$ *which is r-superregular with respect to* $\mathbf{P}$ *and has at least one positive component is a constant vector.*

Proof. Example (a), Section 6 of Chapter 6 can be modified, invoking the convergence of a nonnegative supermartingale, to yield a shorter proof than the direct approach that we now give.

Let the Markov chain be recurrent and consider

$$u_i \geq \sum_j P_{ij} u_j, \quad u_i \geq 0 \qquad \text{for all } i.$$

First we show that if $u_{j_0} > 0$ for some j_0, then $u_j > 0$ for all j. In fact, given j_0 and k there is an n such that $P_{kj_0}^{(n)} > 0$. As in the preceding theorem

$$u_k \geq \sum_j P_{kj}^{(n)} u_j \geq P_{kj_0}^{(n)} u_{j_0} > 0.$$

Thus if $\mathbf{u} \neq \mathbf{0}$ and is nonnegative, then $u_i > 0$ for all i. Now let k be arbitrary but fixed, and set $\xi_i = u_i/u_k$. Then

$$\xi_i \geq \sum_j P_{ij} \xi_j = \sum_{j \neq k} P_{ij} \xi_j + P_{ik}. \tag{5.6}$$

Iterating this inequality yields

$$\begin{aligned}
\xi_i &\geq \sum_{j \neq k} P_{ij} \left(\sum_{s \neq k} P_{js} \xi_s + P_{jk} \right) + P_{ik} \\
&= \sum_{j, s \neq k} P_{ij} P_{js} \xi_s + \sum_{j \neq k} P_{ij} P_{jk} + P_{ik} \\
&= \sum_{j, s \neq k} P_{ij} P_{js} \xi_s + f_{ik}^{(2)} + f_{ik}^{(1)},
\end{aligned}$$

where we have interpreted the last two terms as first passage probabilities. Another application of (5.6) gives

$$\begin{aligned}
\xi_i &\geq \sum_{j, s \neq k} P_{ij} P_{js} \left(\sum_{r \neq k} P_{sr} \xi_r + P_{sk} \right) + f_{ik}^{(2)} + f_{ik}^{(1)} \\
&= \sum_{j, s, r \neq k} P_{ij} P_{js} P_{sr} \xi_r + \sum_{j, s \neq k} P_{ij} P_{js} P_{sk} + f_{ik}^{(2)} + f_{ik}^{(1)} \\
&= \sum_{j, s, r \neq k} P_{ij} P_{js} P_{sr} \xi_r + f_{ik}^{(3)} + f_{ik}^{(2)} + f_{ik}^{(1)}.
\end{aligned}$$

Proceeding in this manner a simple induction proves

$$\xi_i \geq \sum_{n=1}^{m} f_{ik}^{(n)} \qquad \text{for all } m$$

and therefore

$$\xi_i \geq \sum_{n=1}^{\infty} f_{ik}^{(n)} = f_{ik}^* = 1,$$

since the chain is recurrent and irreducible.

Thus $\xi_i = u_i/u_k \geq 1$ or $u_i \geq u_k$. But i and k are arbitrary; hence $u_i = u_k$ for all i, k.

To prove the converse we assume that the chain is nonrecurrent. Let k be arbitrary and set

$$u_i = \begin{cases} f_{ik}^* & \text{if} \quad i \neq k, \\ 1 & \text{if} \quad i = k. \end{cases}$$

(Recall that f_{ik}^* is the probability of ever visiting state k starting from state i.) Then

$$u_i = f_{ik}^* = \sum_{j \neq k} P_{ij} f_{jk}^* + P_{ik} = \sum_j P_{ij} u_j \qquad \text{if} \quad i \neq k$$

and

$$u_k = 1 \geq f_{kk}^* = \sum_{j \neq k} P_{kj} f_{jk}^* + P_{kk} = \sum_j P_{kj} u_j.$$

Thus u is r-superregular. Now if $f_{jk}^* = 1$ for all $j \neq k$, we have

$$f_{kk}^* = \sum_{j \neq k} P_{kj} f_{jk}^* + P_{kk} = \sum_{j \neq k} P_{kj} + P_{kk} = \sum_j P_{kj} = 1,$$

which contradicts the assumption of nonrecurrence of the Markov chain. Hence **u** is a nonconstant bounded r-superregular vector. ∎

With the assistance of Theorem 5.2 it is easy to derive a stronger version of Theorem 3.4 which permits inequalities, namely,

Theorem 5.3.   *For a recurrent irreducible Markov chain the system*

$$v_i \geq \sum_{j=0}^{\infty} v_j P_{ji}, \qquad i = 0, 1, \ldots,$$

$$v_0 = 1, \qquad v_i \geq 0, \quad i = 1, 2, \ldots, \tag{5.7}$$

has a unique solution.

The method of proof as in Theorem 3.4 is to pass to the reverse process of **P** and thereby reduce the consideration of l-superregular vector sequences to that of r-superregular sequences. We then appeal to Theorem 5.2 for the desired conclusion. This device of using the reversed process to transform problems involving left regular concepts to right regular concepts is common and quite powerful.

Proof.   We have seen earlier that $v_0 = 1$, $v_i = {}_0P_{0i}^*$ is a solution of (5.7). In fact, this sequence is a solution with $\geq$ replaced by equality and possessing the property that $v_i > 0$ for all i. Setting

$$Q_{ij} = \frac{v_j P_{ji}}{v_i}, \qquad i, j = 0, 1, \ldots,$$

we have $Q_{ij} \geq 0$ and

$$\sum_{j=0}^{\infty} Q_{ij} = \frac{1}{v_i} \sum_{j=0}^{\infty} v_j P_{ji} = \frac{v_i}{v_i} = 1,$$

so that $\mathbf{Q} = \|Q_{ij}\|$ is a transition probability matrix. Furthermore, as in Theorem 3.4,

$$Q_{ik}^{(n)} = \frac{v_k}{v_i} P_{ki}^{(n)}$$

and the irreducibility and recurrence of $\mathbf{Q}$ follow from that of $\mathbf{P}$. Now if

$$c_i \geq \sum_{j=0}^{\infty} c_j P_{ji}, \qquad i = 0, 1, \ldots,$$

and

$$c_0 = 1, \qquad c_j \geq 0, \quad j = 1, 2, \ldots,$$

then

$$\sum_{j=0}^{\infty} Q_{ij} \frac{c_j}{v_j} = \frac{1}{v_i} \sum_{j=0}^{\infty} c_j P_{ji} \leq \frac{c_i}{v_i},$$

which says that the vector $\{c_j/v_j\}$ is r-superregular with respect to $\mathbf{Q}$. But the preceding theorem asserts that $\{c_j/v_j\}$ is a constant vector. Since $c_0 = v_0 = 1$, we conclude that $c_i = v_i$, $i = 0, 1, \ldots$, which proves the theorem. ∎

6: Stopping Rule Problems

Stopping rule problems occur in many areas of management science as special types of control problems in which the decision maker has only two available actions, "stop" and "continue." The analysis of these problems uses the theory of superregular vectors which motivates us to develop part of this theory here.

Let $\{X_n, n \geq 0\}$ be a discrete state Markov chain with transition matrix $\mathbf{P} = \|P_{ij}\|$. Associated with each state j let $r(j)$ be a nonnegative reward. To achieve some simplicity, yet preserving the underlying ideas, we will assume that $\{r(j)\}$ is a bounded nonnegative sequence, $0 \leq r(j) \leq \|r\| = \sup_j r(j) < \infty$. We sequentially observe $X_0, X_1, \ldots$. At each stage the option is available either to stop or continue observing. If at stage n we stop, the reward $r(X_n)$ is acquired and the game has culminated. If we never stop, then no payment accrues. Our task is to characterize the maximal expected reward and determine an optimal rule for stopping, provided one exists.

Corresponding to a prescribed decision procedure, let T be the time of stopping, with $T = \infty$ signifying the event that we continue forever. Clearly the problem is meaningful only when $T = n$ is determined by the history up to time n and not the future. More precisely, we require for every $n = 0, 1, \ldots$

that the event $\{T = n\}$ be determined only by $X_0, X_1, \ldots, X_n$ (and, of course, on knowledge of the total reward vector $r = \{r(j)\}$ and the transition probabilities P_{ij}). A random variable associated with a Markov process and having this property is called a *Markov time*.

Call a sequence of states $i_0, i_1, \ldots, i_n$ a *stopping sequence* if along this sequence we decide to stop at i_n but not before. A Markov time is fully specified by producing a list of all its stopping sequences, since if at any time n the chain has evolved through $X_0 = i_0, \ldots, X_n = i_n$, in order to decide whether to stop $(T = n)$ or not $(T > n)$ we simply check to see if the sequence $i_0, \ldots, i_n$ is on the list or not.

Stopping at a fixed time $T = m$ is the simplest example of a Markov time. Specifically, any $(m + 1)$-tuple $(i_0, \ldots, i_m)$ of states is a stopping sequence for this time. Also the first time, if any, that the process reaches a given set B of possible states is a Markov time. A stopping sequence in this case is any sequence $(i_0, \ldots, i_v)$ of states for which $i_k \notin B$ for $k = 0, \ldots, v - 1$ and $i_v \in B$. If T_1 and T_2 are Markov times then so is $T = \min\{T_1, T_2\}$. Every stopping sequence for T may be found in the union of the set of stopping sequences for T_i, $i = 1, 2$. It follows from this discussion that if T is any Markov time then $T_m = \min\{T, m\}$ is a bounded Markov time.

If $u = (u(i))$ is a nonnegative vector, the expression $u(X_T)$ may not possess meaning since $T = \infty$ is a possibility. It is convenient to adopt the convention that $u(X_T) = 0$ when $T = \infty$. This is consistent with our prerequisite that never stopping entails no reward. Thus the expected reward $E[r(X_T) | X_0 = i]$ is the expectation in the standard sense, of the random variable Z, where

$$Z = \begin{cases} r(X_T) & \text{if} \quad T < \infty, \\ 0 & \text{if} \quad T = \infty. \end{cases}$$

Computationally,

$$E[r(X_T) | X_0 = i] = \sum_{n=0}^{\infty} r(i_n) \Pr\{T = n \text{ and } X_n = i_n | X_0 = i\}$$

$$= \sum_{(i_0, \ldots, i_n)} r(i_n) \Pr\{X_0 = i_0, \ldots, X_n = i_n | X_0 = i\},$$

where the last sum runs over all stopping sequences $(i_0, \ldots, i_n)$ for T.

Define the optimal reward vector v by

$$v(i) = \sup\{E[r(X_T) | X_0 = i]; T \text{ a Markov time}\}.$$

Since $r(j) \leq \|r\| < \infty$ for all j, then manifestly $v(j) \leq \|r\| < \infty$ for all j. We will establish that v is the smallest right superregular vector satisfying $v(i) \geq r(i)$ for all i. Moreover, if an optimal Markov time exists, then it may be taken to be the first time, if any, that the process is in the set of states i for which $v(i) = r(i)$. That is,

$$T = \begin{cases} \min\{n : v(X_n) = r(X_n)\}, \\ \infty & \text{if} \quad v(X_n) > r(X_n) \quad \text{for all } n. \end{cases}$$

This makes sense. It calls for continuing as long as the optimal reward $v(X_n)$ exceeds the reward $r(X_n)$ available through stopping immediately, and stopping at the first state X_t in which $v(X_t) = r(X_t)$. The stopping sequences are of the form $(i_0, \ldots, i_m)$ where $v(i_k) > r(i_k)$ for $k = 0, \ldots, m - 1$ and $v(i_m) = r(i_m)$.

First we catalog some properties of bounded right superregular vectors.

Lemma 6.1.   (a)   *Every nonnegative constant vector is an r-superregular vector.*

(b)   *If for each α in some set A, $\mathbf{u}_\alpha$ is an r-superregular vector and we define $u(j) = \inf_\alpha u_\alpha(j)$, then $\mathbf{u}$ is r-superregular.*

(c)   *If u is an r-superregular vector, then $u(i) \geq \sum_j P_{ij}^n u(j)$ for any i and $n = 1, 2, \ldots$.*

Proof.   Since $\sum_j P_{ij} = 1$, every nonnegative constant is manifestly an r-regular vector, a fortiori r-superregular, and statement (a) is proved. Let $u(i) = \inf_\alpha u_\alpha(i)$ where each u_α is r-superregular. Certainly u is nonnegative since each u_α is. Also that $u(j) \leq u_\alpha(j)$ for all j and that u_α is r-superregular imply the inequality

$$\sum_j P_{ij} u(j) \leq \sum_j P_{ij} u_\alpha(j) \leq u_\alpha(i),$$

for all i and an arbitrary α. It follows that

$$\sum_j P_{ij} u(j) \leq \inf_\alpha u_\alpha(i) = u(i),$$

which shows that u is r-superregular. This proves (b).

The assertion of (c) is true for $n = 1$ since u is r-superregular by assumption. We proceed by induction. Suppose for n that

$$u(k) \geq \sum_j P_{kj}^n u(j)$$

holds. Now multiply both sides by the nonnegative P_{ik} and sum over k to get

$$\sum_k P_{ik} u(k) \geq \sum_k P_{ik} \sum_j P_{kj}^n u(j).$$

The left-hand side is at most $u(i)$ since u is r-superregular, while the right-hand side is $\sum_j P_{ij}^{n+1} u(j)$. Thus,

$$u(i) \geq \sum_j P_{ij}^{n+1} u(j)$$

is secured advancing the induction. The proof is complete. ∎

The statement of part (c) can be written compactly in the form

$$u(i) \geq E[u(X_n) | X_0 = i]$$

for any time index n. The next theorem indicates that the fixed index n can be replaced any Markov time T while maintaining the inequality.

Theorem 6.1.   *Let u be a bounded r-superregular vector and T a Markov time. Then*

(a) $$u(i) \geq E[u(X_T)|X_0 = i] \text{ for all } i.$$

If S and T are Markov times for which $S \leq T$ (every stopping sequence for T contains a stopping sequence for S), then

(b) $$E[u(X_S)|X_0 = i] \geq E[u(X_T)|X_0 = i] \text{ for all } i.$$

Proof.   We need only prove (b) since (a) follows by taking $S \equiv 0$.

Let α be a constant, $0 < \alpha < 1$. Since u is r-superregular, the function $h(i) = u(i) - \alpha \sum_j P_{ij} u(j)$ is nonnegative. With obvious iterations we may write

$$u(i) = h(i) + \alpha \sum_j P_{ij} u(j)$$

$$= h(i) + \alpha \sum_j P_{ij}[h(j) + \alpha \sum_k P_{jk} u(k)]$$

$$= h(i) + \alpha \sum_j P_{ij} h(j) + \cdots + \alpha^{n-1} \sum_j P_{ij}^{n-1} h(j) + \alpha^n \sum_j P_{ij}^n u(j). \quad (6.1)$$

Part (c) of Lemma 6.1 tells us that $u(i) \geq \sum_j P_{ij}^n u(j)$ and we have further stipulated that $\|u\| < \infty$. It follows, as $n \to \infty$, that the last term goes to zero and (6.1) passes to the equation

$$u(i) = \sum_{n=0}^{\infty} \alpha^n \sum_j P_{ij}^n h(j)$$

$$= \sum_{n=0}^{\infty} \alpha^n E[h(X_n)|X_0 = i]$$

$$= E\left[\sum_{n=0}^{\infty} \alpha^n h(X_n)|X_0 = i\right], \quad (6.2)$$

where h is a nonnegative function.

Invoking the Markov property, we infer on the basis of (6.2) that for any fixed n

$$u(i) = E[h(X_n) + \alpha h(X_{n+1}) + \alpha^2 h(X_{n+2}) + \cdots | X_n = i]$$

$$= E\left[\alpha^{-n} \sum_{k=n}^{\infty} \alpha^k h(X_k)|X_n = i\right]. \quad (6.3)$$

Let T be the given Markov time. The event $\{T = n\}$ is a union of events of the form $\{X_0 = i_0, \ldots, X_n = i_n\}$ where $(i_0, \ldots, i_n)$ is a stopping sequence such that

T calls for stopping at n and not before. Let $I_n(i)$ denote the set of these stopping sequences of length n for which $i_0 = i$. We have

$$E\left[\sum_{k=T}^{\infty} \alpha^k h(X_k) | X_0 = i\right]$$

$$= \sum_{n=0}^{\infty} \sum_{(i_0, \ldots, i_n) \in I_n(i)} E\left[\sum_{k=T}^{\infty} \alpha^k h(X_k) | X_0 = i_0, X_1 = i_1, \ldots, X_n = i_n\right]$$

$$\times \Pr\{X_0 = i_0, \ldots, X_n = i_n | X_0 = i\} \quad \text{(by the law of total probabilities)}$$

$$= \sum_{n=0}^{\infty} \sum_{(i_0, \ldots, i_n) \in I_n(i)} E\left[\sum_{k=n}^{\infty} \alpha^k h(X_k) | X_n = i_n\right]$$

$$\times \Pr\{X_0 = i_0, \ldots, X_n = i_n | X_0 = i\}$$

$$= \sum_{n=0}^{\infty} \sum_{(i_0, \ldots, i_n) \in I_n(i)} \alpha^n u(i_n)$$

$$\times \Pr\{X_0 = i_0, \ldots, X_n = i_n | X_0 = i\} \quad \text{(by virtue of (6.2))}$$

$$= \sum_{n=0}^{\infty} \sum_{i_n} \alpha^n u(i_n) \Pr\{T = n \text{ and } X_T = i_n | X_0 = i\}.$$

(We are able above to replace the Markov time T by the fixed time n since $X_0 = i_0, \ldots, X_n = i_n$ implies $T = n$.)

This last expression is a computation for the expectation of $\alpha^T u(X_T)$, and accordingly we have

$$E[\alpha^T u(X_T) | X_0 = i] = E\left[\sum_{k=T}^{\infty} \alpha^k h(X_k) | X_0 = i\right]. \tag{6.4}$$

Since h is nonnegative, if S is a Markov time that does not exceed T, then

$$\sum_{k=S}^{\infty} \alpha^k h(X_k) \geq \sum_{k=T}^{\infty} \alpha^k h(X_k).$$

Now take the expectation of both sides and compare with (6.4) to deduce

$$E[\alpha^S u(X_S) | X_0 = i] \geq E[\alpha^T u(X_T) | X_0 = i]. \tag{6.5}$$

It remains to justify interchanging the expectation and the limit $\alpha \to 1-$ in (6.5) in order to complete the proof. We next prove the limit relation

$$\lim_{\alpha \to 1-} E[\alpha^T u(X_T) | X_0 = i] = E[u(X_T) | X_0 = i].$$

Let I_A be the indicator of an event A:

$$I_A = \begin{cases} 1 & \text{if } A \text{ occurs,} \\ 0 & \text{if } A \text{ does not occur.} \end{cases}$$

Start with the identities

$$E[\alpha^T u(X_T)|X_0 = i] = E[u(X_T)|X_0 = i] - E[(1 - \alpha^T)u(X_T)|X_0 = i], \quad (6.6)$$

and focus on the second term on the right. A standard partitioning and appropriate analysis leads to the next succession of equations:

$$
\begin{aligned}
E[(1 - \alpha^T)u(X_T)|X_0 = i] &= E[(1 - \alpha^T)u(X_T)I_{0 \le T < \infty}|X_0 = i] \\
&= E[(1 - \alpha^T)u(X_T)I_{0 \le T \le M}|X_0 = i] \\
&\quad + E[(1 - \alpha^T)u(X_T)I_{M < T < \infty}|X_0 = i] \\
&\le (1 - \alpha^M)\|u\| + \|u\| \Pr\{M < T < \infty|X_0 = i\},
\end{aligned}
$$

where M is an arbitrary integer. Let α increase to 1 so that $(1 - \alpha^M)\|u\|$ goes to zero for each fixed M. Therefore

$$\varlimsup_{\alpha \to 1-} E[(1 - \alpha^T)u(X_T)|X_0 = i] \le \|u\| \Pr\{M < T < \infty|X_0 = i\},$$

where M is arbitrary. Since

$$\lim_{M \to \infty} \Pr\{M < T < \infty|X_0 = i\} = 0,$$

it follows that

$$\lim_{\alpha \to 1-} E[(1 - \alpha^T)u(X_T)|X_0 = i] = 0$$

and applying this in (6.6) confirms

$$\lim_{\alpha \to 1-} E[\alpha^T u(X_T)|X_0 = i] = E[u(X_T)|X_0 = i]. \quad (6.7)$$

The same argument works for S. Recalling (6.5) and using (6.7) produces the inequality

$$E[u(X_S)|X_0 = i] \ge E[u(X_T)|X_0 = i],$$

which completes the proof. ∎

This theorem leads to the following simple but important result in our study of optimal stopping.

Theorem 6.2. *Let u be any bounded vector satisfying, for all i,*

$$u(i) \ge r(i) \qquad and \qquad u(i) \ge \sum_j P_{ij}u(j). \quad (6.8)$$

Then for any Markov time T,

$$u(i) \ge E[r(X_T)|X_0 = i] \qquad for\ all\ i.$$

Proof. Since r is a nonnegative vector, the conditions of (6.8) affirm that u is r-superregular. Now application of Theorem 6.1 yields

$$u(i) \geq E[u(X_T)|X_0 = i].$$

But $u(i)$ is never less than $r(i)$ by (6.8), so that

$$E[u(X_T)|X_0 = i] \geq E[r(X_T)|X_0 = i].$$

The last two inequalities together give the desired result. ∎

Remark.   Theorem 6.2 remains valid when the nonnegative functions r and u are not bounded. To see this, create the bounded functions $r_N(i) = \min\{r(i), N\}$ and $u_N(i) = \min\{u(i), N\}$, for an arbitrary positive N. Then $u_N(i) \geq r_N(i)$ and $u_N(i) \geq \sum_j P_{ij} u_N(j)$ by (a) and (b) of Lemma 6.1. Thus the theorem just proved applies and we conclude $u_N(i) \geq E[r_N(X_T)|X_0 = i]$. But $u(i) \geq u_N(i)$ so $u(i) \geq E[r_N(X_T)|X_0 = i]$. Now let N increase. The left side is independent of N while the right converges to $E[r(X_T)|X_0 = i]$, whence $u(i) \geq E[r(X_T)|X_0 = i]$ as claimed.

Any vector u satisfying (6.8) is called an *r-superregular majorant* of $r(i)$. Theorem 6.2 tells us that any r-superregular majorant of $r(i)$ bounds the optimal reward vector v, where

$$v(i) = \sup E[r(X_T)|X_0 = i], \tag{6.9}$$

where the supremum is extended over all Markov times T. Suppose there is available a Markov time T^* such that

$$w(i) = E[r(X_{T^*})|X_0 = i] \tag{6.10}$$

is an r-superregular majorant of $r(i)$. Then T^* is optimal, i.e., achieves the maximum expected reward, because

$$v(i) \geq E[r(X_{T^*})|X_0 = i]$$

by (6.9), while Theorem 6.2 implies

$$w(i) = E[r(X_{T^*})|X_0 = i]$$
$$\geq v(i).$$

Hence

$$v(i) = w(i) = E[r(X_{T^*})|X_0 = i]$$

and T^* is optimal. This program often provides an expeditious method for verifying that a given Markov time is optimal. We illustrate with some concrete examples.

Example 1.   A correct answer for a contestant in a quiz show wins him a dollar and gives him the option of leaving with his total winnings to date or attempting another question. On the other hand, with a wrong answer, he forfeits his cumulative winnings and must quit. What strategy maximizes his mean reward?

Let us suppose the questions are answered independently, each correctly with known probability p and incorrectly with probability $q = 1 - p$. We let X_n denote the cumulative winnings up to trial n, and introduce an artificial state Δ to represent "out of the contest." Then $X_0, X_1, \ldots$ is a success run Markov chain with $X_0 = 0$ and

$$X_{n+1} = \begin{cases} X_n + 1 & \text{with probability} \quad p, \\ \Delta & \text{with probability} \quad q = 1 - p. \end{cases}$$

With the reward function $r(x) = x$ for $x = 0, 1, \ldots$ and $r(\Delta) = 0$ the problem, stated formally, is to find a Markov time T that maximizes $E[r(X_T)|X_0 = 0]$.

Intuitively one might guess that the optimal strategy would be to stop the first time the winnings exceed some critical value ξ. That is, we should examine Markov times of the form

$$T(\xi) = \min\{n : X_n \geq \xi \text{ or } X_n = \Delta\}, \qquad \xi = 1, 2, \ldots.$$

Easily, the mean reward under such a rule is

$$v_\xi(i) = E[r(X_{T(\xi)})|X_0 = i]$$

$$= \begin{cases} \xi p^{\xi - i} & \text{for} \quad i \leq \xi, \\ i & \text{for} \quad i > \xi, \end{cases}$$

and $v_\xi(\Delta) = r(\Delta) = 0$.

Now considering only $X_0 = i = 0$ and looking at the ratio

$$\frac{v_{\xi+1}(0)}{v_\xi(0)} = \frac{(\xi + 1)p^{\xi + 1}}{\xi p^\xi}$$

we see it pays to increase ξ as long as $v_{\xi+1}(0)/v_\xi(0) > 1$. The optimal choice for ξ is, then,

$$\xi^* = \min\{\xi : p(\xi + 1)/\xi \leq 1\}$$
$$= \min\{\xi : p\xi + p \leq \xi\}$$
$$= \min\{\xi : \xi \geq p/(1 - p)\}.$$

In words, ξ^* is the smallest integer greater than or equal to $p/(1 - p)$.

We claim $T^* = T(\xi^*)$ is optimal. According to the comments preceding this example all we need show is that $v(i) = v_{\xi^*}(i)$ is both r-superregular and nowhere less than $r(i) = i$. Checking the second claim, $v(i) = r(i) = i$ for $i \geq \xi^*$, while for $i < \xi^*$, by virtue of the definition of ξ^* we have $p > i/(i + 1)$ and

$$r(i) = i$$

$$= \frac{i}{i + 1} \times \frac{i + 1}{i + 2} \times \cdots \times \frac{\xi^* - 1}{\xi^*} \times \xi^*$$

$$< \xi^* p^{\xi^* - i} = v(i).$$

which verifies $v(i) \geq r(i)$ for all i. Returning to the first claim, if $i \geq \xi^*$, then

$$v(i) = i \geq pv(i + 1) = p(i + 1)$$

by the definition of ξ^*, while if $i < \xi^*$,

$$v(i) = \xi^* p^{\xi^* - i} = pv(i + 1).$$

Thus $v(i)$ is r-superregular.

Completing the argument, $v(i) \geq E[r(X_T)|X_0 = i]$ for any Markov time T, by Theorem 6.2, but $v(i) = E[r(X_{T^*})|X_0 = i]$ by construction. Therefore $T^* = T(\xi^*)$ is optimal.

Example 2.   As a second example, let $X_0, X_1, \ldots$ be a Markov chain having as its state space the set of integers $0, \pm 1, \pm 2, \ldots$ and its transition probabilities

$$P_{i,i+1} = p,$$
$$P_{i,i-1} = q = 1 - p,$$
$$P_{i,j} = 0 \qquad \text{if} \quad |i - j| \neq 1.$$

We suppose $p < \frac{1}{2}$ so that X_n is a random walk having a negative drift. With

$$r(i) = \begin{cases} i & \text{if} \quad i \geq 0, \\ 0 & \text{if} \quad i < 0, \end{cases}$$

we desire a Markov time T that maximizes $E[r(X_T)|X_0 = i]$.

Again, it seems reasonable to suppose that the optimal strategy would assume the form

$$T(a) = \begin{cases} \min\{n : X_n \geq a\} & \text{if} \quad X_n \geq a \quad \text{for some } n, \\ \infty & \text{if} \quad X_n < a \quad \text{for all } n, \end{cases}$$

for some optimally chosen integer a. Consequently, we evaluate

$$\begin{aligned} v_a(i) &= E[r(X_{T(a)})|X_0 = i] \\ &= \begin{cases} a \Pr\{T(a) < \infty | X_0 = i\}, & i < a \\ i, & i \geq a \end{cases} \\ &= \begin{cases} a \Pr\{\max X_n \geq a | X_0 = i\}, & i < a \\ i, & i \geq a \end{cases} \\ &= \begin{cases} a(p/q)^{a-i}, & i < a \\ i, & i \geq a. \end{cases} \quad \text{(See Section 3 of Chapter 3.)} \end{aligned}$$

Looking at the ratios $v_{a+1}(0)/v_a(0)$, an argument analogous to that in the previous example shows the maximizing $a = a^*$ is given by

$$a^* = \min\{a : p(a + 1) \le qa\}$$
$$= \min\{a : a \ge p/(1 - 2p)\}.$$

Again we claim $T^* = T(a^*)$ is optimal. As before, we need only show $v(i) = v_{a^*}(i)$ is both superregular and not less than $r(i)$. Beginning with the second claim, $v(i) = r(i) = i$ for $i \ge a^*$, while for $0 \le i < a^*$, by virtue of the definition of a^* we have $p/q > i/(i + 1)$ and

$$r(i) = i$$

$$= \frac{i}{i + 1} \times \frac{i + 1}{i + 2} \times \cdots \times \frac{a^* - 1}{a^*} \times a^*$$

$$\le a^* \left(\frac{p}{q}\right)^{a^* - i} = v(i)$$

Returning to the first claim, if $i > a^*$, then

$$v(i) = i$$
$$\ge p(i + 1) + q(i - 1) \qquad \text{(because } p < q\text{)}$$
$$= pv(i + 1) + qv(i - 1),$$

while if $i \le a^*$,

$$v(i) = a^* \left(\frac{p}{q}\right)^{a^* - i}$$

$$= pa^* \left(\frac{p}{q}\right)^{a^* - (i + 1)} + qa^* \left(\frac{p}{q}\right)^{a^* - (i - 1)}$$

$$\ge pv(i + 1) + qv(i - 1).$$

Thus $v(i)$ is r-superregular.

Completing the argument, by Theorem 6.2, $v(i) \ge E[r(X_T)|X_0 = i]$ for any Markov time T, but we derived $v(i)$ from $v(i) = E[r(X_{T^*})|X_0 = i]$. Therefore $T^* = T(a^*)$ is optimal.

These two small examples illustrate a general procedure that can be applied in a remarkably large number of problems. The trick is to guess the optimal strategy and then to use Theorem 6.2 to verify that it is indeed optimal. When this shortcut to an answer cannot be taken, the more general approach that we now develop must be used.

A vector f is called the *least r-superregular majorant* of r provided (i) f is an r-superregular majorant of r and (ii) if h is any other r-superregular majorant of r, then $h(i) \ge f(i)$ for all i.

There is at most one least r-superregular majorant of r. Suppose to the contrary there were two, say f_1 and f_2. Then the vector f defined by

$$f(i) = \min\{f_1(i), f_2(i)\} \qquad \text{for all } i,$$

would be an r-superregular majorant of r by Lemma 6.1(b). And $f(i)$ would be strictly smaller than some $f_1(i)$ or $f_2(i)$, thus forcing a contradiction, unless $f_1(i) = f_2(i) = f(i)$ for all i.

Next we prove that at least one least r-superregular majorant of r exists. Let H be the set of all r-superregular majorants of r. Observe first that H is not empty. Indeed the constant vector $h^*(i) = \|r\|$ for all i patently belongs to H by Lemma 6.1(a). Define the vector f by

$$f(i) = \inf\{h(i) : h \in H\} \qquad \text{for all } i,$$

whence $f(i) \geq r(i)$ since each $h(i)$ exceeds $r(i)$, and f is r-superregular by Lemma 6.1(b). Thus f is an r-superregular majorant of r, and it is the smallest such vector by its very definition.

The least r-superregular majorant f can be calculated by successive approximation. Define the vectors $f_0, f_1, \ldots$, recursively by

$$f_0(i) = r(i) \qquad \text{for all } i,$$

and

$$f_{n+1}(i) = \max\left\{r(i), \sum_j P_{ij} f_n(j)\right\} \tag{6.11}$$

Then for every n, $f_n(i) \geq r(i)$. By induction we shall show

$$f_{n+1}(i) \geq f_n(i) \qquad \text{for all } i.$$

Easily

$$f_1(i) = \max\left\{r(i), \sum_j P_{ij} f_0(j)\right\} \geq r(i) = f_0(i).$$

Suppose $f_n(j) \geq f_{n-1}(j)$ for all j. Then

$$\sum_j P_{ij} f_n(j) \geq \sum_j P_{ij} f_{n-1}(j)$$

and a fortiori

$$f_{n+1}(i) = \max\left\{r(i), \sum_j P_{ij} f_n(j)\right\} \geq \sum_j P_{ij} f_{n-1}(j). \tag{6.12}$$

But $f_{n+1}(i) \geq r(i)$ so that, together with (6.12),

$$f_{n+1}(i) \geq \max\left\{r(i), \sum_j P_{ij} f_{n-1}(j)\right\} = f_n(i).$$

Thus, $f_n(i)$ is an increasing sequence in n for each i and we may define the limit

$$f_\infty(i) = \lim_{n \to \infty} f_n(i).$$

It will later emerge that $f_\infty(i)$ is finite, indeed bounded. We claim that $f_\infty(i)$ is the least r-superregular majorant of $r(i)$. First $f_\infty(i) \geq r(i)$ holds since $f_n(i) \geq r(i)$ for each n. Since $f_\infty(i) \geq f_{n+1}(i)$, from the definition (6.11) we have

$$f_\infty(i) \geq \lim_{n \to \infty} \sum_{j=0}^{\infty} P_{ij} f_n(j)$$

$$\geq \lim_{n \to \infty} \sum_{j=0}^{N} P_{ij} f_n(j) = \sum_{j=0}^{N} P_{ij} f_\infty(j),$$

where N is an arbitrary positive integer. Letting N increase indefinitely, we conclude that

$$f_\infty(i) \geq \sum_{j} P_{ij} f_\infty(j).$$

Thus f_∞ is an r-superregular majorant of $r(i)$. To show it is the smallest, let h be an arbitrary r-superregular majorant of r. Then

$$h(i) \geq r(i) = f_0(i) \qquad \text{for all } i. \tag{6.13}$$

We next verify inductively that $h(i) \geq f_n(i)$ for each n. The case $n = 0$ was checked above. Suppose $h(j) \geq f_{n-1}(j)$ for all j. Then $\sum_j P_{ij} h(j) \geq \sum_j P_{ij} f_{n-1}(j)$, and since h is r-superregular, $h(i) \geq \sum_j P_{ij} h(j)$, so that

$$h(i) \geq \sum_{j} P_{ij} f_{n-1}(j).$$

But this last inequality in conjunction with (6.13) shows that

$$h(i) \geq \max\left\{ r(i), \sum_{j} P_{ij} f_{n-1}(j) \right\} = f_n(i).$$

Hence, $h(i) \geq f_n(i)$ for all n and consequently

$$h(i) \geq \lim_{n \to \infty} f_n(i) = f_\infty(i).$$

The claim that $f_\infty = f$ is the least r-superregular majorant of $r(i)$ is hereby validated.

The next result begins the characterization of the optimal mean return function.

Lemma 6.2. *Let f be the least r-superregular majorant of a nonnegative bounded vector r. For each $\varepsilon \geq 0$ let $\Gamma_\varepsilon = \{i : f(i) \leq r(i) + \varepsilon\}$ and let $T(\varepsilon) = \min\{n : X_n \in \Gamma_\varepsilon\}$, $(T(\varepsilon) = \infty$ if $X_n \notin \Gamma_\varepsilon$ for all $n)$. Then for every $\varepsilon > 0$*

$$f(i) - \varepsilon \leq E[r(X_{T(\varepsilon)}) | X_0 = i] \leq f(i).$$

If the Markov chain has a finite state space, then

$$f(i) = E[r(X_{T(0)})|X_0 = i].$$

Proof.   Prescribe $\varepsilon > 0$. Define

$$f_\varepsilon(i) = E[f(X_{T(\varepsilon)})|X_0 = i].$$

We first want to show $f_\varepsilon(i) = f(i)$ for all i. Since f is an r-superregular vector, by Theorem 6.1,

$$f(i) \geq E[f(X_{T(\varepsilon)})|X_0 = i] = f_\varepsilon(i).$$

Thus, we need only prove $f_\varepsilon(i) \geq f(i)$. This will be accomplished if we can show (i) $f_\varepsilon(i) \geq r(i)$ for all i and (ii) $f_\varepsilon(i) \geq \sum_j P_{ij} f_\varepsilon(j)$, since f is the smallest vector satisfying (i) and (ii). For (ii), let $T' = \min\{n : n \geq 1 \text{ and } X_n \in \Gamma_\varepsilon\}$. (Note, in the definition of T', n is restricted to exceed 0.) Then $T' \geq T(\varepsilon)$. Thus, by Theorem 6.1,

$$\begin{aligned} f_\varepsilon(i) &= E[f(X_{T(\varepsilon)})|X_0 = i] \\ &\geq E[f(X_{T'})|X_0 = i]. \end{aligned} \tag{6.14}$$

Now, using the Markov property,

$$E[f(X_{T'})|X_0 = i] = \sum_j P_{ij} E[f(X_{T'})|X_1 = j].$$

But

$$\begin{aligned} E[f(X_{T'})|X_1 = j] &= E[f(X_{T(\varepsilon)})|X_0 = j] \\ &= f_\varepsilon(j). \end{aligned}$$

Taken together we have shown

$$f_\varepsilon(i) \geq \sum_j P_{ij} f_\varepsilon(j).$$

Now to show (i). Suppose $c = \sup_i\{r(i) - f_\varepsilon(i)\} \geq 0$. Then $f_\varepsilon(i) + c$ is r-superregular, since f_ε is by (ii) and a constant vector is regular. Moreover $f_\varepsilon(i) + c \geq r(i)$ so that $f_\varepsilon + c$ exceeds the least r-superregular majorant of $r(i)$; that is,

$$f_\varepsilon(i) + c \geq f(i) \qquad \text{for all } i. \tag{6.15}$$

Take α satisfying $0 < \alpha < \varepsilon$. Since $c = \sup_i\{r(i) - f_\varepsilon(i)\}$, there is a state i_0 for which

$$r(i_0) - f_\varepsilon(i_0) > c - \alpha. \tag{6.16}$$

Because of (6.15)

$$0 \leq f(i_0) - r(i_0) \leq f_\varepsilon(i_0) + c - r(i_0) < \alpha < \varepsilon.$$

Thus, $i_0 \in \Gamma_\varepsilon$ and consequently

$$f_\varepsilon(i_0) = E[f(X_{T(\varepsilon)})|X_0 = i_0] = f(i_0).$$

According to (6.16)

$$c - \alpha < r(i_0) - f(i_0) \leq 0,$$

or

$$c < \alpha.$$

But α is an arbitrary number in $(0, \varepsilon)$. We send α to zero, implying $c \leq 0$. That is, $f_\varepsilon(i) \geq r(i)$ for all i. This verifies requirement (i) attesting that f_ε is an r-superregular majorant of $r(i)$. Hence $f_\varepsilon(i) \geq f(i)$ for all i since f is the least majorant of $r(i)$. But we already pointed out the relations $f_\varepsilon(i) \leq f(i)$ for all i. Thus $f_\varepsilon = f$, and

$$f(i) = E[f(X_{T(\varepsilon)})|X_0 = i].$$

But note that $X_{T(\varepsilon)} \in \Gamma_\varepsilon$ so that $f(X_{T(\varepsilon)}) \leq r(X_{T(\varepsilon)}) + \varepsilon$. Therefore

$$f(i) \leq E[r(X_{T(\varepsilon)}) + \varepsilon | X_0 = i]$$
$$\leq E[r(X_{T(\varepsilon)})|X_0 = i] + \varepsilon.$$

Thus

$$f(i) - \varepsilon \leq E[r(X_{T(\varepsilon)})|X_0 = i] \leq f(i),$$

as was to be shown.

When the state space is finite the same proof will work with $\varepsilon = 0$. We leave this to the reader.

A key by-product of this theorem is the result that $v = f$, that is, that the optimal reward vector is the smallest r-superregular majorant of r. This warrants a formal statement that offers an additional characterization of the optimal mean return.

Theorem 6.3.   *The following are equivalent:*

(a) $v(i) = \sup E[r(X_T)|X_0 = i]$, *where the supremum is over all Markov times T;*

(b) $v(i) = \lim_{n \to \infty} f_n(i) = \sup_n f_n(i)$ *where $f_0(k) = r(k)$ for all k and*

$$f_{n+1}(k) = \max\left\{r(k), \sum_j P_{kj}f_n(j)\right\}, \qquad \text{for all } k;$$

and

(c) $v(i) = \inf h(i)$ *where the infimum is over all vectors h satisfying, for all i,*

$$h(i) \geq r(i) \qquad and \qquad h(i) \geq \sum_j P_{ij}h(j).$$

Proof.   We have already shown that (b) and (c) are equivalent since (c) decribes the least r-superregular majorant of r and (b) gives an algorithm that leads to it. Comparing (c) and (a), if v is the least r-superregular majorant of r then

$v(i) \geq E[r(X_T)|X_0 = i]$ for all Markov times T by Theorem 6.2 while, given a positive ε, we have $v(i) - \varepsilon \leq E[r(X_{T(\varepsilon)})|X_0 = i]$ in the notation of Lemma 6.2.

Since there exist strategies having returns arbitrarily close to $v(i)$ we have $v(i) = \sup E[r(X_T)|X_0 = i]$, and (a) and (c) are also equivalent. ∎

We have characterized the maximal mean reward but said nothing concerning optimal strategies. Unfortunately, there may not exist optimal rules. As an example, suppose $P_{i, i+1} = 1$, $i \geq 1$, so that the Markov chain moves deterministically through successive states. Take $r(i) = 1 - (1/i)$. Then $v(i) = 1$ for all i, but there is no Markov time that will attain this expected reward.

Nevertheless we can show that if there exists an optimal Markov time, then $T(0) = \min\{n : r(X_n) = v(X_n)\}$ is optimal.

Theorem 6.4.   *If there exists an optimal Markov time T^*, e.g.,*

$$v(i) = E[r(X_{T^*})|X_0 = i] \qquad \text{for all } i, \tag{6.17}$$

then $T(0) = \min\{n : v(X_n) = r(X_n)\}$ is also optimal.

Proof.   *Recall that $v(i) \geq r(i)$ for all i. Hence, from (6.17)*

$$\begin{aligned} v(i) &= E[r(X_{T^*})|X_0 = i] \\ &\leq E[v(X_{T^*})|X_0 = i] \leq v(i). \end{aligned}$$

That is,

$$E[r(X_{T^*})|X_0 = i] = E[v(X_{T^*})|X_0 = i].$$

Since $v(i) \geq r(i)$ for all i we must have

$$\Pr\{r(X_{T^*}) > v(X_{T^*})|X_0 = i\} = 0,$$

and

$$\Pr\{r(X_{T^*}) = v(X_{T^*})|X_0 = i\} = 1.$$

That is, $\Pr\{X_{T^*} \in \Gamma_0 | X_0 = i\} = 1$. But $T(0)$ is the first n, if any, for which $X_n \in \Gamma_0$. Hence $T(0) \leq T^*$. From Lemma 6.1

$$E[v(X_{T(0)})|X_0 = i] \geq E[v(X_{T^*})|X_0 = i] = v(i).$$

But $X_{T(0)} \in \Gamma_0$ so that $v(X_{T(0)}) = r(X_{T(0)})$. Hence

$$E[r(X_{T(0)})|X_0 = i] \geq v(i).$$

This shows that $T(0)$ is optimal.

Elementary Problems

1. Consider an irreducible positive recurrent Markov chain with initial state $X_0 = i$. Let $N_n(i)$ be the number of visits to state i in the first n trials. Prove that

$$\lim_{n \to \infty} \frac{E[N_n(i)]}{n} = \frac{1}{\mu_i},$$

where μ_i is the mean recurrence time to state i, i.e., $\mu_i = \sum_{n=1}^{\infty} n f_{ii}^n$.

2. Consider an irreducible but not necessarily recurrent Markov chain. Assume that $_0P_{00}^* = 1$ and $_0P_{0i}^* < \infty$ ($i \geq 1$). Show that the sequence $_0P_{0i}^*$ is l-superregular.

3. Consider the gambler's ruin on $n + 1$ states, where the transition probability matrix was given in Chapter 3. Find all r-regular vectors relative to this Markov chain.

Solution: $u(i) = \alpha \pi_i(C_0) + b \pi_i(C_n)$, a, b arbitrary, using the notation of the example in Chapter 3.

4. Let $\{U_j\}_{j=0}^{\infty}$ be a real solution of the equations

$$\sum_{j=0}^{\infty} P_{ij} U_j - U_i = 1, \qquad i = 0, 1, 2, \dots.$$

Let $E[U_{X_n} | X_0 = i]$ denote the expected value of the r.v. U_{X_n} with the initial condition $X_0 = i$. Prove that

$$\lim_{n \to \infty} \frac{E[U_{X_n} | X_0 = i]}{n} = 1.$$

Hint: Establish the identity $\sum_{j=0}^{\infty} P_{ij}^{(n)} U_j - U_i = n, i = 0, 1, 2, \dots$.

Problems

1. Consider an irreducible positive recurrent Markov chain with initial state $X_0 = i$. Let $N_n(i)$ be the number of visits to state i in the first n trials and let $T_m(i)$ denote the number of trials until the mth visit to state i. Justify the relationship

$$\Pr\{T_m(i) \leq n\} = \Pr\{N_n(i) \geq m\}.$$

2. Assume $\sum_{n=1}^{\infty} n^2 f_{ii}^n < \infty$. Since $T_m(i)$ is a sum of m independent, identically distributed, random variables with mean μ_i and variance σ_i use the central limit theorem and the relationship given in Problem 1 to develop a limit distribution for $N_n(i)$ properly normalized, i.e., find $a_n > 0$ and $b_n > 0$ such that $(N_n(i) - a_n)/b_n$ has a limit normal distribution.

3. Let $\{X_n, n \geq 0\}$ be an irreducible recurrent Markov chain with transition probability matrix $\mathbf{P} = \|P_{ij}\|$ and generalized invariant measure $\{v_i\}$, i.e., $\sum_i v_i P_{ij} = v_j$, $v_j > 0$, $j = 0, 1, \dots$.

Let $\{Y_n, n \geq 0\}$ be the imbedded process obtained by looking at the Markov chain only at times when $X_n = 0$ or 1 (i.e., $Y_0 = X_{n_0}$ where n_0 is the first time the chain is in state 0 or 1; $Y_m = X_{n_m}$ where n_m is the time of mth return of the chain to states 0 or 1). Then $\{Y_n, n \geq 0\}$ is also an irreducible recurrent Markov chain. Let w_0, w_1 denote the stationary probabilities of the embedded Markov chain. Show that $w_1/w_0 = v_1/v_0$.

Hint: Use the interpretation of v_i as given in Theorems 3.3 and 3.4.

4.   Let $\mathbf{P} = \|P_{ij}\|$ and $\mathbf{P}_n = \|P_{ij}(n)\|$, $n = 1, 2, \ldots$, be transition probability matrices of irreducible recurrent Markov chains and let $\{v_i\}$ and $\{v_i^{(n)}\}$, $n = 1, 2, \ldots$, be the corresponding invariant measures normalized so that $v_0 = v_0^{(n)} = 1$ for all n.
   Prove that if $P_{ij}(n) \to P_{ij}$ for all i, j as $n \to \infty$ then $v_i^{(n)} \to v_i$ for all i.

Hint: Prove that $\lim_{n \to \infty} v_j^{(n)} = w_j$ exists and satisfies

$$\sum_{i=0}^{\infty} w_i P_{ij} \leq w_j, \qquad j = 0, 1, 2, \ldots,$$

and $w_0 = 1$. Then use the uniqueness property of Theorem 5.3.

5.   In an irreducible Markov chain prove that any nonnegative r-superregular sequence $\{u(i)\}$ has the property that

$$u(i) \geq f_{ik}^* u(k) \qquad \text{for every } i \text{ and } k,$$

where f_{ik}^* is the probability of reaching state k from state i.

Hint: Consult the proof of Theorem 5.2.

6.   Let $\{u(i)\}$ be a finite nonnegative r-superregular sequence. Define

$$w(i) = u(i) - \sum_j P_{ij} u(j).$$

Show that the set A of all states i for which $w(i) > 0$ is a set of nonrecurrent states.

Hint: Modify slightly the conclusion of Problem 5 for the case at hand to deduce strict inequality and apply the result appropriately.

7.   Consider an irreducible Markov chain. Fix some state; call it 0. Prove that if $f_{i0}^* \geq \alpha > 0$ for every $i \neq 0$ then the chain is recurrent.

Hint: Show that the probability of visiting state 0 only finitely often has probability zero (cf. Theorems 7.1 and 7.2 of Chapter 2).

8.   Show that an irreducible Markov chain is nonrecurrent if and only if there exists a finite-valued l-superregular sequence $\mu(i)$ such that

$$\mu(k) > \sum_i \mu(i) P_{ik} \qquad \text{for some state } k.$$

Hint: (sufficiency)   Use Theorems 5.3, 3.3, and 3.4; (necessity) consult Eq. (5.5).

9.   Consider an electric circuit with m boundary and n interior positions denoted by $B_1, B_2, \ldots, B_m$ and $A_1, A_2, \ldots, A_n$, respectively, as in the accompanying diagram.

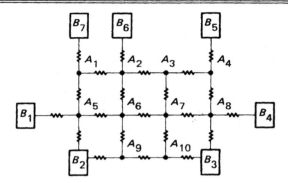

Thus there are $m + n$ positions altogether. The current flowing from position i to j is T_{ij} and the resistance between i and j is R_{ij} when i and j are not both boundary positions. The potential at position i is V_i. Assume that the resistances are all known and the potentials at boundary positions are also given. Ohm's law states that

(i)
$$V_i - V_j = R_{ij} T_{ij}$$

and Kirchoff's first law asserts that

(ii)
$$\sum_j T_{ij} = 0.$$

Using (i) and (ii) prove that V_i satisfies the relations

(iii)
$$V_i = \left(\sum_j R_{ij}^{-1} \right)^{-1} \sum_k V_k R_{ik}^{-1}.$$

By interpreting (iii) show that we may consider a random walk on the positions of the electric network, interpreting the boundary positions as absorbing states and defining transition probabilities as appropriate expressions of the resistance R_{ij}.

10. In Problem 9 order the states of the random walk in such a fashion that states 1, 2, ..., m are all the boundary positions and states $m + 1, m + 2, ..., m + n$ are the interior positions. Prove that the potential at an interior position A_i may be expressed as

$$V_{m+i} = \sum_{k=1}^{m} b_{ik} V_k \qquad \text{for} \quad i = 1, 2, ..., n,$$

where b_{ik} is the probability of absorption at the boundary position B_k having started at the interior position A_i.

11. Let $\| P_{ij} \|$ be a transition probability matrix of a recurrent null or transient irreducible Markov chain. Let $\{u_i\}$ be a I-regular positive vector and let A be a collection of states. We introduce the notation $P_{iA}^{(n)} = \sum_{j \in A} P_{ij}^{(n)}$. Prove that if $\mu(A) = \sum_{i \in A} u_i < \infty$ then $P_{iA}^{(n)} \to 0$ as $n \to \infty$.

Hint: Let B be a finite subset of A such that $\sum_{i \in A - B} u_i < \varepsilon$. Show that $P_{i, A-B}^{(n)} < \varepsilon / u_i$ and use this fact. (The notation $A - B$ designates the set of states in A excluding the states of B.)

Let $\mathbf{N}$ be the matrix with elements $N_{ij} = \sum_{n=0}^{\infty} {}_0 P_{ij}^{(n)} = \sum_{n=0}^{\infty} Q_{ij}^{(n)}$ for $i, j \geq 1$.

(a) Show that $\mathbf{N}$ is the minimal right (and left) nonnegative inverse of $\mathbf{I} - \mathbf{Q}$, where $\mathbf{I}$ is an infinite-dimensional identity matrix.

(b) Show that the stationary distribution $\boldsymbol{\alpha} = \|\alpha_i\|$ of $\{X_n\}$ is proportional to $(1, N_{11}, N_{12}, \ldots)$ by (i) using an algebraic argument involving (a) and (ii) using a renewal argument.

(c) *Age distributions.* Motivated by the current age problem in renewal theory (see Problem 10 of Chapter 5), given that $X_n = j > 0$, we say the process is of age k if $X_{n-k} = 0$ but $X_{n-l} \neq 0$ for $0 < l < k$. Determine the limiting age distribution

$$g_j^{(k)} = \lim_{n \to \infty} \Pr\{X_{n-k} = 0, \, X_{n-l} \neq 0 \text{ for } 0 < l < k \,|\, X_n = j\}, \quad k \geq 1.$$

Answer: $g_j^{(k)} = {}_0 P_{ij}^{(k-1)}/N_{1j}, \, k \geq 1$.

(d) Find the mean limiting age given $X_n = j$. That is, determine $\sum_{k \geq 1} k g_j^{(k)}$.

Answer: $\sum_{r=1}^{\infty} N_{1r} N_{rj}/N_{1j}$.

(e) Let $L(j)$ be the last time that the process visits state j before hitting T_0, with $L(j) = \infty$ should these be no such last visit. Then $L_j = k$ if and only if $T_0 > k$, $X_k = j$, and $X_n \neq j$ for $n = k+1, \ldots, T_0$. Show that

$$\Pr\{L_j = k \,|\, X_0 = 1\} = \Pr\{L_j < \infty \,|\, X_0 = 1\} p_{1j}^{(k)}/N_{1j}, \quad k \geq 0.$$

20. Let $\{X_n\}$ be a Markov chain on the states $0, 1, \ldots$ with transition matrix $\mathbf{P} = \|P_{ij}\|$. Assume that $P_{00} = 1$ and that $E[T_0 | X_0 = j] < \infty$ for all states j where $T_0 = \min\{n \geq 0 : X_n = 0\}$ is the hitting time to state 0. Let $\mathbf{Q}$ be the matrix whose elements are P_{ij} but restricted to $i, j \geq 1$. Thus $\mathbf{P}$ has the form

$$\mathbf{P} = \begin{Vmatrix} 1 & 0 & 0 & \cdots \\ P_{10} & Q_{11} & Q_{12} & \cdots \\ P_{20} & Q_{21} & Q_{22} & \cdots \\ \vdots & \vdots & \vdots & \end{Vmatrix}.$$

Let $\mathbf{N}$ be the matrix with elements $N_{ij} = \sum_{n=0}^{\infty} Q_{ij}^{(n)}$ for $i, j \geq 1$. Next, let $\{\tilde{X}_n\}$ be the Markov chain derived from $\{X_n\}$ by restarting at state 1 whenever state 0 is reached. Then $\{\tilde{X}_n\}$ has transition matrix with elements $\tilde{P}_{01} = 1$ and $\tilde{P}_{ij} = P_{ij}$ for $i \geq 1, j \geq 0$.

Let $\{\tilde{Y}_n\}$ be the time reversal of $\{\tilde{X}_n\}$ with respect to the stationary distribution in Problem 19(b).

(a) Give the transition probabilities for $\{\tilde{Y}_n\}$.

(b) Let $\{Y_n\}$ be the process derived from $\{\tilde{Y}_n\}$ by making state 0 absorbing. Show that

$$\Pr\{T_0 = n \,|\, Y_0 = j\} = P_{1j}^{(n-1)}/N_{1j}, \quad n \geq 1,$$

where T_0 is the absorption time for $\{Y_n\}$. (Compare with Problem 19(c).)

NOTES

This chapter serves merely as a bare introduction to an important and expanding area in probability theory.

A useful book on the subject of potential theory in Markov chains is the comprehensive work of Kemeny, Snell, and Knapp [2].

Bibliographic and historical references on potential theory and boundary theory for Markov chains are found in the above work.

A very modern approach to the study of Markov processes via potential theory is contained in Meyer [3].

Dynkin and Yushkevich [5] have written a nice little book that covers aspects of both Markov chain potential theory and optimal stopping.

A treatment of Brownian motion and certain stable processes in N dimensions with emphasis on potential theoretic concepts is given in Port and Stone [4].

Various classes of optimal stopping problems and gambling systems as part of mathematical statistics are covered in Chow, Robbins and Siegmund [6].

REFERENCES

1. K. L. Chung, "Markov Chains with Stationary Transition Probabilities." Springer-Verlag, Berlin, 1960.
2. J. G. Kemeny, J. L. Snell, and A. W. Knapp, "Denumerable Marker Chains." Van Nostrand–Reinhold, New York, 1966.
3. P. A. Meyer, "Probability and Potentials." Ginn (Blaisdell), Boston, 1966.
4. S. C. Port and C. J. Stone, "Brownian Motion and Classical Potential Theory." Academic Press, New York, 1978.
5. E. B. Dynkin and A. A. Yushkevich, "Markov Processes; Theorems and Problems." Plenum, New York, 1969.
6. Y. S. Chow, H. Robbins, and D. Siegmund, "Great Expectations: The Theory of Optimal Stopping." Houghton-Mifflin, Boston, 1971.

Chapter 12

SUMS OF INDEPENDENT RANDOM VARIABLES AS A MARKOV CHAIN

This chapter is a simplified version of a few topics from the theory of sums of independent random variables. Some additional results on sums of independent random variables appear in Chapter 17 on fluctuation theory.

1: Recurrence Properties of Sums of Independent Random Variables

Let X_1, X_2, ... be a sequence of integer-valued, independent, identically distributed random variables and define $S_n = X_1 + X_2 + \cdots + X_n$ for $n = 1, 2, \ldots$. For completeness we also define $S_0 \equiv 0$.

In this chapter we discuss some aspects of sums of independent random variables S_n, $n = 0, 1, \ldots$, regarding them as successive values of a discrete valued Markov chain of special structure. We barely touch the surface in our treatment of the extensive theory of sums of independent random variables as much of it is beyond the level of this book. For a more complete account of this elegant and rich theory we refer the reader to Spitzer (see references at the close of this chapter).

In Example A, Section 2, Chapter 2, we mentioned the sequence S_n (where X_i are nonnegative integer valued) as an example of a Markov chain. In the present context the state space consists of all integers; positive, negative, and zero. The initial state is prescribed to be zero since we set $S_0 \equiv 0$. The special feature of the Markov chain $\{S_n\}$ is its spatial homogeneous character in that the one-step transition probabilities have the property $\Pr\{S_n = j | S_{n-1} = i\} = P_{ij} = P_{0, j-i} = P_{i-j, 0}$. A simple induction shows that the same is true for the n-step transition probabilities; thus for any $n \geq 1$

$$P_{ij}^n = P_{0, j-i}^n = P_{i-j, 0}^n = \Pr\{S_{n+k} = j | S_k = i\}.$$

We will assume in effect throughout this chapter that the random variable X_1 is "irreducible." By this we shall mean that the Markov chain with transition probability matrix $P_{ij} = \Pr\{S_n = j \mid S_{n-1} = i\}$ is irreducible. Simple criteria which guarantee irreducibility will be indicated later. We will also stipulate, without repeating this at every occasion, that X_1 is nondegenerate, i.e., it has at least two possible values.

We now determine some simple conditions for recurrence and transientness appropriate for the Markov chain generated by $\{S_n\}$. For this purpose we introduce the following quantities

$$G_{ij}^n = \sum_{m=0}^{n} P_{ij}^m, \qquad G_{ij} = \sum_{m=0}^{\infty} P_{ij}^m \leq +\infty,$$

for all integers i, j and $n = 0, 1, 2, \ldots,$ The expression G_{ij} is called the "Green function" and relates to the potential theory development alluded to in Chapter 11.

Lemma 1.1.

$$G_{ij}^n \leq G_{00}^n, \qquad n = 0, 1, 2, \ldots, \tag{1.1}$$

for all integers i, j. In particular, as $n \to \infty$ we have

$$G_{ij} \leq G_{00} \tag{1.2}$$

for all integers i, j.

Proof. Because of spatial homogeneity we have $G_{ij}^n = G_{i-j,0}^n$. Thus it is sufficient to prove $G_{i0}^n \leq G_{00}^n$ for all integers $i = 0, \pm 1, \pm 2, \ldots$ and $n = 0, 1, 2, \ldots$. But

$$G_{i0}^n = \sum_{m=0}^{n} P_{i0}^m = \sum_{m=0}^{n} \sum_{l=0}^{m} f_{i0}^{m-l} P_{00}^l = \sum_{l=0}^{n} P_{00}^l \sum_{m=l}^{n} f_{i0}^{m-l} = \sum_{l=0}^{n} P_{00}^l \sum_{r=0}^{n-l} f_{i0}^r,$$

where f_{i0}^r is the probability of reaching 0 from i for the first time at the rth step. Since

$$\sum_{r=0}^{n} f_{i0}^r \leq 1, \qquad G_{i0}^n \leq \sum_{l=0}^{n} P_{00}^l = G_{00}^n. \quad \blacksquare$$

An elegant and useful criterion for recurrence is the content of the following theorem.

Theorem 1.1. *If*

$$E[|X_k|] = E[|X_1|] = \sum_{j=-\infty}^{\infty} |j| P_{0j} < \infty, \qquad k = 2, 3, \ldots, \tag{1.3}$$

and

$$\mu = E[X_k] = E[X_1] = \sum_{j=-\infty}^{\infty} j P_{0j} = 0, \tag{1.4}$$

then the Markov chain $\{S_n\}$ is recurrent.

Remark. Since $E[X_1] = 0$ and X_1 is a nondegenerate random variable, we infer that there are positive and negative values that X_1 may achieve with positive probability. The irreducibility postulate imposed at the start asserts that the Markov chain generated by S_n, $n = 0, 1, 2, \ldots$, is irreducible (consists of one class) and its state space comprises all integers (positive, negative, and zero). Therefore, in accordance with Corollary 5.1 of Chapter 2, it is enough in verifying the recurrence property to establish recurrence for a single state (say the zero state).

Proof. Now from (1.1) we know that $G_{0j}^n \leq G_{00}^n$, for all integers j. The same inequality is preserved by averaging. Hence

$$\frac{1}{2M+1} \sum_{j=-M}^{M} G_{0j}^n \leq G_{00}^n. \tag{1.5}$$

But

$$\sum_{j=-M}^{M} G_{0j}^n = \sum_{|j| \leq M} \sum_{m=0}^{n} P_{0j}^m = \sum_{m=0}^{n} \sum_{|j| \leq M} P_{0j}^m \geq \sum_{m=0}^{n} \sum_{|j/m| \leq M/n} P_{0j}^m, \tag{1.6}$$

and the last inequality holds trivially since $m \leq n$. Comparing (1.5) and (1.6) we see that

$$G_{00}^n \geq \frac{1}{2M+1} \sum_{m=0}^{n} \sum_{|j/m| \leq M/n} P_{0j}^m. \tag{1.7}$$

Now, by definition,

$$P_{0j}^m = \Pr\{S_m = j \mid S_0 = 0\}. \tag{1.8}$$

Since S_k is the sum of k independent identically distributed random variables with finite mean, $\mu = E[X_1] = E[S_k] = 0$, the weak law of large numbers prevails (see Section 1, Chapter 1, page 19), which asserts specifically that for any prescribed $\varepsilon > 0$,

$$\Pr\left\{\left|\frac{S_m - m\mu}{m}\right| \leq \varepsilon\right\} = \Pr\left\{\left|\frac{S_m}{m}\right| \leq \varepsilon\right\} \to 1, \quad \text{as} \quad m \to \infty. \tag{1.9}$$

From definition (1.8), we have

$$\Pr\{|S_m| \leq m\varepsilon\} = \sum_{|j| \leq [m\varepsilon]} P_{0j}^m.$$

(Here $[h]$ designates the greatest integer which does not exceed h; so $h - 1 < [h] \leq h$.) The limit relation (1.9) can be expressed equivalently in the form

$$H_m(\varepsilon) = \sum_{|j| \leq [m\varepsilon]} P_{0j}^m \to 1, \quad \text{as} \quad m \to \infty. \tag{1.10}$$

Now, choose $M = [n\varepsilon]$ in (1.7), where $\varepsilon > 0$. Then

$$G_{00}^n \geq \frac{1}{2[n\varepsilon] + 1} \sum_{m=0}^{n} \sum_{|j| \leq m[n\varepsilon]/n} P_{0j}^m = \frac{1}{2[n\varepsilon] + 1} \sum_{m=0}^{n} \sum_{|j| \leq [m\varepsilon]} P_{0j}^m$$

$$= \frac{n+1}{2[n\varepsilon] + 1} \frac{1}{n+1} \sum_{m=0}^{n} H_m(\varepsilon). \tag{1.11}$$

It follows from (1.10) that

$$\frac{1}{n+1} \sum_{m=0}^{n} H_m(\varepsilon) \to 1 \qquad \text{as} \quad n \to \infty. \tag{1.12}$$

Further,

$$\lim_{n \to \infty} \frac{n+1}{2[n\varepsilon] + 1} = \lim_{n \to \infty} \frac{n+1}{2n\varepsilon + 1} = \frac{1}{2\varepsilon}.$$

From (1.11) and (1.12) we conclude that

$$\lim_{n \to \infty} G_{00}^n \geq \frac{1}{2\varepsilon}.$$

Since $\varepsilon > 0$ may be chosen arbitrarily small, we have shown that

$$\sum_{k=0}^{\infty} P_{00}^k = \lim_{n \to \infty} G_{00}^n = \infty.$$

Finally, consulting Theorem 5.1 of Chapter 2 we recall that $G_{00} = \infty$ is equivalent to the assertion that the zero state is recurrent. ∎

Notice that we did not use the full force of the hypothesis that X_i has finite mean value. Scrutiny of the proof reveals the fact that the conclusion of the theorem remains valid provided merely that the weak law of large numbers is applicable as expressed in (1.9).

In the following theorem, which is a partial converse of Theorem 1.1, the existence of a mean is more decisively used.

Theorem 1.2. *If*

$$E[|X_i|] = \sum_{j=-\infty}^{\infty} |j| P_{0j} < \infty, \qquad i = 1, 2, \ldots, \tag{1.13}$$

and

$$\mu = E[X_i] = \sum_{j=-\infty}^{\infty} j P_{0j} \neq 0,$$

then the Markov chain $\{S_n\}$ is transient.

Proof.   Let A_n denote the event $\{S_n = 0\}$.

We recall the criterion of recurrence in the form

$$\text{Pr}\{A_n \text{ occurs for infinitely many } n\} = \begin{cases} 1 & \begin{array}{l} \text{if and only if the} \\ \text{Markov chain } \{S_n\} \\ \text{is recurrent,} \end{array} \\ 0 & \begin{array}{l} \text{if and only if the} \\ \text{Markov chain } \{S_n\} \\ \text{is transient} \end{array} \end{cases} \quad (1.14)$$

(see Theorem 7.1 of Chapter 2).

The proof of Theorem 1.2 makes use of the strong law of large numbers (cf. Section 1, Chapter 1), which states that

$$\text{Pr}\left\{ \lim_{n \to \infty} \frac{S_n}{n} = \mu \right\} = 1. \quad (1.15)$$

Now if $\mu \neq 0$ we consider the events

$$C_n = \left\{ \left| \frac{S_n}{n} - \mu \right| > \frac{|\mu|}{2} \right\}, \quad n = 1, 2, \ldots,$$

and let C be the event that C_n occurs for infinitely many n. We will evaluate $\text{Pr}\{C\}$. Any realization of the process for which $\lim_{n \to \infty} S_n/n = \mu$ obviously cannot belong to the event C. But according to (1.15) the realizations of the process fulfilling $\lim_{n \to \infty} S_n/n = \mu$ have probability 1. Therefore $\text{Pr}\{\text{complement of } C\} = 1$ or $\text{Pr}\{C_n \text{ occurs for infinitely many } n\} = 0$. But plainly the event A_n implies the event C_n, i.e., $A_n \subset C_n$. Hence

$$\text{Pr}\{A_n \text{ occurs for infinitely many } n\} \leq \text{Pr}\{C\} = 0$$

Taking cognizance of (1.14) we conclude that the Markov chain $\{S_n\}$ is transient.   ∎

2: *Local Limit Theorems*

Note that if the Markov chain $\{S_n\}$ is recurrent, it can only be null recurrent. This is so because, by spatial homogeneity, we have

$$\pi_i = \lim_{n \to \infty} P_{ii}^n = \lim_{n \to \infty} P_{00}^n = \pi_0, \quad \text{for all } i.$$

Thus, $\pi_0 > 0$ would imply $\sum_{i=-\infty}^{\infty} \pi_i = \infty$, which is impossible. Hence, $\pi_i = 0$ for all i.

Since the Markov chain $\{S_n\}$ is always either transient or null recurrent, we know that for all j

$$P_{0j}^n \to 0, \quad \text{as} \quad n \to \infty.$$

It is of some interest to determine the rate of convergence to zero. Such a result is referred to as a *local limit theorem*. To analyze this problem we introduce the characteristic function

$$\varphi(\theta) = \sum_{v=-\infty}^{\infty} P_{0v} \exp(iv\theta) = E[\exp(iX_k\theta)], \qquad k = 1, 2, \ldots,$$

$$-\pi \le \theta < \pi, \quad (2.1)$$

where the series converges absolutely and uniformly. We claim that

$$[\varphi(\theta)]^n = \sum_{v=-\infty}^{\infty} P_{0v}^n \exp(iv\theta), \qquad -\pi \le \theta < \pi. \quad (2.2)$$

This is so since the X_k, $k = 1, \ldots, n$, are independent, identically distributed random variables and thus

$$P_{0j}^n = \Pr\{S_n = j\},$$

$$\sum_{v=-\infty}^{\infty} P_{0v}^n \exp(iv\theta) = E[\exp(iS_n\theta)] = E[\exp(i\theta(X_1 + \cdots + X_n))]$$

$$= \prod_{k=1}^{n} E[\exp(i\theta X_k)] = \prod_{k=1}^{n} [\varphi_{X_k}(\theta)] = [\varphi_{X_1}(\theta)]^n.$$

Note, further, that

$$\int_{-\pi}^{\pi} e^{i(j-k)\theta} \, d\theta = \begin{cases} 0 & \text{if} \quad j \ne k \\ 2\pi & \text{if} \quad j = k \end{cases} \qquad (\text{here } i = \sqrt{-1}), \quad (2.3)$$

i.e., the functions $(2\pi)^{-1/2}e^{ik\theta}$ (k any integer) are orthonormal. Thus, when we multiply both sides of (2.2) by $(2\pi)^{-1}e^{-ik\theta}$ and integrate with respect to θ over $[-\pi, \pi]$ only the term $v = k$ remains on the right and this gives the formula

$$P_{0k}^n = (2\pi)^{-1} \int_{-\pi}^{\pi} e^{-ik\theta} [\varphi(\theta)]^n \, d\theta. \quad (2.4)$$

Before we can proceed to state and prove the result concerning the rate of convergence of P_{0j}^n to zero as $n \to \infty$, we introduce some additional concepts and discuss their properties. We say that X is a *periodic random variable* if the only values of X that may be achieved with positive probability are contained in the set

$$X = \omega + rc, \qquad r = 0, \pm 1, \pm 2, \ldots,$$

where ω and c are integers and $|c| \ne 1$. We may note that "the Markov chain $\{S_n\}$ is periodic" implies that the random variables X_k are periodic random variables but not conversely. (The student should prove this.) Recall that "$\{S_n\}$ is aperiodic" means that the smallest additive group generated by the integers i for which $\Pr\{X_1 = i\} > 0$ is the group of all integers.

The example

$$X_k = \begin{cases} +1 & \text{with probability } p, \\ -1 & \text{with probability } q \end{cases}$$

is a periodic random variable. In fact, we can represent its possible values in the form

$$X = 1 + 2r, \qquad r = 0, \pm 1, \pm 2, \ldots.$$

(Here $c = 2$, $\omega = 1$.)

Lemma 2.1.   *The X_k are periodic random variables if and only if their characteristic function $\varphi(\theta)$ has the property*

$$|\varphi(\theta_0)| = 1 \qquad (2.5)$$

for some $\theta_0 \neq 0$, $-\pi \leq \theta_0 < \pi$.

Proof.   Suppose that for $\theta_0 = h \neq 0 \,(-\pi \leq h < \pi)$

$$|\varphi(h)| = 1.$$

Then there is a real number w such that $\varphi(h) = e^{iwh}$, and hence

$$1 = e^{-ihw}\varphi(h) = \sum_{j=-\infty}^{\infty} P_{0j} e^{i(j-w)h}$$

$$= \sum_{j=-\infty}^{\infty} P_{0j} \cos(j-w)h + i \sum_{j=-\infty}^{\infty} P_{0j} \sin(j-w)h.$$

The real parts of both sides must be equal; thus

$$1 = \sum_{j=-\infty}^{\infty} P_{0j} \cos(j-w)h.$$

Since $|\cos x| \leq 1$ for all x, we infer that for all j accessible from zero, i.e., for those j for which $P_{0j} > 0$, necessarily

$$\cos(j-w)h = 1.$$

This requires the existence of an integer r for which

$$(j-w)h = 2\pi r, \qquad \text{for some} \quad r = 0, \pm 1, \pm 2, \ldots,$$

i.e., any j accessible from zero may be expressed in the form $j = w + (2\pi/h)r$, $r = 0, \pm 1, \pm 2, \ldots$, and plainly $|c| = |2\pi/h| \neq 1$. Clearly, X_k may take only these j values, i.e., X_k is a periodic random variable.

Conversely, if the possible values of X_k are contained in the set

$$X_k = \omega + rc, \qquad r = 0, \pm 1, \pm 2, \ldots$$

(ω and c are integers, $|c| \neq 1$), then

$$\varphi(\theta) = \sum_{r=-\infty}^{\infty} P_{0, \omega + rc} e^{i(\omega + rc)\theta}$$

and

$$\sum_{r=-\infty}^{\infty} P_{0, \omega + rc} = 1.$$

Now let $\theta_0 = 2\pi/c$. Since c is an integer, $|c| \neq 1$, it follows that $\theta_0 \neq 0$, $-\pi \leq \theta_0 \leq \pi$, and

$$\varphi\left(\frac{2\pi}{c}\right) = \sum_{r=-\infty}^{\infty} P_{0, \omega + rc} e^{i\omega(2\pi/c)} e^{i2\pi r} = e^{i\omega(2\pi/c)} \sum_{r=-\infty}^{\infty} P_{0, \omega + rc} = e^{i\omega(2\pi/c)}.$$

Thus (2.5) is satisfied with $\theta_0 = (2\pi/c) \neq 0$, $-\pi \leq \theta_0 \leq \pi$. ∎

We assume henceforth, unless it is stated explicitly to the contrary, that X_k is a nonperiodic random variable.

Lemma 2.2. *There exists a constant $\lambda > 0$, such that*

$$1 - \operatorname{Re} \varphi(\theta) \geq \lambda\theta^2, \qquad -\pi \leq \theta \leq \pi. \tag{2.6}$$

Proof. Note that

$$1 - \operatorname{Re} \varphi(\theta) = 1 - \sum_{j=-\infty}^{\infty} P_{0j} \cos j\theta = \sum_{j=-\infty}^{\infty} (1 - \cos j\theta) P_{0j}.$$

We employ the identity

$$1 - \cos \alpha = 2 \sin^2 \frac{\alpha}{2} \qquad \text{for any real } \alpha.$$

Thus

$$1 - \operatorname{Re} \varphi(\theta) = 2 \sum_{j=-\infty}^{\infty} \left(\sin^2 \frac{j\theta}{2} \right) P_{0j} \geq 2 \sum_{j=-L}^{L} \left(\sin^2 \frac{j\theta}{2} \right) P_{0j} \tag{2.7}$$

for any positive L.

We next use the familiar inequality

$$|\sin x| \geq \frac{2|x|}{\pi}, \qquad \text{for} \quad -\frac{\pi}{2} \leq x \leq \frac{\pi}{2}. \tag{2.8}$$

One proof of (2.8) proceeds as follows: We claim that the function $(\sin x)/x$ is decreasing over the range $0 \leq x \leq \pi/2$. In fact,

$$\frac{d}{dx}\left(\frac{\sin x}{x}\right) = \frac{x \cos x - \sin x}{x^2}.$$

But $1 \leq \sec^2 x$ and integrating both sides from 0 to x gives $x \leq \tan x$. Since $\cos x > 0$ for $0 \leq x < \pi/2$ we infer that $(d/dx)[(\sin x)/x] < 0$ and so $(\sin x)/x$ is decreasing for $0 \leq x < \pi/2$. Therefore

$$\frac{\sin x}{x} \geq \frac{\sin(\pi/2)}{\pi/2} = \frac{2}{\pi} \quad \text{or} \quad \sin x \geq \frac{2}{\pi} x, \qquad 0 \leq x < \frac{\pi}{2}.$$

Since both sides of the last inequality are odd functions of x, (2.8) follows. Using (2.8) in (2.7) gives

$$1 - \text{Re } \varphi(\theta) \geq 2 \sum_{|j| \leq L} \left(\frac{j\theta}{\pi}\right)^2 P_{0j} = \frac{2}{\pi^2} \theta^2 \sum_{|j| \leq L} j^2 P_{0j}, \qquad (2.9)$$

valid for all θ such that $|j\theta| \leq \pi$. But if $|j| \leq L$, the condition $|j\theta| \leq \pi$ will be satisfied whenever

$$|\theta| \leq \frac{\pi}{L}. \qquad (2.10)$$

By choosing L large enough there must be at least one j such that $|j| \leq L$ and $P_{0j} > 0$. Thus by specifying L large enough such that

$$C = 2\pi^{-2} \sum_{|j| \leq L} j^2 P_{0j} > 0,$$

we have

$$1 - \text{Re } \varphi(\theta) \geq C\theta^2 \qquad (2.11)$$

for all $|\theta| \leq \pi/L$ with $C > 0$.

Thus far we have made no use of the aperiodicity of the random variables $\{X_k\}$. We need this assumption, however, to estimate $1 - \text{Re } \varphi(\theta)$ for $|\theta| > \pi/L$. We know that X_k being nonperiodic random variables is equivalent to the condition that $|\varphi(\theta)| = 1$ is satisfied on $\theta \in [-\pi, \pi]$ only for $\theta = 0$ (Lemma 2.1). But $|\varphi(\theta)| \leq 1$ is always true for any characteristic function. Therefore,

$$1 - \text{Re } \varphi(\theta) \geq 1 - |\varphi(\theta)| > 0 \qquad (2.12)$$

for $\theta \neq 0$, $-\pi \leq \theta \leq \pi$. Since $1 - \text{Re } \varphi(\theta)$ is a continuous function of θ in $[-\pi, \pi]$,

$$m = \min_{\pi \geq |\theta| \geq \pi/L} \{1 - \text{Re } \varphi(\theta)\}$$

exists and is positive in view of (2.12). This can certainly be expressed in the form

$$1 - \text{Re } \varphi(\theta) \geq m \frac{\theta^2}{\pi^2} \qquad (2.13)$$

valid for all θ satisfying $\pi \geq |\theta| \geq \pi/L$. Now let

$$\lambda = \min(C, m/\pi^2)$$

Then (2.6) is satisfied for all $|\theta| \leq \pi$. ∎

Now we are prepared to state the theorem presenting a bound on the rate of convergence of P_{0j}^n to zero.

Theorem 2.1. *If the r.v.'s $\{X_k\}$ are aperiodic (i.e., nonperiodic), then for some constant $A > 0$ (independent of j and n)*

$$P_{0j}^n \leq \frac{A}{\sqrt{n}} \tag{2.14}$$

for all integers j and $n \geq 1$.

Proof. It is clear from (2.4) that

$$P_{0k}^{2n} \leq (2\pi)^{-1} \int_{-\pi}^{\pi} |\varphi(\theta)|^{2n} \, d\theta. \tag{2.15}$$

Note that $|\varphi(\theta)|^2$ is the characteristic function of an integer-valued random variable. Indeed, since

$$\varphi(\theta) = E[\exp(iX_k\theta)] \qquad \text{for any} \quad k = 1, 2, \ldots$$

and

$$\overline{\varphi(\theta)} = E[\exp(-iX_l\theta)] \qquad \text{for any} \quad l = 1, 2, \ldots,$$

we have for $k \neq l$,

$$|\varphi(\theta)|^2 = \varphi(\theta)\overline{\varphi(\theta)} = E[\exp(iX_k\theta)]E[\exp(-iX_l\theta)] = E[\exp(i(X_k - X_l)\theta)],$$

so that $|\varphi(\theta)|^2$ is the characteristic function of the integer-valued random variable $X_k - X_l$ where X_k and X_l are independent and identically distributed. The property that X_k is not a periodic random variable is equivalent to the nonperiodic character of the random variable $X_k - X_l$. This is a direct consequence of Lemma 2.1. Let $\psi(\theta) = |\varphi(\theta)|^2$. We now make use of Lemma 2.2 in the case of the real characteristic function $\psi(\theta)$. The lemma provides us with the inequality

$$1 - \psi(\theta) \geq \lambda\theta^2,$$

valid for all θ such that $-\pi \leq \theta \leq \pi$ for some $\lambda > 0$. We rewrite this relation in the form

$$\psi(\theta) \leq 1 - \lambda\theta^2 \leq \exp(-\lambda\theta^2), \tag{2.16}$$

where the last inequality results from the relation $1 - y \leq e^{-y}$, $y \geq 0$, which follows by integration of the trivial inequality $e^{-\xi} \leq 1$ over $[0, y]$.

From (2.16), by integration we have

$$\int_{-\pi}^{\pi} \psi(\theta)^n \, d\theta \le \int_{-\pi}^{\pi} \exp(-n\lambda\theta^2) \, d\theta = \frac{1}{\sqrt{n}} \int_{-\pi\sqrt{n}}^{\pi\sqrt{n}} \exp(-\lambda\alpha^2) \, d\alpha$$

$$\le (\sqrt{n})^{-1} \int_{-\infty}^{\infty} \exp(-\lambda\alpha^2) \, d\alpha. \quad (2.17)$$

Combining (2.15) and (2.17) yields

$$P_{0k}^{2n} \le \frac{1}{\sqrt{n}} \frac{1}{2\pi} \int_{-\infty}^{\infty} \exp(-\lambda\alpha^2) \, d\alpha = \frac{A_1}{\sqrt{2n}}, \quad (2.18)$$

where

$$A_1 = (2\pi)^{-1} \int_{-\infty}^{\infty} \exp(-\lambda\alpha^2) \, d\alpha.$$

Since $|\varphi(\theta)| \le 1$, we also obtain

$$P_{0k}^{2n+1} \le (2\pi)^{-1} \int_{-\pi}^{\pi} |\varphi(\theta)|^{2n+1} \, d\theta \le (2\pi)^{-1} \int_{-\pi}^{\pi} |\varphi(\theta)|^{2n} \, d\theta$$

$$= (2\pi)^{-1} \int_{-\pi}^{\pi} \{\psi(\theta)\}^n \, d\theta \le \frac{A_1}{\sqrt{2n}} \le \frac{\sqrt{2}\,A_1}{\sqrt{2n+1}}. \quad (2.19)$$

Put $A = \sqrt{2}\,A_1$; then (2.18) and (2.19) assert (2.14), as was to be shown. ∎

It is important to emphasize that the bound exhibited in (2.14) applies whether the Markov chain $\{S_n\}$ is recurrent or transient. It is instructive at this point to consider the example of the classical stochastic model of coin tossing. Specifically,

$$X_k = \begin{cases} 1 & \text{with probability } p, \\ -1 & \text{with probability } q. \end{cases}$$

Then the sequence S_n describes a Markov chain with special transition probabilities

$$P_{ij} = \begin{cases} p & \text{if } j = i + 1, \\ q & \text{if } j = i - 1, \\ 0 & \text{otherwise.} \end{cases}$$

Moreover, it is clear that

$$P_{00}^{2n} = \binom{2n}{n} p^n q^n$$

whose asymptotic behavior [cf. (6.2) of Chapter 2] is

$$P_{00}^{2n} \sim \frac{1}{\sqrt{n\pi}} (4pq)^n. \tag{2.20}$$

This shows that if $p \neq \frac{1}{2}$ then P_{00}^{2n} tends exponentially fast (at a geometric rate) to zero. In the case $p = q = \frac{1}{2}$, we see that the bound in (2.14) is exact.

This example is typical of the general situation. Actually a considerable sharpening of (2.14) is available under the assumption that $E[X_k^2] < \infty$ and $E[X_k] = \mu = 0$. In this case, Theorem 1.1 tells us that the Markov chain $\{S_n\}$ is recurrent and indeed null recurrent. Therefore $\sum_{n=0}^{\infty} P_{00}^n = \infty$ and $P_{0j}^n \to 0$. An application of the central limit theorem proves that

$$\lim_{n \to \infty} \sqrt{n} P_{0j}^n = B, \tag{2.21}$$

where B is a positive finite constant independent of j. The proof of this result is beyond the level of the book. We refer the reader to the works cited in the footnotes†‡ for details. When $E[X_k^2] = \infty$ and $E[|X_k|^{1+\delta}] < \infty$ for some $0 < \delta < 1$ but $E(|X_k|^{1+\zeta}) = \infty$ for $\zeta > \delta$, then frequently the precise asymptotic relation replacing (2.21) becomes

$$\lim_{n \to \infty} n^{1/(1+\delta)} P_{0j}^n = B. \tag{2.22}$$

This is relevant when the central limit theorem is not applicable but rather attraction to a proper stable law occurs. The theory of stable laws is an elaborate area of probability theory of importance in applications particularly to physics and astronomy. We cannot enter into this vast subject in this elementary book. We must be content merely to state its existence and encourage the reader to pursue these topics in later courses on probability theory.

3: Right Regular Sequences for the Markov Chain $\{S_n\}$

For special classes of Markov chains finer results pertaining to the theory of recurrence, occupation time problems, the evaluation of distributions of various functionals of the process, and other results are usually available. In this section we will present a sharpening of the characterization for right regular sequences (Theorem 5.2 of Chapter 11) in the special context of the Markov chain $\{S_n\}$.

If a Markov chain is irreducible and recurrent then the only right regular vector is the constant vector (Theorem 5.2, Chapter 11). In the special case of sums of independent random variables, this theorem can be extended to the

---

† B. V. Gnedenko and A. N. Kolmogorov, "Limit Distributions for Sums of Independent Random Variables" Addison-Wesley, Reading, Massachusetts, 1954.

‡ F. Spitzer, "Principles of Random Walk," D. Van Nostrand, Princeton, New Jersey, 1964.

aperiodic transient case if we require the right regular vector to be bounded. More precisely, we prove the following.

Theorem 3.1.   *If the Markov chain $\{S_n\}$ is irreducible with transition probability matrix $\mathbf{P}$ and if $\{y_j\}$ is right regular, i.e., $\mathbf{y}$ satisfies*

$$y_j \geq 0 \qquad \text{for all} \quad j = 0, \pm 1, \pm 2, \ldots \tag{3.1}$$

and

$$\sum_{j=-\infty}^{\infty} P_{ij} y_j = y_i \qquad \text{for all} \quad i = 0, \pm 1, \pm 2, \ldots \tag{3.2}$$

and $\{y_j\}$ is bounded, then $y_j \equiv$ constant for all j.

Proof.   Assume that $\{y_j\}$ satisfies (3.1) and (3.2) and is bounded. Let k_0 be any state other than 0 that can be reached from state 0, i.e., there exists n such that $P_{0k_0}^n > 0$. We keep k_0 fixed and define

$$z_j = y_j - y_{j-k_0}, \qquad j = 0, \pm 1, \pm 2, \ldots.$$

Now, using spatial homogeneity, we have

$$\sum_{j=-\infty}^{\infty} P_{i,j} y_{j-k_0} = \sum_{k=-\infty}^{\infty} P_{i,k_0+k} y_k = \sum_{k=-\infty}^{\infty} P_{i-k_0,k} y_k = y_{i-k_0},$$

which says that $\{u_j\} = \{y_{j-k_0}\}_{j=-\infty}^{\infty}$ also satisfies (3.2) and therefore $\{z_j\}$ satisfies (3.2). Trivially, $\{z_j\}$ is bounded since $\{y_j\}$ is bounded. Let

$$M = \sup_j z_j < \infty \qquad \text{and} \quad \sup_j |z_j| = M'. \tag{3.3}$$

We select a sequence of integers $\{r_n\}$ for which

$$\lim_{n \to \infty} z_{r_n} = M. \tag{3.4}$$

Since $\{z_j\}$ is bounded we may select a subsequence $\{r_n^{(1)}\}$ from $\{r_n\}$ such that

$$\lim_{n \to \infty} z_{1+r_n^{(1)}}$$

exists. Then we can choose a further subsequence $\{r_n^{(-1)}\}$ of $\{r_n^{(1)}\}$ such that

$$\lim_{n \to \infty} z_{-1+r_n^{(-1)}}$$

exists. Then select another subsequence $\{r_n^{(2)}\}$ of $\{r_n^{(-1)}\}$ such that

$$\lim_{n \to \infty} z_{2+r_n^{(2)}}$$

exists. Continuing in this fashion we determine sequences $\{r_n^{(-2)}\}$, $\{r_n^{(3)}\}$, $\{r_n^{(-3)}\}, \ldots$, each a subsequence of the one preceding it. Now there is a sequence, namely $\{s_n = r_n^{(-n)}\}$ in this case, which from some point on is a subsequence of each of the sequences

$$\{r_n\}, \quad \{r_n^{(1)}\}, \quad \{r_n^{(-1)}\}, \quad \{r_n^{(2)}\}, \quad \{r_n^{(-2)}\}, \quad \{r_n^{(3)}\}, \quad \{r_n^{(-3)}\}, \quad \ldots.$$

In fact, $\{s_n\}$ is a subsequence of $\{r_n^{(\rho)}\}$ at least from $n \geq |\rho|$ on. Because of the construction we know that

$$\lim_{n \to \infty} z_{j+s_n} = z_j^* \tag{3.5}$$

exists for all integers j. (This procedure is called the method of diagonalization.)

Clearly, by the definition of $\{r_n\}$ [see (3.4)] we have

$$z_0^* = \lim_{n \to \infty} z_{s_n} = M \tag{3.6}$$

and by (3.3)

$$z_j^* = \lim_{n \to \infty} z_{j+s_n} \leq M, \qquad j = 0, \pm 1, \pm 2, \ldots.$$

But we observed before that

$$\sum_{j=-\infty}^{\infty} P_{i+s_n, j} z_j = z_{i+s_n},$$

and taking limits on both sides as $n \to \infty$ we have

$$\lim_{n \to \infty} z_{i+s_n} = \lim_{n \to \infty} \sum_{j=-\infty}^{\infty} P_{i+s_n, j} z_j = \lim_{n \to \infty} \sum_{j=-\infty}^{\infty} P_{i, j-s_n} z_j$$
$$= \lim_{n \to \infty} \sum_{l=-\infty}^{\infty} P_{il} z_{l+s_n}. \tag{3.7}$$

We now claim that interchange of limit and summation is permissible. To validate this assertion, we are required to prove that for any prescribed $\varepsilon > 0$ there exists an integer $n(\varepsilon)$ such that

$$\left| \sum_{l=-\infty}^{\infty} P_{il}(z_{l+s_n} - z_l^*) \right| \leq \varepsilon,$$

provided $n \geq n(\varepsilon)$. To this end, we determine L sufficiently large so that

$$\sum_{|l| > L} P_{il} < \frac{\varepsilon}{4M}.$$

Next we choose n_0 sufficiently large so that

$$|z_{l+s_n} - z_l^*| < \frac{\varepsilon}{2} \qquad \text{for all} \quad n \geq n_0 \quad \text{and} \quad l \text{ satisfying } -L \leq l \leq L.$$

Now combining these estimates we have for $n \geq n_0(\varepsilon)$

$$\left| \sum_{l=-\infty}^{\infty} P_{il}(z_{l+s_n} - z_l^*) \right|$$

$$\leq 2M' \sum_{|l|>L} P_{il} + \sum_{|l|\leq L} P_{il}|z_{l+s_n} - z_l^*| \leq \frac{\varepsilon}{2} + \frac{\varepsilon}{2} = \varepsilon.$$

In passing to the limit under the summation sign in (3.7) we obtain

$$z_i^* = \sum_{l=-\infty}^{\infty} P_{il} z_l^*, \tag{3.8}$$

i.e., z_j^* satisfies (3.2), although not necessarily (3.1). From (3.8) by iteration we obtain

$$z_i^* = \sum_{j=-\infty}^{\infty} P_{ij}^n z_j^*, \qquad \text{for any} \quad n \geq 0.$$

Thus, with $i = 0$ and with reference to (3.6), we conclude that

$$\sum_{j=-\infty}^{\infty} P_{0j}^n z_j^* = z_0^* = M, \qquad \text{for any} \quad n \geq 0. \tag{3.9}$$

The left-hand side in (3.9) is a weighted average of numbers which are all $\leq M$. This is only possible if for all j for which $P_{0j}^n > 0$ for some $n \geq 0$ we have $z_j^* = M$. In particular, by the definition of k_0,

$$z_{k_0}^* = M, \qquad z_{2k_0}^* = M, \qquad \ldots, \qquad z_{tk_0}^* = M, \qquad \ldots$$

for any positive integer t. Then by (3.5) n may be chosen large enough such that all the inequalities

$$z_{k_0+s_n} > M - \varepsilon,$$
$$z_{2k_0+s_n} > M - \varepsilon,$$
$$\vdots$$
$$z_{tk_0+s_n} > M - \varepsilon$$

are satisfied simultaneously. Adding these inequalities yields

$$\begin{aligned} t(M - \varepsilon) &< z_{k_0+s_n} + z_{2k_0+s_n} + \cdots + z_{tk_0+s_n} \\ &= (y_{k_0+s_n} - y_{s_n}) + (y_{2k_0+s_n} - y_{k_0+s_n}) + (y_{3k_0+s_n} - y_{2k_0+s_n}) \\ &\quad + \cdots + (y_{tk_0+s_n} - y_{(t-1)k_0+s_n}) \\ &= y_{tk_0+s_n} - y_{s_n}. \end{aligned}$$

Since the y_j are bounded there exists $K > 0$, such that $y_j < K$, for all integers j. Then plainly

$$t(M - \varepsilon) < 2K.$$

Since this must hold for any positive integer t and $\varepsilon > 0$, M necessarily must be negative or zero. Thus $y_j - y_{j-k_0} = z_j \leq 0$ or

$$y_j \leq y_{j-k_0} \tag{3.10}$$

for all integer j and any k_0 that is accessible from zero.

Examination of the preceding analysis shows that the hypothesis (3.1), i.e., $y_j \geq 0$, was never used. Actually only the fact that the $|y_j|$ are bounded was vital. Therefore we could follow the same procedure putting $y_j' = -y_j$ in place of y_j throughout. Of course $\{y_j'\}$ remains bounded and satisfies (3.2). The arguments above now yield the conclusion

$$y_j \geq y_{j-k_0} \qquad \text{for all } j \quad \text{and for all states } k_0 \text{ accessible from 0.} \tag{3.11}$$

Comparing (3.10) and (3.11) we have

$$y_j = y_{j-k_0} \qquad \text{for all } j \quad \text{and for all states } k_0 \text{ accessible from 0.}$$

In particular

$$y_0 = y_{-k_0} \qquad \text{for all } k_0 \text{ accessible from 0.} \tag{3.12}$$

Now for the first time in the proof we will make use of the irreducibility assumption. This assumption guarantees that all states are indeed accessible from state 0. Therefore, (3.12) holds for all k_0 and each component y_j is equal to the constant y_0. ■

As an application of Theorem 3.1, we complete this section by proving the generalized renewal theorem for sums of independent identically distributed integer-valued random variables. Throughout this discussion we shall assume that the sums $\{S_n\}$ are aperiodic, i.e., that the smallest additive group generated by the integers i for which $\Pr\{X_1 = i\} > 0$ is the group of all integers.

The proof presented below is not the most direct, but several of the auxiliary results are of independent interest and the techniques are common to other studies of fluctuation theory for sums of independent r.v.'s.

Theorem 3.2. *If* $S_n = X_1 + X_2 + \cdots + X_n$, $n \geq 1$, *are aperiodic and if* $E[|X_1|] < \infty$, $E[X_1] > 0$, *then*

$$\lim_{j \to \infty} G_{ij} = \lim_{j \to \infty} \sum_{n=0}^{\infty} P_{0, j-i}^n = \frac{1}{E[X_1]}, \tag{3.13}$$

$$\lim_{j \to -\infty} G_{ij} = 0. \tag{3.14}$$

It is convenient to divide the proof into several steps. For any quantity a let $a^+ = \max(a, 0)$ and $a^- = \min(a, 0)$. Let

$$M_n = \min(S_1, S_2, \ldots, S_n).$$

Since obviously M_n is nonincreasing,

$$\lim_{n \to \infty} M_n = \inf(S_1, S_2, \ldots) = M$$

exists and conceivably M could be $-\infty$. However, since $EX_1 > 0$, the strong law of large numbers assures us that

$$\Pr\{M = -\infty\} \le \Pr\{S_n \le 0 \text{ infinitely often}\} = 0.$$

It follows that M is finite with probability 1.

Now

$$
\begin{aligned}
E[M_n] &= E[\min(S_1, \ldots, S_n)] \\
&= E[X_1 + \min(0, X_2, X_2 + X_3, \ldots, X_2 + \cdots + X_n)] \\
&= E[X_1] + E[\min(0, X_2, X_2 + X_3, \ldots, X_2 + \cdots + X_n)]. \quad (3.15)
\end{aligned}
$$

Since X_i are independent and identically distributed, we recognize the last term as $E[M_{n-1}^-]$.

Letting n go to ∞ we obtain

$$E[M] = E[X_1] + E[M^-]. \qquad (3.16)$$

(The student should justify the interchange of the limit operation and expectation.) Obviously

$$E[M] = E[M^+] + E[M^-]$$

and comparison with (3.16) reveals that

$$E[M^+] = E[X_1]. \qquad (3.17)$$

The proof given above is not fully rigorous since we do not know *a priori* that $E[M^-] > -\infty$ and only in that case is (3.16) meaningful.

In order to provide a complete proof we will need the following theorem, of some interest in itself.

Theorem 3.3. *Let Z be a nonnegative integer-valued r.v., i.e. $\Pr\{Z = n\} = p_n$, $n = 0, 1, \ldots, \sum_{n=0}^{\infty} p_n = 1$ with characteristic function $\varphi(\theta)$. Suppose $[\varphi(\theta) - 1]/i\theta = (1/i\theta) \sum_{n=0}^{\infty} p_n(e^{in\theta} - 1)$ converges to α $(0 < \alpha < \infty)$ as $\theta \downarrow 0$. Then $E[Z] = \sum_{n=0}^{\infty} np_n = \alpha$.*

Proof. By taking the real and imaginary part of $(\varphi(\theta) - 1)/i\theta$ we conclude that

$$\lim_{\theta \downarrow 0} \sum_{n=0}^{\infty} p_n \frac{(1 - \cos n\theta)}{\theta} = 0 \quad \text{and} \quad \lim_{\theta \downarrow 0} \sum_{n=0}^{\infty} p_n \frac{\sin n\theta}{\theta} = \alpha. \qquad (3.18)$$

Now for any fixed $0 < \theta < \pi/2$ we determine the largest integer $\hat{k} = k(\theta)$ satisfying $\pi/2 \ge \hat{k}\theta > 0$.

Consider the decomposition

$$\sum_{n=0}^{\infty} p_n \frac{\sin n\theta}{\theta} = \sum_{n=0}^{k} p_n \frac{\sin n\theta}{\theta} + \sum_{n=\hat{k}+1}^{\infty} p_n \frac{\sin n\theta}{\theta}. \tag{3.19}$$

In view of (3.18) the sum $\sum_{n=0}^{\infty} p_n (\sin n\theta)/\theta$ is uniformly bounded for θ sufficiently small.

We estimate $(\sin n\theta)/\theta$ in $\sum_{n=0}^{k} p_n (\sin n\theta)/\theta$ from below using the fact that $(\sin \theta)/\theta$ is decreasing on $0 < \theta < \pi/2$ (see page 79), which yields, in particular,

$$\frac{\sin \theta}{\theta} \geq \frac{\sin \pi/2}{\pi/2} = \frac{2}{\pi} \quad \text{for} \quad 0 < \theta < \frac{\pi}{2}.$$

From the definition of $\hat{k}$, it follows that $0 \leq n\theta \leq \pi/2$ for $0 \leq n \leq \hat{k}$. Applying the above inequality yields

$$\sum_{n=0}^{\hat{k}} p_n \frac{\sin n\theta}{\theta} \geq \frac{2}{\pi} \sum_{n=0}^{\hat{k}} n p_n.$$

On the other hand, we bound the second sum from above to give

$$\left| \sum_{n=\hat{k}+1}^{\infty} p_n \frac{\sin n\theta}{\theta} \right| \leq \frac{1}{\theta} \sum_{n=\hat{k}+1}^{\infty} p_n.$$

But $1 - (\sin n\theta)/n\theta \geq 1 - (1/n\theta) > 1 - 2/\pi = b > 0$ for all n satisfying $n\theta > \pi/2$. From the definition of $\hat{k}$ we know that $n > \hat{k}$ implies $n\theta > \pi/2$. Therefore

$$\frac{1}{\theta} \sum_{n=\hat{k}+1}^{\infty} p_n \leq \frac{1}{b\theta} \sum_{n=\hat{k}+1}^{\infty} p_n \left(1 - \frac{\sin n\theta}{n\theta} \right)$$

$$= \frac{1}{b\theta} \sum_{n=\hat{k}+1}^{\infty} p_n \frac{1}{\theta} \int_0^{\theta} (1 - \cos n\xi) \, d\xi$$

$$\leq \frac{1}{b\theta} \sum_{n=\hat{k}+1}^{\infty} p_n \int_0^{\theta} \frac{1 - \cos n\xi}{\xi} \, d\xi$$

$$= \frac{1}{b\theta} \int_0^{\theta} \sum_{n=\hat{k}+1}^{\infty} p_n \left(\frac{1 - \cos n\xi}{\xi} \right) d\xi \quad \text{(the interchange}$$

$$\text{is justified since all the terms are nonnegative)}$$

$$\leq \frac{1}{b\theta} \int_0^{\theta} \sum_{n=0}^{\infty} p_n \frac{1 - \cos n\xi}{\xi} \, d\xi$$

$$= \frac{1}{b\theta} \int_0^{\theta} \text{Im} \left[\frac{\varphi(\xi) - 1}{i\xi} \right] d\xi.$$

However, $\text{Im}([\varphi(\xi) - 1]/i\xi) \to 0$ as $\xi \downarrow 0$ in accordance with (3.18) and so its average tends to zero, i.e.,

$$\lim_{\theta \downarrow 0} \frac{1}{\theta} \int_0^\theta \text{Im}\left[\frac{\varphi(\xi) - 1}{i\xi}\right] d\xi = 0.$$

Putting together the preceding estimates we see that the second sum of (3.19) tends to zero as $\theta \downarrow 0$. It follows that

$$\sum_{n=0}^{k} np_n \qquad (\hat{k} = k(\theta))$$

is uniformly bounded for $\theta > 0$. Obviously $\hat{k} = k(\theta)$ increases to infinity as $\theta \downarrow 0$ and this implies the convergence of $\sum_{n=1}^{\infty} np_n = E[Z]$. To finish the proof of Theorem 3.3 we must establish that

$$\sum_{n=0}^{\infty} np_n = \lim_{\theta \to 0} \frac{\varphi(\theta) - 1}{i\theta}. \tag{3.20}$$

To this end, we prescribe $\varepsilon > 0$, and determine $N(\varepsilon)$ so that $\sum_{n=N+1}^{\infty} np_n < \varepsilon/2$, which is certainly possible since $\sum_{n=0}^{\infty} np_n$ converges by what was already demonstrated.

Now, consider

$$\frac{\varphi(\theta) - 1}{i\theta} - \sum_{n=0}^{\infty} np_n = \sum_{n=0}^{N} p_n\left(\frac{e^{in\theta} - 1}{i\theta} - n\right) + \sum_{n=N+1}^{\infty} p_n\left(\frac{e^{in\theta} - 1}{i\theta}\right) - \sum_{n=N+1}^{\infty} np_n.$$

Since $|(e^{in\theta} - 1)/i\theta| \leq n$ holds, the second and third sum are each bounded by $\varepsilon/2$. Hence

$$\overline{\lim_{\theta \downarrow 0}} \left|\frac{\varphi(\theta) - 1}{i\theta} - E[Z]\right| \leq \overline{\lim_{\theta \downarrow 0}} \left|\sum_{n=0}^{N} p_n\left[\frac{e^{in\theta} - 1}{i\theta} - n\right]\right| + \varepsilon.$$

For fixed N the right-hand sum tends to zero since each term does, and therefore

$$\overline{\lim_{\theta \downarrow 0}} \left|\frac{\varphi(\theta) - 1}{i\theta} - E[Z]\right| \leq \varepsilon.$$

The left-hand side is a fixed nonnegative number and $\varepsilon > 0$ can be chosen arbitrarily small. The result (3.20) clearly follows. ∎

We are now prepared to give a rigorous proof of the identity (3.17). We introduce the c.f.'s of M_n, X_1, and M_{n-1}^-, i.e.,

$$\varphi_{M_n}(\theta) = \sum_{k=-\infty}^{\infty} e^{ik\theta} \Pr\{M_n = k\},$$

$$\varphi_{X_1}(\theta) = \sum_{k=-\infty}^{\gamma} e^{ik\theta} \Pr\{X_1 = k\}$$

and

$$\varphi_{M_{n-1}^-} = \sum_{k=-\infty}^{0} e^{ik\theta} \Pr\{M_{n-1}^- = k\}.$$

Note that the possible values of M_{n-1}^- are restricted to the set of nonpositive integers. Manifestly, X_1 and

$$\tilde{M}_{n-1}^- = \min(0, X_2, X_2 + X_3, \ldots, X_2 + X_3 + \cdots + X_n)$$

are independent r.v.'s. Moreover $\tilde{M}_{n-1}^-$ and M_{n-1}^- are identically distributed since $\tilde{M}_{n-1}^-$ is defined in terms of $X_2, X_3, \ldots, X_n$ in precisely the same way that M_{n-1}^- is formed from $X_1, \ldots, X_{n-1}$.

Since $M_n = X_1 + \tilde{M}_{n-1}^-$ [this identity is implicit in Eq. (3.15)], we deduce the relation

$$\varphi_{M_n}(\theta) = \varphi_{X_1}(\theta) \varphi_{\tilde{M}_{n-1}^-}(\theta) = \varphi_{X_1}(\theta) \varphi_{M_{n-1}^-}(\theta).$$

From the definitions we plainly have $M_n \to M$ and $M_{n-1}^- \to M^-$ as $n \to \infty$ (convergence is meant in the sense of distributions) and so, by P. Levy's convergence theorem (page 11 of *A First Course*), it follows that

$$\varphi_M(\theta) = \varphi_{X_1}(\theta) \varphi_{M^-}(\theta). \tag{3.21}$$

But

$$\varphi_M(\theta) = \sum_{k=-\infty}^{x} e^{ik\theta} \Pr\{M = k\}$$

$$= \sum_{k=1}^{\infty} e^{ik\theta} \Pr\{M^+ = k\} + \sum_{k=-\infty}^{-1} e^{ik\theta} \Pr\{M^- = k\} + \Pr\{M = 0\}.$$

Since

$$\Pr\{M = 0\} = \Pr\{M^- \geq 0\} + \Pr\{M^+ \leq 0\} - 1$$
$$= \Pr\{M^- = 0\} + \Pr\{M^+ = 0\} - 1,$$

we can write the above expression in the form

$$\varphi_M(\theta) = \sum_{k=0}^{\infty} e^{ik\theta} \Pr\{M^+ = k\} + \sum_{-\infty}^{k=0} e^{ik\theta} \Pr\{M^- = k\} - 1$$
$$= \varphi_{M^+}(\theta) + \varphi_{M^-}(\theta) - 1.$$

Inserting this formula into (3.21) yields

$$\varphi_{M^+}(\theta) - 1 = \varphi_{M^-}(\theta)(\varphi_{X_1}(\theta) - 1). \tag{3.22}$$

Now dividing by $i\theta$ and then letting $\theta \to 0$, we obtain

$$\lim_{\theta \to 0} \frac{\varphi_{X_1}(\theta) - 1}{i\theta} = E[X_1] \quad \text{since} \quad E[|X_1|] < \infty.$$

(A formal proof of this limit relation can be carried out as in the final arguments of Theorem 3.3.) Trivially $\lim_{\theta \to 0} \varphi_{M^-}(\theta) = 1$ and comparison with (3.22) shows that

$$0 \le \lim_{\theta \to 0} \frac{\varphi_{M^+}(\theta) - 1}{i\theta} = E[X_1] < \infty. \tag{3.23}$$

Because M^+ is a nonnegative r.v. we can appeal to Theorem 3.3, which tells us that the limit in (3.23) is $E[M^+]$ and thus $E[M^+] = E[X_1]$ as was claimed. We also need the following lemma.

Lemma 3.1.   *Let e_i ($i \le 0$) be the probability that the process $\{S_n\}$ starting at i enters on the first step the positive states and thereafter never visits a nonpositive state. Define $e_i = 0$ for $i > 0$. Then $\sum_{i=-\infty}^{0} e_i = E[X_1]$.*

Proof.   Clearly, from its very definition

$$e_i = \begin{cases} \Pr\{\inf(S_1, S_2, \ldots) > -i\} = \Pr\{M > -i\}, & i \le 0, \\ 0, & i > 0. \end{cases}$$

Now

$$\sum_{i=-\infty}^{0} e_i = \sum_{i=-\infty}^{0} \Pr\{M > -i\} = \sum_{j=0}^{\infty} \Pr\{M > j\}$$

$$= \sum_{j=0}^{\infty} \sum_{k=j+1}^{\infty} \Pr\{M = k\} = \sum_{k=1}^{\infty} \Pr\{M = k\} \sum_{j=0}^{k-1} 1$$

$$= \sum_{k=1}^{\infty} k \Pr\{M = k\} = \sum_{k=0}^{\infty} k \Pr\{M^+ = k\} = E[M^+] = E[X_1].$$

The proof of Lemma 3.1 is complete.   ■

Let

$$V(i) = \Pr\{S_n \le 0 \text{ for some } n \ge 1 | S_0 = i\},$$

i.e., $V(i)$ is the probability, starting at i, of visiting the nonpositive axis.

Consider the realizations of the process which start at i and in fact visit the nonpositive states. With probability 1, by the law of large numbers, since $E[X_1] = \mu > 0$ by assumption such a path visits these states at most a finite number of times and thus there is a last time that it does. The probability that the last visit occurs at the nth trial is $\sum_{k=-\infty}^{0} P_{ik}^n e_k = \sum_{-\infty}^{\infty} P_{ik}^n e_k$; the last identity follows from the definition of e_k for $k > 0$. Enumerating the contingencies of the last visit to the nonpositive states we get the important identity

$$V(i) = \sum_{n=1}^{\infty} \sum_{k=-\infty}^{\infty} P_{ik}^{(n)} e_k = \sum_{k=-\infty}^{\infty} e_k \left(\sum_{n=1}^{\infty} P_{ik}^{(n)} \right)$$

$$= \sum_{k=-\infty}^{\infty} e_k \left(\sum_{n=0}^{\infty} P_{ik}^{(n)} - \delta_{ik} \right) = \sum_{k=-\infty}^{\infty} G_{ik} e_k - e_i \tag{3.24}$$

(δ_{ik} is the Kronecker delta function). We are now in a position to establish the renewal theorem.

Proof of Theorem 3.2.   From the definition of G_{ij} we find

$$\sum_k P_{ik} G_{kj} = \sum_k P_{ik} \sum_{n=0}^{\infty} P_{kj}^{(n)} = \sum_{n=0}^{\infty} P_{ij}^{(n+1)} = G_{ij} - \delta_{ij}. \tag{3.25}$$

Since $G_{ij} \leq G_{00}$ (Lemma 1.1) we conclude with the help of the diagonal procedure (see the proof of Theorem 3.1) that we may extract a subsequence $j_m \to \infty$ such that

$$\lim_{m \to \infty} G_{ij_m} = \varphi_i$$

exists for all i.

Letting $k = j_m$ tend to $+\infty$ we obtain from (3.25)

$$\sum_{k=-\infty}^{\infty} P_{ik} \varphi_k = \varphi_i \tag{3.26}$$

and $\{\varphi_i\}$ is bounded. [The reader should justify the interchange of limit and sum in passing from (3.25) to (3.26).] Appealing to Theorem 3.1, we know that the regular bounded sequence $\{\varphi_i\}$ is identically constant; call its value α. From (3.24), we have

$$V(-j_m) = \sum_k G_{-j_m, k} e_k - e_{-j_m}.$$

Since $G_{-j_m, k} = G_{-k, j_m}$ and $\sum_k e_k$ converges we conclude that

$$\lim_{m \to \infty} V(-j_m) = \alpha \sum_{k=-\infty}^{\infty} e_k = \alpha E[X_1] \qquad \text{(by Lemma 3.1).}$$

But obviously $\lim_{m \to \infty} V(-j_m) = 1$. Therefore $\alpha = 1/E[X_1]$. If we had another sequence $\tilde{j}_m \to \infty$ with the property that G_{ij_m} converges for all i then the same argument would show that its limit is $(E[X_1])^{-1}$. It follows that (3.13) holds. ∎

The result of (3.14) is proved in a similar manner and will be omitted.

4: The Discrete Renewal Theorem

The discrete random renewal theorem was stated in Section 1 of Chapter 3 and its proof, in a special case, was given in Section 2. Here we give an alternative proof based on Theorem 1.1 and the idea of *coupling* random processes.

Let $X_1, X_2, \dots$ be independent identically distributed nonnegative random variables with $p_k = \Pr\{X_n = k\}$ for $k = 0, 1, \dots$. We assume that $p_0 = 0$, $0 < p_1 < 1$, and $\mu = \sum_{k=0}^{\infty} k p_k < \infty$. Let $S_0 = 0$ and $S_n = X_1 + \cdots + X_n$ for $n \geq 1$ be the times of the successive "renewals" determined by the "lifetimes"

$X_1, X_2, \ldots$ Let $v_n = \Pr\{S_l = n \text{ for some } l \geq 0\}$ be the probability that a renewal takes place at time n.

The renewal theorem that we will prove in this section asserts that $v_n \to 1/\mu$ as $n \to \infty$.

We introduce the so-called "stationary renewal process" associated with $\{X_n\}$. Let $Y_0, Y_1, \ldots$ be nonnegative independent random variables, representing component lifetimes, and let $T_n = Y_0 + \cdots + Y_n$ be the corresponding renewal instants. While $Y_1, Y_2, \ldots$ are assumed to follow the probability distribution $\{p_k\}$, the initial lifetime Y_0 is given the distribution

$$\Pr\{Y_0 = k\} = (p_{k+1} + p_{k+2} + \cdots)/\mu \qquad \text{for} \quad k = 0, 1, \ldots.$$

Let $u_n = \Pr\{T_l = n \text{ for some } l \geq 0\}$ be the probability that a renewal takes place at time n in the stationary renewal process determined by $\{Y_n\}$.

In the stationary renewal process, renewals occur at a constant rate (probability) exactly equal to $u_n \equiv 1/\mu$. In the ordinary renewal process we wish to show that renewals occur asymptotically at rate $\lim v_n = 1/\mu$. To establish this result we couple the two processes together.

Let $U_0 = X_0 - Y_0 = -Y_0$ and $U_n = U_{n-1} + (X_n - Y_n)$ measure the discrepancy between the two processes. Let $N = \min(n \geq 0 : U_n = 0)$ be the first instant that the same renewal simultaneously takes place in both processes. Then $\Pr\{N < \infty\} = 1$ because $E[X_n - Y_n] = 0$ and by Theorem 1.1, the process $\{U_n - U_0\}$ is recurrent, whence

$$\Pr\{N < \infty\} = \sum_{l=0}^{+\infty} \Pr\{U_0 = -l \text{ and } U_n - U_0 \text{ visits } l\}$$

$$= \sum_{l=0}^{+\infty} \Pr\{U_0 = -l\} = 1.$$

Since N is finite we may couple the two renewal processes by assuming $X_n = Y_n$ for $n > N$. The probabilistic properties of the individual renewal processes $\{X_n\}$ and $\{Y_n\}$ are not affected by this coupling. Then

$$v_n = \Pr\{S_l = n \text{ for some } l \geq 0\}$$
$$= \Pr\{S_l = n \text{ for some } l > N\} + \Pr\{S_l = n \text{ for some } l \leq N\}.$$

Because of the coupling we have $S_l = T_l$ for $l > N$, whence

$$v_n = \Pr\{T_l = n \text{ for some } l > N\} + \Pr\{S_l = n \text{ for some } l \leq N\}$$
$$= \Pr\{T_l = n \text{ for some } l \geq 0\} + \Pr\{S_l = n \text{ for some } l \leq N\}$$
$$- \Pr\{T_l = n \text{ for some } l \leq N\}.$$

Now $\Pr\{T_l = n \text{ for some } l \geq 0\} = 1/\mu$ for all n. Thus to complete the demonstration that $v_n \to 1/\mu$ as $n \to \infty$, we need only show

$$\lim_{n \to \infty} \Pr\{S_l = n \text{ for some } l \leq N\} = 0$$

and

$$\lim_{n \to \infty} \Pr\{T_l = n \text{ for some } l \leq N\} = 0.$$

But because $S_l \leq S_N$ for $l \leq N$, we have

$$\lim_{n \to \infty} \Pr\{S_l = n \text{ for some } l \leq N\} \leq \lim_{n \to \infty} \Pr\{S_N \geq n\} = 0,$$

and similarly,

$$\lim_{n \to \infty} \Pr\{T_l = n \text{ for some } l \leq N\} \leq \lim_{n \to \infty} \Pr\{T_N \geq n\} = 0.$$

This completes the proof that $v_n \to 1/\mu$ as $n \to \infty$.

Remark. The assumption that $0 < p_1 < 1$ implies that the state space for the process $\{U_n\}$ comprises all integers and that this chain is aperiodic. The weaker assumption that the greatest common divisor of the set $\{k : p_k > 0\}$ is 1 will suffice.

Elementary Problems

1. Let $X_1, X_2, \ldots$ be a sequence of independent, identically distributed, integer-valued random variables. Define the partial sums $S_0 = 0$ and $S_k = X_1 + X_2 + \cdots + X_k$ for $k \geq 1$. The subscript $n \geq 1$ is called a ladder index if $S_n > S_j$ for $j = 0, 1, \ldots, n - 1$. Call the event that n is a ladder index $\mathscr{E}$. Define $Y_0 = 0$ and Y_N as the time (i.e., the index) of the last occurrence of $\mathscr{E}$ where the present trial is the Nth. Let W denote the time of first occurrence of $\mathscr{E}$. Suppose $\mathscr{E}$ occurs at trial n. Prove that the number of trials until the next occurrence of $\mathscr{E}$ is independent of n and distributed as W.

2. Under the hypothesis of Elementary Problem 1 prove the identity

$$\sum_{n=0}^{\infty} t^n E[x^{Y_n}] = \frac{1 - F(t)}{1 - t} \frac{1}{1 - F(xt)},$$

where $F(t) = \sum_{n=1}^{\infty} \Pr\{W = n\}t^n$.

Hint: Use the relation

$$\Pr\{Y_n = k\} = \Pr\{Y_k = k\} \Pr\{Y_{n-k} = 0\}.$$

3. Under the hypothesis of Elementary Problem 1 prove the exponential representation

$$U(t) = \exp\left[\sum_{k=1}^{\infty} \frac{t^k}{k} (E[Y_k] - E[Y_{k-1}])\right],$$

where $U(t) = 1/(1 - F(t))$.

Hint: With the aid of Elementary Problem 2 derive the differential equation

$$\frac{U'(t)}{U(t)} = \sum_{n=1}^{\infty} E[Y_n - Y_{n-1}]t^{n-1}$$

and solve it.

Problems

1. Consider a sample space consisting of the n cyclic permutations of $(a_1, a_2, \ldots, a_n)$ with each permutation having probability $1/n$. For a given point $\mathbf{x} = (a_k, a_{k+1}, \ldots, a_{k+n-1})$, where by definition $a_{n+l} = a_l$, $l = 1, \ldots, n-1$, let $N(\mathbf{x})$ be the number of partial sums among $\{a_k, a_k + a_{k+1}, a_k + a_{k+1} + a_{k+2}, \ldots, a_k + \cdots + a_{k+n}\}$ which are zero. Let $M(\mathbf{x})$ be the number of distinct partial sums. Show that if $\sum_{i=1}^{n} a_i = 0$, then $E[1/N] = M(\mathbf{x})/n$ for any $\mathbf{x}$.

2. Let $X_1, X_2, \ldots, X_n, \ldots$ be independent random variables uniformly distributed on $[0, 1]$. Show that for $0 \le a \le n-1$

$$\Pr\{X_1 + \cdots + X_n \le a\} = \sum_{j=0}^{[a]} (-1)^j \binom{n}{j} \frac{(a-j)^n}{n!}.$$

$[a]$ denotes, as usual, the greatest integer smaller than or equal to a.

Hint: Use induction with respect to n.

3. (Refer to Problem 2.) Establish the identity

$$1 = \sum_{j=0}^{n} (-1)^j \binom{n}{j} \frac{(n-j)^n}{n!}.$$

4. In Problem 2 let r be the index such that $X_1 + \cdots + X_{r-1} \le a$ but $X_1 + \cdots + X_r > a$. Show that

$$E(r) = \sum_{j=0}^{[a]} (-1)^j \frac{(a-j)^j}{j!} e^{a-j}.$$

Hint: Verify the identity $E(r) = \sum_{n=0}^{\infty} \Pr\{r > n\} = \sum_{n=0}^{\infty} \Pr\{X_1 + \cdots + X_n \le a\}$ and then use the result of Problem 2.

5. Let $X_1, X_2, \ldots$ be independent, identically distributed, integer-valued random variables. Define

$$S_0 = 0 \quad \text{and} \quad S_k = X_1 + X_2 + \cdots + X_k \quad \text{for} \quad k = 1, 2, \ldots.$$

Let $f_{00}^{(n)}$ be the probability that the process $\{S_k\}$ first returns to the origin at the nth step. Let γ_n be the probability that at the nth step the process will occupy a new state, i.e., $\gamma_n = \Pr\{S_n \ne S_0, S_n \ne S_1, S_n \ne S_2, \ldots, S_n \ne S_{n-1}\}$. Prove that

$$\gamma_n = \Pr\{S_1 \ne 0, S_2 \ne 0, \ldots, S_n \ne 0\}.$$

In the case $E[|X_i|] < \infty$ and $E[X_i] = 0$ express γ_n in terms of $\{f_{00}^{(k)}\}_{k=1}^{\infty}$.

6. Define S_k as in Problem 5. Let R_n be the number of distinct states visited in the first n steps by the process $\{S_k\}_{k=0}^{\infty}$. Compute the expected value of R_n in terms of $\{\gamma_k\}$ defined in the preceding problem.

Solution: $E[R_n] = \sum_{i=1}^{n} \gamma_i$.

7. Under the conditions of Problem 5 prove that

$$\lim_{n \to \infty} \frac{E[R_n]}{n} = 1 - f_{00}^*,$$

where

$$f_{00}^* = \sum_{k=1}^{\infty} f_{00}^k.$$

8. We retain the notation of Problem 5. Let $L_n = j$ if and only if $S_j > S_i$, $0 \le i < j$, and $S_j \ge S_i$ for $j < i \le n$, i.e., L_n is the first index of S_i where $\max_{0 \le i \le n} S_i$ is achieved. Prove the identity

$$\Pr\{L_n = j\} = \Pr\{L_j = j\} \Pr\{L_{n-j} = 0\}.$$

9. Define X_t with X_t a real number and $t = 0, 1, 2, \ldots$ as follows: $X_0 = 0$,

$$X_t = \begin{cases} X_{t-1} + (\frac{1}{2})^t & \text{with probability } \frac{1}{2}, \\ X_{t-1} - (\frac{1}{2})^t & \text{with probability } \frac{1}{2}. \end{cases}$$

Show that in the limit as $t \to \infty$, the distribution of X_t tends to the uniform distribution on $(-1, 1)$.

Hint: The distribution of X_t tends to the distribution of $Y = \sum_{k=1}^{\infty} Y_k(\frac{1}{2})^k$ where the Y_k are identically and independently distributed with values ± 1 equally likely. Let $f(s) = \prod_{k=1}^{\infty} \cos(s/2^k)$, which satisfies the functional equation

$$f(2s) = (\cos s) f(s).$$

Show that the only continuous solution of the latter equation satisfying $f(0) = 1$ is $(\sin s)/s$, which is the characteristic function of the uniform distribution on $(-1, 1)$.

10. Consider a random walk on the integer lattice of the positive quadrant in two dimensions. If at any step the process is at (m, n), it moves at the next step to $(m + 1, n)$ or $(m, n + 1)$ with probability $\frac{1}{2}$ each. Let the process start at $(0, 0)$. Let Γ be any curve connecting neighboring lattice points (extending from the Y axis to the X axis) in the first quadrant. Show that $EY_1 = EY_2$, where Y_1 and Y_2 denote the number of steps to the right and up, respectively, before hitting the boundary Γ. The diagram describes an example of Γ.

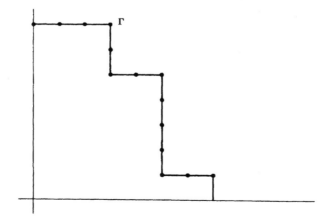

Hint: (a)   First consider the case when the curve Γ consists of two segments AB and BC. AB is the horizontal line extending from coordinates $(0, 1)$ to $(N, 1)$ and BC the vertical segment from $(N, 1)$ to $(N, 0)$. In this case verify

$$E[Y_2] = \tfrac{1}{2} + \sum_{k=1}^{N-1} (\tfrac{1}{2})^{k+1}, \qquad E[Y_1] = \sum_{k=1}^{N-1} k(\tfrac{1}{2})^{k+1} + (\tfrac{1}{2})^N N,$$

and that these quantities are equal.

   (b)   Any region bounded by the X and Y axes and a curve Γ can be broken into blocks as in case (a) above. Use an appropriate form of the addition law of expectations and the result proved in case (a).

Problems 11–16 are based on the following model

   Consider a random walk $\mathbf{X}_n = (A_n, B_n)$ in the plane where the possible states are all points with integer coordinates in the two-dimensional space. Assume that the probability of transition in one step from any state to any of the four neighboring states is $\tfrac{1}{4}$. Let T be the time that the random walk starting at the origin first hits the 45° line. Let X_T denote the point where the random walk hits the 45° line. Define

$$Q_{0j} = \Pr\{\mathbf{X}_T = (j, j) \mid X_0 = (0; 0)\}$$

This is the transition probability of a one-dimensional random walk which is the same as the original random walk observed only at times when it hits the 45° line. Define the characteristic function of this random walk as

$$\psi(\theta) = \sum_{m=-\infty}^{\infty} Q_{0m} e^{im\theta}, \qquad -\infty < \theta < \infty.$$

11.   Define $U_0 = V_0 = 0$, $U_n = A_n + B_n$, and $V_n = A_n - B_n$, $n = 1, 2, \ldots$. Prove that the sequence of random variables $\{U_n\}$ is independent of the sequence $\{V_n\}$. (The variables $\{U_n, V_n\}$ produce a change of coordinate system so that the 45° lines comprise one of the system of coordinates.)

Hint: (a)   Show first that

$$\Pr\{U_m - U_{m-1} = \pm r \mid U_0 = V_0 = 0\} = \begin{cases} \tfrac{1}{2} & \text{if } |r| = 1 \quad (m = 1, 2, \ldots), \\ 0 & \text{if } |r| \neq 1, \end{cases}$$

and

$$\Pr\{V_n - V_{n-1} = \pm s \mid U_0 = V_0 = 0\} = \begin{cases} \tfrac{1}{2} & \text{if } |s| = 1, \\ 0 & \text{if } |s| \neq 0, \quad n = 1, 2, \ldots. \end{cases}$$

   (b)   Show next that

$$\Pr\{U_m - U_{m-1} = \pm 1, V_n - V_{n-1} = \pm 1 \mid U_0 = V_0 = 0\} = \tfrac{1}{4}$$

and

$$\Pr\{U_m - U_{m-1} = r, V_n - V_{n-1} = s \mid U_0 = V_0 = 0\} = 0$$

if either $|r| \neq 1$ or $|s| \neq 1$ $(n, m = 1, 2, \ldots)$.
   Use (a) and (b) to prove that the sequence $\{U_m - U_{m-1}\}$ is independent of the sequence $\{V_n - V_{n-1}\}$ and from this conclude that $\{U_m\}$ is independent of $\{V_n\}$.

12. Prove that the random variable T is independent of the sequence $\{U_n\}$.

13. Note that $T = k$ implies $V_k = 0$. Establish the formula

$$\psi(\theta) = \sum_{k=1}^{\infty} \Pr\{T = k \,|\, X_0 = (0, 0)\} \sum_{l=-\infty}^{\infty} e^{il\theta} \Pr\{U_k = 2l \,|\, X_0 = (0, 0)\}.$$

14. Using the fact that U_n describes a one-dimensional random walk show that

$$\psi(\theta) = \sum_{k=1}^{\infty} \Pr\{T = k \,|\, X_0 = (0, 0)\} (\cos \theta/2)^k.$$

15. Show that the generating function of T is

$$E[s^T] = 1 - \sqrt{1 - s^2}.$$

16. Prove the formula

$$\psi(\theta) = 1 - |\sin \theta/2|, \qquad -\infty < \theta < \infty.$$

NOTES

The source of inspiration for this chapter is the elegant book by Spitzer [1]. Our discussion is a simplified version of a scant few topics from this important work.

Excellent textbooks on advanced probability which contain much material on limit theorems for sums of independent random variables are those by Gnedenko and Kolmogorov [2], Loéve [3], and Renyi [4]. A modern comprehensive treatment of limit theorems in probability theory and their ramifications is Volume II of Feller [5]. Petrov's book [6] is an excellent source of recent developments without assuming identical distributions for the summands.

REFERENCES

1. F. Spitzer, "Principles of Random Walk." Van Nostrand, Princeton, New Jersey, 1964.
2. B. V. Gnedenko and A. N. Kolmogorov, "Limit Distributions for Sums of Independent Random Variables." Addison-Wesley Publ. Co., Reading, Massachusetts, 1954.
3. M. Loéve, "Probability Theory," 2nd ed. Van Nostrand, Princeton, New Jersey, 1955.
4. A. Renyi, "Wahrscheinlichkeitsrechnung, mit einem Anhaug über Informationstheorie." Deutscher Verlag der Wissenschaften, Berlin, 1962.
5. W. Feller, "An Introduction to Probability Theory and Its Applications," Vol. II. Wiley, New York, 1966.
6. V. V. Petrov, "Sums of Independent Random Variables." Springer, New York, 1975.

Chapter 13

ORDER STATISTICS, POISSON PROCESSES, AND APPLICATIONS

In this chapter we present a variety of applications of some of the methods of Poisson processes and sums of independent random variables.

1: Order Statistics and Their Relation to Poisson Processes

Let $Y_1, Y_2, \ldots, Y_n$ be n independent, identically distributed, random variables with continuous, strictly increasing, cumulative distribution function $F(y)$. We define random variables $Y_1^*, \ldots, Y_n^*$ by

$$Y_i^* = \text{the } i\text{th smallest among } Y_1, Y_2, \ldots, Y_n.\dagger$$

In particular,

$$Y_1^* = \min\{Y_1, Y_2, \ldots, Y_n\} \quad \text{and} \quad Y_n^* = \max\{Y_1, Y_2, \ldots, Y_n\}.$$

Clearly, we have

$$Y_1^* \le Y_2^* \le \cdots \le Y_n^*.$$

Y_i^* is called the *i*th-*order statistic* of the sample $(Y_1, \ldots, Y_n)$ and $(Y_1^*, \ldots, Y_n^*)$ is called the set of *order statistics of size n* associated with the sample $(Y_1, \ldots, Y_n)$.

This chapter will be concerned with the distribution theory of the order statistics of a sample, their relationships to Poisson processes, and other applications. First, however, we will make a significant simplification without loss of generality.

Let us set

$$X_i = F(Y_i) \qquad (i = 1, \ldots, n)$$

---

† In case of ties, we make an arbitrary choice from among those Y_i that qualify. Actually the event of at least two equal Y_i values occurs with zero probability and for our purposes can be ignored.

and compute the distribution of X_i:

$$\Pr\{X_i < x\} = \Pr\{F(Y_i) < x\} = \Pr\{Y_i < F^{-1}(x)\}$$
$$= F[F^{-1}(x)] = x \quad \text{for} \quad 0 \le x \le 1, \quad i = 1, \ldots, n, \quad (1.1)$$

where F^{-1}, the inverse function of F, is uniquely defined by our assumptions about F. Further, since $0 \le F(y) \le 1$,

$$\Pr\{X_i < x\} = \begin{cases} 0 & \text{if} \quad x < 0, \\ 1 & \text{if} \quad x > 1, \end{cases} \quad \text{for} \quad i = 1, \ldots, n. \quad (1.2)$$

Thus, by (1.1) and (1.2), X_i is uniformly distributed over $[0, 1]$ for every $i = 1, \ldots, n$ regardless of the form the continuous, strictly increasing function F takes. Notice that the order relationship among the $\{Y_i^*\}$ is preserved by the transformation $X_i = F(Y_i)$. This means that instead of investigating the order statistics $\{Y_i^*\}$ corresponding to the general sample $(Y_1, \ldots, Y_n)$ we may study the order statistics $\{X_i^*\}$ taken from the uniform distribution on $[0, 1]$.

Therefore from now on we restrict ourselves to studying the order statistics

$$X_1^* \le X_2^* \le \cdots \le X_n^* \quad (1.3)$$

based on a sample of independent uniformly distributed random variables

$$X_1, X_2, \ldots, X_n$$

on the interval $[0, 1]$.

The fact that (1.3) holds for the random variables $X_1^*, X_2^*, \ldots, X_n^*$ clearly indicates that they are not independent. We will first determine the joint distribution of the order statistics $X_1^*, \ldots, X_n^*$, or rather its probability density function, denoted by $f^*(x_1, \ldots, x_n)$. That this exists will be clear from the proof. Prescribing $0 < x_1 < x_2 < \cdots < x_n < 1$ and sufficiently small increments $h_1, h_2, \ldots, h_n$ such that $[x_1, x_1 + h_1], [x_2, x_2 + h_2], \ldots, [x_n, x_n + h_n]$ comprise nonoverlapping intervals, we obtain

$$\int_{x_n}^{x_n + h_n} \cdots \int_{x_1}^{x_1 + h_1} f^*(x_1, \ldots, x_n) \, dx_1 \cdots dx_n$$

$$= \Pr\{x_i \le X_i^* < x_i + h_i, i = 1, 2, \ldots, n\}$$

$$= \sum_{\substack{\text{all permutations} \\ \sigma \text{ of } (1, 2, \ldots, n)}} \Pr\{x_i \le X_{i_\sigma} < x_i + h_i, \ i = 1, 2, \ldots, n\}$$

$$= \sum_\sigma \prod_{i=1}^n \Pr\{x_i \le X_{i_\sigma} < x_i + h_i\} \quad \text{(by independence)}$$

$$= \sum_\sigma \prod_{i=1}^n h_i = n! \, h_1 h_2 \cdots h_n, \quad (1.4)$$

where we have used the independence and uniform distribution of the X_i $(i = 1, \ldots, n)$ and the fact that there are $n!$ permutations of the indices $1, 2, \ldots, n$.

Now if we let each h_i shrink to zero it follows that the joint probability density function of the order statistics $X_1^*, \ldots, X_n^*$ is

$$f^*(x_1, \ldots, x_n) = n! \qquad \text{for} \quad 0 \leq x_1 \leq x_2 \leq \cdots \leq x_n \leq 1,$$
$$= 0 \qquad \text{elsewhere.} \tag{1.5}$$

The same proof shows that if the original $X_1, \ldots, X_n$ were taken from a uniform distribution over the interval $[a, b]$, the corresponding joint probability density function of the order statistics would be

$$f^*(x_1, \ldots, x_n) = \frac{n!}{(b - a)^n} \qquad \text{for} \quad a \leq x_1 \leq x_2 \leq \cdots \leq x_n \leq b,$$

$$= 0 \qquad \text{elsewhere.} \tag{1.6}$$

We encountered the density function (1.5) in our discussion of Poisson processes (see page 126 of *A First Course*). Specifically, let $\{Y(t), 0 \leq t \leq 1\}$ be a Poisson process, in particular for every $t \in [0, 1]$, $Y(t)$ is a discrete random variable with probability density function

$$p_k(t) = e^{-\lambda t} \frac{(\lambda t)^k}{k!} \qquad \text{for} \quad k = 0, 1, 2, \ldots,$$

$$= 0 \qquad \text{elsewhere,}$$

where $\lambda > 0$ is a fixed parameter. Assume that $Y(0) = 0$. Now, under the condition that $Y(1) = n$ (a positive integer) there will be exactly n time points in the interval $[0, 1]$ at which $Y(t)$ makes a "jump." The exact location in $[0, 1]$ of these jumps, depending on chance, define the random variables

$$T_1, T_2, \ldots, T_n \qquad (T_1 < T_2 < \cdots < T_n)$$

with values in $[0, 1]$.

We make the following assertion: Under the condition that

$$Y(1) = n$$

the random variables $T_1, T_2, \ldots, T_n$ are distributed as a set of order statistics of size n taken from a uniform distribution over $[0, 1]$.

The proof of this statement is quite immediate in view of the results already obtained. In fact, the evaluation of the conditional density function of $T_1, T_2, \ldots, T_n$ under the condition $Y(1) = n$ was given in Theorem 2.3 of Chapter 4. Comparison with (1.5) shows that these formulas agree and our assertion is hereby established.

The identification of order statistics and the conditioned occurrence of events of Poisson processes simplifies the derivation of other properties of order statistics.

For example, as earlier, let $X_1^*, X_2^*, \ldots, X_n^*$ denote order statistics based on a sample of size n from the uniform distribution. We claim that the joint distribution of $X_1^*, X_2^*, \ldots, X_{k-1}^*$ under the condition that $X_k^* = c_k, \ldots, X_n^* = c_n$ will be that of $k - 1$ order statistics associated with the sample $X_1, \ldots, X_{k-1}$, where each r.v. follows a uniform law on $[0, c_k]$. To verify this fact, we pass to the formulation of the problem in terms of events of a Poisson process.

Let $0 \le T_1 \le T_2 \le \cdots \le T_n \le 1$ denote the times at which the events of a Poisson process $Y(t)$ occur under the condition $Y(1) = n$. Suppose we impose the further conditions $T_k = c_k, T_{k+1} = c_{k+1}, \ldots, T_n = c_n$ and seek to determine the joint distribution of $T_1, T_2, \ldots, T_{k-1}$. Because $Y(t)$ is a process with independent increments it is clear that the only information pertinent to $T_1, T_2, \ldots, T_{k-1}$ can be summarized in the assertion $T_k = c_k$ or in the equivalent statement $Y(c_k - \varepsilon) = k - 1$ for ε positive and sufficiently small. Under this last condition we know that $T_1, \ldots, T_{k-1}$ are distributed as the order statistics of a $k - 1$ sample from a uniform distribution on $[0, c_k]$. Thus the joint conditional distribution of $X_1^*, X_2^*, \ldots, X_{k-1}^*$ under the condition $X_k^* = c_k, \ldots, X_n^* = c_n$ is the joint distribution of a size $k - 1$ order statistic taken from a uniform distribution over $[0, c_k]$. Thus the conditional density is given by

$$f^*(x_1, \ldots, x_{k-1} | c_k, \ldots, c_n) = \frac{(k-1)!}{c_k^{k-1}} \qquad \text{for} \quad 0 \le x_1 \le \cdots \le x_{k-1} \le c_k,$$

$$= 0 \qquad \text{elsewhere.} \tag{1.7}$$

By exactly the same reasoning we may deduce the fact that the conditional density function of $X_{k+1}^*, \ldots, X_n^*$, given that the values of the first k order statistics are

$$X_1^* = c_1, \ldots, X_k^* = c_k,$$

is equal to the joint density function of $n - k$ order statistics based on a sample of $n - k$ independent observations each following a uniform distribution over $[c_k, 1]$:

$$f^*(x_{k+1}, \ldots, x_n | c_1, \ldots, c_k)$$

$$= \frac{(n-k)!}{(1 - c_k)^{n-k}} \qquad \text{for} \quad c_k \le x_{k+1} \le \cdots \le x_n \le 1,$$

$$= 0 \qquad \text{elsewhere.} \tag{1.8}$$

Formulas (1.7) and (1.8) exhibit the feature that these joint conditional density functions are dependent on c_k only and independent of c_i, $i = k + 1, \ldots, n$ and $1, 2, \ldots, k - 1$, respectively. This means that the joint conditional density functions of $X_1^*, \ldots, X_{k-1}^*$ and $X_{k+1}^*, \ldots, X_n^*$ under the same single condition $X_k^* = c_k$ have the form

$$f^*(x_1, \ldots, x_{k-1} | c_k) = \frac{(k-1)!}{c_k^{k-1}} \qquad \text{for} \quad 0 \le x_1 \le \cdots \le x_{k-1} \le c_k,$$

$$= 0 \qquad \text{elsewhere.} \tag{1.9}$$

and

$$f^*(x_{k+1}, \ldots, x_n | c_k) = \frac{(n-k)!}{(1-c_k)^{n-k}} \qquad \text{for} \quad c_k \leq x_{k+1} \leq \cdots \leq x_n \leq 1,$$

$$= 0 \qquad\qquad \text{elsewhere}, \qquad (1.10)$$

respectively.

Formulas (1.7) and (1.8) in conjunction with (1.9) and (1.10) also show that the sets of variables $X_1^*, \ldots, X_{k-1}^*$ and $X_{k+1}^*, \ldots, X_n^*$ under the condition $X_k^* = c_k$ are (conditionally) independent.

More generally, the two sets of random variables

$$X_1^*, \ldots, X_i^* \quad \text{and} \quad X_{k+1}^*, \ldots, X_n^* \quad (i < k)$$

will be conditionally independent given the values of the remaining variables $X_{i+1}^*, \ldots, X_k^*$.

By the same reasoning we may obtain the joint density function of any number of consecutive order statistics. Thus, the joint conditional density function of $X_1^*, \ldots, X_i^*, X_{k+1}^*, \ldots, X_n^* (i < k)$, given $X_{i+1}^* = x_{i+1}, \ldots, X_k^* = x_k$, is

$$f^*(x_1, \ldots, x_i, x_{k+1}, \ldots, x_n | x_{i+1}, \ldots, x_k) = \frac{f^*(x_1, \ldots, x_n)}{f^*(x_{i+1}, \ldots, x_k)}. \qquad (1.11)$$

On the other hand, by the above assertion of independence, the left-hand side is also equal to

$$f^*(x_1, \ldots, x_i | x_{i+1}, \ldots, x_k) f^*(x_{k+1}, \ldots, x_n | x_{i+1}, \ldots, x_k)$$

$$= f^*(x_1, \ldots, x_i | x_{i+1}) f^*(x_{k+1}, \ldots, x_n | x_k)$$

$$= \frac{i!}{x_{i+1}^i} \frac{(n-k)!}{(1-x_k)^{n-k}}.$$

From this expression, (1.11), and (1.5) we obtain for $0 \leq i < k \leq n$

$$f^*(x_{i+1}, \ldots, x_k) = \frac{n!}{i!(n-k)!} x_{i+1}^i (1 - x_k)^{n-k}$$

$$\text{for} \quad 0 \leq x_{i+1} \leq \cdots \leq x_k \leq 1,$$

$$= 0 \quad \text{otherwise}. \qquad (1.12)$$

In particular, (1.12) with $i + 1 = k$ gives the marginal density function of X_k^*:

$$f^*(x_k) = \frac{n!}{(k-1)!(n-k)!} x_k^{k-1}(1 - x_k)^{n-k} \qquad \text{for} \quad 0 \leq x_k \leq 1,$$

$$= 0 \qquad\qquad \text{elsewhere}, \qquad (1.13)$$

which is the density function of a beta distribution.

The order statistics $X_1^*, X_2^*, \ldots, X_n^*$ partition the interval $[0, 1]$ into $n + 1$ disjoint intervals with lengths

$$U_1 = X_1^*, \quad U_2 = X_2^* - X_1^*, \quad \ldots, \quad U_n = X_n^* - X_{n-1}^*, \quad U_{n+1} = 1 - X_n^*.$$

Then $U_1, U_2, \ldots, U_n, U_{n+1}$ are obviously not independent random variables since $\sum_{i=1}^{n+1} U_i = 1$. By executing the transformation of variables $(x_1^*, \ldots, x_n^*)$ into $(u_1, \ldots, u_n)$, we have

$$
\begin{aligned}
u_1 &= x_1^*, \\
u_2 &= -x_1^* + x_2^*, \\
&\;\;\vdots \\
u_n &= \qquad\qquad -x_{n-1}^* + x_n^*,
\end{aligned}
\tag{1.14}
$$

and calculating the Jacobian of this transformation, which in this case is identically equal to 1, we may determine the joint density function $g(u_1, \ldots, u_n)$ of the random variables $\{U_1, \ldots, U_n\}$. Thus

$$g(u_1, \ldots, u_n) = n! \quad \text{for} \quad u_i \geq 0 \quad (i = 1, \ldots, n), \quad \sum_{i=1}^{n} u_i \leq 1,$$

$$= 0 \quad \text{otherwise.} \tag{1.15}$$

We can express this by saying that the random variables $\{U_1, \ldots, U_n\}$ are "uniformly" distributed over the region

$$u_i \geq 0 \quad (i = 1, \ldots, n) \qquad \sum_{i=1}^{n} u_i \leq 1.$$

Equivalently this also determines the distribution of the random variables $\{U_1, \ldots, U_n, U_{n+1}\}$ in the region

$$u_i \geq 0, \quad (i = 1, \ldots, n, n + 1) \qquad \sum_{i=1}^{n+1} u_i = 1.$$

We will now demonstrate that the joint distribution of $\{U_1, \ldots, U_{n+1}\}$ is the same as that of

$$\frac{Y_1}{S}, \quad \ldots, \quad \frac{Y_n}{S}, \quad \frac{Y_{n+1}}{S},$$

where $S = Y_1 + \cdots + Y_n + Y_{n+1}$ and the $Y_i, i = 1, \ldots, n, n + 1$, are independent exponentially distributed random variables with parameter $\lambda > 0$. This result can be demonstrated by defining a related Poisson process and analyzing the problem in these terms. For the sake of variety, we present a proof involving direct computation.

For this purpose we write the joint density function of $(Y_1, \ldots, Y_{n+1})$,

$$f(y_1, \ldots, y_{n+1}) = \lambda^{n+1} \exp\left(-\lambda \sum_{i=1}^{n+1} y_i\right) \quad \text{for} \quad y_i \geq 0, \quad i = 1, \ldots, n + 1,$$

$$= 0 \qquad\qquad \text{elsewhere,}$$

and make the transformation

$$v_1 = \frac{y_1}{y_1 + \cdots + y_{n+1}},$$

$$v_2 = \frac{y_2}{y_1 + \cdots + y_{n+1}},$$

$$\vdots$$

$$v_n = \frac{y_n}{y_1 + \cdots + y_{n+1}},$$

$$v_{n+1} = y_1 + \cdots + y_{n+1}.$$

The inverse of this transformation is

$$y_1 = v_1 v_{n+1},$$
$$y_2 = v_2 v_{n+1},$$
$$\vdots$$
$$y_n = v_n v_{n+1},$$
$$y_{n+1} = v_{n+1}[1 - (v_1 + \cdots + v_n)].$$

From this the Jacobian can be computed:

$$J = \begin{vmatrix} v_{n+1} & 0 & 0 & \cdots & 0 & v_1 \\ 0 & v_{n+1} & 0 & \cdots & 0 & v_2 \\ 0 & 0 & v_{n+1} & \cdots & 0 & v_3 \\ \vdots & \vdots & \vdots & \ddots & \vdots & \vdots \\ 0 & 0 & 0 & & v_{n+1} & v_n \\ -v_{n+1} & -v_{n+1} & -v_{n+1} & \cdots & -v_{n+1} & 1 - v_1 - v_2 - \cdots - v_n \end{vmatrix}$$

$$= \begin{vmatrix} v_{n+1} & 0 & 0 & \cdots & 0 & v_1 \\ 0 & v_{n+1} & 0 & \cdots & 0 & v_2 \\ 0 & 0 & v_{n+1} & \cdots & 0 & v_3 \\ \vdots & \vdots & \vdots & \ddots & \vdots & \vdots \\ & & & & v_{n+1} & v_n \\ 0 & 0 & 0 & \cdots & 0 & 1 \end{vmatrix} = v_{n+1}^n.$$

Hence, the joint density function of the variables

$$\frac{Y_1}{S}, \quad \frac{Y_2}{S}, \quad \cdots, \quad \frac{Y_n}{S}, \quad S$$

is

$$f(v_1, \ldots, v_n, v_{n+1}) = \lambda^{n+1} \exp(-\lambda v_{n+1}) v_{n+1}^n$$

$$\text{for } v_i \geq 0, \quad i = 1, \ldots, n+1, \quad \sum_{i=1}^{n} v_i = 1,$$

$$= 0 \quad \text{otherwise.}$$

From this representation we may infer that S and the random vector $(Y_1/S, \ldots, Y_n/S)$ are independent and possess the respective marginal density functions

$$f(v_{n+1}) = \frac{\lambda^{n+1}}{n!} \exp(-\lambda v_{n+1}) v_{n+1}^n \quad \text{for} \quad v_{n+1} \geq 0,$$

$$= 0 \quad \text{elsewhere,}$$

and

$$f(v_1, \ldots, v_n) = n! \quad \text{for} \quad v_i \geq 0, \quad i = 1, \ldots, n, \quad \sum_{i=1}^{n} v_i \leq 1,$$

$$= 0 \quad \text{elsewhere.} \tag{1.16}$$

Since (1.16) agrees with (1.15) and since

$$\frac{Y_1}{S} + \frac{Y_2}{S} + \cdots + \frac{Y_n}{S} + \frac{Y_{n+1}}{S} = 1,$$

the assertion about the equality of the distributions of

$$(U_1, \ldots, U_{n+1}) \quad \text{and} \quad \left(\frac{Y_1}{S}, \ldots, \frac{Y_{n+1}}{S} \right)$$

is proved.

2: The Ballot Problem

We now intend to present several applications of Poisson processes and related order statistics to various random variables connected with empirical distribution functions. To this end, we first develop some results familiar under the name of the *ballot problem*, which are of considerable interest and utility in their own right.

The ballot problem can be stated as follows:

In a ballot where c votes are cast, candidates A and B receive a and b number of votes, respectively, $a + b = c$. Throughout the counting of the votes the lead may continually change hands. The ballot problem (in its simplest version) consists of the question: Assuming that $a \geq b$, what is the probability that candidate A will always lead (at least by one vote) throughout the counting of the votes?

A direct solution of the ballot problem runs as follows. Consider a fixed arrangement of the a A's and the b B's on a circle. For the given arrangement we will determine the number of starting positions from which, say, going clockwise one full circle, A will always lead in the counting.

To find these positions, we delete successively all adjacent pairs AB going around in the circle perhaps several times. We are finally left with $a - b$ A's. A little reflection will convince oneself that the places left are exactly the starting positions from which A always leads. It follows that the probability that A always leads where the possible observation is a prescribed arrangement or one of its cyclic permutations is $(a - b)/(a + b)$. This quantity is independent of the choice of the prescribed sequence. It follows that the probability that candidate A will always lead throughout the vote counting is $(a - b)/(a + b)$.

The above analysis is very elegant. However, since we have in mind other generalizations we now formulate the ballot problem in a more general context of drawing cards from an urn and analyze its structure.

An urn contains a cards each labeled 0 and b cards each labeled 2; $a + b = c$, $a > b$. The cards are selected one by one in a random way from the urn without replacement until the last card is drawn. Let v_i be the random variable equal to the number on the ith card drawn; $i = 1, \ldots, c$. Then the ballot problem will be solved by finding

$$\Pr\{v_1 + v_2 + \cdots + v_r < r \text{ for } r = 1, 2, \ldots, c\}. \tag{2.1}$$

This assertion follows from the observation that if among the first r drawings there are α 0's and β 2's ($\alpha + \beta = r$) then the condition $v_1 + \cdots + v_r < r$ means that $\alpha \cdot 0 + \beta \cdot 2 < \alpha + \beta$, which reduces to $\beta < \alpha$. Obviously, (2.1) is the probability that this inequality holds for $r = 1, 2, \ldots, c$.

To find the probability (2.1) first notice that $(v_1, v_2, \ldots, v_c)$ is an arrangement of a 0's and b 2's and any one of the $c!/a!b!$ possible arrangements is equally likely to occur. That means that for every r ($r = 1, \ldots, c$) and every set $i_1, \ldots, i_r$ of distinct members of $\{1, \ldots, c\}$ the joint distribution of $(v_{i_1}, v_{i_2}, \ldots, v_{i_r})$ is the same as that of $(v_1, \ldots, v_r)$. This fact is expressed by saying that the random variables $v_1, \ldots, v_r$ are *interchangeable*. (Independent, identically distributed random variables are, of course, interchangeable.) Then, since

$$v_1 + v_2 + \cdots + v_c = a \cdot 0 + b \cdot 2 = 2b, \tag{2.2}$$

we have

$$\sum_{i=1}^{c} E[v_i] = E[v_1 + v_2 + \cdots + v_c] = 2b.$$

Since the r.v.'s $v_1, v_2, \ldots, v_n$ are interchangeable, each v_i has the same marginal distribution and therefore

$$cE[v_i] = 2b \quad \text{or} \quad E[v_i] = \frac{2b}{a + b} \quad \text{for} \quad i = 1, 2, \ldots, c. \tag{2.3}$$

Using this fact we now prove by induction with respect to c that

$$\Pr\{v_1 + \cdots + v_r < r \text{ for } r = 1, \ldots, c \,|\, v_1 + v_2 + \cdots + v_c = 2b\}$$

$$= 1 - \frac{2b}{c} = \frac{a - b}{a + b}. \tag{2.4}$$

First we show that (2.4) holds for $c = 1$. But $c = 1$ implies that $a = 1$ and $b = 0$, since $a > b \geq 0$ is assumed in the statement of the problem. Then clearly

$$\Pr\{v_1 < 1\} = \Pr\{v_1 = 0\} = 1.$$

The assertion of (2.4) is trivially true for $2b = c$, since in this case $v_1 + \cdots + v_c = 2b = c$.

Now, assume (2.4) holds for all $c \leq n - 1$ and $0 \leq 2b < c$. We wish to prove that (2.4) also holds for $c = n$ and $0 \leq 2b < c$.

Let b' be an integer such that $0 \leq b' \leq b$; then

$$\Pr\{v_1 + \cdots + v_r < r \text{ for } r = 1, \ldots, c \,|\, v_1 + \cdots + v_{2b} = 2b'\}$$
$$= \Pr\{v_1 + \cdots + v_r < r \text{ for } r = 1, \ldots, 2b \,|\, v_1 + \cdots + v_{2b} = 2b'\} \tag{2.5}$$

because the inequality $v_1 + \cdots + v_r < r$ is always satisfied for $r = 2b + 1, \ldots, c$ by the condition (2.2). But the right-hand side of (2.5) is the same type of expression as the left-hand side of (2.4), subject to a condition of the form (2.2). In fact, in (2.4) just replace c by $2b$ and $2b$ by $2b'$. Using the induction hypotheses with c and b replaced by $2b$ and b', respectively, yields

$$\Pr\{v_1 + \cdots + v_r < r \text{ for } r = 1, \ldots, 2b \,|\, v_1 + \cdots + v_{2b} = 2b'\} = 1 - \frac{2b'}{2b}. \tag{2.6}$$

To complete our induction proof write

$$\Pr\{v_1 + \cdots + v_r < r \text{ for } r = 1, \ldots, n\}$$

$$= \sum_{b'=0}^{b} \Pr\{v_1 + \cdots + v_r < r \text{ for } r = 1, \ldots, n \,|\, v_1 + \cdots + v_{2b} = 2b'\}$$

$$\times \Pr\{v_1 + \cdots + v_{2b} = 2b'\}$$

$$= \sum_{b'=0}^{b} \left(1 - \frac{2b'}{2b}\right) \Pr\{v_1 + \cdots + v_{2b} = 2b'\}, \tag{2.7}$$

by (2.6). But

$$\sum_{b'=0}^{b} 2b' \, \Pr\{v_1 + \cdots + v_{2b} = 2b'\} = E[v_1 + \cdots + v_{2b}] = 2b \frac{2b}{n},$$

by (2.3). Hence from (2.7)

$\Pr\{v_1 + \cdots + v_r < r \text{ for } r = 1, \ldots, n\}$

$$= \sum_{b'=0}^{b} \Pr\{v_1 + \cdots + v_{2b} = 2b'\} - \frac{1}{2b} \sum_{b'=0}^{b} 2b' \Pr\{v_1 + \cdots + v_{2b} = 2b'\}$$

$$= 1 - \frac{1}{2b} 2b \frac{2b}{n} = 1 - \frac{2b}{n}.$$

So (2.4) holds for $c = n$ and $0 \leq 2b < n$. This completes the induction proof.

A generalization of the ballot problem consists of having the urn contain c cards with numbers $k_1, \ldots, k_c$ such that

$$k_i = \text{nonnegative integer}, \qquad i = 1, \ldots, c,$$

$$k_1 + \cdots + k_c = k, \qquad 0 \leq k \leq c.$$

If v_i again denotes the number on the ith drawing from the urn without replacement, then (2.4) generalizes to

$$\Pr\{v_1 + \cdots + v_r < r \text{ for } r = 1, \ldots, c\} = 1 - \frac{k}{c}. \tag{2.8}$$

The proof of this more general statement can be carried out by repeating almost verbatim the proof of (2.4). We omit the details. (The student should check the steps for himself.)

A further generalization leads to the following problem. Let $v_1, \ldots, v_n$ be nonnegative interchangeable random variables with

$$v_1 + \cdots + v_n = y \qquad \text{(fixed number)}. \tag{2.9}$$

Let $\tau_1, \ldots, \tau_n$ be the order statistics based on n independent observations each uniformly distributed over $[0, t]$. Assume further that the random variables v_i $(i = 1, \ldots, n)$ are independent of the variables τ_j $(j = 1, \ldots, n)$.

Imagine the step function (see Fig. 1)

$$f(x) = \begin{cases} 0 & \text{if } 0 \leq x < \tau_1, \\ v_1 + \cdots + v_r & \text{if } \tau_r \leq x < \tau_{r+1}, \quad r = 1, \ldots, n-1, \\ v_1 + \cdots + v_n & \text{if } \tau_n \leq x < t. \end{cases}$$

Then we pose the problem, What is the probability that the graph of $f(x)$ will not cross the 45° line. The analytic statement of this problem with its solution is

$$\Pr\{v_1 + \cdots + v_r \leq \tau_r \text{ for } r = 1, \ldots, n\} = \begin{cases} 1 - \dfrac{y}{t} & \text{if } 0 \leq y \leq t, \\ 0 & \text{if } y > t. \end{cases} \tag{2.10}$$

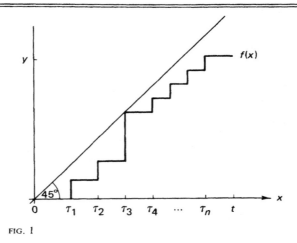

FIG. 1

We will prove (2.10) by induction. For $n = 1$,

$$\Pr\{v_1 \le \tau_1\} = \begin{cases} 1 - \dfrac{y}{t} & \text{if} \quad 0 \le y \le t, \\ 0 & \text{if} \quad y > t, \end{cases}$$

since $v_1 = y$ and τ_1 is uniformly distributed on $[0, t]$.

Assume now that (2.10) holds for $n - 1$ in place of n. Imposing the conditions

$$v_1 + \cdots + v_{n-1} = z \qquad (0 \le z \le y)$$

and

$$\tau_n = u \qquad (0 \le u \le t), \tag{2.11}$$

let us compute the conditional probability

$$\Pr\{v_1 + \cdots + v_r \le \tau_r \text{ for } r = 1, \ldots, n \,|\, v_1 + \cdots + v_{n-1} = z, \tau_n = u\}, \tag{2.12}$$

Assume first that $y \le u \le t$. Then by (2.9), the inequality

$$v_1 + \cdots + v_n \le \tau_n$$

will certainly prevail under the condition (2.11). Hence (2.12) equals

$$\Pr\{v_1 + \cdots + v_r \le \tau_r \text{ for } r = 1, \ldots, n - 1 \,|\, v_1 + \cdots + v_{n-1} = z, \tau_n = u\}$$

$$= \begin{cases} 1 - \dfrac{z}{u} & \text{if} \quad 0 \le z \le u, \\ 0 & \text{if} \quad z \ge u, \end{cases}$$

by the induction hypothesis, since $(\tau_1, \ldots, \tau_{n-1})$ under the conditions (2.11) are order statistics taken from $n - 1$ independent random variables uniformly distributed over $[0, u]$.

Now, remembering that v_i is independent of τ_n for all $i = 1, \ldots, n$, consider the density function

$$\varphi(z) = \frac{d}{dz} \Pr\{v_1 + \cdots + v_{n-1} \leq z\}.$$

Then for $y \leq u \leq t$

$$\Pr\{v_1 + \cdots + v_r \leq \tau_r \text{ for } r = 1, \ldots, n \,|\, \tau_n = u\}$$

$$= \int_0^y \Pr\{v_1 + \cdots + v_r \leq \tau_r \text{ for } r = 1, \ldots, n \,|\, v_1 + \cdots + v_{n-1} = z, \tau_n = u\}$$

$$\times \varphi(z) \, dz$$

$$= \int_0^y \left(1 - \frac{z}{u}\right)\varphi(z) \, dz = 1 - \frac{1}{u}\int_0^y z\varphi(z) \, dz$$

$$= 1 - \frac{1}{u} E[v_1 + \cdots + v_{n-1}] = 1 - \frac{1}{u}\frac{n-1}{n}y \qquad (2.13)$$

because (2.9) and the interchangeability of $v_1, \ldots, v_n$ imply

$$E[v_i] = \frac{y}{n} \quad \text{for} \quad i = 1, \ldots, n.$$

For $u < y$, however,

$$\Pr\{v_1 + \cdots + v_r \leq \tau_r \text{ for } r = 1, \ldots, n \,|\, \tau_n = u\} = 0, \qquad (2.14)$$

since under condition (2.11),

$$v_1 + \cdots + v_n \leq \tau_n$$

cannot hold. Now by (1.13) the probability density function of τ_n is

$$\psi(u) = \begin{cases} n\left(\dfrac{u}{t}\right)^{n-1}\dfrac{1}{t} & \text{if } 0 \leq u \leq t, \\ 0 & \text{otherwise.} \end{cases}$$

Then by (2.13) and (2.14) for $0 \leq y \leq t$

$$\Pr\{v_1 + \cdots + v_r \leq \tau_r \text{ for } r = 1, \ldots, n\}$$

$$= \int_0^t \Pr\{v_1 + \cdots + v_r \leq \tau_r \text{ for } r = 1, \ldots, n \,|\, \tau_n = u\}\psi(u) \, du$$

$$= \int_y^t \left(1 - \frac{1}{u}\frac{n-1}{n}y\right)n\left(\frac{u}{t}\right)^{n-1}\frac{du}{t} = 1 - \frac{y}{t}.$$

For $y > t$, obviously,

$$\Pr\{v_1 + \cdots + v_r \leq \tau, \text{ for } r = 1, \ldots, n\} = 0.$$

This proves (2.10).

We are now prepared to develop some applications of the preceding ideas to order statistics.

3: Empirical Distribution Functions

An important class of problems connected with order statistics concerns the empirical cumulative distribution function of a random variable. If X is a random variable with distribution function $F(x)$ and

$$(X_1^*, X_2^*, \ldots, X_n^*)$$

is the set of order statistics corresponding to a sample of size n from $F(x)$, then the *empirical cumulative distribution function* $F_n(x)$ of X is a random variable defined as

$$F_n(x) = \begin{cases} 0 & \text{if } x < X_1^*, \\ \dfrac{k}{n} & \text{if } X_k^* \leq x < X_{k+1}^*, \quad k = 1, \ldots, n-1, \\ 1 & \text{if } x \geq X_n^*. \end{cases} \tag{3.1}$$

Let X be a random variable with a continuous strictly increasing cumulative distribution function $F(x)$ and empirical cumulative distribution $F_n(x)$. We want to determine the probability

$$\Pr\{F_n(x) < \gamma F(x) \text{ for } -\infty < x < \infty\}. \tag{3.2}$$

As we have seen at the start of this chapter there is no loss of generality in assuming that X is uniformly distributed over $[0, 1]$. Indeed, replace X by $F(X)$, whose corresponding observations become $F(X_1)$, $F(X_2)$, $\ldots$, $F(X_n)$. Then (3.2) reduces to

$$\Pr\{F_n(x) < \gamma x \text{ for } 0 \leq x \leq 1\}, \tag{3.3}$$

where $F_n(x)$ is now the empirical distribution function associated with a uniform distribution.

Figure 2 shows a typical realization of $F_n(x)$ for the uniform distribution. We will prove that

$$\Pr\{F_n(x) < \gamma x \text{ for } 0 < x \leq 1\} = \begin{cases} 0 & \text{if } \gamma \leq 1, \\ 1 - \dfrac{1}{\gamma} & \text{if } \gamma > 1. \end{cases} \tag{3.4}$$

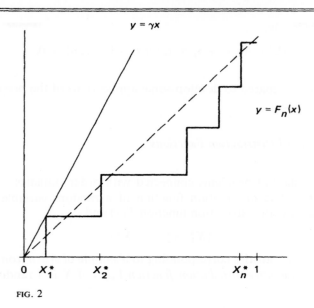

FIG. 2

If $\gamma \leq 1$, then (3.4) is obviously zero since the condition $F_n(1) < \gamma$ is violated. The result when $\gamma > 1$ is a corollary of (2.10). It is only necessary to apply the substitution $t = 1$, $y = 1/\gamma$. In fact, it is clear (consult Fig. 2) that the event

$$F_n(x) < \gamma x \qquad \text{for} \quad 0 < x \leq 1$$

will occur if and only if

$$X_k^* > \frac{k}{n\gamma}, \qquad k = 1, 2, \ldots, n, \tag{3.5}$$

where X_k^* $(k = 1, 2, \ldots, n)$ are the order statistics of size n whose underlying distribution is uniform over $[0, 1]$. The random variables $v_1, v_2, \ldots, v_n$ should be identified with the fixed numbers $v_i = 1/n\gamma$ $(i = 1, \ldots, n)$, which are manifestly interchangeable. Then (3.5) may be rewritten as

$$v_1 + v_2 + \cdots + v_k < X_k^*, \qquad k = 1, 2, \ldots, n,$$

where $v_1 + v_2 + \cdots + v_n = 1/\gamma = y$.

The validation of (3.4) is now complete by virtue of (2.10). Note that the right-hand side of (3.4) is independent of n.

The following problem connected with coin tossing also involves the empirical cumulative distribution function. Let X be a random variable with continuous, strictly increasing, cumulative distribution function $F(x)$. Let $(X_1, \ldots, X_n)$ and $(Y_1, \ldots, Y_n)$ be two independent random samples of size n from the distribution of X. Let $(X_1^*, \ldots, X_n^*)$ and $(Y_1^*, \ldots, Y_n^*)$ be the corresponding order statistics and form the empirical cumulative distribution functions

$$F_n(x) = \begin{cases} 0 & \text{if } x < X_1^*, \\ \dfrac{k}{n} & \text{if } X_k^* \le x < X_{k+1}^*, \quad k = 1, \ldots, n-1, \\ 1 & \text{if } x \ge X_n^*, \end{cases}$$

(3.6)

$$G_n(y) = \begin{cases} 0 & \text{if } y < Y_1^*, \\ \dfrac{l}{n} & \text{if } Y_l^* \le y < Y_{l+1}^*, \quad l = 1, \ldots, n-1, \\ 1 & \text{if } y \ge Y_n^*. \end{cases}$$

We may again assume that $F(x)$ is the uniform distribution over $[0, 1]$. As Fig. 3 shows, $F_n(x) = G_n(x) = 0$ for $x < \min(X_1^*, Y_1^*)$ and $F_n(x) = G_n(x) = 1$ for $x > \max(X_n^*, Y_n^*)$. In the interval $I = [\min(X_1^*, Y_1^*), \max(X_n^*, Y_n^*)]$ one of the two graphs may lie entirely below the other or, as in Fig. 3, they may equal or cross each other several times. We will proceed to determine the probability of the former event happening, that is,

$$\Pr\{F_n(x) \ne G_n(x) \text{ on any subinterval of } I\}. \tag{3.7}$$

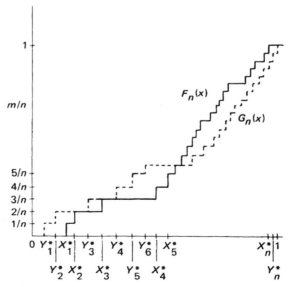

FIG. 3

This probability may be interpreted as follows. Regard $(X_1, \ldots, X_n, Y_1, \ldots, Y_n)$ as one sample and form the corresponding order statistics. A possible version of this could be

$$(Y_1^*, Y_2^*, X_1^*, X_2^*, Y_3^*, X_3^*, Y_4^*, Y_5^*, Y_6^*, X_4^*, X_5^*, \ldots, X_n^*, Y_n^*)$$

as depicted in Fig. 3.

Because both samples are taken from the same distribution any such arrangement is equally likely to occur.

At any rate we denote the order statistics as

$$(Z_1^*, Z_2^*, \ldots, Z_{2n}^*), \tag{3.8}$$

where half of the Z_k^*'s are the X^*'s and the other half are the Y^*'s. For every k we may compute the ratio

$$\rho_k = \frac{\text{number of } X_i^*\text{'s} \le Z_k^*}{\text{number of } Y_j^*\text{'s} \le Z_k^*}.$$

Clearly $\rho_{2n} = 1$. Since the graph of $F_n(x)$ will be entirely below the graph of $G_n(x)$ for all $x \in I$ if and only if $\rho_k < 1$ for all $k = 2, 3, \ldots, 2n - 1$, the expression (3.7) is equivalent to

$$\Pr\{\rho_k < 1 \text{ for } k = 2, 3, \ldots, 2n - 1 \text{ or } \rho_k > 1 \text{ for } k = 2, 3, \ldots, 2n - 1\}. \tag{3.9}$$

Another graphical representation of the event in (3.9) can be described as follows. Consider the step function (see Fig. 4) that makes a horizontal unit jump whenever we encounter an X_i^* in the arrangement of order statistics (3.8) and a vertical unit jump whenever we come upon a Y_j^*. This step function will lie strictly above the 45° line (except for the end point, where it will always coincide with the 45°

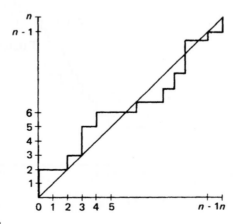

FIG. 4

line) if and only if $\rho_k < 1$ for all $k = 2, 3, \ldots, 2n - 1$. Hence, (3.9) [and also (3.7)] expresses the probability that the step function of Fig. 4 will be entirely on one side of the 45° line (except for the end points).

This problem can be interpreted as a coin-tossing game as follows. Suppose we conduct a series of $2n$ tosses with a fair coin. Assuming that at the end of the series the cumulative number of heads equals the cumulative number of tails (both equal to n), we may ask: What is the probability that the number of heads observed will always lead the number of tails observed as the game proceeds, or vice versa, that is, the number of tails will lead the number of heads throughout?

If we make the correspondence of horizontal jumps to heads and vertical jumps to tails, we thereby associate with every possible outcome of the series a step function as in Fig. 4. Hence the probability that heads leads tails throughout the series of $2n$ tosses or vice versa, provided they are tied at the end, equals the quantity in (3.9).

We will now proceed to compute this probability for which we have given several interpretations. For this purpose we will refer for convenience to the typical random step functions of Fig. 4. These step functions lead from $(0, 0)$ to (n, n) in $2n$ steps, n of which must be vertical. These are obviously altogether $\binom{2n}{n}$ such step functions. To count those among them that do not have points in common with the 45° line (except the end points) it is sufficient, for reasons of symmetry, to count those step functions that remain entirely below the 45° line and double their number. Every step function, however, that remains entirely below the 45° line goes to the point $(1, 0)$ in the first jump and then proceeds to (n, n). Obviously, there are $\binom{2n - 1}{n}$ step functions that lead to (n, n) from $(1, 0)$; we want to count only those that remain entirely under the 45° line. To obtain this number we shall count the number of step functions that lead from $(1, 0)$ to (n, n) and have at least one point in common with the 45° line, then we shall subtract this number from $\binom{2n - 1}{n}$. We show first that every step function leading from $(1, 0)$ to (n, n) that has at least one point in common with the 45° line and ends with a vertical jump (see Fig. 5) corresponds to a step function leading from $(1, 0)$ to (n, n) crossing the 45° line and ending with a horizontal jump. To see this, take the step function that touches the 45° line and ends in a vertical jump; let (k, k) be the point where it last touches the 45° line before reaching (n, n). Reflect the portion of the step function between (k, k) and (n, n) symmetrically with respect to the 45° line (see broken lines in Fig. 5). This process clearly establishes a one-to-one correspondence between the two kinds of step functions, those hitting the 45° line and ending with a vertical jump and those ending with a horizontal jump after hitting the 45° line. Thus we only have to count the step functions leading from $(1, 0)$ to (n, n) that cross the 45° line and end with a horizontal jump. But obviously *every* step function

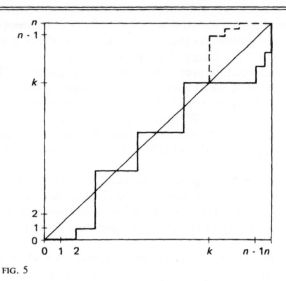

FIG. 5

leading from $(1, 0)$ to (n, n) that ends with a horizontal jump must pass through $(n - 1, n)$, and there are exactly $\binom{2n - 2}{n}$ such step functions. By virtue of the reflection principle, we infer that the number of step functions that lead from $(1, 0)$ to (n, n) and cross or touch the $45°$ line with a final vertical step is also $\binom{2n - 2}{n}$. Then the number of step functions that lead from $(1, 0)$ to (n, n) and always remain below the $45°$ line [except for (n, n)] is

$$\binom{2n - 1}{n} - 2\binom{2n - 2}{n} = \frac{1}{n - 1}\binom{2n - 2}{n}.$$

This is also the number of step functions that lead from $(0, 0)$ to (n, n) and remain entirely below the $45°$ line (except at the end points).

Hence, the probability that a step function (in Fig. 4) chosen at random will have no common point with the $45°$ line (except at the end points) is

$$\frac{2 \cdot \dfrac{1}{n - 1}\binom{2n - 2}{n}}{\binom{2n}{n}} = \frac{1}{2n - 1}. \tag{3.10}$$

This is also the probability expressed by (3.7) and (3.9). The result (3.10) can also be derived by a straightforward application of the ballot theorem.

Finally, the result of (3.10) can be obtained by appealing directly to the results on coin tossing developed in Chapter 10. In view of the identifications with coin tossing we are dealing with the first Markov chain of Section 4,

Chapter 10. In this formulation the probability in question is precisely the probability that first passage from state 1 to state 0 occurs at trial $2n - 1$, given that the cumulative number of heads equals the cumulative number of tails at trial $2n$.

The calculation of the probability that first passage from state 1 to state 0 occurs at trial $2n - 1$ with no further conditioning was carried out in Section 6 of Chapter 10 and shown to be $(1/(2n - 1))2^{-2n}\binom{2n}{n}$. Therefore, the conditional first passage probability is manifestly $f_{00}^{2n}/\mu_{2n} = 1/(2n - 1)$, as it should be.

4: Some Limit Distributions for Empirical Distribution Functions

The following lemma is interesting in itself and will be used in establishing a limit distribution for empirical distribution functions.

Lemma 4.1. *Let* $\mathbf{x} = (x_1, \ldots, x_n)$ *be a vector in n-dimensional Euclidean space. Assume that*

$$x_1 + x_2 + \cdots + x_n = 0$$

and that $\sum_{k=i+1}^{j} x_k \neq 0$ *for* $1 \leq i < j \leq n$.

Let $x_{k+n} = x_k$ *and* $\mathbf{x}(k) = (x_k, x_{k+1}, \ldots, x_n, \ldots, x_{n+k-1})$ *be a cyclic permutation of the components of* $\mathbf{x}$ *for* $k = 1, 2, \ldots, n$. *Then for each* $r = 0, 1, \ldots, n - 1$ *exactly one of the* $\mathbf{x}(k), k = 1, \ldots, n$, *is such that r among the successive partial sums of its components are positive.*

Proof. Let $s_k = x_1 + x_2 + \cdots + x_k$ for $k = 1, \ldots, n$ and $s_0 = 0$, $s_{k+n} = s_k$ for $k = 0, 1, \ldots, n$. Then the $s_k, k = 1, \ldots, n$, are distinct, since if we had

$$s_i = s_j \qquad \text{for} \qquad 1 \leq i < j \leq n,$$

then $x_1 + \cdots + x_i = x_1 + \cdots + x_i + x_{i+1} + \cdots + x_j$, i.e., $x_{i+1} + \cdots + x_j = 0$, which contradicts our assumption. Clearly, the successive partial sums of the components of $\mathbf{x}(k)$ are

$$s_k - s_{k-1}, \quad s_{k+1} - s_{k-1}, \quad \ldots, \quad s_n - s_{k-1}, \quad s_{n+1} - s_{k-1}, \quad \ldots,$$
$$s_{k+n-1} - s_{k-1}. \tag{4.1}$$

This sequence is identical with the sequence

$$s_k - s_{k-1}, \quad s_{k+1} - s_{k-1}, \quad \ldots, \quad s_n - s_{k-1}, \quad s_1 - s_{k-1}, \quad \ldots, \quad s_{k-1} - s_{k-1}. \tag{4.2}$$

Now let

$$s_1^*, \quad s_2^*, \quad \ldots, \quad s_n^*$$

be the unique relabeling of $s_1, s_2, \ldots, s_n$ for which

$$s_1^* > s_2^* > \cdots > s_n^*.$$

Such a relabeling can be achieved since the s_k $(k = 1, \ldots, n)$ are distinct.

The number of positive members of the sequence (4.2) (which is the same as (4.1)) will then be the same as the number of positive members in the sequence

$$s_1^* - s_{k-1}, \quad s_2^* - s_{k-1}, \quad \ldots, \quad s_n^* - s_{k-1}, \tag{4.3}$$

as this is simply a rearrangement of (4.2). Now for any $r = 0, 1, \ldots, n - 1$ there will be exactly one k $(1 \le k \le n)$ such that exactly r members in (4.3) (and hence also in (4.1)) will be positive. To see this just choose k such that

$$s_{k-1} = s_{r+1}^*.$$

Then $s_1^* - s_{k-1} > 0$, $s_2^* - s_{k-1} > 0$, $\ldots$, $s_r^* - s_{k-1} > 0$, but $s_{r+1}^* - s_{k-1} = 0$, $s_{r+2}^* - s_{k-1} < 0, \ldots, s_n^* - s_{k-1} < 0$. ∎

This lemma has a geometric setting. Omit the assumption $s_n = 0$. Plot the points $(0, 0)$, $(1, s_1)$, $\ldots$, (n, s_n) in a two-dimensional Cartesian coordinate system. Join the neighboring points with straight lines as in Fig. 6. The resulting broken line is called the sum polygon of the vector $\mathbf{x} = (x_1, \ldots, x_n)$. In quite the same way we can obtain sum polygons for the vector $\mathbf{x}(k)$ whose components are a cyclic permutation of those of $\mathbf{x}$. The straight line segment joining the points $(0, 0)$ and (n, s_n) is called the chord of the sum polygon. Consider the point of intersection P of this chord with the vertical line through (k, s_k). The ordinate

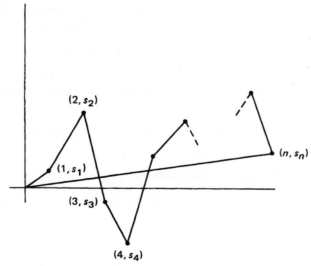

FIG. 6

of P is equal to $(k/n)s_n$ from elementary geometric considerations. Hence, the vertical distance from the vertex (k, s_k) to the chord of the polygon is equal to $s_k - (k/n)s_n$.

Now the vector whose components are

$$\left(x_1 - \frac{1}{n}s_n, x_2 - \frac{1}{n}s_n, \ldots, x_n - \frac{1}{n}s_n \right)$$

clearly satisfies the conditions of the lemma including the requirement that the nth partial sum vanish. The conclusion of the theorem asserts that among the n cyclic permutations of the sum polygons of $\mathbf{x}$ there is for each $r = 0, 1, \ldots, n - 1$ exactly one which will have r of its vertices above its chord. Particularly, the one for $r = 0$ is obtained by the cyclic permutation starting with $(1 +$ the index k where $\max_k[s_k - ks_n/n]$ is achieved). The case of $r = n - 1$ corresponds to $(1 +$ the index k where $\min_k(s_k - (k/n)s_n)$ is attained), etc.

We close this chapter with an application of Lemma 4.1 to the analysis of certain r.v.'s connected with empirical distribution functions. Let, as usual, $F_n(x)$ denote the empirical distribution function of a sample of size n from the uniform distribution on $(0, 1)$. We introduce the two random variables U_n and V_n:

$$U_n = \begin{cases} \text{cumulative length of all segments} \\ \text{of } x \text{ values for which } F_n(x) > x \end{cases} \qquad (4.4)$$

(see Fig. 7). In the particular realization depicted above, U_n equals the sum of all the darkened line segments.

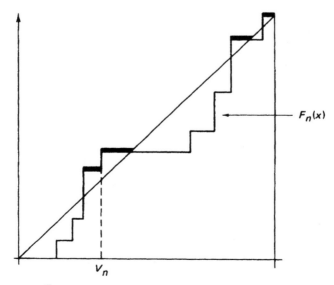

FIG. 7

We define

$$V_n = \inf\left\{x \mid F_n(x) - x = \max_{0 \le x \le 1}[F_n(x) - x]\right\}. \tag{4.5}$$

The quantity $\max_{0 \le x \le 1}[F_n(x) - x] = D_n^+$, commonly called the one-sided Kolmogorov–Smirnov statistic, is an important function of the observations used in performing statistical tests to decide whether the observed sample comes from an underlying uniform distribution.

Our objective in this section is to determine the distribution law of U_n and V_n. The remarkable result is that for all n each is uniformly distributed over $[0, 1]$.

To prove this assertion we consider a Poisson process $X(t)$, $0 \le t \le 1$, with parameter 1. Now divide the interval $(0, 1)$ into $r + 1$ parts

$$\left(0, \frac{1}{r+1}\right), \quad \left(\frac{r}{r+1}, \frac{2}{r+1}\right), \quad \ldots, \quad \left(\frac{r}{r+1}, 1\right),$$

where $r + 1$ is a prime number greater than n. (The reason for this specification will be clear later.) The increments

$$X\left(\frac{1}{r+1}\right), \quad X\left(\frac{2}{r+1}\right) - X\left(\frac{1}{r+1}\right), \quad \ldots, \quad X(1) - X\left(\frac{r}{r+1}\right)$$

are independent and identically distributed Poisson random variables. We denote these increments by $W_1, W_2, \ldots, W_{r+1}$, respectively, and define

$$Y_i = W_i - \frac{n}{r+1}, \qquad i = 1, \ldots, r+1,$$

which are obviously independent and identically distributed. We form the sequence of partial sums

$$S_k = Y_1 + \cdots + Y_k, \qquad k = 1, 2, \ldots, r+1,$$

and note that

$$\Pr\{S_i = 0\} = 0 \qquad (i = 1, 2, \ldots, r). \tag{4.6}$$

This is true because $S_i = 0$ implies $(r + 1)X(i/(r + 1)) = ni$. But this cannot hold owing to the hypothesis that $r + 1 > n$, i and does not divide n or i since $r + 1$ is prime. Hence (4.6) is valid.

A similar argument shows that, for any $j > i$, $\Pr\{S_j - S_i = 0\} = 0$. This means that with probability 1 none of the partial sums $S_j - S_i$ ($0 \le i < j < r + 1$, $S_0 = 0$) is zero.

Let

$N_r =$ number of positive terms in the sequence $S_1, S_2, \ldots, S_r$,

$L_r =$ smallest j for which $S_j = \max(0, S_1, \ldots, S_r)$.

The hypotheses of Lemma 4.1 are fulfilled for the sequence $x_i = S_i - S_{i-1}$, $i = 1, \ldots, r$, whenever no $S_j - S_i$ $(j > i)$ is zero and $S_{r+1} = 0$. This event occurs with probability 1. According to the lemma

$$\Pr\{L_r = m \,|\, S_{r+1} = 0\} = \Pr\{N_r = m \,|\, S_{r+1} = 0\} = \frac{1}{r+1} \qquad (m = 0, 1, \ldots, r).$$

(4.7)

Now under the condition $X(1) = n$, $X(t)/n$ is distributed like $F_n(t)$, $0 \leq t \leq 1$. (This is the identification made in Section 1.) Hence, we can define U_n, V_n for $X(t)/n$, $0 \leq t \leq 1$, under the condition $X(1) = n$. We claim that

$$\left| U_n - \frac{N_r}{r+1} \right| < \frac{A}{r+1}, \qquad \left| V_n - \frac{L_r}{r+1} \right| < \frac{B}{r+1}, \qquad (4.8)$$

where A and B are constants which depend on n but not on r.

A complete justification of the first inequality of (4.8) goes as follows. If

$$S_{k+1} = X[(k+1)/(r+1)] - n(k+1)/(r+1) > 0,$$

and if no jump occurs in the interval $(k/(r+1), (k+1)/(r+1))$, then certainly for the whole segment $k/(r+1) \leq t \leq (k+1)/(r+1)$, we have

$$X(t) - nt > 0. \qquad (4.9)$$

Since $X(t)$ has precisely n jumps under the condition $X(1) = n$ there are at most n intervals each of length not exceeding $1/(r+1)$ for which $S_i > 0$ does not imply that

$$X(t) - nt > 0, \qquad \frac{i}{r+1} \leq t \leq \frac{i+1}{r+1}$$

holds. But $N_r/(r+1)$ is equal to the number of positive S_i multiplied by a length $1/(r+1)$. In view of the preceding analysis we may conclude that $N_r/(r+1)$, can differ from U_n by a quantity no larger than $n/(r+1)$. Thus $A \leq n$. A similar argument leads to the second relation of (4.8). Thus under the condition $X(1) = n$, both absolute values in (4.8) converge in probability to zero as $r \to \infty$. Since $N_r/(r+1)$ and $L_r/(r+1)$ are asymptotically uniformly distributed over $(0, 1)$ as $r \to \infty$ ($r + 1$ tends to ∞ through prime values), this completes the proof of the following theorem.

Theorem 4.1. *Let $F_n(x)$ denote the empirical distribution function based on a sample of size n from the uniform distribution on $[0, 1]$. Consider the random variables U_n and V_n defined in (4.4) and (4.5). Then*

$$\Pr\{U_n \leq x\} = \Pr\{V_n \leq x\} = x \qquad (0 \leq x \leq 1).$$

Elementary Problems

1. Let $X_1, X_2, \ldots, X_n$ be a sequence of independent, identically distributed random variables with continuous distribution function $F(x)$. Denote by X_{nk} the kth smallest among the variables

$$X_1, X_2, \ldots, X_n.$$

Thus

$$X_{n1} \leq X_{n2} \leq \cdots \leq X_{nn}.$$

Find the distribution function $F_{nk}(x)$ of X_{nk}.

Solution:

$$F_{nk}(x) = \sum_{l=k}^{n} \binom{n}{l} [F(x)]^l [1 - F(x)]^{n-l}.$$

2. Using the notation of Problem 1, show that

$$F_{nk}(x) = k \binom{n}{k} \int_{1-F(x)}^{1} t^{n-k}(1-t)^{k-1} \, dt.$$

Hint: Integrate by parts.

3. Using the notation of Problem 1, find

$$\Pr\{X_{nk} > y, X_{n+1,k} \leq x\} \qquad \text{for} \quad x \leq y.$$

Solution:

$$\binom{n}{k-1} [F(x)]^k [1 - F(y)]^{n-k+1}.$$

4. Let $X_i^*, i = 1, 2, 3, \ldots, n$, be the order statistics from a uniform $(0, 1)$ distribution. Show that $\log X_k^*$ has the same distribution as $-\sum_{j=k}^{n} \theta_j/j$, where the θ_j are independent random variables with the exponential distribution with parameter 1.

Hint: Use the relationship between Poisson processes and order statistics of the uniform distribution.

5. Prove that $(X_i^*/X_{i+1}^*)^i, i = 1, 2, \ldots, n$, where X_{n+1}^* is defined to be 1, are independent uniform $(0, 1)$ random variables, where $X_1^*, \ldots, X_n^*$ are order statistics corresponding to a sample of size n from the uniform distribution.

6. Let $X_0, X_1, X_2, \ldots$ be independent identically distributed positive random variables having the continuous distribution function $F(x)$. Let N be the value of the first subscript n for which $X_n > X_0$. Find $\Pr\{N = k\}$ for $k = 1, 2, \ldots$ and $E[N]$.

(*Meaning of the problem:* $X_0, X_1, \ldots$ are offers for the car you are trying to sell. $E[N]$ is the average time you must wait before you receive an offer better than your first offer.)

Problems

1. If $X_1, X_2, \ldots, X_n$ are independent observations from the uniform $(0, 1)$ distribution find the distribution of $P = \prod_{i=1}^{n} X_i$.

Hint: Either calculate directly or introduce the new variables

$$X_i = \exp(-Y_i), \qquad i = 1, 2, \ldots, n.$$

Answer:

$$\Pr\{P \le p\} = \int_0^p \int_t^1 \int_{\xi_2}^1 \cdots \int_{\xi_{n-1}}^1 \frac{d\xi_2 \, d\xi_3 \cdots d\xi_n}{\xi_2 \xi_3 \cdots \xi_n} \, dt$$

$$= \frac{(-1)^n}{n!} \int_0^p t(\log t)^n \, dt.$$

2. Let $X_1, X_2, \ldots, X_k$ be independent identically distributed random variables with distribution function $F(x)$ and density function $f(x)$. Let $X_1^* \le X_2^* \le \cdots \le X_k^*$ be the corresponding order statistics. Let $Y_1, Y_2, \ldots$ be a sequence of independent, identically distributed random variables from the same distribution as the $\{X_i\}$. Define N as the integer-valued random variable for which

$$Y_i \le X_k^*, \quad i = 1, 2, \ldots, N - 1, \quad \text{but} \quad Y_N > X_k^*.$$

Similarly, define M as the random variable for which

$$Y_i \le X_{k-1}^*, \quad i = 1, 2, \ldots, M - 1, \quad \text{but} \quad Y_M > X_{k-1}^*.$$

Find the distributions of N and M, respectively.

Answer:

$$\Pr\{N = n\} = \frac{k}{(n + k)(n + k - 1)},$$

$$\Pr\{M = n\} = \frac{2k(k - 1)}{(n + k)(n + k - 1)(n + k - 2)}.$$

Problems 3–10 are based on the following structure.

Let $X_1, \ldots, X_{n-1}$ be independent random variables each uniformly distributed over the interval $(0, 1)$. Let

$$X_1^* \le X_2^* \le \cdots \le X_{n-1}^*$$

be the ordered values of the X_i. These order statistics partition the unit interval into n disjoint intervals whose lengths are

$$U_1 = X_1^*, \quad U_2 = X_2^* - X_1^*, \quad \ldots, \quad U_n = 1 - X_{n-1}^*.$$

Let

$$U_1^* \le U_2^* \le \cdots \le U_n^*$$

denote the ordered values of the U_i.

It is implicitly established in the text that
(a)   The random vector

$$\left(\frac{Y_1}{S_n}, \frac{Y_2}{S_n}, \ldots, \frac{Y_n}{S_n}\right),$$

where Y_i are independent r.v.'s exponentially distributed with parameter 1 and the r.v.'s $S_n = Y_1 + \cdots + Y_n$, are independent.

(b)   The random vectors $(U_1, \ldots, U_n)$ and

$$\left(\frac{Y_1}{S_n}, \frac{Y_2}{S_n}, \ldots, \frac{Y_n}{S_n}\right)$$

are identically distributed.

(c)   If S is a random variable distributed as S_n and independent of $(U_1, U_2, \ldots, U_n)$, then the random vectors $(SU_1, SU_2, \ldots, SU_n)$ and $(Y_1, \ldots, Y_n)$ are identically distributed.

3.   Compute $E[U_1^{i_1} U_2^{i_2} \cdots U_n^{i_n}]$, where $i_1, i_2, \ldots, i_n$ are nonnegative integers.

Answer:

$$\frac{n! \, i_1! \cdots i_n!}{(n + i_1 + \cdots + i_n)!}$$

4.   Prove that

$$\frac{X_1^*}{X_2^*}, \quad \frac{X_2^*}{X_3^*}, \quad \ldots, \quad \frac{X_n^*}{1}$$

are independent random variables.

Hint: Consult Elementary Problem 5.

5.   Let $S_k = Y_1 + \cdots + Y_k$. Show that

$$\frac{S_1}{S_2}, \quad \frac{S_2}{S_3}, \quad \ldots, \quad \frac{S_{n-1}}{S_n}$$

are independent random variables.

6.   If i, j, k, and l are distinct indices show that

$$\frac{U_i}{U_j}, \quad \frac{U_k}{U_l}$$

are independent random variables.

7.   Find the distribution of

$$\frac{U_1 + \cdots + U_r}{U_{r+1} + \cdots + U_{r+s}}.$$

Answer: Ratio of two independent r.v.'s each following a gamma distribution.

8. Show that

$$F(t_1, \ldots, t_n) = \Pr\{U_1 > t_1, U_2 > t_2, \ldots, U_n > t_n\}$$
$$= \begin{cases} 1 & \text{if } t < 0, \\ (1 - t)^n & \text{if } 0 \le t \le 1, \\ 0 & \text{if } t > 1, \end{cases}$$

where $t = t_1 + \cdots + t_n$.

9. Let $Y_1^* \le Y_2^* \le \cdots \le Y_n^*$ be the ordered values of $Y_1, \ldots, Y_n$. Find the joint distribution of

$$Z_1 = nY_1^*, \quad Z_2 = (n - 1)(Y_2^* - Y_1^*), \quad \ldots, \quad Z_n = (Y_n^* - Y_{n-1}^*).$$

Hint: Cf. Elementary Problem 5.

Answer: Independent exponential random variables with parameter 1.

10. Show that $[U_1, U_2, \ldots, U_n]$ and $[nU_1^*, (n - 1)(U_2^* - U_1^*), \ldots, U_n^* - U_{n-1}^*]$ have the same joint distribution.

11. Using the notation of Elementary Problem 1, we define the rank of X_j in the set $X_1, X_2, \ldots, X_n$ to be r if $X_j = X_{nr}$. Now, let R_j be the rank of X_j in the set $X_1, X_2, \ldots, X_j$. Show that the random variables $R_1, R_2, \ldots, R_n$ are independent and

$$\Pr\{R_n = r\} = \frac{1}{n} \quad \text{for} \quad r = 1, 2, \ldots, n.$$

Hint: Prove that

$$\Pr\{R_1 = r_1, R_2 = r_2, \ldots, R_n = r_n\} = \frac{1}{n!}$$

for all choices $r_1 = 1$; $r_2 = 1, 2$; $r_3 = 1, 2, 3$; $\ldots$; $r_n = 1, 2, \ldots, n$.

12. Let $X_1, X_2, \ldots$ be independent, identically distributed, positive random variables with continuous distribution function $F(x)$. Prove that

$$\Pr\{X_k > \max(X_1, X_2, \ldots, X_{k-1})\} = \frac{1}{k}.$$

13. Let $X_1, X_2, \ldots$ be independent identically distributed r.v.'s and define

$$M_n = \max(X_1, X_2, \ldots, X_n).$$

Let $N_n(\lambda, \mu)(0 < \lambda < \mu)$ be the number of terms in the sequence

$$X_{[n\lambda]}, \quad X_{[n\lambda]+1}, \quad X_{[n\lambda]+2}, \quad \ldots, \quad X_{[n\mu]},$$

where $X_i = M_i$. Here, as customary, $[n\lambda]$ denotes the greatest integer less than or equal to $n\lambda$ and $[n\mu]$ has an analogous connotation. Show that the probability generating function of $N_n(\lambda, \mu)$ is

$$E[z^{N_n(\lambda, \mu)}] = \exp\left\{\sum_{i=[\lambda n]}^{[\mu n]} \log\left[1 - \frac{(1 - z)}{i}\right]\right\}.$$

Hint: Introduce the r.v.'s

$$C(i) = \begin{cases} 1 & \text{if} \quad X_i = M_i, \\ 0 & \text{otherwise.} \end{cases}$$

Use Problem 11 to deduce that $C(i)$ are independent r.v.'s and then calculate

$$\prod_{i=[n\lambda]}^{[n\mu]} E[z^{C(i)}].$$

14. (*continuation*).   Show that the limit distribution of $N_n(\lambda, \mu)$ as $n \to \infty$ is a Poisson law with parameter $\log(\mu/\lambda)$.

15. Let $W_{1,n\mu}$ = the index of the last maximum among the sequence $X_1, X_2, \ldots, X_{[n\mu]}$. Prove that $W_{1,n\mu}/n\mu$ has a limit distribution as $n \to \infty$ that is uniform on $(0, 1)$.

16. Let $W_1, W_2, \ldots$ be r.v.'s independent and uniformly distributed on $(0, 1)$. Let N be the number of indices i satisfying $t < \prod_{j=1}^{i} W_j < 1$, where t is a fixed number between 0 and 1. Find the distribution function of N.

Answer: N follows a Poisson law with parameter $-\log t$.

17. Let $X_1, X_2, \ldots, X_n$ be n observations from a continuous distribution function $F(x)$. Let N_n denote the number of indices k for which $X_k > \max(X_1, X_2, \ldots, X_{k-1})$, $k = 1, 2, \ldots, n$. Show that the generating function of N_n is

$$\sum_{r=1}^{n} \Pr\{N_n = r\}s^r = \binom{s+n-1}{n} = \frac{(s+n-1)(s+n-2)\cdots(s+1)s}{n!}.$$

By convention, X_1 satisfies the requirement of $N_1 = 1$.

18. In the previous problem, establish the relation

$$\lim_{n\to\infty} \frac{E[N_n]}{\log n} = 1.$$

19. Consider a sequence $X_1, X_2, \ldots$ of independent, identically distributed r.v.'s with continuous distribution $F(x)$. A value X_k is called "outstanding" if

$$X_k > \max(X_1, X_2, \ldots, X_{k-1}).$$

(By convention X_1 is outstanding.)
    Let N_n be the number of outstanding values among the sequence $X_1, X_2, \ldots, X_n$.
    Let V_k be the index of the kth outstanding value.
    Prove the relation

$$\Pr\{V_k > n\} = \Pr\{N_n < k + 1\}.$$

20. (*continuation*).   Prove the formula

$$\Pr\{N_n = r\} = \frac{1}{n} \sum_{1 < k_1 < k_2 < \cdots < k_{r-1} < n} \frac{1}{(n - k_{r-1})} \frac{1}{(n - k_{r-2})} \cdots \frac{1}{(n - k_1)}.$$

21. Let $X_1, X_2, \ldots, X_m$ and $Y_1, Y_2, \ldots, Y_n$ be collections of m and n independent r.v.'s with the same continuous distribution function $F(x)$. Let N be the number of Y_i in the second sample not exceeding X_k^*, the kth-order statistic from the first sample. Show that the probability density function of N is

$$\Pr\{N = l\} = \frac{\binom{m}{k}\binom{n}{l}}{\binom{m+n}{k+l}} \frac{k}{k+l} \qquad \text{for} \quad l = 0, 1, \ldots, n.$$

Hint: Use the result of Problem 2.

22. Consider a random sample of size n taken from a population with continuous density function $f(x)$. Take another random sample independent of the first one and also of size n from the same population. Let U_r^n be the random variable associated with the number of values in the second sample which exceed the rth smallest value in the first sample. Similarly, let V_s^n be the random variable associated with the number of values in the second sample which exceed the sth largest value in the first sample.

Prove that

$$\Pr\{U_r^n = x\} = \frac{\binom{n-x+r-1}{r-1}\binom{n-r+x}{x}}{\binom{2n}{n}}$$

$$= \frac{1}{2}\frac{\binom{n-1}{r-1}\binom{n}{x}}{\binom{2n-1}{n-r+x}} \qquad \text{for} \quad x = 0, 1, \ldots, n$$

and $\Pr\{U_r^n = x\} = \Pr\{V_s^n = x\}$ where $s = n - r$.

Hint: Use the result of Problem 2.

23. Let $X_1, X_2, \ldots, X_m$ and $Y_1, Y_2, \ldots, Y_n$ be two independent random samples from distributions $F(x)$ and $G(y)$, respectively. Assume that F and G are continuous distribution functions and

$$F = G^\delta, \qquad \delta > 0.$$

Let

$$X_{1,m} \leq X_{2,m} \leq \cdots \leq X_{m,m}$$

and

$$Y_{1,n} \leq Y_{2,n} \leq \cdots \leq Y_{n,n}$$

be the order statistics of the corresponding samples.

Find the probability

$$\Pr\{X_{m-i,m} \leq Y_{n,n} < X_{m-i+1,m} \mid Y_{n,n} = t\}$$

Answer:

$$\binom{m}{i}[G(t)]^{\delta(m-i)}(1 - [G(t)]^\delta)^i.$$

24. Under the conditions of Problem 23 find the probability that exactly i of the X's are larger than all of the Y's.

Answer:

$$\binom{m}{i} \frac{n}{\delta} \frac{\Gamma((n/\delta) + m - i)\Gamma(i + 1)}{\Gamma((n/\delta) + m + 1)}.$$

25. Under the conditions of Problem 23 find the probability that exactly i of the X's are smaller than all of the Y's.

Answer:

$$\binom{m}{i} n! \sum_{r=0}^{m-i} (-1)^r \binom{m - i}{r} \frac{1}{[\delta(i + r) + 1][\delta(i + r) + 2] \cdots [\delta(i + r) + n]}.$$

26. Let $X_1, X_2, \ldots, X_m$ and $Y_1, Y_2, \ldots, Y_n$ be two independent random samples from the distributions $F(x)$ and $G(y)$, respectively. Assume that $F(x)$ and $G(y)$ are strictly increasing and continuous and let

$$p = \Pr\{Y < X\},$$

where X and Y are independent and distributed according to F and G, respectively. Define

$$U = \text{number of pairs } (X_i, Y_j) \text{ such that } Y_j < X_i.$$

Show that

$$E[U] = mnp.$$

Hint: Write

$$U = \sum_{j=1, i=1}^{m, n} U_{ij}, \quad \text{where} \quad U_{ij} = \begin{cases} 1 & \text{if } Y_j < X_i, \\ 0 & \text{otherwise.} \end{cases}$$

27. In the preceding problem show that

$$\text{Var}(U) = mn\{(m - 1)\alpha + (n - 1)\beta - (m + n - 1)p^2 + (2m - 1)p - (m - 1)\},$$

where

$$\alpha = \int_0^1 [L(t)]^2 \, dt \quad \text{and} \quad \beta = 1 - 2 \int_0^1 tL(t) \, dt,$$

for $L(t) = F(G^{-1}(t))$ and G^{-1} the inverse function of G.

28. Let $\xi_1, \xi_2, \ldots, \xi_n$ be n independent random variables uniformly distributed on $[0, t]$. Find the expectation and the variance of their minimum.

Answer: Expectation $= t/(n + 1)$; variance $= nt^2/[(n + 1)^2(n + 2)]$.

29. Let $X_1, X_2, \ldots, X_n$ be independent r.v's, possessing the density functions $f_1(x), f_2(x), \ldots, f_n(x)$ and the corresponding cumulative distribution functions $F_1(x), F_2(x), \ldots, F_n(x)$ respectively.
   Define

$$Z = \min(X_1, X_2, \ldots, X_n)$$

and

$$N = \text{first index where } \min(X_1, \ldots, X_n) \text{ is achieved.}$$

Let

$$H_k(z) = \Pr\{N = k, Z \le z\}.$$

Prove that

$$H_k(z) = \int_{-\infty}^{z} \left\{ \prod_{j \ne k} [1 - F_j(t)] \right\} f_k(t)\, dt.$$

30. Players are all of equal skill and in a contest the probability is $\frac{1}{2}$ that a specified one of the two contestants will be the victor. A group of 2^n players are paired off against each other at random. The 2^{n-1} winners are again paired off randomly, and so on, until one winner remains. If at the start we designate two players as A and B, what is the probability that they will ever contest each other?

Answer: $p_n = (\frac{1}{2})^{n-1}$.

31. Players A and B match pennies N times in a fair game. What is the probability of never being even?

Answer:

$$\Pr(\text{no tie}) = \begin{cases} \dbinom{N-1}{n} \dfrac{1}{2^{N-1}} & \text{if } N = 2n+1, \\[2ex] \dbinom{N}{n} \Big/ 2^N & \text{if } N = 2n. \end{cases}$$

32. Let $X_1, X_2, \ldots$ be a sequence of random variables with X_k uniformly distributed over the interval $(X_{k-1}, 1)$, $X_0 \equiv 0$. Prove that

$$\prod_{i=1}^{\infty} E[X_i] = \prod_{i=1}^{\infty} [1 - (\tfrac{1}{2})^i].$$

Hint: Prove by induction that the density of X_i is

$$f(x) = \frac{[-\log(1-x)]^{i-1}}{(i-1)!}, \qquad i \ge 2.$$

33. Under the conditions of Problem 32, show that $-\sum_{i=1}^{n} E[\log X_i]$ is uniformly bounded in n.

Hint: Verify that

$$-\sum_{i=1}^{n} E[\log X_i] = -\int_0^{\infty} \log(1 - e^{-\omega}) e^{-\omega} \sum_{i=0}^{n-1} \frac{\omega^i}{i!}\, d\omega$$

$$\le -\int_0^{\infty} \log(1 - e^{-\omega})\, d\omega = \sum_{k=1}^{\infty} \frac{1}{k^2}.$$

34. Under the conditions of Problem 32 show that

$$E\left[\prod_{i=k}^{\infty} X_i \,\Big|\, X_{k-1} = \xi \right] = e^{\xi - 1}.$$

Hint: Let $\varphi(\xi) = E[\prod_{i=k}^{\infty} X_i \mid X_{k-1} = \xi]$. (Why is φ independent of k?) Use the result of Problem 33 to show that $\varphi(\xi) \ne 0$ for $0 \le \xi < 1$.

Next derive the functional equation

$$\varphi(x) = \frac{1}{1-x} \int_x^1 \xi \varphi(\xi) \, d\xi, \qquad \varphi(1) = 1,$$

and solve. There are two solutions, one of which is continuous at 1, the other not. It has to be proved that $\varphi(\xi) = 0 \ (0 \le \xi < 1), \ \varphi(1) = 1$ is not the desired solution.

35. Under the conditions of Problem 32 show that $E[\prod_{i=1}^{\infty} X_i] \ne \prod_{i=1}^{\infty} E[X_i]$ and find which is larger.

Answer:

$$E\left[\prod_1^\infty X_i\right] > \prod_1^\infty E[X_i].$$

*36. If n points are chosen "at random" (i.e., according to the uniform distribution) on a line of length L, show that if $0 < d < L/(n-1)$, the probability is $\{[L - (n-1)d]/L\}^n$ that no two points will be closer together than the distance d.

Hint: Establish and interpret the relation

$$\frac{n!}{L^n} \int_0^{x_2-d} dx_1 \int_d^{x_3-d} dx_2 \int_{2d}^{x_4-d} dx_3 \cdots \int_{(n-2)d}^{x_n-d} dx_{n-1} \int_{(n-1)d}^{L} dx_n = \frac{[L - (n-1)d]^n}{L^n}.$$

*37. Suppose n points are chosen independently according to a uniform distribution on the circumference of a circle. Show that the probability that a specified j of the n arcs determined by consecutive points will be longer than α is

$$\pi_j = \begin{cases} \left(1 - \dfrac{j\alpha}{t}\right)^{n-1} & \text{if} \quad 0 \le j\alpha \le t, \\ 0 & \text{otherwise,} \end{cases}$$

where t is the circumference of the circle.

Hint: Because of circular symmetry select any one of the n points as the origin and assume that the other $n-1$ points were chosen at random independently in the interval $[0, t]$. Let

$$0 \le X_1^* \le X_2^* \le \cdots \le X_{n-1}^* \le t$$

be the order statistics that mark the distances from the origin of the first, second, $\ldots$, $(n-1)$th point, respectively. Use the method of Problem 16.

*38. In Problem 37, show that the probability that exactly k of the n arcs between consecutive points will be longer than α is

$$P_k = \binom{n}{k} \sum_{j=k}^{[t/\alpha]} (-1)^{j-k} \binom{n-k}{n-j} \left(1 - \frac{j\alpha}{t}\right)^{n-1}.$$

Hint: Let V_k be the probability that a given k arcs are each longer than α and the remaining $n-k$ arcs are shorter. Note that

$$P_k = \binom{n}{k} V_k, \qquad k = 0, 1, \ldots, n.$$

Next establish the formula

$$\pi_j = \sum_{k=j}^{n} \binom{n-j}{n-k} V_k,$$

where π_j is defined in Problem 37. Show that inversion gives the expression

$$V_k = \sum_{j=k}^{n} (-1)^{j-k} \binom{n-k}{n-j} \pi_j.$$

$$\left(\text{Recall that } \binom{n}{r} = 0 \text{ for } r > n. \right)$$

*39. Let n impulses arrive at a counter at times $t_1, \ldots, t_n$ where $t_1, t_2, \ldots, t_n$ are distributed as order statistics from a uniform $(0, 1)$ distribution. Following each time the counter *registers* an impulse, it has a "dead time," of length τ, in which it will not register, even if it receives an impulse. The inverval $(0, \tau)$ is also a "dead time." Find the probability that the counter registers the first k impulses which it receives (i.e., $t_i - t_{i-1} \geq \tau$, $i = 1, 2, \ldots, k$, $t_0 = 0$).

Hint: Use the method of Problem 36.

Answer:

$$(1 - k\tau)^n, \qquad k\tau \leq 1;$$
$$0, \qquad\qquad k\tau > 1.$$

*40 (*continuation*). In Problem 39, find the density $f(y)$ of the r.v. Y, which is the time at which the nth particle arrives, given that the counter registers the first k particles it receives ($n > k$).

Answer:

$$f(y) = \frac{n(y - k\tau)^{n-1}}{(1 - k\tau)^n} \qquad \text{for} \quad k\tau \leq y \leq 1.$$

*41. Let $Z_i = (X_i, Y_i)$, $i = 1, \ldots, n$, denote n pairs of real random variables all independently and identically distributed with continuous distribution functions $F(x)$. A vector $Z_i = (X_i, Y_i)$ is said to be *admissible* if there exists no other $Z_j = (X_j, Y_j)$ for which $X_j \geq X_i$, $Y_j \geq Y_i$. Let I_n denote the number of admissible vectors in the sequence $Z_1, Z_2, \ldots, Z_n$.

Without loss of generality assume that the Z_i are labeled in such a way that

$$X_1 \leq X_2 \leq \cdots \leq X_n.$$

Prove that

$$I_n = \sum_{i=1}^{n} U_i,$$

where U_i are *independent* random variables defined as

$$U_i = \begin{cases} 1 & \text{if the rank } Y_i \text{ in the set } Y_i, Y_{i+1}, \ldots, Y_n \text{ is } n - i + 1, \\ 0 & \text{otherwise.} \end{cases}$$

Hint: Use the result of Problem 11.

*42. Under the conditions of Problem 41, find the probability generating function $g(t)$ of I_n (cf. Problem 17).

Answer:

$$g(t) = \frac{1}{n!} \frac{\Gamma(t + n)}{\Gamma(t)}.$$

*43. Let $X_1, X_2, \ldots$ be a sequence of independent, identically distributed random variables with distribution function $F(x)$ and density function $f(x)$. Let n_1 be the integer-valued random variable, which is the first subscript such that $X_{n_1} > X_1$, then let n_2 be the first subscript $> n_1$ with $X_{n_2} > X_{n_1}$; in general, let n_r be the first subscript $> n_{r-1}$ with $X_{n_r} > X_{n_{r-1}}$. Find the distribution of X_{n_r}.

Hint: Condition on the values of $X_1, X_{n_1}, X_{n_2}, \ldots, X_{n_{r-1}}$ and thereby prove the formula

$$\Pr\{X_{n_r} \le z\} = \int_{-\infty}^{z} f(x_r)\,dx_r \int_{-\infty}^{x_r} \frac{f(x_{r-1})\,dx_{r-1}}{1 - F(x_{r-1})}$$

$$\times \int_{-\infty}^{x_{r-1}} \frac{f(x_{r-2})\,dx_{r-2}}{1 - F(x_{r-2})} \cdots \int_{-\infty}^{x_1} \frac{f(x_0)\,dx_0}{1 - F(x_0)}.$$

Then evaluate it. Alternatively, employ induction on r.

Solution: The density function of X_{n_r} is

$$f_{X_{n_r}}(z) = \frac{[-\log(1 - F(z)]^r}{r!} f(z), \qquad z > 0.$$

*44. Let $X_1, X_2, \ldots$ be a sequence of independent, identically distributed, random variables with distribution function $F(x)$ and density function $f(x)$. Let n_i be the r.v.'s as defined in Problem 43. Find the distribution of the random variable $n_r - n_{r-1}$.

Answer:

$$\Pr\{n_r - n_{r-1} = i\} = \sum_{k=0}^{i-1} (-1)^k \binom{i-1}{k} \frac{1}{(k+2)^r}.$$

Problems 45–49 are variations on the ballot problems.

*45. As a modification of the ballot problem in the text, find the probability that A never falls behind during the vote counting.

Hint: Consider the more general setup of c cards in an urn. The cards are labeled with numbers $k_1, k_2, \ldots, k_c$. Let $n_0, n_1, n_2, \ldots$ be the number of zeros, ones, twos, ... among $k_1, k_2, \ldots, k_c$. Suppose $k_1 + \cdots + k_c = k$ $(0 \le k \le c)$. Let v_i $(i = 1, 2, \ldots, c)$ be the r.v. whose value is equal to the number on the ith card drawn. Drawings are made one at a time without replacements. We introduce the notation

$$P_j(c, k; n_0, n_1, \ldots) = \Pr\{v_1 + \cdots + v_r < r + j \text{ for } r = 1, 2, \ldots, c\}.$$

By inserting an extra card labeled 0 into the urn develop the recursion formula

$$P_{j-1}(c+1, k; n_0+1, n_1, n_2, \ldots) = \frac{n_0+1}{c+1} P_j(c, k; n_0, n_1, \ldots)$$

$$+ \sum_{i=1}^{j-1} \frac{n_i}{c+1} P_{j-i}(c, k-i; n_0+1, n_1, \ldots, n_{i-1}, n_{i+1}, \ldots).$$

With the method of the text, where $n_0 = a$, $n_2 = b$ $(a + b = c)$, show that the probability that A never falls behind during the vote counting is precisely

$$\Pr\{v_1 + \cdots + v_r < r + 1 \text{ for } r = 1, 2, \ldots, c\}.$$

Answer: $(a - b + 1)/(a + 1)$.

*46. Consider an urn containing a white and b black balls $(a > b)$. We draw all the balls at random without replacement one after the other. Simultaneously we perform a random walk on the nonnegative integers. We start at the point 1 and move to the right or left by one unit, if the ball just drawn is white or black, respectively. What is the probability that state 0 will be reached at some time during the $a + b$ drawings.

Answer: $b/(a + 1)$.

*47. Let a and b be positive integers satisfying $a > b$. Prove the following identities:

(i) $\dfrac{b}{a+b} + \sum_{k=1}^{b-1} (2k+1)^{-1} \binom{2k+1}{k}$

$\times \dfrac{a(a-1)(a-2)\cdots(a-k+1)b(b-1)\cdots(b-k)}{(a+b)(a+b-1)\cdots(a+b-2k)}$

$= \dfrac{b}{a+1};$

(ii) $\dfrac{a-b}{a+b} + \sum_{r=1}^{b} \dfrac{a(a-1)\cdots(a-r+1)b(b-1)\cdots(b-r+1)}{(a+b)(a+b-1)\cdots(a+b-2r+1)} \binom{2r}{r}$

$\times \dfrac{a-b}{a+b-2r} = \dfrac{b}{a+1}.$

Hint: (i) Consider the setup of Problem 46. Break up the event that a visit to the origin occurs at some time in terms of the first time a visit to the origin occurs.

(ii) This is derived by considering the last time of equality of white and black draws.

The results of the ballot problem, Problem 46, and the calculation of formula (6.12) of Chapter 10 are to be used.

*48. In the setup of Problem 46, prove that the probability that a visit to the origin is

$$p_n(a, b) = \frac{b(b-1)\cdots(b-n+1)}{(a+1)(a+2)\cdots(a+n)},$$

starting the random walk at the point n. Here n is a positive integer not exceeding b.

Hint: Find a recursion formula for $p_n(a, b)$. Then use induction on n coupled with the result of Problem 34 that $p_1(a, b) = b/(a + 1)$.

*49.   Let $F_n(x)$ be the empirical cumulative distribution function of n observations from the uniform distribution on $(0, 1)$. Compute the probability

$$P_n(a, \gamma) = \Pr\{F_n(x) < a + \gamma x \text{ for all } 0 \le x \le 1\},$$

where $(n - 1)/n \le a \le 1, \gamma > 0, \gamma + a > 1$.

Answer: $1 - ((1 - a)/\gamma)^n$.

*50.   Under the same conditions as in Problem 49, except that $a < (n - 1)/n$, prove the relation

$$P_n(a, \gamma) = \int_{(1-a)/\gamma}^{1} P_{n-1}\left(\frac{n}{n - 1} a, \frac{n}{n - 1} \gamma t\right) nt^{n-1} dt.$$

Hint: Condition on the largest observation.

*51.   Let

$$C_n(a, \gamma, i) = \binom{n}{i}\left(\frac{n - i}{n\gamma} - \frac{a}{\gamma}\right)^{n-i}\left[1 - \left(\frac{n - i}{n\gamma} - \frac{a}{\gamma}\right)\right]^{i-1}\left(\frac{\gamma + a - 1}{\gamma}\right).$$

Show that

$$P_n(a, \gamma) = 1 - \sum_{i=0}^{k} C_n(a, \gamma, i),$$

where k is the integer defined by

$$\frac{k}{n} \le 1 - a < \frac{k + 1}{n}.$$

Hint: Consider the complementary event that $F_n(x) \ge a + \gamma x$ occurs for some x. This happens if and only if

$$F_n\left(\frac{n - r}{n\gamma} - \frac{a}{\gamma}\right) \ge \frac{n - r}{n} \qquad \text{for some} \quad r = 1, 2, \ldots, k.$$

Introduce the event A_i that the index r where the inequality

$$F_n\left(\frac{n - r}{n\gamma} - \frac{a}{\gamma}\right) \ge \frac{n - r}{n}$$

first occurs is $r = i$ ($i = 0, 1, \ldots, k$). Compute $\sum_{i=0}^{k} \Pr\{A_i\}$.

An alternative method is to use the result of Problem 49 and induction on n.

*52.   Consider the ballot problem where candidate A scores a votes and candidate B scores b votes ($a > b$). Suppose the ballots are drawn one at a time and denote by α_r and β_r the number of votes recorded for A and B, respectively, among the first r votes. Let $\Delta_{a,b}$ be the number of times that A strictly leads, i.e., $\Delta_{a,b}$ = the number of subscripts r satisfying the condition $\alpha_r > \beta_r$ ($r = 1, 2, \ldots, a + b$). Let $\Delta_{a,b}^*$ be the number of subscripts satisfying $\alpha_r \ge \beta_r$ ($r = 1, 2, \ldots, a + b$).

Prove that

$$\Pr\{\Delta_{a,b} = a - b + r\}$$

$$= \begin{cases} \dfrac{a-b-1}{\dbinom{a+b}{a}} \displaystyle\sum_{m=0}^{[r/2]} \dfrac{\dbinom{2b-2m}{b-m}\dbinom{a-b-2+2m}{m}}{(b+m+1)(a-b+m-1)} & \text{if } a > b + 1, \\[30pt] \dfrac{1}{2b+1} & \text{if } a = b + 1, \end{cases}$$

and

$$\Pr\{\Delta_{a,b}^* = a - b + r\}$$

$$= \begin{cases} \dfrac{a-b+1}{\dbinom{a+b}{a}} \displaystyle\sum_{m=0}^{[(r-1)/2]} \dfrac{\dbinom{2b-2m-2}{b-m-1}\dbinom{a-b+2m}{m}}{(b-m)(a-b+m+1)}, & r = 1, 2, \ldots, 2b-1, \\[30pt] \dfrac{a-b+1}{a+1}, & r = 2b, \\[12pt] 0, & \text{otherwise.} \end{cases}$$

NOTES

The point of view adopted in this chapter concerning order statistics and Poisson processes follows Rényi [1].

   The combinatorial methods of the last half of this chapter are mostly based on work of L. Takács which has not appeared in book form yet.

   Substantial discussions on order statistics from the point of view of classical statistics can be found in Wilks [2]. See also references therein.

   A recent summary of the asymptotic behavior of order statistics is included in the book by Galambos [3].

REFERENCES

1. A. Rényi, "Wahrscheinlichkeitsrechnung, mit einem Anhang uber Informationstheorie." Deutscher Verlag der Wissenschaften, Berlin, 1962.
2. S. S. Wilks, "Mathematical Statistics." Wiley, New York, 1962.
3. J. Galambos, "The Asymptotic Theory of Extreme Order Statistics." Wiley, New York, 1978.

Chapter 14

CONTINUOUS TIME MARKOV CHAINS

The purpose of this chapter is to introduce the reader to some of the more advanced topics and concepts which arise in the study of the theory of continuous parameter Markov chains. As before we shall consider only the case of stationary transition probabilities.

1: Differentiability Properties of Transition Probabilities

Let X_t be a continuous time discrete state Markov process with transition probability matrix $\| P_{ij}(t) \|_{i,j=0}^{\infty}$. Thus

$$\Pr\{X(t + s) = j \,|\, X(s) = i\} = P_{ij}(t).$$

In addition to the usual assumptions on the transition matrix function $\| P_{ij}(t) \|$, i.e.,

(a) $P_{ij}(t) \geq 0, \qquad t > 0,$

(b) $\sum_j P_{ij}(t) = 1, \qquad t > 0,$

(c) $\sum_k P_{ik}(t) P_{kj}(h) = P_{ij}(t + h), \qquad t, h > 0,$

we also assume that the $P_{ij}(t)$ are continuous for $t > 0$ and that

(d) $\displaystyle \lim_{t \to 0} P_{ij}(t) = \begin{cases} 1 & \text{if} \quad i = j, \\ 0 & \text{if} \quad i \neq j \end{cases}$

(see also Problem 3). Such a transition matrix function is often called "standard." It turns out that conditions (a)–(d) imply a great deal more than might be expected. One of these results is that the $P_{ij}(t)$ are differentiable for every $t \geq 0$. We will prove here only the much simpler result that the $P_{ij}(t)$ are differentiable (i.e., have right-hand derivatives) at $t = 0$. We begin with $P_{ii}(t)$.

Theorem 1.1. *For every i*

$$- P_{ii}'(0) = \lim_{t \to 0} \frac{1 - P_{ii}(t)}{t}$$

exists but may be infinite.

Proof. First we show that $P_{ii}(t) > 0$ for all $t > 0$. In fact, (d) asserts that for each i there is $\varepsilon > 0$ such that $0 \le t \le \varepsilon$ implies $P_{ii}(t) > 0$. Now (c) can easily be iterated to give

$$P_{ij}(t_1 + \cdots + t_n) = \sum_{k_1 \ldots, k_{n-1}'} P_{ik_1}(t_1) P_{k_1 k_2}(t_2), \ldots, P_{k_{n-1} j}(t_n). \tag{1.1}$$

Letting $t_1 = \cdots = t_n = t/n$, $i = j$, and taking only that term on the right corresponding to $k_1 = k_2 = \cdots = k_{n-1} = i$, we obtain

$$P_{ii}(t) \ge [P_{ii}(t/n)]^n. \tag{1.2}$$

For n sufficiently large $t/n \le \varepsilon$; hence $P_{ii}(t/n) > 0$ and so $P_{ii}(t) > 0$. Let $-\log P_{ii}(t) = \varphi(t)$, this being well defined since $P_{ii}(t) > 0$. The inequality

$$P_{ii}(t + s) \ge P_{ii}(t) P_{ii}(s)$$

holds, as can be proved in the same manner as (1.2). Taking the logarithm on both sides yields the subadditivity inequality for $\varphi(t)$:

$$\varphi(t + s) \le \varphi(t) + \varphi(s).$$

Also $\varphi(t) \ge 0$ since $0 < P_{ii}(t) \le 1$. We put

$$q_i = \sup_{t > 0} \frac{\varphi(t)}{t};$$

then $0 \le q_i \le \infty$ since $\varphi(t) \ge 0$ for $t > 0$. If $q_i < \infty$, there exists $t_0 > 0$ such that $\varphi(t_0)/t_0 \ge q_i - \varepsilon$. For each t, we write $t_0 = nt + \delta$, where $0 \le \delta < t$. Then

$$\varphi(t_0) \le \varphi(nt) + \varphi(\delta) \le \varphi((n - 1)t) + \varphi(t) + \varphi(\delta) \le \cdots \le n\varphi(t) + \varphi(\delta),$$

and so

$$q_i - \varepsilon \le \frac{\varphi(t_0)}{t_0} \le \frac{n\varphi(t) + \varphi(\delta)}{t_0} = \frac{nt}{t_0} \frac{\varphi(t)}{t} + \frac{\varphi(\delta)}{t_0}.$$

Hence

$$q_i - \varepsilon \le \varliminf_{t \to 0} \left[\frac{nt}{t_0} \frac{\varphi(t)}{t} + \frac{\varphi(\delta)}{t_0} \right].$$

But as $t \to 0$, $nt/t_0 \to 1$ and $\varphi(\delta) \to 0$ (since $P_{ii}(\delta) \to 1$ as $\delta \to 0$); hence

$$\varliminf_{t \to 0} \left[\frac{nt}{t_0} \frac{\varphi(t)}{t} + \frac{\varphi(\delta)}{t_0} \right] = \varliminf_{t \to 0} \frac{\varphi(t)}{t}.$$

Now by the definition of q_i,

$$\overline{\lim_{t \to 0}} \frac{\varphi(t)}{t} \leq q_i.$$

Combining the last three inequalities, we have

$$q_i - \varepsilon \leq \underline{\lim_{t \to 0}} \frac{\varphi(t)}{t} \leq \overline{\lim_{t \to 0}} \frac{\varphi(t)}{t} \leq q_i.$$

Since ε was arbitrary, we have

$$\underline{\lim_{t \to 0}} \frac{\varphi(t)}{t} = \overline{\lim_{t \to 0}} \frac{\varphi(t)}{t} = q_i.$$

If $q_i = \infty$, we can replace $q_i - \varepsilon$ by M, an arbitrarily large constant, and then obtain

$$M \leq \underline{\lim_{t \to 0}} \frac{\varphi(t)}{t}; \qquad \text{thus} \qquad \infty = \lim_{t \to 0} \frac{\varphi(t)}{t}.$$

In either case we have

$$\lim_{t \to 0} \frac{\varphi(t)}{t} = q_i.$$

Now

$$\lim_{t \to 0} \frac{1 - P_{ii}(t)}{t} = \lim_{t \to 0} \frac{1 - e^{-\varphi(t)}}{\varphi(t)} \frac{\varphi(t)}{t} = q_i. \quad \blacksquare$$

Theorem 1.2. *For every i and j, $i \neq j$,*

$$P'_{ij}(0) = \lim_{t \to 0} \frac{P_{ij}(t)}{t}$$

exists and is finite.

Proof. For each fixed $h > 0$, $\| P_{ij}(h) \|$ is a transition probability matrix of a Markov chain $\{X_{nh}\}$; clearly $P_{ij}^n(h) = P_{ij}(nh)$. We now define $_jP_{ii}^0(h) = 1$ and

$$_jP_{ii}^n(h) = \Pr\{X_{nh} = i; X_{vh} \neq j, 1 \leq v < n | X_0 = i\},$$

$$f_{ij}^n(h) = \Pr\{X_{nh} = j; X_{vh} \neq j, 1 \leq v < n | X_0 = i\}.$$

Then

$$P_{ij}(nh) \geq \sum_{v=0}^{n-1} {}_jP_{ii}^v(h) P_{ij}(h) P_{jj}[(n - v - 1)h] \tag{1.3}$$

since each term on the right corresponds to a possible way of going from i to j in n steps (relative to $\| P_{ij}(h) \|$) and these paths are mutually exclusive but not necessarily exhaustive. The term $_jP_{ii}^v(h)P_{ij}(h)$ is the probability of the event that the last visit to i before visiting j occurs at trial v. (The relation (1.3) also appeared in our discussion of ratio theorems in Chapter 11.) Furthermore

$$P_{ii}(vh) = \ _jP_{ii}^v(h) + \sum_{m=1}^{v-1} f_{ij}^m(h)P_{ji}[(v-m)h]$$

for similar reasons. The first term is the probability of visiting i at trial v without entering state j in the intervening trials. The sum terms consider the cases where state j is entered at some intermediate trial. Since

$$\sum_{m=1}^{v-1} f_{ij}^m(h) \le 1,$$

we have

$$_jP_{ii}^v(h) \ge P_{ii}(vh) - \max_{1 \le m < v} P_{ji}[(v-m)h]. \tag{1.4}$$

Now by property (d) it follows that for every $\varepsilon > 0$ and any preassigned i, j ($i \ne j$) there exists t_0 such that

$$\max_{0 \le t \le t_0} P_{ji}(t) < \varepsilon, \qquad \min_{0 \le t \le t_0} P_{ii}(t) > 1 - \varepsilon, \qquad \min_{0 \le t \le t_0} P_{jj}(t) > 1 - \varepsilon.$$

Hence if $nh < t_0$ and $v \le n$, it follows from (1.4) that

$$_jP_{ii}^v(h) > 1 - 2\varepsilon.$$

Using this estimate in (1.3) we obtain

$$P_{ij}(nh) \ge (1 - 2\varepsilon) \sum_{v=0}^{n-1} P_{ij}(h)(1 - \varepsilon) \ge (1 - 3\varepsilon)nP_{ij}(h)$$

or

$$\frac{P_{ij}(nh)}{nh} \ge (1 - 3\varepsilon)\frac{P_{ij}(h)}{h} \qquad \text{if} \quad nh < t_0. \tag{1.5}$$

Put

$$q_{ij} = \lim_{t \to 0} \frac{P_{ij}(t)}{t}.$$

Then (1.5) shows that $q_{ij} < \infty$. In fact, if $q_{ij} = \infty$, we could find h arbitrarily small for which $P_{ij}(h)/h$ is arbitrarily large; choosing n' so that $t_0/2 < n'h < t_0$ we would conclude on the basis of (1.5) that $P_{ij}(n'h)/n'h$ would be arbitrarily large, but at the same time

$$\frac{P_{ij}(n'h)}{n'h} < \frac{\varepsilon}{n'h} < \frac{2\varepsilon}{t_0}.$$

This contradiction implies that $q_{ij} < \infty$.

The remainder of the proof is purely analytic and is a consequence of (1.5). From the definition of q_{ij} there exists $t_1 < t_0$ such that

$$\frac{P_{ij}(t_1)}{t_1} < q_{ij} + \varepsilon.$$

Since $P_{ij}(t)$ is continuous, we can find h_0 so small that $t_1 + h_0 < t_0$ and

$$\frac{P_{ij}(t)}{t} < q_{ij} + \varepsilon \qquad \text{for} \quad t \in I = [t_1 - h_0, t_1 + h_0]. \tag{1.6}$$

Now for any $h < h_0$ we determine an integer n_k such that $n_h h \in I$; thus, using (1.5) and (1.6) we find

$$(1 - 3\varepsilon)\frac{P_{ij}(h)}{h} \leq \frac{P_{ij}(n_h h)}{n_h h} < q_{ij} + \varepsilon, \qquad h < h_0,$$

from which we conclude that

$$(1 - 3\varepsilon)\overline{\lim_{h \to 0}}\,\frac{P_{ij}(h)}{h} \leq q_{ij} + \varepsilon,$$

Since ε is arbitrary, it follows that

$$\overline{\lim_{h \to 0}}\,\frac{P_{ij}(h)}{h} \leq q_{ij}.$$

The theorem now follows from the definition of q_{ij}. ∎

As an example, in the case of birth and death processes, we have

$$q_i = \lambda_i + \mu_i, \quad q_{ij} = \begin{cases} \lambda_i, & j = i + 1, \\ 0, & j \neq i - 1, i + 1, i, \\ \mu_i, & j = i - 1, \end{cases} \qquad i = 0, 1, \ldots.$$

We claim that generally

$$\sum_{j \neq i} q_{ij} \leq q_i \qquad \text{for all } i.$$

In fact, since

$$\sum_{j \neq i} P_{ij}(h) = 1 - P_{ii}(h)$$

we have for any finite N

$$\sum_{j=1, j \neq i}^{N} P_{ij}(h) \leq 1 - P_{ii}(h).$$

Dividing by h and then letting $h \to 0$ leads to the inequality

$$\sum_{j=1, j \neq i}^{N} q_{ij} \leq q_i.$$

Since N is arbitrary and all the terms are positive, the assertion follows.

2: Conservative Processes and the Forward and Backward Differential Equations

A continuous time Markov chain is said to be "conservative" if

$$\sum_{j \neq i} q_{ij} = q_i < \infty \qquad \text{for all } i.$$

Notice that a birth and death process is *conservative*. We now prove that for a conservative Markov chain not only are all the $P_{ij}(t)$ differentiable, if $q_i < \infty$ ($i \geq 0$), but they satisfy a set of differential equations known as the backward Kolmogoroff equations. (For the special case of the birth and death process see Section 5 of Chapter 4.) We remind the reader, however, that the differentiability of the $P_{ij}(t)$ follows directly from (a)–(d); it is only that the proof under the assumption of conservativeness is quite simple. Indeed,

$$P_{ij}(s + t) - P_{ij}(t) = \sum_k P_{ik}(s)P_{kj}(t) - P_{ij}(t)$$

$$= \sum_{k \neq i} P_{ik}(s)P_{kj}(t) + [P_{ii}(s) - 1]P_{ij}(t).$$

Dividing by s and letting $s \to 0+$, we obtain formally the *backward equations*.

$$P'_{ij}(t) = \sum_{k \neq i} q_{ik}P_{kj}(t) - q_i P_{ij}(t), \qquad \text{for all } i. \tag{2.1}$$

To derive these equations rigorously, we must show that

$$\lim_{s \to 0+} \frac{1}{s} \sum_{k \neq i} P_{ik}(s)P_{kj}(t) = \sum_{k \neq i} q_{ik}P_{kj}(t).$$

Now

$$\varliminf_{s \to 0} \frac{1}{s} \sum_{k \neq i} P_{ik}(s)P_{kj}(t) \geq \varliminf_{s \to 0} \frac{1}{s} \sum_{k=1, k \neq i}^{N} P_{ik}(s)P_{kj}(t) = \sum_{k=1, k \neq i}^{N} q_{ik}P_{kj}(t)$$

for any $N > 0$, and so

$$\varliminf_{s \to 0} \frac{1}{s} \sum_{k \neq i} P_{ik}(s)P_{kj}(t) \geq \sum_{k \neq i} q_{ik}P_{kj}(t). \tag{2.2}$$

On the other hand, for $N > i$,

$$\sum_{k \neq i} P_{ik}(s)P_{kj}(t) \leq \sum_{k=1, k \neq i}^{N} P_{ik}(s)P_{kj}(t) + \sum_{k=N+1}^{\infty} P_{ik}(s)$$

$$= \sum_{k=1, k \neq i}^{N} P_{ik}(s)P_{kj}(t) + 1 - P_{ii}(s) - \sum_{k=1, k \neq i}^{N} P_{ik}(s).$$

Dividing by s and taking $\overline{\lim}_{s\to 0+}$ of both sides, we obtain

$$\overline{\lim_{s\to 0}} \frac{1}{s} \sum_{k\neq i} P_{ik}(s)P_{kj}(t) \leq \sum_{k=1,k\neq i}^{N} q_{ik}P_{kj}(t) + q_i - \sum_{k=1,k\neq i}^{N} q_{ik}.$$

Letting $N \to \infty$ and using the conservative nature of the system, we see that

$$\overline{\lim_{s\to 0+}} \frac{1}{s} \sum_{k\neq 1} P_{ik}(s)P_{kj}(t) \leq \sum_{k\neq i} q_{ik}P_{kj}(t).$$

Comparing this inequality with (2.2) we conclude that

$$\lim_{s\to 0} \frac{1}{s} \sum_{k\neq i} P_{ik}(s)P_{kj}(t)$$

exists and equals

$$\sum_{k\neq i} q_{ik}P_{kj}(t).$$

In a similar fashion we can formally derive a set of equations called the *forward equations*. We write

$$P_{ij}(s+t) - P_{ij}(s) = \sum_k P_{ik}(s)P_{kj}(t) - P_{ij}(s)$$

$$= \sum_k P_{ik}(s)[P_{kj}(t) - \delta_{kj}].$$

Dividing by t and letting $t \to 0$, we obtain formally

$$P'_{ij}(s) = \sum_{k\neq j} P_{ik}(s)q_{kj} - P_{ij}(s)q_j, \qquad \text{for all } i, j, \tag{2.3}$$

the *forward equations*. A discussion of the scope and validity of these equations is rather more involved than in the case of the backward equations, and we shall not enter into this question.

Both sets of equations assume a very simple form in matrix notation. Indeed, consider the infinite matrix $\mathbf{A} = \|a_{ij}\|$ defined by

$$a_{ij} = \begin{cases} q_{ij}, & i \neq j, \\ -q_i, & i = j, \end{cases}$$

called the *infinitesimal matrix* of the process. The backward equations may be compactly expressed by the matrix differential equation

$$\mathbf{P}'(t) = \mathbf{A}\mathbf{P}(t)$$

and the forward equations are of the form

$$\mathbf{P}'(t) = \mathbf{P}(t)\mathbf{A},$$

where

$$\mathbf{P}(t) = \|P_{ij}(t)\|.$$

3: Construction of a Continuous Time Markov Chain from Its Infinitesimal Parameters

An interesting and important question in the theory of continuous parameter Markov chains is the following. Suppose that we are given a set of nonnegative numbers $\{q_{ij}\}$ with the property

$$\sum_{j \neq i} q_{ij} \leq q_i \qquad \text{for all } i.$$

For uniformity of notation, we sometimes write $q_{ii} = -q_i$ as is done above. Is there a Markov chain, i.e., a standard transition matrix function $\| P_{ij}(t) \|$, for which

$$P'_{ij}(0) = q_{ij}, \qquad i \neq j,$$

and

$$P'_{ii}(0) = -q_i?$$

The problem becomes more specific if we assume that

$$\sum_{j \neq i} q_{ij} = q_i < \infty \qquad \text{for all } i,$$

for in that case we know that any Markov chain associated with the $\{q_{ij}\}$ must at least satisfy the backward equations.

The practical importance of these questions rests on the fact (as we have seen specifically in the case of birth and death processes; see Chapter 4) that quite often a continuous time Markov chain is defined in a manner that enables one to derive the backward equations. One then attempts to solve these equations in order to calculate the complete transition probability function of the process.

At the present time, definitive results in the general case are not available. It is known that under the condition

$$\sum_{j \neq i} q_{ij} = q_i, \qquad \text{all } i,$$

there exists at least one associated transition matrix function $\mathbf{P}(t)$, and if there exists more than one, then there is an infinity of them. Of course more is known for special forms of $\mathbf{A} = \| q_{ij} \|$, e.g., in the birth and death case. In that special case, a complete classification of all processes possessing a prescribed infinitesimal matrix is known. These processes differ mostly in boundary behavior, i.e., in the nature of the path functions at ∞. We remind the reader in the special example of birth and death processes of the condition (4.5) of Chapter 4 that $\mathbf{A}$ must satisfy in order to assure a unique process.

For the general, continuous time, Markov chain the problem of classifying the infinitesimal matrix $\mathbf{A}$ and its associated processes is quite complicated, and we can only refer the interested reader to the literature (see references at the close of this chapter).

Let i be such that $0 < q_i < \infty$. We now present some interpretations of the elements of $\mathbf{A}$ analogous to the interpretation ascribed to the quantities $\lambda_i + \mu_i$ and $p_i = \lambda_i/(\mu_i + \lambda_i)$ in the case of birth and death processes. Recall that we formally proved that $(\lambda_i + \mu_i)^{-1}$ is the expected waiting time in state i and $\lambda_i/(\mu_i + \lambda_i)$ is the probability of a transition to state $i + 1$ from i, given that a transition occurs. The results for the case of general, continuous time Markov chains are analogous.

Let $t > 0$ be fixed and $n > 0$ be an arbitrary positive integer. Suppose that the process is started in state i. Then consider

$$\Pr\left\{X(\tau) = i \text{ for } \tau = 0, \frac{t}{n}, \frac{2t}{n}, \frac{3t}{n}, \dots, t\right\} = \left[P_{ii}\left(\frac{t}{n}\right)\right]^n.$$

Since

$$\frac{1 - P_{ii}(t)}{t} = q_i + o(1) \qquad (o(1) \to 0 \text{ as } t \to 0+),$$

we have

$$\left[P_{ii}\left(\frac{t}{n}\right)\right]^n = \left[1 - \frac{t}{n}q_i + o\left(\frac{t}{n}\right)\right]^n = \exp\left\{n \log\left[1 - \frac{tq_i}{n} + o\left(\frac{t}{n}\right)\right]\right\}.$$

Using the expansion for the logarithm of the form $\log(1 - x) = -x + \theta(x)x^2$ valid for $|x| \le \frac{1}{2}$ and $|\theta| \le 1$ with x identified as $-tq_i/n + o(t/n)$ and then letting $n \to \infty$, we obtain

$$\lim_{n \to \infty}\left[P_{ii}\left(\frac{t}{n}\right)\right]^n = \exp(-q_i t).$$

But (see Chapter 1, page 33, *A First Course*) we may consider

$$\lim_{n \to \infty} \Pr\left\{X(\tau) = i, \tau = 0, \frac{t}{n}, \frac{2t}{n}, \dots, \frac{n-1}{n}t, t\right\}$$

as just

$$\Pr\{X(\tau) = i, \quad \text{for all} \quad 0 \le \tau \le t\}.$$

(this entails the tacit assumption that the process is separable), which therefore shows that $\exp(-q_i t)$ is the probability of remaining in state i for at least a length of time t.

In other words, the waiting time distribution in state i is an exponential distribution with parameter q_i. This confirms rigorously for general continuous time Markov chains what we have heuristically shown in the special case of birth and death processes (cf. *A First Course*, page 132).

A state i for which $0 \le q_i < \infty$ is called *stable*. It is *absorbing* if $q_i = 0$, which obviously implies that once state i is entered the process remains there permanently. Indeed,

$$\Pr\{X(\tau) = i, 0 \le \tau < t \mid X(0) = i\} = \exp(-q_i t) = 1$$

for all t. On the other hand, if $q_i > 0$, then the waiting time in state i is a random variable whose distribution function is that of a bona fide exponential and therefore some transition out of state i occurs in finite time.

A state i for which $q_i = \infty$ is called an *instantaneous state*. The expected waiting time in such a state is zero. The name "instantaneous" is appropriate since the sojourn time in state i is zero, i.e., when entering an immediate transition out of the state occurs.

The theory of continuous time Markov chains with instantaneous states is exceptionally complicated, particularly with reference to the path behavior of the process. What is even worse is that Markov chains can be constructed with only instantaneous states. It is worthwhile to appreciate the technical problems inherent in such examples but at the same time it is comforting to know that almost all continuous time Markov chains arising in applications have only stable states. Actually, in most cases of interest the process under study is usually defined by specifying the infinitesimal parameters as the known data. To complete the theory, it is then necessary to establish the existence of a process (i.e., to determine the sample paths and their probability laws) possessing the prescribed infinitesimal matrix.

Most elementary textbooks and discussions of continuous time Markov chains avoid this aspect of the problem (as we do) and concentrate primarily on the distribution theory of the process and the calculation of various probabilistic quantities of interest. The computation of the transition probability function for all t is traditionally accomplished by deriving the backward differential equations and, hopefully, solving them. This approach was the basis of our treatment of birth and death processes (see Chapter 4).

Throughout what follows we restrict attention exclusively to continuous time Markov chains with only stable states. Our next task is to provide an intuitive meaning for the parameters q_{ij}. In fact, if the process is conservative the elements q_{ij}/q_i ($i \neq j$) can be interpreted as the conditional probabilities that a transition out of state i occurs to state j. To see this we consider

$$R_{ij}(h) = \Pr\{X(h) = j \,|\, X(0) = i, X(h) \neq i\}, \qquad j \neq i,$$

and compute $\lim_{h \to 0} R_{ij}(h)$. This is the probability of a transition from state i to state j, given that a transition occurs. Now by the very meaning of the symbols we have

$$R_{ij}(h) = \frac{P_{ij}(h)}{1 - P_{ii}(h)}.$$

Dividing numerator and denominator by h, letting $h \downarrow 0$, and using the results of Theorems 1.1 and 1.2 produces the desired formula:

$$\lim_{h \downarrow 0} R_{ij}(h) = \frac{q_{ij}}{q_i}, \qquad j \neq i.$$

The sum (with respect to j) equals 1 since the process is conservative.

We remarked above that for any infinitesimal parameters $q_{ij} \geq 0 \, (i \neq j)$ and $q_i \, (0 < q_i < \infty) \, (i \geq 0)$ satisfying $q_i \geq \sum_{j \neq i} q_{ij}$ there may exist one or infinitely many continuous time Markov chains with the same infinitesimal matrix $\mathbf{A}$. In the case of a conservative infinitesimal matrix (i.e., $q_i = \sum q_{ij}$ for all i) there is one special process (the *minimal process*) for which the sample paths can be simply described. The construction of the minimal process in the case of birth and death processes was indicated in Section 4 of Chapter 4. The method is the same for the general case. We review briefly the essential ideas of this construction in the general context. A typical realization starting in any state, say i, is of the following form. We take an observation from the exponential distribution with parameter (q_i). This determines the initial waiting time in state i. At the end of this wait the particle moves to state j with probability $q_{ij}/q_i \, (j \neq i)$. In the new state just entered, say j', it waits a random length of time (exponentially distributed with parameter $q_{j'}$); concluding its sojourn time in state j', it then moves to a new state j'' with probability $q_{j'j''}/q_{j'} \, (j'' \neq j')$; it waits there a random time duration whose distribution law is an appropriate exponential and then moves again, etc. By this inductive procedure we build up all possible realizations of the process. By using rather deep measure theory methods one can determine the transition probability function $\| \tilde{P}_{ij}(t) \|$ having the prescribed infinitesimal matrix.

Another way of describing the minimal process is as follows: The transition probability matrix $\| \tilde{P}_{ij}(t) \|$ is determined as the probabilities of the various transitions, given that only a finite number of steps have occurred. More specifically we introduce the quantity $P_{ij}(t; N)$, which represents the probability of a transition from i to j in time t, given that at most N transitions occur. In particular, consistent with the meaning of the infinitesimal parameters, we have

$$
P_{ij}(t; 0) = \begin{cases} \exp(-q_i t), & i = j, \\ 0, & i \neq j, \end{cases}
$$

and we can further develop a recursion formula connecting $P_{ij}(t; N)$ and $P_{ij}(t; N - 1)$ as follows.

We consider first the term $P_{ii}(t; N)$. Two possibilities arise according to whether a transition occurs before time t or not. The waiting time in state i is exponentially distributed with parameter q_i so that with probability $\exp(-q_i t)$ no transition occurs. Suppose that a transition first occurs in the time span τ to $\tau + d\tau$ [the probability of this event is $q_i \exp(-q_i \tau) \, d\tau$] and that the state then changes to state $j \neq i$. (The probability of this last event is q_{ij}/q_i.) The probability of returning to state i in the remaining time span $t - \tau$ in at most $N - 1$ transitions is $P_{ji}(t - \tau; N - 1)$. By the law of total probabilities we have

$$
P_{ii}(t; N) = \exp(-q_i t) + \sum_{j \neq i} \frac{q_{ij}}{q_i} \int_0^t P_{ji}(t - \tau; N - 1) q_i \exp(-q_i \tau) \, d\tau.
$$

By a similar enumeration of the alternative contingencies we derive the identity

$$P_{ij}(t; N) = \sum_{k \neq i} q_{ik} \int_0^t P_{kj}(t - \tau; N - 1) \exp(-q_i \tau) \, d\tau, \qquad i \neq j.$$

It can be proved that

$$\lim_{N \to \infty} P_{ij}(t; N) = \tilde{P}_{ij}(t)$$

(see references at the close of this chapter).

At this point we feel compelled to apologize to the reader for introducing a wealth of concepts of importance in the theory of Markov processes and doing very little with them. These concepts are fraught with many subtleties and pathologies well beyond the level of this textbook. We can only hope the student will pursue these problems further in the many excellent treatises of the subject.

4: Strong Markov Property

A rigorous discussion of the sojourn times for a Markov process, or even for continuous time Markov chains, is intimately related to the so-called *strong Markov property*. Although a complete discussion of this subject requires the use of the language of measure theory, the central idea can be explained in much simpler terms.

To elaborate the relevant ideas we need to introduce a special type of random variable associated with a stochastic process, known variously as a *random variable independent of the future*, or a *Markov time*, or *stopping time* (we shall use "Markov time"). Let σ be a nonnegative random variable associated with a given continuous parameter process $\{X_t\}$, $0 \leq t < \infty$; in other words, associate with each sample function X_t a nonnegative (and possibly infinite) number, which we denote by $\sigma(X_t)$. The random variable σ is said to be a Markov time relative to $\{X_t\}$ if it has the following property:

If X_t and Y_t are two sample functions of the process such that $X_\tau = Y_\tau$ for $0 \leq \tau \leq s$ and $\sigma(X_t) < s$, then $\sigma(X_t) = \sigma(Y_t)$.

In descriptive language, which is useful in understanding the essential properties of such random variables, we can say that the random variable σ scans the sample function starting from $t = 0$ up until some time σ_0 which depends only upon the values of the sample function under consideration in the interval $[0, \sigma_0]$, and the time σ_0 at which σ stops scanning is the value $\sigma(X_t)$.[†]

---

† Strictly speaking, this description is slightly more restrictive than the precise definition given earlier, owing to the fact that we had $\sigma(X_t) < s$ rather than $\sigma(X_t) \leq s$. This restriction can be avoided by saying that for each $\varepsilon > 0$ the value of $\sigma(X_t)$, if σ stops scanning at σ_0, lies in the interval $(\sigma_0 - \varepsilon, \sigma_0)$.

As an example of a Markov time, suppose that we are given a Markov chain that starts in state j_0, i.e., $X(0) \equiv j_0$. If we set

$$\sigma(X_t) = \inf\{\tau \mid X_\tau = i\},$$

where i is a fixed state, and the infimum is taken as $+\infty$ if there is no τ for which $X_\tau = i$, then σ is a Markov time. The formal proof is simple. In fact, suppose for a fixed sample function X_t that $\sigma(X_t) < s$. This means that there is $t' < s$ for which $X_{t'} = i$. Now if $X_\tau = Y_\tau$ for $0 \leq \tau \leq s$, then $Y_{t'} = i$ and certainly $\sigma(Y_t) \leq \sigma(X_t)$. Hence by symmetry $\sigma(X_t) = \sigma(Y_t)$. The random variable σ is called the first entrance time into the state i. Similarly, the first entrance time into any finite collection C of states not containing $X(0) = j_0$, defined by

$$\sigma(X_t) = \inf\{\tau \mid X_\tau \in C\},$$

is a Markov time, the proof being identical with that of the preceding example. Trivially $\sigma \equiv$ constant is a Markov time. The reader should have no difficulty in constructing other examples of Markov times.

Intuitively speaking, the Markov property of stationary Markov processes asserts the following: If we know the values of $X(s)$ for $0 \leq s_1 < s_2 < \cdots < s_n = t_0$ ($t_0 > 0$ fixed), the probability distribution of

$$X(t_0 + t_1), \quad X(t_0 + t_2), \quad \ldots, \quad X(t_0 + t_k) \qquad (0 < t_1 < t_2 < \cdots < t_k) \qquad (4.1)$$

depends only on the value of $X(t_0)$. More exactly, the probability distribution of (4.1) under the condition that we know the path $X(s)$ at times $0 \leq s_1 < s_2 < \cdots < s_n = t_0$, or even more generally that we know the complete history of $X(s)$ up to time t_0 ($0 \leq s \leq t_0$), coincides with the probability distribution of

$$X(t_1), \quad X(t_2), \quad \ldots, \quad X(t_k), \qquad \text{given } X(0).$$

In other words, we can calculate the probability law of (4.1) by translating the time scale so that $t_0 = 0$ and take the initial point as the value of $X(t_0)$.

It seems intuitively plausible that the same relationship should hold if we replace the fixed time value t_0 by a "Markov time" σ. More explicitly, suppose we wish to compute the probability law of

$$X(\sigma + \tau), \qquad \text{given} \quad X(\sigma) = x, \qquad (4.2)$$

i.e., the probability law of $X(\tau)$, $\tau > 0$ units after the occurrence of σ, knowing the value of X at time σ. Now the random variable σ involves the values of the path only up to and inclusive of the time σ but not beyond σ, although the value of σ is not necessarily fixed and may vary from path to path. In other words it is a random time.

It seems reasonable that we may invoke the Markov property at the random time σ. It would then follow that the probability law of (4.2) coincides with the probability law of

$$X(\tau), \qquad \text{given} \quad X(0) = x. \qquad (4.3)$$

This fact is not a direct consequence of the statement of the Markov property since the original formulation requires fixed times. The assertion that (4.2) and (4.3) are indeed governed by the same probability law is the sense of the strong Markov property. More precisely, if for any Markov time σ, the probability distribution of

$$X(t_1 + \sigma), \quad X(t_2 + \sigma), \quad \ldots, \quad X(t_k + \sigma) \qquad (t_1 < t_2 < \cdots < t_k), \quad (4.4)$$

given $X(s), s \leq \sigma$, and $X(\sigma) = x$, is identical with the probability distribution of

$$X(t_1), \quad X(t_2), \quad \ldots, \quad X(t_k) \qquad (t_1 < \cdots < t_k),$$

given $X(0) = x$, then the Markov process is said to possess the *strong Markov property*. There are examples of Markov processes which are not strong Markov.

Such a result as is implied by (4.4) is exceedingly important for purposes of calculating various probabilistic quantities of interest. In fact, one of the basic techniques in analyzing stochastic processes and in computing probabilistic quantities is to write suitable recursion relations, usually in terms of the first time or last time some specified event happens. As an example, suppose we attempt to compute $P_{ij}(t)$ by decomposing the implied event in terms of the event "the first time state j is entered."

Let σ_{ij} be the time at which state j is first entered starting from state i. We pointed out before that first entrance times to a finite set of states are Markov times and hence particularly σ_{ij} is a Markov time. Let

$$F_{ij}(s) = \Pr\{\sigma_{ij} \leq s\}.$$

Of course, at the instant σ_{ij} when the particle enters state j, the probability distribution of its future history is the same as if we translated time so that $\sigma_{ij} = 0$; the initial state is j and the Markov chain is governed by the transition probability matrix $P_{kl}(t)$ in the usual way thereafter. This principle is only correct provided the strong Markov property holds. Then the relation

$$P_{ij}(t) = \int_0^t P_{jj}(t - s) \, dF_{ij}(s) \qquad (4.5)$$

is valid and represents a continuous time analog of formula (5.9) of Chapter 2. The reader may interpret $dF_{ij}(s)$ as the "density function" $f_{ij}(s) \, ds$ of σ_{ij} when a density exists. The formula (4.5) results by the law of total probabilities where $dF_{ij}(s)$ is the probability that $s \leq \sigma_{ij} \leq s + ds$ and $P_{jj}(t - s)$ is the transition probability that if the start is at instant σ_{ij} (the particle is then necessarily in state j), then $t - s$ time units later the process is again at j.

Numerous other intuitive renewal relations of the type (4.5) can be written out [the counterparts of (2.1)–(2.3) of Chapter 11]; we emphasize again that such relations are generally correct provided the strong Markov property prevails.

The fact that sojourn times of successive visits to a given state (or to two different prescribed states) enjoy the property that they are independent random variables is an immediate consequence of the strong Markov property. (The reader should supply a formal proof.)

Because of the fundamental role of the strong Markov property in the analysis of Markov processes we close by stating the most welcome result: *Any continuous time, conservative, Markov chain with only stable states is strong Markov.* More generally, the strong Markov property prevails for any Markov time σ for which $X(\sigma) \neq \infty$ and a "nice" realization of the process is used.

From a practical point of view this means that almost all intuitive renewal relations for these processes are fully rigorous and can be used without fear of error. The proof of the above stated result and a more thorough discussion of the strong Markov property for continuous time Markov chains and other foundational questions can be found in the book by Chung. (See Notes and References at the close of the chapter.)

Problems

1. Let $\mathbf{P}(t) = \| P_{ij}(t) \|_{i, j=0}^{N}$ denote the transition probability matrix of a continuous time finite state Markov chain. Show that $\det[\mathbf{P}(t)] > 0$ for all $t > 0$.

2. Let $\mathbf{P}$ be a 2×2 stochastic matrix, i.e., $\mathbf{P} = \begin{Vmatrix} \alpha & \beta \\ \gamma & \delta \end{Vmatrix}$, where $\alpha, \beta, \gamma, \delta \geq 0$, $\alpha + \beta = 1$, $\gamma + \delta = 1$. Prove there exists a continuous family of 2×2 stochastic matrices $\mathbf{P}(t)$, $t > 0$, such that $\mathbf{P}(1) = \mathbf{P}$ if and only if

$$\det \mathbf{P} = \alpha\delta - \gamma\beta > 0 \qquad \text{and} \qquad \alpha + \delta > 1.$$

Hint: Use the fact that, under the hypotheses, $\mathbf{P}$ can be expressed in the form $e^{\mathbf{A}}$ where $\mathbf{A}$ has the structure

$$\mathbf{A} = \begin{Vmatrix} -a & a \\ b & -b \end{Vmatrix} \quad (a, b > 0).$$

3. Show that if $\mathbf{P}(t) = \| P_{ij}(t) \|_{i, j=0}^{\infty}$ satisfies postulates (a)–(d) of Section 1 then $\mathbf{P}(t)$ is continuous for all $t > 0$.

Hint: Consult Section 8 of Chapter 4.

4. Consider a continuous time irreducible Markov chain with a finite number of states: $1, 2, \ldots, N$. Let q_{ij}, $i, j = 1, \ldots, N$ denote the infinitesimal parameters of the process. Assume that $q_{ij} = q_{ji}$ for all $i, j = 1, 2, \ldots, N$. Let $P_i(t)$ be the probability that the process is in state i at time t for some prescribed initial state. Define

$$E(t) = - \sum_{i=1}^{N} P_i(t) \log P_i(t),$$

where $x \log x$ is taken as zero for $x = 0$. Prove that $E(t)$ is a nondecreasing function of $t \geq 0$.

Hint: First prove

$$P'_i(t) = \sum_{j=1}^{N} q_{ij}[P_j(t) - P_i(t)].$$

Using this formula then prove

$$\frac{dE(t)}{dt} = \frac{1}{2} \sum_{i,j=1}^{N} q_{ij}[P_j(t) - P_i(t)][\log P_j(t) - \log P_i(t)].$$

5. Let X_t $(t > 0)$ be a conservative continuous time Markov chain, that is $\sum_{j \neq i} q_{ij} = q_i$, and assume that $q_{ij} > 0$, $i \neq j$, and $0 < q_i < \infty$. Show that X_t is recurrent if there exists a sequence $z = (z_0, z_1, z_2, \ldots)$ such that (i) $z_n \to \infty$ as $n \to \infty$ and (ii) $-q_i z_i + \sum_{j \neq i} q_{ij} z_j \leq 0$, $i \geq 1$. (Recurrence means that every state is visited infinitely often or equivalently each state is occupied a total infinite length of time.)

Hint: Use the embedded Markov chain. This is the Markov chain $\{Y_n\}$, $n = 0, 1, 2, \ldots$, with stationary transition probability matrix $\mathbf{P} = \| P_{ij} \|$ where

$$P_{ij} = \frac{q_{ij}}{q_i}, \qquad i \neq j,$$

$$P_{ii} = 0.$$

The embedded Markov chain $\{Y_n\}$ simply records the sequence of states through which the process X_t passes without regard to the length of time required for the transitions. Consult Theorem 4.2 of Chapter 3.

6. Under the set-up of Problem 5 show that X_t is nonrecurrent if there exists a bounded nonconstant sequence $z = (z_0, z_1, z_2, \ldots)$ such that

$$-z_i q_i + \sum_{j \neq i} q_{ij} z_j = 0, \qquad i \geq 1.$$

Hint: Cf. Section 4, of Chapter 3.

7. For an irreducible aperiodic recurrent Markov chain, show that

$$\lim_{n \to \infty} (P_{00}^n)^{1/n} = 1.$$

Hint: Use the result of subadditive functions developed in the course of the proof of Theorem 1.1.

8. Consider an irreducible finite state (say N states) continuous time Markov chain. Let $\mathbf{A}$ be its infinitesimal matrix. Show that $\mathbf{A}$ has rank $N - 1$.

9. Consider a two-state, continuous time, conservative Markov chain, and let

$$\mathbf{A} = \begin{Vmatrix} a_{11} & a_{12} \\ a_{21} & a_{22} \end{Vmatrix}$$

be the infinitesimal matrix of the process. Show that, in this case, the backward equations become

$$P''_{ij}(t) - (\text{trace } \mathbf{A})P'_{ij}(t) + (\det \mathbf{A})P_{ij}(t) = 0, \qquad i, j = 1, 2.$$

Solve these equations for $\mathbf{A} = \left\|\begin{matrix} -1 & 1 \\ 1 & -1 \end{matrix}\right\|$.

10. Consider a symmetric random walk in r dimensions, i.e., in a single step the probability is $1/2r$ of visiting any of the neighboring states. Let χ_n be the number of self-avoiding walks in n steps, i.e., the number of n-step paths with the property that the same state is not occupied twice. Prove the simple inequality

$$\chi_{n+m} \leq \chi_n \chi_m,$$

and using this relation show that $\lim_{n \to \infty} \chi_n^{1/n}$ exists.

Hint: Use the fact that $\psi_n = \log \chi_n$ is subadditive, i.e., $\psi_{n+m} \leq \psi_n + \psi_m$; see also page 139.

11. Let $\mathbf{P}_t = e^{\mathbf{A}t}$ be the transition probabilities of a finite state Markov chain having infinitesimal matrix $\mathbf{A}$. Show that $A_{ij} = 0$ whenever $|i - j| > 1$ if and only if for all sets of states $i \leq k$ and $j \leq l$ the determinant

$$\det \left\|\begin{matrix} P_{ij}(t) & P_{il}(t) \\ P_{kj}(t) & P_{kl}(t) \end{matrix}\right\| \geq 0$$

is nonnegative.

Hint: A matrix $\mathbf{A}$ is said to be totally positive of order 2, written $\mathbf{TP}_2$, if

$$\det \left\|\begin{matrix} a_{ij} & a_{il} \\ a_{kj} & a_{kl} \end{matrix}\right\| \geq 0$$

whenever $i \leq k$ and $j \leq l$. Show by direct calculation that if $\mathbf{A}$ is an infinitesimal matrix with $a_{ij} = 0$ when $|i - j| > 1$, then $\mathbf{A}$ is $\mathbf{TP}_2$. Then, for n sufficiently large, $\mathbf{I} + (t/n)\mathbf{A}$ is $\mathbf{TP}_2$. Next, verify for 2×2 matrices U and V that $\det(UV) = (\det U)(\det V)$. Thus, products of $\mathbf{TP}_2$ matrices are $\mathbf{TP}_2$, whence $(\mathbf{I} + (t/n)\mathbf{A})^n$ is $\mathbf{TP}_2$ and finally $\mathbf{P}(t) = \lim_{n \to \infty} (\mathbf{I} + (t/n)\mathbf{A})^n$ is $\mathbf{TP}_2$.

For the converse, evaluate the following for $j > i + 1$:

$$0 \leq \det \left\|\begin{matrix} P_{i,i+1}(t) & P_{ij}(t) \\ P_{i+1,i+1}(t) & P_{i+1,j}(t) \end{matrix}\right\|$$

$$= \det \left\|\begin{matrix} a_{i,i+1}t + o(t) & a_{ij}t + o(t) \\ 1 - a_{i+1,i+1}t + o(t) & a_{i+1,j}t + o(t) \end{matrix}\right\|$$

$$= -a_{ij}t + O(t^2).$$

This implies $a_{ij} = 0$ if $j > i + 1$. A similar argument holds for $j < i - 1$.

12. Let $\{X(t) : t \geq 0\}$ be a finite state Markov chain with generator $\mathbf{Q} = \|q_{ij}\|$. Set $q_i = -q_{ii} > 0$. Define $\sigma_0 = 0$ and let σ_n be the time of the nth change of state of $\{X(t)\}$.

Formally $\sigma_n = \inf\{t \ge \sigma_{n-1} : X(t) \ne X(\sigma_{n-1^+})\}$. Let $\xi_n = X(\sigma_{n^+})$ for $n = 0, 1, \ldots$. Then (cf. Chapter 4) $\{\xi_n\}$ is a discrete time Markov chain whose transition matrix $\mathbf{P} = \| P_{ij} \|$ is given by

$$P_{ij} = q_{ij}/q_i, \qquad i \ne j$$
$$= 0, \qquad i = j.$$

Suppose $\{X(t)\}$ has a single recurrent class with stationary distribution $\boldsymbol{\alpha} = \| \alpha_i \|$ and that $\{\xi_n\}$ has stationary measure $\boldsymbol{\pi} = \| \pi_j \|$. Show that $\alpha_i q_i$ is proportional to π_i.

13. (*continuation*). Suppose that $\{X(t)\}$ is a continuous time Markov chain whose state space is $\{0, \pm 1, \pm 2, \ldots\}$ and let $\{\xi_n\}$ be defined as in Problem 12. Show that $\{X(t)\}$ and $\{\xi_n\}$ have the same recurrent classes, but they need not be positive recurrent together. (Suppose $q_{i,i+1} = q_{i,i-1} = \frac{1}{2}q_i$. Then $\{\xi_n\}$ is null recurrent, but $\{X(t)\}$ is positive recurrent when $\sum_i 1/q_i < \infty$.)

14. Let $\{X(t)\}$ be a continuous time Markov chain on $\{0, 1, \ldots, N\}$ with absorption at 0 and N. Let $\mathbf{A}$ be the generator over the transient states. Let m_{ij} be the mean duration spent in j starting from i. Show $\mathbf{M} = -\mathbf{A}^{-1}$ where $\mathbf{M} = \| m_{ij} \|$.

15. (*continuation*). Let T_j be the total sojourn time in state j and define $m_{ij} = E[T_j | X(0) = i]$ for $i, j = 1, \ldots, N - 1$. Show that

$$\Pr\{T_j = 0 | X(0) = i\} = 1 - m_{ij}/m_{jj};$$

and

$$\Pr\{0 < T_j \le t | X(0) = i\} = (m_{ij}/m_{jj})[1 - \exp(t/m_{jj})], \qquad t > 0.$$

16. (*continuation*). Let ζ be the time to absorption in state 0 or N. Formally $\zeta = \sum_{j=1}^{n-1} T_j$. Show that, conditioned on $X(0) = i$, ζ has probability density function $f_i(t) = -\mathbf{e}_i' Q \exp(Qt)\mathbf{e}_i$, where $\mathbf{e}_i = \| \delta_{ij} \|$ is the vector with a 1 in the ith position and 0 otherwise.

17. Consider an irreducible continuous time Markov chain on the states $\{0, \ldots, M\}$. Let $(\alpha_0, \ldots, \alpha_n)$ be the stationary distribution. Define

$$\rho_i = \inf\{t > 0 : X(t) \ne i\},$$
$$E_i = \inf\{t > \rho_i : X(t) = i\}.$$

Then ρ_i is called the *i-residence* and E_i the *i-excursion*.
   Show that

$$P_{ii}(t) = e^{-q_i t} + \int_0^t P_{ii}(t - s) \, dH_i(s),$$

where $H_i(s) = \Pr\{E_i \le s\}$ and $\mathbf{Q} = \| q_{ij} \|$ is the infinitesimal matrix, with $q_i = -\sum_{j \ne i} q_{ij}$.

18. Let $\{X(t)\}$ be a continuous time Markov chain with state space $S = \{0, 1, \ldots\}$ and generator $\mathbf{Q}$. Let $\mathbf{y}$ be a nonzero solution of $\mathbf{Qy} = \lambda \mathbf{y}$. Show that $Y(t) = e^{-\lambda t} y(X(t))$ is a martingale with respect to $\{X(t)\}$.

NOTES

This chapter incorporates a bare few topics with expanded detail from Chung [1]. See also references therein.

The bulk of the ideas of this chapter developed in the context of general Markov processes can be found in Dynkin [2] and Doob [3].

Kingman's interesting book [4] incorporates valuable results concerning recurrence phenomena and the transition probabilities of general continuous time Markov chains.

REFERENCES

1. K. L. Chung, "Markov Chains with Stationary Transition Probabilities," Chapters 3 and 4. Springer–Verlag, Berlin, 1960.
2. E. B. Dynkin, "Theory of Markov Processes." Academic Press, New York, 1965.
3. J. L. Doob, "Stochastic Processes." Wiley, New York, 1953.
4. J. F. C. Kingman, "Regenerative Phenomena." Wiley, New York, 1972.

Chapter 15

DIFFUSION PROCESSES

1: General Description

A continuous time parameter stochastic process which possesses the (strong) Markov property and for which the sample paths X(t) are (almost always)† continuous functions of t is called a *diffusion process*. A number of alternative characterizations of diffusion processes are set forth later in this section.

Apart from their intrinsic interest, diffusion processes are of value for a manifold of purposes.

1. Many physical, biological, economic, and social phenomena are either well approximated or reasonably modeled by diffusion processes. These include examples from molecular motions of enumerable particles subject to interactions, security price fluctuations in a perfect market, some communication systems with noise, neurophysiological activity with disturbances, variations of population growth, changes in species numbers subject to competition and other community relationships, gene substitutions in evolutionary development, etc.

2. Many functionals, including first-passage probabilities, mean absorption times, occupation time distributions, boundary behavior properties, and stationary distributions, can be calculated explicitly for one-dimensional diffusions. The calculations largely reduce to solving second-order differential

---

† The phase "almost always" means with probability 1. For our purposes, without loss in generality we may and do assume that all sample paths are continuous.

equations with simple boundary conditions (see Sections 3 and 4). In principle, these formulations and calculations can be extended to multidimensional diffusions but then involve partial differential equations whose explicit resolution is usually formidable.

3. By transforming the time scale and renormalizing the state variable, many Markov processes can be approximated by diffusion processes. The concept of rescaling in a quite general model is described in intuitive form at the close of this section, and a variety of examples of these approximations will be set forth in Sections 2 and 10.

Consider a diffusion process $\{X(t), t \geq 0\}$ whose state space is an interval I with endpoints $l < r$, so that I is necessarily of the form (l, r), $(l, r]$, $[l, r)$, or $[l, r]$ (a parenthesis signifies that the interval is open at that end, while a square bracket means it is closed). We allow the possibilities $l = -\infty$ and/or $r = +\infty$. Such a process is said to be *regular* if starting from any point in the interior of I any other point in the interior of I may be reached with positive probability. This may be expressed more precisely through the concept of *hitting time* random variables. For any point z in the interval I, let T_z denote the random variable equal to the first time the process attains the value z. In the event that z is never reached, by convention we set $T_z = \infty$. The process is regular if

$$\Pr\{T_z < \infty \,|\, X(0) = x\} > 0,$$

whenever $l < x$, $z < r$. *Henceforth, without further mention, we shall consider only regular diffusion processes.*

Most Markov processes, including the Poisson process, birth and death processes, etc., satisfy the property

$$\lim_{h \downarrow 0} \frac{1}{h} \Pr\{|X(h) - x| > \varepsilon \,|\, X(0) = x\} = \lambda(x, \varepsilon),$$

with $\lambda(x, \varepsilon)$ nonnegative and possibly positive for ε small. In fact, recall that for the Poisson process we have

$$\lim_{h \downarrow 0} \frac{1}{h} \Pr\{X(h) - i = 1 \,|\, X(0) = i\} = \lambda, \qquad i = 0, 1, \ldots, \tag{*}$$

where λ is the mean rate of the occurrence of events. The sample realizations of the Poisson process are discontinuous step functions having jumps of unit increase.

In contrast to (*) every diffusion process satisfies the following condition: For every $\varepsilon > 0$,

$$\lim_{h \downarrow 0} \frac{1}{h} \Pr\{|X(t + h) - x| > \varepsilon \,|\, X(t) = x\} = 0 \qquad \text{for all } x \text{ in } I. \tag{1.1}$$

This relation asserts that large displacements, of order exceeding a fixed ε, are very unlikely over sufficiently small time intervals. This fact can be viewed as a formalization of the property that the sample paths of the process are continuous. Theorem 1.1 asserts a partial converse: A Markov process for which (1.1) holds in an appropriate uniform sense is a diffusion process. The theorem is followed by Lemma 1.1 which establishes the necessary condition (1.1) as a consequence of moment properties that are usually satisfied in practice.

Almost all diffusion processes that have appeared to date as models of physical or biological phenomena are characterized by two basic conditions which augment (1.1) and describe the mean and variance of the infinitesimal displacements. Let $\Delta_h X(t)$ be the increment in the process accrued over a time interval of length h. Thus, $\Delta_h X(t) = X(t + h) - X(t)$. These key conditions affirm the existence of the limits

$$\lim_{h \downarrow 0} \frac{1}{h} E[\Delta_h X(t) | X(t) = x] = \mu(x, t) \tag{1.2}$$

and

$$\lim_{h \downarrow 0} \frac{1}{h} E[\{\Delta_h X(t)\}^2 | X(t) = x] = \sigma^2(x, t) \tag{1.3}$$

whenever $l < x < r$. The functions $\mu(x, t)$ and $\sigma^2(x, t)$ are termed the *infinitesimal parameters* of the process, and, in particular, $\mu(x, t)$ is called the *drift parameter*, *infinitesimal mean*, or *expected infinitesimal displacement* and $\sigma^2(x, t)$ the *diffusion parameter*, or *infinitesimal variance*.

Generally, $\mu(x, t)$ and $\sigma^2(x, t)$ are continuous functions of x and t, and a regular process typically has $\sigma^2(x, t)$ positive for all $l < x < r$ and $t > 0$. (But see Problem 26.) Indeed, some texts (especially those with emphasis on practical models) define $\{X(t), t \geq 0\}$ to be a diffusion if $X(t)$ is a Markov process satisfying (1.1) and if (1.2) and (1.3) give continuous functions of x and t.

It is tacitly assumed in relations (1.2) and (1.3) that the displayed moments exist. A more general version presents these relations in terms of truncated moments where $\Delta_h X(t)$ is replaced by

$$\Delta_{h,\varepsilon} X(t) = \begin{cases} \Delta_h X(t) & \text{if } |\Delta_h X(t)| \leq \varepsilon, \\ 0 & \text{otherwise.} \end{cases}$$

Formally, the conditions (1.2) and (1.3) are replaced by

$$\lim_{h \downarrow 0} \frac{1}{h} E[\Delta_{h,\varepsilon} X(t) | X(t) = x] = \mu(x, t), \tag{1.2'}$$

$$\lim_{h \downarrow 0} \frac{1}{h} E[\{\Delta_{h,\varepsilon} X(t)\}^2 | X(t) = x] = \sigma^2(x, t), \tag{1.3'}$$

respectively, and of course (1.1) must be adjoined to (1.2′) and (1.3′) to be assured that the process is a diffusion.

The use of $\Delta_{h,\varepsilon}X(t)$ in place of $\Delta_h X(t)$ adds only technical complexity and essentially no new ideas. Moreover, in most (if not all) practical examples of diffusions, the indicated moments exist. Accordingly, we stipulate henceforth that the relations (1.2) and (1.3) hold. Moreover we concentrate primarily on the time homogeneous case where the functions $\mu(x, t) = \mu(x)$ and $\sigma^2(x, t) = \sigma^2(x)$ are independent of t.

The motivation for the name "infinitesimal mean" for $\mu(x)$ is clear, since

$$E[\Delta_h X(t) | X(t) = x] = \mu(x)h + o_1(h)$$

where $o_1(h)$ accumulates the remainder terms of order less than h as $h \to 0$. For the variance we have

$$\text{var}[\Delta_h X(t) | X(t) = x] = E[(\Delta_h X(t))^2 | X(t) = x] - \{E[\Delta_h X(t) | X(t) = x]\}^2$$
$$= \sigma^2(x)h + o_2(h) - [\mu(x)h + o_1(h)]^2$$
$$= \sigma^2(x)h + o_3(h)$$

where

$$o_3(h) = o_2(h) - [\mu(x)h + o_1(h)]^2$$

is again a remainder term of order smaller than h.

In addition to the infinitesimal relations (1.2) and (1.3), the following higher-order infinitesimal moment relations are usually satisfied:

$$\lim_{h \downarrow 0} \frac{E[|\Delta_h X(t)|^r | X(t) = x]}{h} = 0, \qquad r = 3, 4, \dots . \tag{1.4}$$

Sometimes (and frequently in the physical literature) a diffusion process is defined to be a Markov process obeying the infinitesimal moment conditions (1.2) and (1.3), and (1.4) is stipulated to hold only for *some* $r, r > 2$. In fact, subject to smoothness assumptions on $\mu(x)$ and $\sigma^2(x)$, a process satisfying (1.2), (1.3), and (1.4) for some $r > 2$ can be realized having only continuous paths. To formalize this approach is arduous and unnatural in some ways, and therefore the modern setup is to define a diffusion as we have done, as a strong Markov process with continuous sample paths.

MULTIVARIATE DIFFUSION PROCESSES

Although we concentrate on one-dimensional diffusions, it is often relevant to consider a vector diffusion process, e.g., a diffusion on the state space $I = E^n$ (Euclidean n-space) or an open region of E^n. Let $\mathbf{X}(t) = (X_1(t), X_2(t), \dots, X_n(t))$ be a vector process. The analog of the infinitesimal relations (1.2) and (1.3) are

$$\lim_{h \downarrow 0} \frac{1}{h} E[X_i(t + h) - X_i(t) | \mathbf{X}(t) = \mathbf{x} = (x_1, x_2, \dots, x_n)] = \mu_i(\mathbf{x}, t),$$

$$i = 1, 2, \dots, n,$$
$$\tag{1.2''}$$

and

$$\lim_{h\downarrow 0} \frac{1}{h} E[\{X_i(t+h) - X_i(t)\}\{X_j(t+h) - X_j(t)\} | \mathbf{X}(t) = \mathbf{x}] = \sigma_{ij}(\mathbf{x}, t), \quad (1.3'')$$

$$i, j = 1, \ldots, n.$$

The matrix $\|\sigma_{ij}(\mathbf{x}, t)\|_1^n$ is required to be positive definite, i.e.,

$$\sum_{i,j=1}^{n} \sigma_{ij}(\mathbf{x}, t) a_i a_j > 0 \quad \text{for all nontrivial real } n\text{-tuples } (a_1, a_2, \ldots, a_n) \quad \text{and}$$

$$\mathbf{x} \in I$$

consistent with the positive value of

$$\sum_{i,j=1}^{n} a_i a_j [X_i(t+h) - X_i(t)][X_j(t+h) - X_j(t)] = \left\{ \sum_{\nu=1}^{n} a_\nu [X_\nu(t+h) - X_\nu(t)] \right\}^2.$$

The higher-order moments are again taken to be negligible compared to h, e.g.,

$$\lim_{h\downarrow 0} \frac{1}{h} E[\{X_i(t+h) - X_i(t)\}^4 | \mathbf{X}(t) = \mathbf{x}] = 0, \quad \text{for} \quad i = 1, \ldots, n.$$

CONSERVATIVE PROCESSES AND DIFFUSIONS WITH KILLING

A process is called a *diffusion with killing* if the sample paths behave like those of a regular diffusion until a possibly random, possibly infinite time ζ when the process is killed. There are many natural models, especially in biological and physical contexts, for which this generalization is appropriate; see Section 2, Example G and Section 10.

We write a diffusion with killing in the form $\{X(t), 0 \leq t < \zeta\}$. Since we allow $\zeta = \infty$, this includes the case where no killing occurs. If $\zeta = \infty$ from all starting points $X(0) = x$, the process is said to be conservative, and we write $\{X(t), t \geq 0\}$. That is, a regular diffusion $\{X(t)\}$ on I is *conservative* if

$$\Pr\{X(t) \in I \,|\, X(0) = x\} = \Pr\{\zeta > t \,|\, X(0) = x\} = 1$$

for all $t \geq 0$ and x in I.

Many authors append a distinguished point Δ (called "the cemetery") to the interval I, thereby enlarging the state space to $I \cup \{\Delta\}$, with the prescription that $X(t) = \Delta$ for $t \geq \zeta$. This construction serves as a technical convenience, ensuring that the process is always conservative on the augmented state space $I \cup \{\Delta\}$. However, the convention needlessly encumbers the notation in applications, where it is preferred simply to allow $X(t)$ to be undefined for $t \geq \zeta$.

For a diffusion with killing, at each point x there is a probability $k(x) dt + o(dt)$ that the process ceases (is killed) over the infinitesimal time duration $(t, t + dt)$, and probability $1 - k(x) dt + o(dt)$ that no killing occurs. The above prescription is merely descriptive. Its rigorous formulation requires the concept of stochastic multiplicative functionals. We shall say more on this in Section 12. Notice that the killing rate can depend on the position and even on time, if we allow $k(x, t)$ to be a function of time t as well as position x. In symbols,

$$\lim_{h \downarrow 0} \frac{1}{h} \Pr\{t < \zeta < t + h \,|\, X(t) = x\} = k(x, t).$$

There is a canonical construction of the process governed by the infinitesimal rates (1.2) and (1.3) with infinitesimal killing rate $k(x, t)$ that is expressed in terms of a conservative diffusion process having the same mean and diffusion coefficients.

HITTING TIMES

Hitting times of points and sets play a fundamental role in the study of one-dimensional diffusion processes. Formally we define the *hitting time* of the process $\{X(t), 0 \leq t < \zeta\}$ to the level z by

$$T_z = \infty \qquad\qquad \text{if} \quad X(t) \neq z \quad \text{for} \quad 0 \leq t < \zeta$$

$$= \inf\{t \geq 0; X(t) = z\} \qquad \text{otherwise.}$$

For typographical convenience we write $T(z)$ interchangeably for T_z, the meaning being clear from the context.

We use the notation

$$T^* = T_{a,b} = T(a, b) = \min\{T(a), T(b)\} = T(a) \wedge T(b)$$

for the hitting time to a or b, the first time $X(t) = a$ or $X(t) = b$. For processes starting at $X(0) = x$ in (a, b), this is the same as the *exit time* of the interval (a, b):

$$T(a, b) = \inf\{t \geq 0; X(t) \notin (a, b)\}, \qquad X(0) = x \quad \text{in } (a, b).$$

WHEN IS A STOCHASTIC PROCESS RECOGNIZED AS A DIFFUSION?†

It is of value to have verifiable sufficient conditions under which a Markov process is (or can be realized as) a diffusion process. To this end, we introduce the concept of a *standard‡ process*. A strong Markov process $\{X(t), t \geq 0\}$ is called a standard process if the sample paths possess the following regularity properties:

(i) $X(t)$ is right continuous; i.e., for all $s \geq 0$

$$\lim_{t \downarrow s} X(t) = X(s);$$

---

† The remainder of this section can be skipped at first reading.

‡ The definition of "standard" given here is for a process without killing.

(ii)   left limits of $X(t)$ exist; i.e.,

$$\lim_{t\uparrow s} X(t) \text{ exists} \qquad \text{for all} \quad s > 0;$$

and

(iii) $X(t)$ is continuous from the left through Markov times; i.e., if $T_1 \leq T_2 \leq \cdots$ are Markov times (see Sections 3, 7, and 8 of Chapter 6) converging to $T \leq \infty$, then $\lim_{n\to\infty} X(T_n) = X(T)$ whenever $T < \infty$. (To be precise, equality holds in the almost sure sense in which $\Pr\{T < \infty$ and $\lim_{n\to\infty} X(T_n) \neq X(T)\} = 0$.)

The third condition is sometimes referred to as *quasi-left continuity*. It does not imply that the sample paths are left continuous, but only that jumps cannot be predicted. As an example, the Poisson process is a standard process that exhibits jump discontinuities. The random times $T_n = T - 1/n$, where T is the time of the first jump, are not Markov times.

In particular, a sample path of a standard process can at worst exhibit discontinuities of the first kind (jumps).

A deep and remarkable result is that every strong Markov process $\{X(t), t \geq 0\}$ *continuous in probability* (see below) and subject to mild regularity conditions, possesses an equivalent version $\{\tilde{X}(t), t \geq 0\}$ which is a standard process. (The processes $\tilde{X}(t)$ and $X(t)$ are *equivalent* provided that $\tilde{X}(t)$ and $X(t)$ have the same finite-dimensional distributions.) Recall that a stochastic process is continuous in probability if for any $\varepsilon > 0$ and $s \geq 0$

$$\lim_{t\to s} \Pr\{|X(t) - X(s)| > \varepsilon\} = 0. \tag{1.5}$$

Property (1.5) is quite weak and is satisfied in most stochastic models, and thus for all practical purposes we can assume that the Markov process at hand is a standard process.

A sufficient condition that a standard process be a diffusion is the fulfillment of the *Dynkin condition*:

$$\frac{1}{h} Pr\{|X(t + h) - X(t)| > \varepsilon \,|\, X(t) = x\} \to 0 \qquad \text{when} \quad \varepsilon > 0, \tag{1.6}$$

as $h \downarrow 0$, where the convergence prevails uniformly for x restricted to any compact subinterval of (l, r) and t traversing any finite interval $[0, N]$. (We emphasize that uniformity is crucial.) The following theorem asserts the sufficiency of this condition.

Theorem 1.1.   *Let $\{X(t), t \geq 0\}$ be a standard process (see (i)–(iii) above) and suppose the Dynkin condition holds. Then $\{X(t), t \geq 0\}$ is a diffusion process.*

Proof.   We give the proof assuming that the state space I is a closed bounded interval so that the convergence in the Dynkin condition (1.6) holds uniformly for all x.

Take an arbitrary number $\delta > 0$, a fixed integer $N > 0$, and a sequence of integers $k = 1, 2, \ldots$. Consider the events

$$A_N(k, i) = \left\{ \left| X\left(\frac{iN}{k}\right) - X\left(\frac{(i-1)N}{k}\right) \right| > \tfrac{1}{2}\delta \right\}, \qquad i = 1, \ldots, k.$$

That is, $A_N(k, i)$ consists of all sample paths that move more than $\tfrac{1}{2}\delta$ between times $(i - 1)N/k$ and iN/k. Consider the event

$$B_\delta = \left\{ \sup_{0 < t < N} |X(t) - X(t-)| > \delta \right\},$$

consisting of all realizations exhibiting a jump exceeding δ in the interval $0 < t < N$. Since the sample paths all have left limits and are right continuous, we claim that

$$B_\delta \subset \bigcup_{j=1}^{\infty} \bigcap_{k=j}^{\infty} \bigcup_{i=1}^{k} A_N(k, i). \tag{1.7}$$

Indeed, any sample path in B_δ must have a point t $(0 < t < N)$, where $|X(t) - X(t-)| > \delta$. For all k sufficiently large and appropriate i satisfying $(i - 1)N/k < t \leq iN/k$, we can guarantee the validity of (1.7) by relying on the property that $X(t)$ is right continuous and has a limit from the left at t. Now, from (1.7) we deduce

$$\Pr\{B_\delta\} \leq \Pr\left\{ \bigcup_{j=1}^{\infty} \bigcap_{k=j}^{\infty} \bigcup_{i=1}^{k} A_N(k, i) \right\}$$

$$= \lim_{j \to \infty} \Pr\left\{ \bigcap_{k=j}^{\infty} \bigcup_{i=1}^{k} A_N(k, i) \right\}$$

$$\leq \lim_{k \to \infty} \Pr\left\{ \bigcup_{i=1}^{k} A_N(k, i) \right\}$$

$$\leq \lim_{k \to \infty} \sum_{i=1}^{k} \Pr\{A_N(k, i)\}. \tag{1.8}$$

But with $h = N/k$, the uniformity in the Dynkin condition (1.6) implies

$$\Pr\{A_N(k, i)\} \leq \varepsilon(\delta, k)(N/k),$$

where $\varepsilon(\delta, k) \to 0$ as $k \to \infty$. Adding this over i, we obtain

$$\sum_{i=1}^{k} \Pr\{A_N(k, i)\} \leq N\varepsilon(\delta, k). \tag{1.9}$$

Sending $k \to \infty$, and comparing (1.8) and (1.9), we deduce that

$$\Pr\{B_\delta\} \leq N \lim_{k \to \infty} \varepsilon(\delta, k) = 0. \tag{1.10}$$

This holds for every $\delta > 0$. Consequently

$$\Pr\left\{\sup_{0 < t < N} |X(t) - X(t-)| > 0\right\} \leq \sum_{\substack{\delta > 0 \\ \text{rational}}} \Pr\left\{\sup_{0 < t < N} |X(t) - X(t-)| > \delta\right\} = 0.$$

Thus, with probability 1, each path is left continuous, and therefore continuous, for $0 < t < N$. But N is arbitrary and a standard process is right continuous, so $X(t)$ is continuous at all $t \geq 0$. The proof is complete. ∎

The following lemma is valuable and will be cited on several occasions:

Lemma 1.1. *If a standard process satisfies the infinitesimal moment condition*

$$\lim_{h \downarrow 0} \frac{1}{h} E[|\Delta_h X(t)|^p \,|\, X(t) = x] = 0 \tag{1.11}$$

for some $p > 2$ uniformly for x in any compact subinterval of (l, r), and t in any finite interval $[0, N]$, then the Dynkin condition (1.6) is satisfied.

Proof. A direct application of the Chebyshev inequality gives

$$\frac{1}{h} \Pr\{|\Delta_h X(t)| > \varepsilon \,|\, X(t) = x\} \leq \frac{E[|\Delta_h X(t)|^p \,|\, X(t) = x]}{h\varepsilon^p} \tag{1.12}$$

and the desired result is clear. ∎

Remark. The inequality in (1.12) applies for any $p > 0$ but its utility for diffusions ordinarily requires $p > 2$, with $p = 4$ often being the most convenient choice.

Applications of Theorem 1.1 and Lemma 1.1 will be made in the later sections.

Another criterion frequently used to check that a one-dimensional stochastic process $X(t)$ (not necessarily possessing the Markov property) has a continuous path realization is the *condition of Kolmogorov* now stated.

Let $\{X(t), t \geq 0\}$ be a stochastic process obeying the bound

$$E[|X(t) - X(s)|^\gamma] \leq C|\varphi(t) - \varphi(s)|^{1+\alpha} \qquad \text{for all} \quad s, t \geq 0, \tag{1.13}$$

where α, γ, and C are positive constants independent of s and t and φ is a continuous nondecreasing function. Then there exists an equivalent version $\tilde{X}(t)$ possessing continuous paths.

For a d-dimensional process the Kolmogorov condition must be modified to

$$E[|X(t) - X(s)|^\gamma] \leq C|\varphi(t) - \varphi(s)|^{d+\alpha}$$

with α, γ, and C positive as before and the exponent on the right now exceeding d.

CHARACTERIZATIONS OF DIFFUSION PROCESSES

It is useful to conclude this section by briefly and informally reviewing several alternative characterizations of diffusion processes.

(a) Definition of a Diffusion in Terms of the Infinitesimal Coefficients

We have already remarked with a view to applications that the infinitesimal parameters $\mu(x, t)$ and $\sigma^2(x, t)$, assumed to be smooth with $\sigma^2(x, t) > 0$ for $l < x < r$, summarize the basic information at hand. When appropriate higher-order moment restrictions as in (1.4) are in force, then apart from boundary behavior the probability laws governing the realizations of the diffusion are determined from the nature of the infinitesimal mean and variance coefficients.

The classification of all possible diffusions with prescribed infinitesimal coefficients depends on a delineation of the motion on or off the boundary. Some of the problems in these constructions will be exemplified in our discussions of the boundary behavior for the diffusions covered in Sections 6–8. Our primary approach in dealing with diffusions will be to operate in terms of the infinitesimal coefficients and the moment conditions (1.2)–(1.4).

(b) The Stroock–Varadhan Martingale Characterization of Diffusions

Let $X(t)$ be a time homogeneous diffusion with drift $\mu(x)$ and variance $\sigma^2(x)$ for $l < x < r$.

Consider for each real λ, the stochastic process

$$Y_\lambda(t) = \exp\left[\lambda X(t) - \lambda \int_0^t \mu(X(s))\, ds - \tfrac{1}{2}\lambda^2 \int_0^t \sigma^2(X(s))\, ds \right] \tag{1.14}$$

defined for $t > 0$. A very useful result affirms under proper regularity and boundary constraints that $\{Y_\lambda(t)\}$ constitutes a martingale with respect to the family of σ-fields $\mathscr{F}_t$ determined by the process $X(\tau)$ up to time t. When $X(t)$ is standard Brownian motion so that $\mu(x) \equiv 0$ and $\sigma^2(x) \equiv 1$, then plainly

$$Y_\lambda(t) = e^{[\lambda X(t) - \lambda^2 t/2]}$$

which is a familiar martingale process encountered in Chapter 7. Subject to suitable technical requirements a converse theorem holds as follows: *If $Y_\lambda(t)$ is a continuous time martingale process for each real λ, then $\{X(t)\}$ is a diffusion.*

A related martingale characterization, quite useful in some contexts, is the following. Subject to suitable technical requirements, $\{X(t), t \geq 0\}$ is a diffusion process with infinitesimal coefficients $\mu(x)$ and $\sigma^2(x)$ if and only if for every bounded and twice continuously differentiable function $f(x)$ the process

$$Z_f(t) = f(X(t)) - f(X(0)) - \int_0^t [\tfrac{1}{2}\sigma^2(X(s))f''(X(s)) + \mu(X(s))f'(X(s))]\, ds$$

is a martingale.

When $X(t)$ is standard Brownian motion and $f(x) = x$, we recover the well-known fact that $Z_f(t) = X(t)$ is a martingale. When $f(x) = x^2$ then $Z_f(t) = X(t)^2 - t$ is also a martingale. (See Section 5 of Chapter 7.)

The above martingale characterizations are dealt with further in Sections 11, 12, and 16.

(c) Stochastic Differential Equations

A useful tool in modeling physical, biological, and economic phenomena is the concept and solution of stochastic differential equations. The procedure is to extend the method of successive approximations used to solve ordinary differential equations to the stochastic context. The solutions obtained are diffusions. An introductory account of these ideas is set forth in Section 14 and further elaborated in Sections 15 and 16.

(d) Total Positivity Characterization of One-Dimensional Diffusions

Consider a diffusion process with state space a segment (l, r) of the real line. For ease of exposition assume the process is governed by a transition density

$$p(t, x, y)\, dy \sim \Pr\{y < X(t) \le y + dy \,|\, X(0) = x\}.$$

Choose $2r$ points inside the state space satisfying

$$x_1 < x_2 < \cdots < x_r, \qquad y_1 < y_2 < \cdots < y_r, \tag{1.15}$$

and form the determinant

$$\det \| p(t, x_i, y_j) \| = \det \begin{Vmatrix} p(t, x_1, y_1) & p(t, x_1, y_2) & \cdots & p(t, x_1, y_r) \\ p(t, x_2, y_1) & p(t, x_2, y_2) & \cdots & p(t, x_2, y_r) \\ \vdots & & & \vdots \\ p(t, x_r, y_1) & p(t, x_r, y_2) & \cdots & p(t, x_r, y_r) \end{Vmatrix}. \tag{1.16}$$

It is a remarkable fact that when $X(t)$ is a one-dimensional diffusion, then the quantities of (1.16) are always positive.

The converse, under mild regularity requirements imposed on $p(t, x, y)$, is also valid. Thus, if $X(t)$ is a strong Markov process and (1.16) is positive for each t and all choices of (1.15) ($r = 2$ suffices), then $X(t)$ is equivalent to a diffusion.

The assumption of the existence of a density is not necessary in that the determinantal inequalities can be reexpressed in terms of the transition probability distribution.

The determinantal expression (1.16) has a probabilistic interpretation which we elaborate in Problem 21.

SOME COMMENTS ON APPLIED STOCHASTIC MODELING AND RELATED DIFFUSIONS

Many stochastic models evolving in discrete time have the structure

$$Z_{k+1} = f(Z_k, s_k) + \xi_k, \qquad k = 0, 1, 2, \ldots, \tag{1.17}$$

or more generally

$$Z_{k+1} = g(Z_k, s_k, \xi_k). \tag{1.18}$$

(A broad spectrum of examples of (1.17) and (1.18) are dispersed throughout the chapter.) Thus, the process variable Z_{k+1} at generation $k + 1$ is influenced by two types of forces. The first contribution depends on the current state variable Z_k and other systematic or randomly varying parameters $\{s_k\}$ whose effects may be temporally correlated. The $\{s_k\}$ factors are coupled to Z_k through the function f. The second contribution embodied in ξ_k is generally regarded as a "noise" perturbation. In ecological, genetic, and other biological contexts, the process $\{s_k\}$ is sometimes referred to as *stochastic* or *deterministic environmental effects* while $\{\xi_k\}$ is regarded as *sampling* or *demographic* effects.

The analysis of $\{Z_k\}$ following (1.17) is often facilitated by passing to an analogous diffusion (or stochastic differential equation) model which is in many instances more tractable, while retaining the relevant qualitative consequences.

In applied stochastic modeling, there arise families of discrete or continuous time Markov chains $Z^{(N)}(t)$ indexed by a natural parameter N. Depending on the problem and objectives, it is frequently natural to rescale the time and state variables, producing the transformed process

$$X^{(N)}(t) = a(N)[Z^{(N)}(b(N)t) - c(N)] \tag{1.19}$$

where $c(N)$ is a centering constant, $a(N)$ performs a scaling of the state variable, and $b(N)$ performs the required time scaling. Accordingly, a unit of time in the $X^{(N)}$ process corresponds to an epoch of about $b(N)$ time units in the original $Z^{(N)}$ process.

By judicious rescalings, we often uncover the convergence phenomenon

$$X^{(N)}(t) \to X(t) \tag{1.20}$$

(convergence in the sense defined in (1.24) and (1.25)) where $X(t)$ is a diffusion. In this way properties of certain functionals of the process $X^{(N)}(t)$ can be obtained from the calculation of the corresponding functionals for the $X(t)$ process. We amply illustrate the concept and mechanism of (1.19) and (1.20) in terms of concrete examples in Sections 2 and 10.

CONVERGENCE TO DIFFUSIONS

Consider the sequence of discrete time stochastic processes $X^{(N)} = \{X_k^{(N)}\}$ with state space confined to a closed bounded interval of the real line adapted† to

---

† That is, $X_k^{(N)}$ is measurable with respect to $\mathscr{F}_k^{(N)}$ for every $k \geq 0$.

an increasing sequence $\{\mathscr{F}_k^{(N)}, k \geq 0\}$ of σ fields. It is usual to take $\mathscr{F}_k^{(N)}$ to be the σ field determined by the process values $\{X_i^{(N)}, i \leq k\}$; cf. Chapter 6. The processes $X^{(N)}$ need not possess the Markov property.

The conditional moments of $\Delta X_n^N = X_{n+1}^{(N)} - X_n^{(N)}$ are assumed to satisfy

$$E[\Delta X_n^N | \mathscr{F}_n^N] = h_N \mu(X_n^{(N)}) + \varepsilon_{1,n}^N,$$
$$E[(\Delta X_n^N)^2 | \mathscr{F}_n^N] = h_N \sigma^2(X_n^{(N)}) + \varepsilon_{2,n}^N, \tag{1.21}$$

and

$$E[(\Delta X_n^N)^4 | \mathscr{F}_n^N] = \varepsilon_{3,n}^N, \tag{1.22}$$

where $h_N > 0$ and $h_N \to 0$ as $N \to \infty$ and the error terms are uniformly small to the extent that for any positive t

$$\sum_{n < [t/h_N]} E[|\varepsilon_{i,n}^N|] \to 0 \quad \text{for} \quad i = 1, 2, 3 \quad \text{as} \quad N \to \infty. \tag{1.23}$$

When the processes $\{X_n^{(N)}\}$ are Markov chains, then the conditioning in (1.21) and (1.22) refer to the state of $X_n^{(N)}$.

Define the continuous time processes

$$X^{(N)}(t) = X_{[t/h_N]}^{(N)},$$

where $[t/h_N]$ is the largest integer not exceeding t/h_N.

Under sufficient smoothness requirements we have the convergence

$$X^{(N)}(t) \text{ converges in distribution to } X(t) \quad \text{for each } t \tag{1.24}$$

and also for any set of time points $0 \leq t_1 < t_2 < \cdots < t_r$, the finite dimensional distributions of

$$\{X^{(N)}(t_1), X^{(N)}(t_2), \ldots, X^{(N)}(t_r)\} \quad \text{converge to those of}$$
$$\{X(t_1), X(t_2), \ldots, X(t_r)\} \quad \text{as} \quad N \to \infty \tag{1.25}$$

where $X(t)$ is a diffusion process whose infinitesimal drift and variance coefficients are the functions $\mu(x)$ and $\sigma^2(x)$ of (1.21).

The resemblance of (1.21) and (1.22) to the infinitesimal moment relations (1.2)–(1.4) is manifest. The criteria (1.21) and (1.22) tell us that if a sequence of processes closely obeys the moment conditions then there exists a diffusion that serves to approximate the processes $\{X_n^{(N)}\}$ with N large. The proofs of (1.24) and (1.25) are technical and will not be presented.

2: Examples of Diffusion

A. BROWNIAN MOTION

Brownian motion is a regular diffusion process on the interval $(-\infty, +\infty)$ with $\mu(x) = 0$ and $\sigma^2(x) = \sigma^2$, a constant, for all x. We may compute these infinitesimal parameters from the knowledge that $\Delta_h X = X(h) - X(0)$ is normally distributed with mean zero and variance $\sigma^2 h$, whence

$$E[\Delta_h X \,|\, X(0) = x] = 0 \quad \text{and} \quad E[(\Delta_h X)^2 \,|\, X(0) = x] = \sigma^2 h.$$

We will now verify that the Dynkin condition is satisfied in the stronger form such that (1.11) applies with $p = 4$. Precisely, we have

$$E[(\Delta_h X)^4 \,|\, X(0) = x] = \frac{1}{\sqrt{2\pi h}\,\sigma} \int_{-\infty}^{+\infty} y^4 \exp\left(-\frac{1}{2}\frac{y^2}{\sigma^2 h}\right) dy \qquad \left(\text{set } \frac{y^2}{\sigma^2 h} = 2\xi\right)$$

$$= \frac{4\sigma^4 h^2}{\sqrt{\pi}} \int_0^\infty \xi^{3/2} e^{-\xi}\, d\xi = 3h^2 \sigma^4$$

and (1.11) is confirmed.

Adding the trend μt to a Brownian motion $X(t)$ produces a *Brownian motion with drift* $X(t) + \mu t$. In this case, the drift parameter is μ, while the variance parameter remains σ^2.

B. ABSORBED AND REFLECTED BROWNIAN MOTION

Absorbed and reflected Brownian motion are regular diffusion processes defined on the common state space $I = [0, \infty)$. Starting from a point $X(0) = x$ in the interior of the interval, that is, $x > 0$, both processes act like Brownian motion until the level zero is first reached. Therefore, the infinitesimal parameters are $\mu(x) = 0$ and $\sigma^2(x) = \sigma^2$ for $0 < x < \infty$.

Here we have two diffusion processes with the same state space and the same infinitesimal parameters, which shows that these parameters do not uniquely define a diffusion process. The infinitesimal parameters govern the process evolution only while the process is in the interior of the state space I. To fully define a diffusion process, in addition one must adjoin boundary conditions to specify the behavior at any endpoints of I that the process may reach. Often physical considerations dictate the choice of the boundary conditions in a natural way. A more detailed discussion of absorbed and reflected Brownian motion was covered in Chapter 7 on Brownian motion (see Section 4).

C. ORNSTEIN–UHLENBECK PROCESS

This diffusion process has the entire real line $I = (-\infty, \infty)$ as its state space and $\mu(x) = -\alpha x$, $\sigma^2(x) = \sigma^2$, as its infinitesimal parameters, where α and σ are arbitrary positive constants. The infinitesimal drift parameter reflects a restoring force directed towards the origin and of a magnitude proportional to the distance.

If Brownian motion represents the position of a particle, the derivative of Brownian motion should represent the particle's velocity. But, as mentioned in Chapter 7, this derivative does not exist at any point in time. The Ornstein–

Uhlenbeck process is an alternative model which overcomes this defect by directly modeling the velocity of the particle as a function of time. Two factors are considered to affect this velocity during a small time duration. First, the frictional resistance of the surrounding medium is assumed to reduce the magnitude of the velocity by a proportional amount. Second, there is a change in velocity caused by the random collisions with neighboring particles. These two factors lead to the specifications $\mu(x) = -\alpha x$ and $\sigma^2(x) = \sigma^2$, respectively.

In Chapter 7, Section 1, we sketched how Brownian motion was approximated by means of a discrete random walk. The Ornstein–Uhlenbeck process may be approximated by the Ehrenfest urn model depicting diffusion of particles through a permeable membrane. (See Example B of Section 2, Chapter 2.) In this model, if there are i particles in urn A, the probability that there will be $i + 1$ after one time unit is $1 - (i/2N)$ and the probability of $i - 1$ is $i/2N$, where $2N$ is the aggregate number of particles in both urns. We might expect an appropriate limiting process, in which the time between transitions becomes small and the number of particles becomes large, to be a diffusion process, where the changes over an interval of time would vary continuously. Let Δt be the time between transitions and let $X_N(t)$ be the number of particles in urn A at time t. Then, with $X_N(t) = x$ and $\Delta X = X_N(t + \Delta t) - X_N(t)$, the probability law is

$$\Pr\{\Delta X = \pm 1 \,|\, X_N(t) = x\} = \frac{1}{2} \pm \frac{N - x}{2N}.$$

We let N increase and Δt decrease while maintaining $N\,\Delta t = 1$, and measure fluctuations of the rescaled process about its limiting mean value N in units of order $1/\sqrt{N}$. A transition occurs every $1/N$ time units. The full definition of the approximating process is

$$Y_N(\tau) = \frac{X_N([N\tau]) - N}{\sqrt{N}}.$$

A unit of time in the limiting process corresponds roughly to N transitions of the original process and a unit change in the limiting process corresponds roughly to a fluctuation of order $\sqrt{N}$ in the urn composition. Let $\Delta Y = \Delta Y_N(\tau, \Delta t) = Y_N(\tau + 1/N) - Y_N(\tau)$ determine the displacement over the time interval $\Delta t = 1/N$. Then

$$\Pr\left\{\Delta Y = \pm\frac{1}{\sqrt{N}}\,\bigg|\, Y_N(\tau) = y\right\} = \Pr\{\Delta X = \pm 1\,|\, X_N([N\tau]) = x = N + y\sqrt{N}\}$$

$$= \frac{1}{2} \pm \frac{N - (N + y\sqrt{N})}{2N}$$

$$= \frac{1}{2} \mp \frac{y}{2\sqrt{N}}.$$

We now compute for the urn process $\{Y_N(\tau)\}$ what are roughly equivalent to the infinitesimal parameters for the Ornstein–Uhlenbeck process. Accordingly, we have

$$E[\Delta Y | Y_N(0) = y] = \frac{1}{\sqrt{N}}\left(\frac{1}{2} - \frac{y}{2\sqrt{N}}\right) - \frac{1}{\sqrt{N}}\left(\frac{1}{2} + \frac{y}{2\sqrt{N}}\right) = -\frac{y}{N} = -y\,\Delta t,$$

so $\lim_{\Delta t \to 0} E[\Delta Y | Y_N(0) = y]/\Delta t = -y$ uniformly for y in bounded intervals. We now compute

$$\frac{1}{\Delta t} E[\{\Delta Y\}^2 | Y_N(0) = y] = \frac{1}{\Delta t}\left[\frac{1}{N}\left(\frac{1}{2} - \frac{y}{2\sqrt{N}}\right) + \frac{1}{N}\left(\frac{1}{2} + \frac{y}{2\sqrt{N}}\right)\right] = \frac{1}{N\,\Delta t} = 1.$$

A similar computation leads to the relation

$$\frac{1}{\Delta t} E[\{\Delta Y\}^4 | Y_N(0) = y] = \frac{1}{N} \to 0 \qquad \text{as} \quad N = \frac{1}{\Delta t} \to \infty$$

uniformly for the y variable restricted to bounded sets.

Taking account of the discussion of Section 1, the evidence is compelling that the processes $\{Y_N(\tau), \tau \ge 0\}$ will converge in an appropriate sense to the Ornstein–Uhlenbeck process, and this is indeed the case.† This suggests that the Ornstein–Uhlenbeck process can be used to calculate quantities that will be approximately correct for the Ehrenfest urn model when the number of particles is large.

If $\{V(t), t \ge 0\}$ is an Ornstein–Uhlenbeck process with parameters $\mu(x) = -\alpha x$ and $\sigma^2(x) = \sigma^2$, then conditioned on $V(t) = x$, the distribution of $V(t + s)$ is normal with mean

$$E[V(t + s) | V(t) = x] = xe^{-\alpha s}$$

and variance

$$\text{var}[V(t + s) | V(t) = x] = \left(\frac{1 - e^{-2\alpha s}}{2\alpha}\right)\sigma^2.$$

This will be derived in Section 5. For now let us check the consistency of this distribution with the desired infinitesimal parameters. With $h = \Delta t$ and $\Delta V = V(t + h) - V(t)$, we have

$$\lim_{h \downarrow 0} \frac{1}{h} E[\Delta V | V(t) = x] = \lim_{h \downarrow 0} \frac{1}{h}\{E[V(t + h) | V(t) = x] - x\}$$

$$= \lim_{h \downarrow 0} \frac{x}{h}(e^{-\alpha h} - 1)$$

$$= -\alpha x,$$

† We give no formal proof here.

and

$$\lim_{h\downarrow 0} \frac{1}{h} E[(\Delta V)^2 | V(t) = x] = \lim_{h\downarrow 0} \frac{1}{h} \{\text{var}[\Delta V | V(t) = x] + E[\Delta V | V(t) = x]^2\}$$

$$= \lim_{h\downarrow 0} \frac{1}{h} \left[\left(\frac{1 - e^{-2\alpha h}}{2\alpha} \right) \sigma^2 + \alpha^2 x^2 h^2 \right]$$

$$= \sigma^2.$$

(a) *Elementary Transformations of Processes*

Before constructing other important diffusion processes, we determine the form and infinitesimal parameters of new processes built from certain transformations applied to given ones. A continuous strictly increasing function g may be used to transform an arbitrary stochastic process $\{X(t)\}$ into a new process defined by $Y(t) = g(X(t))$. If $\{X(t)\}$ is a continuous path Markov process, i.e., a diffusion, then so is $\{Y(t)\}$ since g was assumed continuous and monotone, and hence, $\{Y(t)\}$ will be a Markov process having continuous paths. If, in addition, $\{X(t)\}$ has infinitesimal parameters $\mu(x)$ and $\sigma^2(x)$ given by (1.2) and (1.3) and g has two uniformly continuous derivatives g' and g'', then $Y(t)$ will also have infinitesimal parameters.

In Theorem 2.1 below we determine the infinitesimal parameters of $Y(t) = g[X(t)]$. A complete proof requires one to operate in terms of truncated moments, but this development is not worth the effort, ab initio and is not given here. We develop the pertinent formulas without full formal rigor.

Theorem 2.1. *Let $\{X(t), t \geq 0\}$ be a regular diffusion whose state space is an interval I having endpoints l and r, and suppose $\{X(t)\}$ has infinitesimal parameters $\mu(x)$ and $\sigma^2(x)$. Let g be a strictly monotone function on I with continuous second derivative $g''(x)$ for $l < x < r$. Then $Y(t) = g[X(t)]$ defines a regular diffusion process on the interval with endpoints $g(l)$ and $g(r)$, and $\{Y(t)\}$ has infinitesimal parameters*

$$\mu_Y(y) = \tfrac{1}{2}\sigma^2(x)g''(x) + \mu(x)g'(x)$$

$$\sigma_Y^2(y) = \sigma^2(x)[g'(x)]^2 \tag{*}$$

where $y = g(x)$.

Remark. Extensions and elaborations on Theorem 2.1 are covered in the later developments of Sections 14 and 16. These transformations of diffusions are subsumed in what is known as the Ito transformation formula. We present a direct discussion producing the infinitesimal parameters (*) which incorporates the main steps in their validation.

Proof. We consider only the case where g is strictly increasing. The strictly decreasing case is similar.

For g twice continuously differentiable, the Taylor expansion furnishes the representation

$$g(x + \Delta x) = g(x) + \Delta x\, g'(x) + \tfrac{1}{2}(\Delta x)^2 g''(x) + \tfrac{1}{2}(\Delta x)^2 [g''(\xi) - g''(x)]$$

with $x \le \xi \le x + \Delta x$. Substituting $X(t) = x$ and $\Delta X = X(t + h) - X(t)$, we have

$$
\begin{aligned}
g(X(t + h)) = g(X(t)) &+ \Delta X g'(X(t)) + \tfrac{1}{2}(\Delta X)^2 g''(X(t)) \\
&+ \tfrac{1}{2}(\Delta X)^2 [g''(\xi(\omega)) - g''(X(t))],
\end{aligned}
\tag{2.1}
$$

where $\xi(\omega)$ lies between $X(t)$ and $X(t + h)$ and ω signifies the particular realization of the process at hand and serves to emphasize that $\xi(\omega)$ is random. We can write (2.1) in the form

$$Y(t + h) - Y(t) = \Delta X\, g'(X(t)) + \tfrac{1}{2}(\Delta X)^2 g''(X(t)) + \tfrac{1}{2}(\Delta X)^2 [g''(\xi(\omega)) - g''(X(t))].$$
$$\tag{2.2}$$

We condition on $X(t) = x$ so that $g(x) = g(X(t)) = Y(t) = y$, and take expectations in (2.2). Then dividing by h leads to

$$\lim_{h \downarrow 0} \frac{1}{h} E[Y(t + h) - Y(t)\,|\, Y(t) = y] = \mu(x)g'(x) + \tfrac{1}{2}\sigma^2(x)g''(x)$$

$$+ \tfrac{1}{2} \lim_{h \downarrow 0} \frac{1}{h} E[(\Delta X)^2 \{g''(\xi(\omega)) - g''(X(t))\}].$$

The stipulated continuity of g'' ensures the convergence of $g''(\xi(\omega))$ to $g''(X(t))$, and this in conjunction with the convergence of $h^{-1}E[(\Delta X)^2]$ produces† the limit

$$\lim_{h \downarrow 0} \frac{1}{h} E[(\Delta X)^2 \{g''(\xi(\omega)) - g''(X(t))\}] = 0,$$

whence

$$\mu_Y(y) = \lim_{h \downarrow 0} \frac{1}{h} E[Y(t + h) - Y(t)\,|\, Y(t) = y]$$

$$= \mu(x)g'(x) + \tfrac{1}{2}\sigma^2(x)g''(x)$$

as stated.

The infinitesimal variance of $Y(t)$ is ascertained following a similar procedure. We square (2.2) to obtain

$$[Y(t + h) - Y(t)]^2 = (\Delta X)^2 [g'(X(t))]^2 + R_h \tag{2.3}$$

---

† A rigorous derivation requires consideration of the infinitesimal truncated moments mentioned in (1.2′) and (1.3′).

where the remainder R_h contains only $(\Delta X)^3$ and higher-order terms. We know by (1.4) that $\lim_{h \downarrow 0} h^{-1} E[|\Delta X|^r | X(t) = x] = 0$ for $r \geq 3$. A straightforward argument on (2.3) then establishes

$$\lim_{h \downarrow 0} \frac{1}{h} E[\{Y(t + h) - Y(t)\}^2 | Y(t) = y] = \sigma^2(x)[g'(x)]^2.$$

The condition (1.1) for $Y(t) = g[X(t)]$ is readily verified. Thus $Y(t)$ determines a diffusion with infinitesimal parameters as displayed in the statement of the theorem. ∎

D. GEOMETRIC BROWNIAN MOTION

We apply the transformation device of Theorem 2.1 to define geometric Brownian motion. Let $\{X(t), t \geq 0\}$ be a Brownian motion process with drift μ and diffusion σ^2. The process defined by $Y(t) = e^{X(t)}$ is sometimes called *geometric Brownian motion*. The state space is the interval $(0, \infty)$. If $t_0 < t_1 < \cdots < t_n$ are time points, the successive *ratios*

$$Y(t_1)/Y(t_0), \ldots, Y(t_n)/Y(t_{n-1})$$

are independent random variables, so that, roughly speaking, for geometric Brownian motion, the percentage changes over nonoverlapping time intervals are independent. With $y = g(x) = e^x$ we have $g'(x) = g''(x) = y$, and hence the infinitesimal parameters for geometric Brownian motion are

$$\mu_Y(y) = (\mu + \tfrac{1}{2}\sigma^2)y \qquad \text{and} \qquad \sigma_Y^2(y) = \sigma^2 y^2.$$

Geometric Brownian motion is often used to model prices of assets, say, shares of stock, that are traded in a perfect market. Prices are nonnegative and exhibit long-run exponential growth (or decay), two properties shared by geometric Brownian motion. More recently, geometric Brownian motion has featured in describing certain population growth processes.

E. THE BESSEL PROCESS

In Chapter 7, the Bessel process was defined as the Euclidean distance from the origin of an n-dimensional Brownian motion and shown to be a Markov process. We shall use Theorem 2.1 to determine the appropriate infinitesimal parameters. First let

$$Z(t) = X_1(t)^2 + \cdots + X_n(t)^2$$

where $\{X_i(t), t \geq 0\}$ are independent standard Brownian motion processes. We shall condition on $X_i(t) = x_i$, $i = 1, \ldots, n$, and write

$$X_i(t + \Delta t) = x_i + \Delta X_i \qquad \text{and} \qquad Z(t + \Delta t) = z + \Delta Z,$$

where

$$z = x_1^2 + \cdots + x_n^2.$$

Then

$$\Delta Z = [X_1(t + \Delta t)]^2 - x_1^2 + \cdots + [X_n(t + \Delta t)]^2 - x_n^2$$
$$= 2(x_1 \Delta X_1 + \cdots + x_n \Delta X_n) + [(\Delta X_1)^2 + \cdots + (\Delta X_n)^2].$$

Since $\Delta X_1, \ldots, \Delta X_n$ are independent and normally distributed with zero means and variances Δt, it follows that

$$E[\Delta Z | Z(t) = z] = n \Delta t$$

and

$$E[(\Delta Z)^2 | Z(t) = z] = 4(x_1^2 + \cdots + x_n^2) \Delta t + o(\Delta t)$$
$$= 4z \Delta t + o(\Delta t),$$

where $o(\Delta t)$ represents the expectations of terms of the form $(\Delta X_i)^4$ and of higher orders. We may use the known normal distribution of ΔX_i to conclude that these terms in total are of order less than Δt. Similarly, by a more tedious but straightforward computation (cf. Example A) we find

$$E[(\Delta Z)^4 | Z(t) = z] = O((\Delta t)^2)$$

and so condition (1.11) holds for the specification $p = 4$. We conclude therefore that $Z(t)$ is a diffusion with the infinitesimal parameters

$$\mu(z) = n \quad \text{and} \quad \sigma^2(z) = 4z.$$

The Bessel process is $Y(t) = g[Z(t)]$ for $g(z) = \sqrt{z}$. If $y = g(z)$, then

$$\mu(z) = n, \quad \sigma^2(z) = 4y^2,$$

$$g'(z) = 1/(2\sqrt{z}) = 1/(2y), \quad \text{and} \quad g''(z) = -1/(4z^{3/2}) = -1/(4y^3).$$

We apply Theorem 2.1 to obtain for the infinitesimal parameters of the Bessel process

$$\mu_Y(y) = (n - 1)/(2y) \quad \text{and} \quad \sigma_Y^2(y) = 1.$$

The transition density for the Bessel process was evaluated in Section 6 of Chapter 7 and is reproduced in Example 3 of Section 6 which follows.

For $n = 1$, the Bessel process can be identified with reflecting Brownian motion, for which $\mu_Y(y) = 0$, and $\sigma_Y^2(y) = 1$ as indicated earlier.

F. WRIGHT–FISHER (HAPLOID) GENETIC MODELS AND ASSOCIATED DIFFUSION APPROXIMATIONS†

Consider a population of constant size N individuals composed of two types A and a. Suppose the current state (the number of A-types) is i and therefore the other $N - i$ individuals are of a-type. The next generation is produced subject

† Consult also Chapter 2, pp. 55–58.

to the influence of mutation, selection, and sampling forces. We stipulate that mutation converts at birth an A-type to an a-type and an a-type to an A-type with probabilities α and β, respectively. Given the parental population comprised of i A-types and $N - i$ a-types, the expected fraction of A-types after mutation is $(i/N)(1 - \alpha) + (1 - i/N)\beta$ and of the a-types is $(i/N)\alpha + (1 - i/N)(1 - \beta)$. We next stipulate that the relative survival abilities of the two types A and a in contributing to the next generation are in the ratio of $1 + s$ to 1 where s is small and positive. Thus, type A is selectively superior to type a. Taking account of these mutation and selection forces, the expected fraction of mature A-types before reproduction is

$$p_i = \frac{(1 + s)[i(1 - \alpha) + (N - i)\beta]}{(1 + s)[i(1 - \alpha) + (N - i)\beta] + [i\alpha + (N - i)(1 - \beta)]} \tag{2.4}$$

The Wright–Fisher model postulates that the composition of the next generation is determined through N binomial trials, where the probability of producing an A-type offspring on each trial is p_i as given in (2.4). Thus the population process $\{X(t) =$ number of A-types in the tth generation$\}$ evolves as a Markov chain governed by the transition probability matrix

$$P_{ij} = \binom{N}{j} p_i^j (1 - p_i)^{N-j} \tag{2.5}$$

Note that the average change $p_i - i/N$ of the fraction of A-types at the end of a generation cycle expresses the mean changes accountable to selection and mutation as in a deterministic infinite population process. The statistical fluctuations due to the finite population size are reflected in the probabilistic transition matrix (2.5) and its inherent sampling characteristics. Even for this explicit Markov chain it is rarely possible to compute in analytic form relevant probabilistic functionals.

However, for N large the process can be approximated by a number of diffusion processes depending on the relative orders of magnitude of the parameters α, β, and s. We provide some basis for this statement in the remainder of this section. For instructional purposes we consider a series of cases all of independent interest.

(a) *Wright–Fisher Gene Frequency Diffusion Model Involving Only Mutation Effects*

$$\alpha = \frac{\gamma_1}{N}, \quad \beta = \frac{\gamma_2}{N} \quad \text{with} \quad \gamma_1 > 0, \quad \gamma_2 > 0, \quad s = 0.$$

(The intensity of mutation of $A \to a$ in the population, $N\alpha = \gamma_1$, is positive and finite, and similarly for $N\beta = \gamma_2$.) Consider the associated process

$$Y_N(\tau) = \frac{X([N\tau])}{N}. \tag{2.6}$$

One unit $\tau = 1$ of the $Y_N(\tau)$ process corresponds to the lapse of N generations in the $X(t)$ process and, of course, $Y_N(\tau)$ represents the fraction of the A-type in the population at generation time $[N\tau]$. The assumptions $\alpha = \gamma_1/N$ and $\beta = \gamma_2/N$ have a convenient intuitive interpretation: The rate of mutation is constant per unit of time of the scaled process, which is N generations in the X process.

We adopt the notation (with $h = 1/N$)

$$\Delta Y_N(\tau, h) = Y_N\left(\tau + \frac{1}{N}\right) - Y_N(\tau) = \frac{X([N\tau] + 1) - X([N\tau])}{N}$$

and compute

$$E\left[\Delta Y_N(\tau, h) | Y_N(\tau) = \xi = \frac{i}{N}\right] = E\left[\left\{Y_N\left(\tau + \frac{1}{N}\right) - Y_N(\tau)\right\} \Big| Y_N(\tau) = \frac{i}{N}\right].$$

(2.7)

Under the conditioning as indicated, $N Y_N(\tau + 1/N)$ is distributed binomially with parameters (p_i, N). Thus the expectation in (2.7) is

$$p_i - \frac{i}{N} = \left[\frac{i}{N}(1 - \alpha) + \frac{(N - i)}{N}\beta\right] - \frac{i}{N}$$

$$= -\alpha\frac{i}{N} + \left(1 - \frac{i}{N}\right)\beta = \frac{1}{N}\left[-\gamma_1\frac{i}{N} + \left(1 - \frac{i}{N}\right)\gamma_2\right].$$ (2.8)

For $h = 1/N$, after combining (2.7) and (2.8) and letting $i/N \to \xi$ as $N \to \infty$, we obtain

$$\lim_{h \downarrow 0+} \frac{1}{h} E[\Delta Y_N(\tau, h) | Y_N(\tau) = \xi] = -\gamma_1\xi + (1 - \xi)\gamma_2,$$ (2.9)

and the convergence holds uniformly for $0 \leq \xi \leq 1$.

Next we shall compute

$$\frac{1}{h} E[\{\Delta Y_N(\tau, h)\}^2 | Y_N(\tau) = \xi] = N E[\{\Delta Y_N(\tau, h)\}^2 | Y_N(\tau) = \xi].$$ (2.10)

Again using elementary moment formulas for the binomial distribution (2.10) is calculated to be equal to

$$N\left\{\frac{1}{N^2}[Np_i(1 - p_i) + N^2 p_i^2] - 2\frac{i}{N}p_i + \left(\frac{i}{N}\right)^2\right\}.$$

Inserting $\alpha = \gamma_1/N$, $\beta = \gamma_2/N$ and applying a little manipulation simplifies the above to

$$\frac{i}{N}\left(1 - \frac{i}{N}\right) + o\left(\frac{1}{N}\right)$$

where $O(1/N)$ is of order at most $1/N$. Thus, we obtain (with $h = 1/N$)

$$\lim_{N \to \infty} \frac{1}{h} E[\{\Delta Y_N(\tau, h)\}^2 | Y_N(\tau) = \xi] = \xi(1 - \xi) \qquad (2.11)$$

(the convergence taking place uniformly for $0 \le \xi \le 1$).

A more tedious but straightforward evaluation leads to the relation

$$\lim_{N \to \infty} \frac{1}{h} E[\{\Delta Y_N(\tau, h)\}^4 | Y_N(\tau) = \xi] = 0 \qquad (2.12)$$

and again the convergence is uniform with respect to $0 \le \xi \le 1$.

Indeed, rewriting (2.12) in terms of X yields

$$\frac{1}{N^3} E[\{X([N\tau] + 1) - X([N\tau])\}^4 | X([N\tau]) = N\xi]$$

$$= \frac{1}{N^3} (E[\{X([N\tau] + 1)\}^4 | X([N\tau]) = N\xi]$$

$$- 4N\xi E[\{X([N\tau] + 1)\}^3 | X([N\tau]) = N\xi]$$

$$+ 6(N\xi)^2 E[\{X([N\tau] + 1)\}^2 | X([N\tau]) = N\xi]$$

$$- 4(N\xi)^3 E[X([N\tau] + 1) | X([N\tau]) = N\xi]$$

$$+ (N\xi)^4)$$

(and abbreviating $q_\xi = N^{-1} E[X([N\tau] + 1) | X([N\tau]) = N\xi]$)

$$= \frac{1}{N^3} \{N^4 (q_\xi^4 - 4\xi q_\xi^3 + 6\xi^2 q_\xi^2 - 4\xi^3 q_\xi + \xi^4)$$

$$+ 6q_\xi(1 - q_\xi)N^3(q_\xi^2 - 2\xi q_\xi + \xi^2) + o(N^3)\}$$

$$= \frac{1}{N^3} [N^4 (q_\xi - \xi)^4 + 6N^3(q_\xi - \xi)^2 + o(N^3)].$$

Recalling from (2.8) and (2.9) that $q_\xi - \xi = O(1/N)$ we achieve the desired result.

Comparison of (2.9), (2.11), and (2.12) suggests strongly that the processes $Y_N(\tau) = X([N\tau])/N$ converge as $N \to \infty$ to a diffusion process $Y(\tau)$ whose state space is the unit interval $[0, 1]$ with drift coefficient $\mu(\xi) = -\gamma_1 \xi + (1 - \xi)\gamma_2$ and variance coefficient $\xi(1 - \xi)$. The above results can indeed be rigorously validated.

In the population genetics literature the process $Y(\tau)$ has been used to compute important functionals of $X(t)$ with quite good accuracy. Several of these calculations will be illustrated later.

(b) *Wright–Fisher Gene Frequency Diffusion Model Involving Selection*

Consider $\alpha = \beta = 0$ (no mutation) but with the presence of selection differences such that $s = \sigma/N$ with σ finite. Consider again the process $Y_N(\tau)$ as in (2.6). We make the same computations as in (2.7). These give

$$E\left[\Delta Y_N(\tau, h)| Y_N(\tau) = \frac{i}{N}\right] = p_i - \frac{i}{N} = \frac{(1 + s)i/N}{(1 + s)i/N + (1 - i/N)} - \frac{i}{N}$$

$$= \frac{(1 + \sigma/N)i/N}{(1 + \sigma/N)(i/N) + (1 - i/N)} - \frac{i}{N} \text{(inserting } s = \sigma/N)$$

$$= \frac{(1/N)\sigma(i/N)(1 - i/N)}{1 + (\sigma/N)i/N} \quad \text{(expanding the denominator)}$$

$$= \frac{1}{N}\sigma\xi(1 - \xi) + o\left(\frac{1}{N}\right) \qquad \left(\xi = \frac{i}{N}\right) \tag{2.13}$$

so that for $h = 1/N$ we derive on the basis of (2.13) the limit relation

$$\lim_{h\downarrow 0} NE[\Delta Y_N(\tau, h)| Y_N(\tau) = \xi] = \sigma\xi(1 - \xi). \tag{2.14}$$

In a similar manner we obtain

$$\lim_{h\downarrow 0} NE[\{\Delta Y_N(\tau, h)\}^2| Y_N(\tau) = \xi] = \xi(1 - \xi) \tag{2.15}$$

and also (2.12) holds.

In view of (2.14) and (2.15) we expect the associated limiting diffusion $Y(\tau)$ to have coefficients

$$\mu(\xi) = \sigma\xi(1 - \xi) \qquad \text{and} \qquad \sigma^2(\xi) = \xi(1 - \xi). \tag{2.16}$$

(c) *Two-Types Growth Model*

The scaling used to create the Y process from the X process in (a) converted the actual number of A-types into the frequency of the A-types in the population. The assumption was that the number of A-types was of order N and that fluctuations of order N were of primary interest. For a large population, the frequency of A-types might be relatively low ($o(N)$); hence scaling by N could conceal the fact that the number of A-types is large, for example of order $\sqrt{N}$. In the previous cases we dealt with a sequence of processes $Y_N(\tau)$ $(N \to \infty)$ tracing the fluctuations of the fraction of the A-type in the population. Where the A- and a-types are abundant, fluctuations in $Y_N(\tau)$ are visible provided observations are made approximately each N generations. In what follows, we consider a scaling by $\sqrt{N}$. Such a model is useful in determining whether introduction of a new type into a population on a lower order of magnitude than N (the population size) will result in extinction or spread of the new type.

Consider now the case where the number of A-types is of the order $z\sqrt{N}$ ($0 < z < \infty$) so that for N large the fraction of A-types is virtually zero. To study the fine stochastic structure of this situation we consider the sequence of processes

$$Z_N(\tau) = \frac{X([\sqrt{N}\,\tau])}{\sqrt{N}}, \quad Z_N(0) = z, \quad 0 < z < \infty. \tag{2.17}$$

This process focuses on the fluctuations of the number of A-types ranging in the order of magnitude $\sqrt{N}$. We have adjusted the time scale so that fluctuations in this process are pronounced provided observations are made at time epochs of approximately $\sqrt{N}$ generations. Consider the conditions on the mutation parameters as in Model (a), $\alpha = \gamma_1/N$ and $\beta = \gamma_2/N, s = 0$. We set $h = 1/\sqrt{N}$ and introduce

$$\Delta_N Z(\tau, h) = Z_N\left(\tau + \frac{1}{\sqrt{N}}\right) - Z_N(\tau) = \frac{X([\sqrt{N}\,\tau] + 1) - X([\sqrt{N}\,\tau])}{\sqrt{N}}.$$
$$\tag{2.18}$$

The computations of the moments of (2.18) run parallel to the preceding cases. With $i = [z\sqrt{N}]$, substitution for p_i yields

$$\sqrt{N}\,E[\Delta_N Z(\tau, h)|Z_N(\tau) = z] = \sqrt{N}(\sqrt{N}\,p_i - z)$$

$$= \sqrt{N}\left[\frac{\gamma_1}{\sqrt{N}}\left(1 - \frac{i}{N}\right) + \left(1 - \frac{i}{N}\right)\frac{\gamma_2}{\sqrt{N}} - z\right] \to \gamma_2$$

$$\text{as} \quad N \to \infty. \tag{2.19}$$

A similar evaluation of the second approximating infinitesimal moment leads to

$$\sqrt{N}\,E\left[\left\{\Delta_N Z\left(\tau, \frac{1}{\sqrt{N}}\right)\right\}^2 \Big| Z_N(0) = z\right] \to z \qquad \begin{array}{l}\text{as } N \to \infty \text{ uniformly in} \\ \text{any compact region of} \\ \text{the } z \text{ variable,}\end{array} \tag{2.20}$$

and finally

$$\sqrt{N}\,E\left[\left\{\Delta_N Z\left(\tau, \frac{1}{\sqrt{N}}\right)\right\}^4 \Big| Z_N(0) = z\right] \to 0 \qquad \begin{array}{l}\text{as } N \to \infty \text{ uniformly for} \\ \text{compact regions of the} \\ z \text{ variable}\end{array} \tag{2.21}$$

obtains. From interpretation of (2.19)–(2.21), it is apparent that the appropriate limiting diffusion $Z(t) = \lim Z_N(t)$ has state space $I = (0, \infty)$ with

$$\mu(z) \equiv \gamma_2 = \text{constant} \qquad \text{and} \qquad \sigma^2(z) = z. \tag{2.22}$$

This diffusion can be used to study the fine fluctuations of the number of A-types when they are of the order $\sqrt{N}$. The variable $Z(t)$ reaching ∞ means practically that the number of A-types has overgrown order $\sqrt{N}$, and $Z(t)$ shrinking to 0 is interpreted as meaning that the A-type has become more rare than order $\sqrt{N}$.

(d) *Different Orders of Mutation Rates*

Here $\alpha \sim \gamma_1/N^d$, $\beta \sim \gamma_2/N$, $0 < d < 1$ with $0 < \gamma_1, \gamma_2 < \infty$, $s = 0$. Notice that the mutation rate of A $\rightarrow$ a is of a substantially larger order of magnitude than the reverse mutation rate. In this case we consider the sequence of processes

$$W_N(\tau) = \frac{X([N^d\tau])}{N^d}. \tag{2.23}$$

Notice that the right time scale to reflect changes involves observations about N^d generations apart. The level of the A-type where fluctuations are distinguishable is that the number of A-types be of the order N^d. Straightforward calculations paraphrasing the analysis of the previous cases show that

$$N^d E\left[\frac{X([N^d\tau] + 1) - X([N^d\tau])}{N^d}\middle| X([N^d\tau]) = xN^d\right] \rightarrow \gamma_2 - \gamma_1 x,$$

$$N^d E\left[\left\{\frac{X([N^d\tau] + 1) - X([N^d\tau])}{N^d}\right\}^2\middle| X([N^d\tau]) = xN^d\right] \rightarrow x, \tag{2.24}$$

and the higher-order infinitesimal moments go to zero. This suggests that the appropriate limiting process $W(\tau)$ for $W_N(\tau)$ is a diffusion with

$$\mu(x) = \gamma_2 - \gamma_1 x \qquad \text{and} \qquad \sigma^2(x) = x. \tag{2.25}$$

This process is sometimes called a Laguerre diffusion since Laguerre orthogonal polynomials are involved. This diffusion features also in models of population growth and branching processes. Further analysis of this case is deferred to Section 13.

(e) *The Growth Process for Rare Types*

Here $\beta \sim \gamma_2/N$, α fixed but not small, $0 < \alpha < 1$, and $s = 0$.

No associated limiting diffusion is possible, but if the A-type is exceptionally rare, then the process of the actual number of A-types $X_N[\tau]$ with time scale one unit corresponding to one generation is described by a limiting process $\{U(\tau), \tau \geq 0\}$ composed of a branching process part plus an immigration process.

To see this, we consider the generating function of $X_N(\tau + 1)$ conditioned on $X_N(\tau) = i$:

$$\sum_{k=0}^{N} P_{ik}^{(N)} s^k = \sum_{k=0}^{N} \Pr\{X_N(\tau + 1) = k \mid X_N(\tau) = i\} s^k$$

$$= [p_i s + (1 - p_i)]^N = [1 + (s - 1)p_i]^N$$

$$= \left\{ 1 + (s - 1)\left[\frac{i}{N}(1 - \alpha) + \beta\left(1 - \frac{i}{N}\right) \right] \right\}^N$$

$$= \left[1 + \frac{(s - 1)i(1 - \alpha)}{N} + \frac{\gamma_2(s - 1)}{N}\left(1 - \frac{i}{N}\right) \right]^N, \qquad (2.26)$$

and as $N \to \infty$ the generating function clearly approaches [since $(1 + a/N)^N \to e^a$]

$$\sum_{k=0}^{\infty} \Pr\{U(\tau + 1) = k \mid U(\tau) = i\} s^k = (e^{(s-1)(1-\alpha)})^i e^{\gamma_2(s-1)}. \qquad (2.27)$$

Thus the process $U(\tau)$ behaves as an ordinary branching process whose offspring distribution is Poisson with parameter $1 - \alpha$. In addition, in each generation a number of A type individuals following a Poisson distribution with parameter γ_2 immigrates into the system.

(f) The Ornstein–Uhlenbeck Process Again

Here $0 < \alpha, \beta < 1$ are not small mutation rates and there are no selection effects; $s = 0$. As $N \to \infty$ the Wright–Fisher process (2.5) tends to a deterministic limit in the sense that the fraction of the A-type converges so that

$$U_N(t) = \frac{X(t)}{N} \to \frac{\beta}{\alpha + \beta} \qquad \text{with probability 1.}$$

It is possible to make a fine analysis of the fluctuation behavior of $X(t)$ about its mean $N\beta/(\alpha + \beta)$. To do this we introduce the sequence of random processes

$$V_N(t) = \frac{X_N(t) - N\beta/(\alpha + \beta)}{\sqrt{N\alpha\beta/(\alpha + \beta)^2}}.$$

It can be established that $V_N(t)$ for N large is approximated by an Ornstein–Uhlenbeck process having parameters $\mu(x) = -(\alpha + \beta)x$ and $\sigma^2(x) = 1$; cf. Example C. We shall not enter into details.

To sum up, we have highlighted six different processes associated with the Wright–Fisher model all relevant for different ranges of the parameter values. Uses of the approximating diffusions in calculating functionals of the process will be elaborated in Section 4.

(g)   A Model with Selection Parameters Varying Stochastically in Time (Extension of Case b)†

Fluctuations of selection intensities may be caused by random changes in environment or genetic background affecting both large and small populations.

Consider a population of N individuals reproducing in discrete generations manifesting two types A and a with fitness coefficients as indicated.

$$\text{fitnesses in generation } t: \quad 1 + \overset{\text{A}}{\sigma^{(t)}} \quad 1 + \overset{\text{a}}{\rho^{(t)}}.$$

The selection intensities $\{\sigma^{(t)}, \rho^{(t)}; t = 0, 1, 2, \ldots\}$ are assumed to fluctuate over time in a random and or systematic manner reflecting a changing ecological or genetic environment. Population size is held constant at N individuals per generation.

The fluctuations in the number of A-types (and a-types) over successive generations are generated in accordance with the standard Wright–Fisher Markov chain process. Specifically, the probability law governing the number of A types in generation t, given that the selection parameters are $(\sigma^{(t)}, \rho^{(t)})$ and the population consists of i A-types (and accordingly $N - i$ a-types) in generation $t - 1$, is that of a binomial distribution with parameters $(N, p_i^{(t)})$, where

$$p_i^{(t)} = (1 + \sigma^{(t)})i/[(1 + \sigma^{(t)})i + (1 + \rho^{(t)})(N - i)]. \tag{2.28}$$

Thus, the stochastic model is structured by superimposing a binomial sampling scheme where the average changes are determined by (2.28). Conditioned on an outcome of the environmental process, i.e., for each realization of the selection values, we construct a time inhomogeneous Markov chain on the state space $\{0, 1, \ldots, N\}$ with transition probability law from generation $t - 1$ to t given by

$$P_{i,j}^{(t-1,t)} = \binom{N}{j}(p_i^{(t)})^j(1 - p_i^{(t)})^{N-j},$$

where $p_i^{(t)}$ is determined in (2.28).

In essence, there are three sources of variation affecting the changes in the population numbers of the A-types over successive generations:

(i)   Sampling variance.   Statistical or sampling fluctuation stemming from small population size. Formally, its effects are expressed in the binomial sampling mechanism.

(ii)   Within generation selection variance.   The randomness of the selection intensities in a given generation is reflected in $(\sigma^{(t)}, \rho^{(t)})$, which is a random vector for generation t usually having a known or estimable distribution

† This case involves a rather detailed analysis and might be skipped on first reading.

function. Randomness here is tied to the ecological and genetic background of the immediate time epoch.

(iii) Between generation selection variance. The correlations and general dependence relationships of the selection process over time, that is, $\{\sigma^{(t)}, \rho^{(t)}; t \geq 0\}$, may constitute a complex stochastic process embodying a hierarchy of dependence associations. The simplest assumption would have $S_t = \{\sigma^{(t)}, \rho^{(t)}\}$, $t = 0, 1, 2, \ldots$, mutually independent, identically distributed random vectors.

Our objective is to evaluate more completely the relative effects resulting from the two different random factors: random sampling of gametes (induced by the small constant population size) and random fluctuation of selection intensities ascribable mainly to temporal environmental changes. When population size is finite and constant, then ultimate fixation of the A- or a-type necessarily occurs. We focus on two main problems. First the determination of the probability of fixation of the A-type (the so-called absorption probabilities) in terms of the known information on the selection process and the initial numbers of A-types; second, the calculation of the expected time to fixation. The main emphasis is to contrast qualitatively and quantitatively the findings on these problems for the constant selection environment with the case of a fluctuating pattern of selection intensities. We will again use diffusion approximations.

We make the following assumptions for $\sigma^{(t)}$ and $\rho^{(t)}$:

$$E[\sigma^{(t)}] = \frac{\gamma}{N} + o\left(\frac{1}{N}\right),$$

$$E[\rho^{(t)}] = \frac{\delta}{N} + o\left(\frac{1}{N}\right),$$

$$E[(\sigma^{(t)})^2] = \frac{v_1}{N} + o\left(\frac{1}{N}\right), \tag{2.29}$$

$$E[(\rho^{(t)})^2] = \frac{v_2}{N} + o\left(\frac{1}{N}\right),$$

$$E[\sigma^{(t)}\rho^{(t)}] = \frac{r}{N} + o\left(\frac{1}{N}\right),$$

and that all terms higher than second degree are of smaller order than $1/N$.

There are now two stochastic processes involved here describing the changes in A type frequencies: the $X(t)$ process, in which randomness is due to both the sampling procedure and the variable selection forces; and, the process $\hat{X}(t) = Np_i$, which describes the number of A-types as a result of the selection process alone (with no sampling). We shall first compute the drift and variance of the diffusion associated with the $\hat{X}(t)$ process.

Let $\hat{Y}_N(\tau) = \hat{X}([N\tau])/N = x$. Then

$$\hat{\mu}(x) = \lim_{N \to \infty} E\left[\frac{\hat{Y}_N(\tau + 1/N) - \hat{Y}_N(\tau)}{1/N}\middle| \hat{Y}_N(\tau) = x\right]$$

$$= \lim_{N \to \infty} E[\hat{X}([N\tau] + 1) - \hat{X}([N\tau])|\hat{X}([N\tau]) = Nx]$$

$$= \lim_{N \to \infty} E[N(p_x - x)],$$

where

$$p_x - x = \frac{x(1 + \sigma)}{x(1 + \sigma) + (1 - x)(1 + \rho)} - x = \frac{x(1 - x)(\sigma - \rho)}{1 + \rho + x(\sigma - \rho)}.$$

We expand $p_x - x$ in a power series and discard terms higher than second order, in line with our assumption on σ and ρ. This gives

$$\hat{\mu}(x) = \lim_{N \to \infty} Nx(1 - x)\left[\frac{\gamma - \delta}{N} + \frac{v_2}{N} - \frac{r}{N} - \frac{x}{N}(v_1 + v_2 - 2r) + o\left(\frac{1}{N}\right)\right]$$

$$= x(1 - x)[\gamma - \delta + v_2 - r - x(v_1 + v_2 - 2r)]$$

$$= x(1 - x)[\Gamma + V(\tfrac{1}{2} - x)] \qquad (2.30)$$

where $\Gamma = \gamma - \delta + \frac{1}{2}(v_2 - v_1)$ and $V = v_1 + v_2 - 2r$.

For the variance term,

$$\hat{\sigma}^2(x) = \lim_{N \to \infty} E\left[\frac{\{\hat{Y}_N(\tau + 1/N) - \hat{Y}_N(\tau)\}^2}{1/N}\middle| \hat{Y}_N(\tau) = x\right]$$

$$= \lim_{N \to \infty} NE[(p_x - x)^2|\hat{X}([N\tau]) = Nx]$$

$$= x^2(1 - x)^2 V.$$

Thus, we have determined that

$$\hat{\mu}(x) = x(1 - x)[\Gamma + V(\tfrac{1}{2} - x)], \qquad \hat{\sigma}^2(x) = x^2(1 - x)^2 V.$$

There is a simple relationship between the infinitesimal parameters of the processes induced by $\hat{X}(\tau)$ and $X(\tau)$. For the drift and variance, the random effects due to the sampling effects and the selection process are *additive*. In fact, the drift and variance due to sampling alone are 0 and $x(1 - x)$, respectively. We find for the $X(t)$ process,

$$\mu(x) = \hat{\mu}(x) + 0 = x(1 - x)[\Gamma + V(\tfrac{1}{2} - x)],$$
$$\sigma^2(x) = \hat{\sigma}^2(x) + x(1 - x) = x(1 - x)[1 + Vx(1 - x)]. \qquad (2.31)$$

In fact, a more general statement concerning the additivity of stochastic effects is true. Suppose we incorporate mutation into the last mentioned model. That is, each generation consists of mutation, followed by selection, and finally binomial sampling. The process $Z(t)$, which equals the number of A types, has a transition matrix with entries

$$P_{ij} = \binom{N}{j} p_i^j (1 - p_i)^{N-j}$$

where

$$p_i = \frac{[i(1 - \alpha) + \beta(N - i)](1 + \sigma)}{[i(1 - \alpha) + \beta(N - i)](1 + \sigma) + [N - i(1 - \alpha) - \beta(N - i)](1 + \rho)}.$$

The parameters α and β and the random variables $\sigma = \sigma^{(t)}$ and $\rho = \rho^{(t)}$ are defined as before. We construct the scaled process $W_N(\tau) = Z([N\tau])/N$. The drift and variance of the diffusion induced by $W_N(\tau)$ (as $N \to \infty$) are the sum of the drifts and variances, respectively, for each of the forces (mutation, fluctuating selection, sampling) acting on the population. Moreover, this result is independent of the order in which selection and mutation are performed. (The student should verify all this.)

The parameters Γ and V carry the following motivation. Suppose the frequency of A, equal to x, was low. Then the fraction of A in the next generation due to selection would be

$$x' = \frac{x(1 + \sigma)}{x(1 + \sigma) + (1 - x)(1 + \rho)} \approx x\left(\frac{1 + \sigma}{1 + \rho}\right)$$

($\sigma = \sigma^{(t)}$ and $\rho = \rho^{(t)}$ depending on the generation time).

After n generations, we would have

$$X_n \approx X_0\left(\prod_{k=1}^n \frac{1 + \sigma^{(k)}}{1 + \rho^{(k)}}\right).$$

To determine the asymptotic behavior of this quantity we compute the expectation of $\log\{\prod_{k=1}^N [(1 + \sigma^{(k)})/(1 + \rho^{(k)})]\}$, and take the limit as $N \to \infty$ to adjust the time scale with the order terms of (2.29). So,

$$E\left[\log \prod_{k=1}^N \left(\frac{1 + \sigma^{(k)}}{1 + \rho^{(k)}}\right)\right] = \sum_{k=1}^N E\left[\log\left(\frac{1 + \sigma^{(k)}}{1 + \rho^{(k)}}\right)\right].$$

Since the vectors $(\sigma^{(k)}, \rho^{(k)})$ are identically distributed,

$$\sum_{k=1}^N E\left[\log\left(\frac{1 + \sigma^{(k)}}{1 + \rho^{(k)}}\right)\right] = NE\left[\log\left(\frac{1 + \sigma}{1 + \rho}\right)\right]$$

$$= NE[\log(1 + \sigma) - \log(1 + \rho)]$$

$$= NE[\sigma - \tfrac{1}{2}\sigma^2 - \rho + \tfrac{1}{2}\rho^2 + \text{higher-order terms}]$$
(by expanding the log)

$$= N\left[\frac{\gamma}{N} - \frac{v_1}{2N} - \frac{\delta}{N} + \frac{v_2}{2N} + o\left(\frac{1}{N}\right)\right] \quad \text{by (2.29)}.$$

Hence,

$$\lim_{N\to\infty} E\left[\log\left\{\prod_{k=1}^{N}\left(\frac{1+\sigma^{(k)}}{1+\rho^{(k)}}\right)\right\}\right] = \gamma - \delta + \tfrac{1}{2}(v_2 - v_1) = \Gamma. \qquad (2.32)$$

Clearly, if $\Gamma > 0$, then the A-type increases when rare, and in the circumstance $\Gamma < 0$, it goes extinct.

If we compute the variance of $\log \prod_{k=1}^{N} [(1 + \sigma^{(k)})/(1 + \rho^{(k)})]$, we have

$$\text{var}\left[\log \prod_{k=1}^{N}\left(\frac{1+\sigma^{(k)}}{1+\rho^{(k)}}\right)\right] = \text{var}\left[\sum_{k=1}^{N} \log\left(\frac{1+\sigma^{(k)}}{1+\rho^{(k)}}\right)\right]$$

$$= N \text{ var}\left(\log \frac{1+\sigma^{(1)}}{1+\rho^{(1)}}\right).$$

Since $\{E[\log((1 + \sigma)/(1 + \rho))]\}^2 = o(1/N)$, we find

$$\text{var}\left[\log \prod_{k=1}^{N}\left(\frac{1+\sigma^{(k)}}{1+\rho^{(k)}}\right)\right] = NE\left[\left\{\log\left(\frac{1+\sigma^{(k)}}{1+\rho^{(k)}}\right)\right\}^2\right] + o(1)$$

$$= NE[(\sigma - \rho)^2] - o(1)$$

$$= v_1 + v_2 - 2r + o(1)$$

and therefore

$$\lim_{N\to\infty} \text{var}\left[\log \prod_{k=1}^{N}\left(\frac{1+\sigma^{(k)}}{1+\rho^{(k)}}\right)\right] = v_1 + v_2 - 2r = V. \qquad (2.33)$$

Thus, the quantities Γ and V play an important role in describing the behavior of the system when the A-type is rare.

G. A POPULATION MODEL OF MUTANT TYPES

The population maintains constant size of N individuals. The times between successive changes in the population are exponentially distributed with parameter λ. When a change occurs, an individual is chosen at random to die, and independently an individual is chosen at random to bear an offspring. Each newborn may mutate with probability μ and create a new type. An individual is called a p-mutant if in its chain of ancestors (including itself) p mutations have occurred.

Let $X(t)$ be the number of 0-mutants at time t and suppose the population composition at time t has $X(t) = k$. Consider the time duration $(t, t + h)$ with h small. The number $X(t)$ will increase by 1 if an event occurs entailing the death of a 1 or higher mutant type and the birth of a new 0-mutant which does not mutate. At the end of the time interval $(t, t + h)$ we have

$$X(t + h) = k + 1 \quad \text{with probability} \quad \lambda h \left(\frac{N - k}{N}\right) \frac{k}{N} (1 - \mu) + o(h).$$

$$(2.34)$$

In a similar manner we find that

$$X(t + h) = k - 1 \quad \text{with probability} \quad \lambda h \frac{k}{N} \left[\frac{k}{N} \mu + \frac{N - k}{N}\right] + o(h)$$

$$(2.35)$$

$$X(t + h) = k \quad \text{with probability}$$

$$1 - \lambda h + \lambda h \left[\frac{N - k}{N} \left(\frac{N - k}{N} + \frac{k}{N} \mu\right) + \frac{k}{N} \frac{k}{N} (1 - \mu)\right] + o(h).$$

Of course $X(t + h) = l$ $(l \neq k, k - 1, k + 1)$ occurs with probability $o(h)$. The expected time between successive events is $1/\lambda$. Thus the average duration of one generation (i.e., replacement of N individuals or equivalently the occurrence of N events) is N/λ. Consider $Y_N(t) = X(t)/N$, the proportion of 0-mutants at time t. We shall speed up the rate of events ($\lambda \to \infty$) and also let $N \to \infty$ in such a manner that $Y_N(t)$ will behave approximately as a diffusion process where one unit of time in the limit process corresponds to about N generations of the original process. For this objective we let $\lambda = N^2$.

Let

$$\Delta_h Y_N(t) = Y_N(t + h) - Y_N(t) = \frac{X(t + h) - X(t)}{N}.$$

Next we compute, using (2.34) and (2.35) with $Nx = k$ and $h = 1/N$, the mean expected change

$$\frac{1}{h} E[\Delta_h Y_N(t) | Y_N(t) = x] = \frac{\lambda}{hN} \left[\frac{N - k}{N} \frac{k}{N} (1 - \mu) h - \frac{k}{N} \left(\frac{k}{N} \mu + \frac{N - k}{N}\right) h\right] + \frac{o(h)}{h}.$$

$$(2.36)$$

Let $\mu = \theta/N$, with θ fixed, and recall that $\lambda = N^2$. The right side of (2.36) then reduces to $-\mu N[k/N] + o(1)$, which tends to $-\theta x$. A similar calculation leads to the relation

$$\lim_{h \downarrow 0} \frac{1}{h} E\left[\{\Delta_h Y_N(t)\}^2 | Y_N(t) = \frac{k}{N} = x\right] = 2x(1 - x) \qquad (2.37)$$

and we also obtain

$$\lim_{h \downarrow 0} \frac{1}{h} E[\{\Delta_h Y_N(t)\}^4 | Y_N(t) = x] = 0. \qquad (2.38)$$

The computations in this model are quite simple since the difference between $X(t + h)$ and $X(t)$ effectively can be only 0 or ± 1 for h sufficiently small.

A First Passage Process for 2-Mutant Types

We next examine the same process as above which is stopped, however, when the first 2-mutant is formed.

Let $Z(t)$ be the number of 1-mutants existing at time t and suppose that a 2-mutant type has as yet not appeared. The transition probabilities over the time epoch $(t, t + h)$ are readily evaluated by merely recalling the mechanisms for changes. We claim that

$$P_{k,k+1}(t, t + h) = \Pr\{Z(t + h) = k + 1 \,|\, Z(t) = k\}$$

$$= \lambda h \frac{N - k}{N} \left[\frac{N - k}{N} \mu + \frac{k}{N} (1 - \mu) \right] + o(h). \quad (2.39)$$

Indeed, for $Z(t) = k$ to pass to $Z(t + h) = k + 1$ requires a death of one of the 0-mutants (the present state involves $N - k$ such individuals) and then either replication of a new 1-mutant or a 0-mutant that mutates without the process ceasing, that is, without the formation of a 2-mutant. These contingencies lead to the formula (2.39). By similar considerations we obtain

$$P_{k,k-1}(t, t + h) = \lambda h \left[\frac{k}{N} \frac{N - k}{N} (1 - \mu) \right] + o(h),$$

$$P_{k,k}(t, t + h) = 1 - \lambda h + \lambda h \left[\left(\frac{N - k}{N} \right)^2 (1 - \mu) \right.$$

$$\left. + \frac{k}{N} \left\{ \frac{N - k}{N} \mu + \frac{k}{N} (1 - \mu) \right\} \right] + o(h),$$

$$P_{k,j}(t, t + h) = o(h), \quad j \neq k, \; j \neq k - 1, \; j \neq k + 1, \quad (2.40)$$

and finally

$$\Pr\{\text{a 2-mutant is created and thus the process stops in } (t, t + h) \,|\, X(t) = k\}$$

$$= \lambda h \frac{k}{N} \mu + o(h).$$

Let $\mu = \theta/N$. If $Z(t)$ is of the order N and events occur rapidly, then double mutants are created almost instantly. On the other hand, if $Z(t)$ is small and λ is small, the first creation time of a double mutant may involve a very long time. The correct scaling of events to get balance requires $\lambda = N^{3/2}$ and only if $Z(t)$ is of the order $\sqrt{N}$ is there a bona fide nontrivial probability distribution for the creation time of the first 2-mutant. With these normalizations $\lambda = N^{3/2}$, $\mu = \theta/N$, $Z(t) = x\sqrt{N}$, let

$$Z_N(t) = \frac{Z(t)}{\sqrt{N}}.$$

Then

$$\frac{1}{h} E[\Delta_k Z_N(t) | Z_N(t) = x, \text{ i.e., } Z(t) = x\sqrt{N}]$$

$$= \frac{1}{h} hN^{3/2} \frac{1}{\sqrt{N}} \left\{ \left(1 - \frac{x}{\sqrt{N}}\right)\left[\left(1 - \frac{x}{\sqrt{N}}\right)\frac{\theta}{N} + \frac{x}{\sqrt{N}}\left(1 - \frac{\theta}{N}\right)\right] \right.$$

$$\left. - \frac{x}{\sqrt{N}}\left(1 - \frac{x}{\sqrt{N}}\right)\left(1 - \frac{\theta}{N}\right) \right\} + o(1). \tag{2.41}$$

The limit as $N \to \infty$ is θ. An analogous calculation gives

$$\lim_{h \downarrow 0} \frac{1}{h} E[\{\Delta_h Z_N(t)\}^2 | Z_N(t) = x] = 2x.$$

The infinitesimal killing rate at $Z_N(t) = x = k/\sqrt{N}$ is

$$\lim_{h \downarrow 0} \frac{1}{h} \lambda h \frac{k}{N} \mu = \theta x.$$

Thus it is plausible that the approximating process can be identified with a diffusion involving killing (see Section 1) whose infinitesimal parameters are

$$\mu(x) = \theta, \qquad \sigma^2(x) = 2x, \qquad k(x) = \theta x.$$

We shall return to this class of examples in Section 10.

3: Differential Equations Associated with Certain Functionals

We emphasized in the introductory section the ease of ascertaining distributional properties of a broad spectrum of natural functionals defined on diffusion processes (especially in the one-dimensional case). The next two sections will amply illustrate these claims. Formal justifications of the methods rely on semigroup operator arguments (Sections 11 and 12). We at first discuss three basic problems and then provide, in Section 4, examples in which they arise. A number of important extensions of the ideas and methods are set forth at the close of this section.

We assume in this section, unless stated otherwise, that $\{X(t), t \geq 0\}$ is a time homogeneous diffusion process satisfying the following conditions:

1. The state space is an interval I of the form $[l, r]$, $(l, r]$, $[l, r)$, or (l, r), where $-\infty \leq l < r \leq \infty$.

2. The process is regular in the interior of I; i.e.,

$$\Pr\{T(y) < \infty | X(0) = x\} > 0, \qquad l < x, \quad y < r,$$

where $T(y)$ is the first time, if any, the process reaches the value y (the hitting time of y; cf. p. 162).

3. The process has infinitesimal parameters $\mu(x)$ and $\sigma^2(x)$, for $l < x < r$, where $\Delta X = X(h) - X(0)$ and

$$\mu(x) = \lim_{h \downarrow 0} \frac{1}{h} E[\Delta X \,|\, X(0) = x]$$

and

$$\sigma^2(x) = \lim_{h \downarrow 0} \frac{1}{h} E[(\Delta X)^2 \,|\, X(0) = x].$$

4. The infinitesimal parameters $\mu(x)$ and $\sigma^2(x)$ are continuous functions of x and $\sigma^2(x) > 0$ for $l < x < r$.

Let a and b be fixed, subject to $l < a < b < r$, and *let $T(y) = T_y$ be the hitting time of y.* Throughout this section we let

$$T^* = T_{a,b} = \min\{T(a), T(b)\} = T(a) \wedge T(b)$$

be the first time the process reaches either a or b.

This section concentrates on three problems.

Problem A.   Find

$$u(x) = \Pr\{T(b) < T(a) \,|\, X(0) = x\}, \qquad a < x < b,$$

the probability that the process reaches b before a.

Problem B.   Find

$$v(x) = E[T^* \,|\, X(0) = x], \qquad a < x < b,$$

the mean time to reach either a or b.

Problem C.   For a bounded and continuous function g, find

$$w(x) = E\left[\int_0^{T^*} g(X(s))\, ds \,\Big|\, X(0) = x\right], \qquad a < x < b.$$

Since the sample paths of the diffusion process are continuous, the integral $A = \int_0^{T^*} g(X(s))\, ds$ is defined. If $g(x)$ represents a cost rate incurred whenever the process is in state x, then A would be the total cost up to the time when either a or b was first reached. If $g(x) = 1$ for all x, then $A = T^*$, the time to reach a or b, so that Problem B is a special case of Problem C.

Under the stated conditions (1)–(4), it can be shown that $v(x)$ and $w(x)$ are finite, that $u(x)$, $v(x)$, and $w(x)$ possess two bounded derivatives for $a < x < b$, and that these functions satisfy the following differential equations:

Equation A

$$0 = \mu(x)\frac{du}{dx} + \frac{1}{2}\sigma^2(x)\frac{d^2u}{dx^2} \qquad \text{for} \quad a < x < b, \quad u(a) = 0, \quad u(b) = 1;$$

$$(3.1)$$

Equation B

$$-1 = \mu(x)\frac{dv}{dx} + \frac{1}{2}\sigma^2(x)\frac{d^2v}{dx^2} \qquad \text{for} \quad a < x < b, \qquad v(a) = v(b) = 0; \quad (3.2)$$

Equation C

$$-g(x) = \mu(x)\frac{dw}{dx} + \frac{1}{2}\sigma^2(x)\frac{d^2w}{dx^2} \qquad \text{for} \quad a < x < b, \quad w(a) = w(b) = 0. \quad (3.3)$$

The explicit solutions to these differential equations appear in (3.10) to (3.12).

A straightforward heuristic justification of the above differential equations is now described. Consider Problem A. The boundary conditions $u(a) = 0$, $u(b) = 1$ are obvious, since $u(x)$ is the probability of reaching b before a, starting from x. Now consider $a < x < b$, and choose a time duration h sufficiently small that the probability of reaching a or b before time h is negligible. At time h, conditioning on the position of $X(h)$, the probability of reaching b before a is $u(X(h))$. Invoking the law of total probabilities gives

$$u(x) = E[u(X(h))|X(0) = x] + o(h)$$

where the error term $o(h)$ is of smaller order than h. We now write $\Delta X = X(h) - x$ and assume we can expand in a Taylor series

$$u(X(h)) = u(x + \Delta X)$$
$$= u(x) + \Delta X\, u'(x) + \tfrac{1}{2}(\Delta X)^2 u''(x) + \cdots,$$

for which the fourth and further terms are of order smaller than $(\Delta X)^2$ and may be neglected. Since $E[\Delta X|X(0) = x] = \mu(x)h + o(h)$ and

$$E[(\Delta X)^2|X(0) = x] = \sigma^2(x)h + o(h),$$

we have

$$u(x) = E[u(X(h))|X(0) = x] + o(h)$$
$$= E[u(x + \Delta X)|X(0) = x] + o(h)$$
$$= u(x) + E[\Delta X|X(0) = x]u'(x) + \tfrac{1}{2}E[(\Delta X)^2|X(0) = x]u''(x) + o(h)$$
$$= u(x) + \mu(x)hu'(x) + \tfrac{1}{2}\sigma^2(x)hu''(x) + o(h).$$

We may subtract $u(x)$ from both sides, divide by h, and let h decrease to zero to conclude

$$0 = \mu(x)u'(x) + \tfrac{1}{2}\sigma^2(x)u''(x), \qquad a < x < b,$$

which is the desired differential equation for Problem A.

We shall motivate the derivation of (3.3) in the same vein. Again, the boundary conditions should be clear. Choose a short time duration h as before. At time h, conditioning on $X(h)$, the expectation of the total integral is the expectation of the contribution up to time h, $\int_0^h g(X(s))\,ds$, plus the expectation of the contribution over the remaining time. Conditioned on $X(h) = z$, the conditional mean of the second part is

$$E\!\left[\int_h^{T^*} g(X(\tau))\,d\tau \,\middle|\, X(h) = z\right] = E\!\left[\int_0^{T^*} g(X(\tau))\,d\tau \,\middle|\, X(0) = z\right]$$

$$\text{(by the Markov property and stationarity)}$$

$$= w(z) \qquad \text{(by the definition of } w).$$

Thus, for $a < x < b$, we have

$$w(x) = E\!\left[\int_0^h g(X(s))\,ds \,\middle|\, X(0) = x\right] + E[w(X(h))\,|\,X(0) = x]. \qquad (3.4)$$

Since the sample paths and g are continuous we have the approximation

$$E\!\left[\int_0^h g(X(s))\,ds \,\middle|\, X(0) = x\right] = g(x)h + o(h),$$

and, as before

$$E[w(X(h))\,|\,X(0) = x] = E[w(x + \Delta X)\,|\,X(0) = x]$$

$$= w(x) + \mu(x)w'(x)h + \tfrac{1}{2}\sigma^2(x)w''(x)h + o(h),$$

so that (3.4) becomes

$$w(x) = g(x)h + w(x) + \mu(x)w'(x)h + \tfrac{1}{2}\sigma^2(x)w''(x)h + o(h).$$

Then, subtracting $w(x)$ from both sides, dividing by h and sending h to zero will produce the differential equation for Problem C.

We turn to the solutions of the three problems. After recalling the assumption that $\sigma^2(x) > 0$ for $l < x < r$, let

$$s(x) = \exp\left\{-\int^x [2\mu(\xi)/\sigma^2(\xi)]\,d\xi\right\} \qquad \text{for } l < x < r. \qquad (3.5)$$

We use indefinite integrals here and in what follows for reasons that will later become clear. Next we introduce the fundamentally important *scale function* of the process

$$S(x) = \int^x s(\eta)\,d\eta = \int^x \exp\left\{-\int^\eta [2\mu(\xi)/\sigma^2(\xi)]\,d\xi\right\}d\eta \qquad \text{for } l < x < r \quad (3.6)$$

and *speed density*

$$m(x) = 1/[\sigma^2(x)s(x)] \qquad \text{for} \quad l < x < r. \tag{3.7}$$

Equations A–C each involve the differential operator L defined by

$$Lf(x) = \mu(x)f'(x) + \tfrac{1}{2}\sigma^2(x)f''(x),$$

for $f(x)$ a twice continuously differentiable function on (a, b). A modern approach to their solution is to express this operator as successive differentiations with respect to the scale and speed measure. To this end, first verify that $s'(x)/s(x) = -2\mu(x)/\sigma^2(x)$. Then, following a classical approach, introduce $1/s(x)$ as an integrating factor and thereby separate the variables, achieving

$$Lf(x) = \frac{1}{2}\left(\frac{1}{1/[\sigma^2(x)s(x)]}\right)\frac{d}{dx}\left[\frac{1}{s(x)}\frac{df(x)}{dx}\right]. \tag{3.8}$$

To obtain a more succinct and meaningful expression for L, write $s(x) = dS(x)/dx$ in the differential form $dS = s(x)\,dx$ and similarly write the speed density as a differential of a *speed measure* M in the form $dM = m(x)\,dx$. In terms of these differentials, the operator L in (3.8) is simply

$$Lf(x) = \frac{1}{2}\frac{d}{dM}\left[\frac{df(x)}{dS}\right]. \tag{3.9}$$

The above expression is called the *canonical representation of the differential infinitesimal operator* associated with the diffusion process. In terms of this canonical representation, the differential equation for **Problem A** becomes

$$\frac{1}{2}\frac{d}{dM}\left[\frac{du(x)}{dS}\right] = 0 \qquad \text{for} \quad a < x < b, \quad u(a) = 0, \quad u(b) = 1.$$

The solution follows directly from two successive integrations.

Denoting the constants of integration by α and β, we integrate once to obtain $du(x)/dS(x) = \beta$, and then again to get $u(x) = \alpha + \beta S(x)$, $a \le x \le b$. The boundary condition $u(a) = 0$ determines $\alpha = -\beta S(a)$, and then $u(b) = 1$ yields $\beta = 1/[S(b) - S(a)]$. We summarize the foregoing analysis in an explicit form.

Solution A

$$u(x) = \frac{S(x) - S(a)}{S(b) - S(a)} \qquad \text{for} \quad a \le x \le b. \tag{3.10}$$

Note that $u(x)$ is unchanged if $S(x)$ is replaced by $S^*(x) = \alpha + \beta S(x)$ for any constants α and $\beta \ne 0$. Thus if $S(x)$ is a scale function, so is $S^*(x)$. It follows that we may specify the scale function $S(x)$ using indefinite integrals since the resulting $u(x)$ does not depend on the lower limits of integration [Section (3.6)].

Remark 3.1. The scale function can be used to rescale the state space (l, r) in terms of the probabilities of achieving various levels, and this use motivates the name. Fix a point x_0 as the origin and determine the scale function by performing a translation, if necessary, causing $S(x_0) = 0$. Then form the process $Y(t) = S(X(t))$ on the interval $(S(l), S(r))$. Since S is strictly monotone and twice continuously differentiable, an appeal to Theorem 2.1 establishes that the infinitesimal parameters of the $\{Y(t)\}$ process are

$$\mu_Y(y) = \tfrac{1}{2}\sigma^2(x)S''(x) + \mu(x)S'(x) = 0,$$

and

$$\sigma_Y^2(y) = \sigma^2(x)[S'(x)]^2 = \sigma^2(x)s^2(x), \qquad \text{where} \quad y = S(x).$$

The scale measure for the $\{Y(t)\}$ process is accordingly $S_Y(y) = y$, or what is equivalent, $S_Y(y) = \alpha + \beta y$, with $\alpha, \beta \neq 0$ constants. A process $\{Y(t)\}$ whose scale function is linear is said to be in *natural* or *canonical scale* since the hitting probabilities

$$\Pr\{T_a(Y) < T_b(Y) \mid Y(0) = y\} = (b - y)/(b - a) \qquad \text{for} \quad a < y < b,$$

are manifestly proportional to actual distances.

We proceed to Problem C since Problem B is a special case. In the canonical representation, the differential equation is written

$$\frac{1}{2}\frac{d}{dM}\left[\frac{dw(x)}{dS(x)}\right] = -g(x) \qquad \text{for} \quad a < x < b,$$

subject to the boundary conditions

$$w(a) = w(b) = 0.$$

Upon the first integration, we obtain

$$\frac{dw(\eta)}{dS} = -2\int_a^\eta g(\xi)\, dM(\xi) + \beta$$

$$= -2\int_a^\eta g(\xi)m(\xi)\, d\xi + \beta,$$

and after the second,

$$w(x) = -2\int_a^x \left[\int_a^\eta g(\xi)m(\xi)\, d\xi\right] dS(\eta) + \beta[S(x) - S(a)] + \alpha.$$

Then $w(a) = 0$ implies $\alpha = 0$, and $w(b) = 0$ requires

$$\beta = \frac{2}{S(b) - S(a)} \int_a^b \left[\int_a^\eta g(\xi)m(\xi)\, d\xi\right] dS(\eta).$$

Using $u(x) = [S(x) - S(a)]/[S(b) - S(a)]$, we obtain

$$w(x) = 2\left\{u(x)\int_a^b\left[\int_a^\eta g(\xi)m(\xi)\,d\xi\right]dS(\eta) - \int_a^x\left[\int_a^\eta g(\xi)m(\xi)\,d\xi\right]dS(\eta)\right\}.$$

Changing the order of integration followed by elementary algebraic manipulations reduces this to a symmetric form.

Solution C

$$w(x) = 2\left\{u(x)\int_x^b [S(b) - S(\xi)]m(\xi)g(\xi)\,d\xi \right.$$

$$\left. + [1 - u(x)]\int_a^x [S(\xi) - S(a)]m(\xi)g(\xi)\,d\xi\right\}. \tag{3.11}$$

As mentioned earlier, the special case $g(\xi) \equiv 1$ yields the solution to Problem B.

Solution B

$$v(x) = 2\left\{u(x)\int_x^b [S(b) - S(\xi)]m(\xi)\,d\xi \right.$$

$$\left. + [1 - u(x)]\int_a^x [S(\xi) - S(a)]m(\xi)\,d\xi\right\}. \tag{3.12}$$

Remark 3.2. Consider for a moment a process in natural scale wherein $S(x) = x$ and $s(x) = S'(x) = 1$. This can be achieved by a suitable transformation of the state space; see Remark 3.1. Calculate the mean time to exit the interval $(x - \varepsilon, x + \varepsilon)$ starting at x. Inserting $a = x - \varepsilon$, $b = x + \varepsilon$, $u(x) = \frac{1}{2}$, and $S(x) = x$ in (3.12), we obtain

$$E[T_{x-\varepsilon, x+\varepsilon}| X(0) = x] = \int_x^{x+\varepsilon} (x + \varepsilon - \xi)m(\xi)\,d\xi + \int_{x-\varepsilon}^x (\xi - x + \varepsilon)m(\xi)\,d\xi,$$

whence

$$\lim_{\varepsilon\downarrow 0}\frac{1}{\varepsilon^2} E[T_{x-\varepsilon, x+\varepsilon}| X(0) = x] = \lim_{\varepsilon\downarrow 0}\frac{1}{\varepsilon^2}\int_x^{x+\varepsilon}(x + \varepsilon - \xi)m(\xi)\,d\xi$$

$$+ \lim_{\varepsilon\downarrow 0}\frac{1}{\varepsilon^2}\int_{x-\varepsilon}^x (\xi - x + \varepsilon)m(\xi)\,d\xi$$

$$= m(x).$$

Thus the speed density $m(x)$ can be construed as the speed at which the clock of the process runs when located at the state point x. Equivalently, if we regard the clock of a standard Brownian motion as standard, and provided the process is in natural scale, the quantity $m(x)\varepsilon^2$ is of the order of the expected time the process spends in the interval $(x - \varepsilon, x + \varepsilon)$ given $X(0) = x$ before departure thereof.

Remark 3.3. It is of interest to calculate the expected amount of time prior to T^* that the process spends in an interval $[\xi, \xi + \Delta)$. Shrinking Δ to $d\xi$ gives the *expected local time* at ξ. To make these ideas more precise, write the solution to Problem C,

$$w(x) = E\left[\int_0^{T^*} g(X(s))\,ds \,|\, X(0) = x\right], \qquad a \le x \le b, \tag{3.13}$$

in the form

$$w(x) = \int_a^b G(x, \xi) g(\xi)\,d\xi, \tag{3.14}$$

where

$$G(x, \xi) = \begin{cases} 2\dfrac{[S(x) - S(a)][S(b) - S(\xi)]}{S(b) - S(a)} \dfrac{1}{\sigma^2(\xi)s(\xi)}, & a \le x \le \xi \le b \\[3mm] 2\dfrac{[S(b) - S(x)][S(\xi) - S(a)]}{S(b) - S(a)} \dfrac{1}{\sigma^2(\xi)s(\xi)}, & a \le \xi \le x \le b. \end{cases} \tag{3.15}$$

The function $G(x, \xi)$ is called the *Green function* of the process on the interval $[a, b]$. Determining the mean time prior to T^* that the process spends in the interval $[\xi, \xi + \Delta)$ is equivalent to evaluating

$$w(x) = E\left[\int_0^{T^*} g(X(s))\,ds \,|\, X(0) = x\right]$$

for

$$g(x) = \begin{cases} 1, & \xi \le x < \xi + \Delta, \\ 0 & \text{otherwise.} \end{cases} \tag{3.16}$$

Following the format of (3.14), this is

$$w(x) = \int_\xi^{\xi + \Delta} G(x, \eta)\,d\eta. \tag{3.17}$$

While the function $g(x)$ as defined in (3.16) does not satisfy the continuity assumption of Problem C, nevertheless, (3.17) can be established by introducing suitable approximating continuous functions.

Shrinking Δ to $d\xi$, we see from (3.17) that $G(x, \xi)\,d\xi$ measures the mean time prior to T^* that the process spends in the infinitesimal interval $[\xi, \xi + d\xi)$ given $X(0) = x$.

DIGRESSION: THE GREEN FUNCTION OF A SECOND-ORDER DIFFERENTIAL OPERATOR†

In terms of the differential operator $L = \mu(x)\,d/dx + \frac{1}{2}\sigma^2(x)\,d^2/dx^2$, the solution to the second-order differential equation

$$Lw(x) = -g(x), \quad a < x < b, \qquad \text{with} \quad w(a) = w(b) = 0$$

is given by

$$w(x) = \int_a^b G(x, \xi)g(\xi)\,d\xi,$$

so that $-G$, considered as an integral operator, is the inverse to the differential operator L.

We now review some material relating to the more general second-order differential equation

$$Ly \equiv p(x)y'' + q(x)y' + r(x)y = -f(x) \qquad \text{for} \quad a < x < b, \qquad (3.18)$$

with $y(x)$ obeying associated boundary conditions and $p(x) > 0$ on $[a, b]$. We assume the coefficients p, q, and r are as smooth as required.

To be specific, we concentrate on the boundary conditions

$$y(a) = y(b) = 0. \qquad (3.19)$$

(Other choices of boundary conditions can be handled by similar means). If the homogeneous equation $Ly = 0$ admits no nonzero solution fulfilling the boundary conditions, then (3.18) can be inverted in the form of an integral operator

$$y(x) = \int_a^b G(x, \xi)f(\xi)\,d\xi, \qquad (3.20)$$

where $G(x, \xi)$ is commonly referred to as the Green function of the boundary value problem. The following properties characterize $G(x, \xi)$:

(i)　For each ξ in (a, b) the function $y(x) = G(x, \xi)$ solves the equation $Ly = 0$ for x in each interval (a, ξ) and (ξ, b), so that $LG(x, \xi) = 0$.

(ii)　For each ξ in (a, b) the function $y(x) = G(x, \xi)$ satisfies the boundary conditions (3.19).

(iii)　The derivative of G exhibits a suitable jump discontinuity

$$\frac{\partial G}{\partial x}(\xi+, \xi) - \frac{\partial G}{\partial x}(\xi-, \xi) = \frac{1}{p(\xi)}.$$

The Green function $G(x, \xi)$ is obtained as follows. Let $y_1(x)$ $[y_2(x)]$ be a solution of $Ly = 0$ satisfying the initial conditions $y_1(a) = 0$ and $y_1'(a) > 0$ $[y_2(b) = 0, y_2'(b) < 0]$. More generally, y_1 is determined up to a constant factor

---

† This digression is not a prerequisite to what follows.

as a solution of $Lu = 0$ satisfying the left boundary condition, while analogously the solution y_2 obeys the right-hand boundary condition. These exist and are easily determined in most examples. The functions $y_1(x)$ and $y_2(x)$ are linearly independent since we assumed that there was no nonzero solution of $Ly = 0$ obeying both boundary conditions. Let

$$W(\xi) = W(y_1, y_2) = y_1(\xi)y_2'(\xi) - y_2(\xi)y_1'(\xi) \tag{3.21}$$

be the Wronskian of the functions. The following argument shows that $W(\xi)$ is nonzero for $a \le \xi \le b$. Differentiate W and use the fact that y_i satisfies $Ly_i = 0$, to produce

$$\frac{d}{d\xi} W(y_1(\xi), y_2(\xi)) = y_1 y_2'' - y_2 y_1'' = \frac{-q}{p}(y_1 y_2' - y_2 y_1') = \frac{-q}{p} W. \tag{3.22}$$

Solving this differential equation, we infer immediately that

$$W(\xi) = W(\xi_0) \exp\left[\int_{\xi_0}^{\xi} \frac{-q(\eta)}{p(\eta)} \, d\eta\right];$$

so if W vanishes at some point, then W is identically zero. Consequently

$$\left(\frac{y_1}{y_2}\right)' = \frac{W}{(y_2)^2} \equiv 0$$

wherever y_2 does not vanish, which means that y_1 is a multiple of y_2, and they are not linearly independent contrary to a previous assumption. Thus $W(\xi) \ne 0$ for $a \le \xi \le b$.

Now form

$$G(x, \xi) = \begin{cases} -\dfrac{y_1(x)y_2(\xi)}{W(\xi)p(\xi)} & \text{for} \quad a \le x \le \xi \le b, \\[3mm] -\dfrac{y_1(\xi)y_2(x)}{W(\xi)p(\xi)} & \text{for} \quad a \le \xi \le x \le b, \end{cases} \tag{3.23}$$

which is well defined since $W(\xi)p(\xi) \ne 0$ for all $a \le \xi \le b$. For $f(x)$ continuous, we claim that the function

$$w(x) = \int_a^b G(x, \xi)f(\xi) \, d\xi = -y_1(x) \int_x^b \frac{y_2(\xi)f(\xi)}{W(\xi)p(\xi)} \, d\xi - y_2(x) \int_a^x \frac{y_1(\xi)f(\xi)}{W(\xi)p(\xi)} \, d\xi \tag{3.24}$$

satisfies the boundary conditions (3.19) and solves $Lw = -f$. That w satisfies (3.19) is verified using the boundary values of y_1 and y_2. Next, we evaluate Lw. Notice that

$$w'(x) = -y_1'(x) \int_x^b \frac{y_2(\xi)f(\xi)}{W(\xi)p(\xi)} \, d\xi - y_2'(x) \int_a^x \frac{y_1(\xi)f(\xi)}{W(\xi)p(\xi)} \, d\xi. \tag{3.25}$$

Next we calculate

$$w''(x) = -y_1''(x) \int_x^b \frac{y_2(\xi)f(\xi)}{W(\xi)p(\xi)} d\xi - y_2''(x) \int_a^x \frac{y_1(\xi)f(\xi)}{W(\xi)p(\xi)} d\xi$$
$$+ \frac{y_1'(x)y_2(x)f(x)}{W(x)p(x)} - \frac{y_2'(x)y_1(x)f(x)}{W(x)p(x)}$$
$$= -y_1''(x) \int_x^b \frac{y_2(\xi)f(\xi)}{W(\xi)p(\xi)} d\xi - y_2''(x) \int_a^x \frac{y_1(\xi)f(\xi)}{W(\xi)p(\xi)} d\xi - \frac{f(x)}{p(x)}. \qquad (3.26)$$

Combining Eqs. (3.24)–(3.26) in the obvious manner, we obtain

$$Lw(x) = pw'' + qw' + rw = -f(x).$$

Returning to the diffusion process operator for which $p(x) = \frac{1}{2}\sigma^2(x)$, $q(x) = \mu(x)$ and $r(x) \equiv 0$, identify

$$Ly = \frac{1}{2}\sigma^2(x)y'' + \mu(x)y'. \qquad (3.27)$$

Take an interval (a, b) properly contained in (l, r) and construct the Green function for the differential operator (3.27) with boundary conditions (3.19) according to the recipe of (3.23). Explicitly take

$$y_1(x) = \frac{[S(x) - S(a)]}{[S(b) - S(a)]}$$

where $S(x)$ is the scale function defined in (3.6). Clearly $y_1(x)$ vanishes at a and $y_1'(a) > 0$. Take $y_2(x) = [S(b) - S(x)]/[S(b) - S(a)]$ exhibiting the properties $y_2(b) = 0$, $y_2'(b) < 0$. The Wronskian of y_1 and y_2 reduces to

$$W(\xi) = \left(\frac{S(\xi) - S(a)}{S(b) - S(a)}\right)\left(\frac{-s(\xi)}{S(b) - S(a)}\right) - \left(\frac{S(b) - S(\xi)}{S(b) - S(a)}\right)\left(\frac{s(\xi)}{S(b) - S(a)}\right)$$
$$= -\frac{s(\xi)}{S(b) - S(a)}$$

with

$$\frac{dS(\xi)}{d\xi} = s(\xi) = \exp\left[-\int^\xi \frac{2\mu(\eta)}{\sigma^2(\eta)} d\eta\right]. \qquad (3.28)$$

Thus for the case at hand

$$G(x, \xi) = \begin{cases} \dfrac{2[S(x) - S(a)]}{S(b) - S(a)} [S(b) - S(\xi)] \dfrac{1}{\sigma^2(\xi)s(\xi)}, & a \le x \le \xi \le b, \\[4mm] \dfrac{2[S(b) - S(x)]}{S(b) - S(a)} [S(\xi) - S(a)] \dfrac{1}{\sigma^2(\xi)s(\xi)}, & a \le \xi \le x \le b. \end{cases}$$
$$(3.29)$$

Frequently, in the literature the speed measure density $1/\sigma^2(\xi)s(\xi)$ is separated out and the Green function is identified as

$$G^*(x, \xi) = \begin{cases} \dfrac{2[S(x) - S(a)]}{S(b) - S(a)} [S(b) - S(\xi)], & a \leq x \leq \xi \leq b, \\ \dfrac{2[S(b) - S(x)]}{S(b) - S(a)} [S(\xi) - S(a)], & a \leq \xi \leq x \leq b. \end{cases} \tag{3.30}$$

SOME FURTHER FUNCTIONALS

Extending the analysis of Problem C, we consider a hierarchy of functionals of the form

$$U(x) = E\left[f\left(\int_0^T g(X(\tau))\, d\tau \right) \middle| X(0) = x \right], \qquad a < x < b,$$

$$= E_x\left[f\left(\int_0^T g(X(\tau))\, d\tau \right) \right], \tag{3.31}$$

where $T = T(a) \wedge T(b)$ is the hitting time to $\{a, b\}$, f is twice continuously differentiable, and g is piecewise smooth. Note that we introduce the subscript notation on E to indicate the initial point $X(0) = x$ of the process realization.

For $f(x) \equiv x$, we recover the functional of Problem C. The evaluation of (3.31) for the choice $f(x) = x^n$ produces the nth moment of the random variable $Z = \int_0^T g(X(\tau))\, d\tau$. The calculation with $f_\lambda(x) = e^{\lambda x}$, where feasible, yields the moment generating function of Z.

By the definition of T we plainly have the boundary conditions

$$U(a) = U(b) = f(0). \tag{3.32}$$

We derive a differential equation for $U(x)$ by paraphrasing the method of solution of Problem C. To this end, assume $g(x)$ is bounded. For h sufficiently small, the Taylor expansion leads to

$$U(x) = E_x\left[f\left(\int_0^h g(X(\tau))\, d\tau + \int_h^T g(X(\tau))\, d\tau \right) \right]$$

$$= E_x\left[f\left(\int_h^T g(X(\tau))\, d\tau \right) + f'\left(\int_h^T g(X(\tau))\, d\tau \right) \int_0^h g(X(\tau))\, d\tau \right] + O(h^2). \tag{3.33}$$

If $g(\xi)$ is continuous at x, then (3.33) has the form

$$E_x\left[f\left(\int_h^T g(X(\tau))\, d\tau \right) + hg(x)f'\left(\int_h^T g(X(\tau))\, d\tau \right) \right] + o(h).$$

Invoking the law of total probabilities and the Markov property, we can reduce the first two terms above to

$$E_x\left[E_{X(h)}\left[f\left(\int_h^T g(X(\tau))\,d\tau\right)\right] + hg(x)E_{X(h)}\left[f'\left(\int_h^T g(X(\tau))\,d\tau\right)\right]\right]$$

$$= E_x[U(X(h))] + hg(x)E_x[V(X(h))] \qquad (3.34)$$

where we define

$$V(x) = E_x\left[f'\left(\int_0^T X(\tau)\,d\tau\right)\right]. \qquad (3.35)$$

Expanding about x since $X(h)$ is close to $X(0) = x$ for h small (because the sample paths are continuous by the basic characterization of diffusion processes), we obtain

$$E_x[U(X(h))] = U(x) + U'(x)E_x[X(h) - x] + \tfrac{1}{2}U''(x)E_x[(X(h) - x)^2] + o(h)$$

and

$$E_x[V(X(h))] = V(x) + O(h). \qquad (3.36)$$

Substituting from the infinitesimal relations, subtracting out $U(x)$, and then dividing by h and sending h to zero produces the differential equation

$$\tfrac{1}{2}\sigma^2(x)U''(x) + \mu(x)U'(x) + g(x)V(x) = 0. \qquad (3.37)$$

The solution of (3.37) in conjunction with (3.32) depends on knowing $V(x)$.

For the case $f(x) = x^n$, then $f'(x) = nx^{n-1}$. In accordance with (3.37), the nth moment of Z with $X(0) = x$; that is,

$$U_n(x) = E_x\left[\left(\int_0^T g(X(\tau))\,d\tau\right)^n\right]$$

solves

$$\tfrac{1}{2}\sigma^2(x)U_n''(x) + \mu(x)U_n'(x) + nU_{n-1}(x)g(x) = 0 \qquad (3.38)$$

subject to the boundary conditions $U_n(a) = U_n(b) = 0$.

Note that $U_1(x)$ is exactly the expectation of the functional of Problem C, whose determination is given in (3.12).

In the case $f_\lambda(x) = e^{\lambda x}$, then (3.38) becomes

$$\tfrac{1}{2}\sigma^2(x)U''(x) + \mu(x)U'(x) + \lambda U(x)g(x) = 0.$$

We next indicate briefly the calculation, of some practical interest, of the function

$$R(x) = E_x\left[\int_0^T \exp\left\{-\int_0^t k(X(\tau))\,d\tau\right\}g(X(t))\,dt\right] \qquad (3.39)$$

for $a < x < b$, where T as before stands for the first passage time to a or b. We assume g and k are continuous at x. The function $k(x)$ is referred to as the killing rate. In fact, where k is positive the multiplier $\exp\{-\int_0^t k(X(\tau))\,d\tau\}$, depending on the previous history of the process, serves as a discount factor on the yield $g(X(t))$ aggregated up to the random time T.

The decomposition induced by the process after the lapse of a slight time duration gives

$$R(x) = E_x\left[\int_0^h \exp\left\{-\int_0^t k(X(\tau))\,d\tau\right\}g(X(t))\,dt\right.$$

$$+ \int_h^T \left(\exp\left\{-\int_0^h k(X(\tau))\,d\tau\right\} - 1\right)\exp\left\{-\int_h^t k(X(\tau))\,d\tau\right\}g(X(t))\,dt$$

$$\left. + \int_h^T \exp\left\{-\int_h^t k(X(\tau))\,d\tau\right\}g(X(t))\,dt\right].$$

The usual Taylor expansion combined with the law of total probabilities, Markov property, and obvious manipulations, lead to

$$\tfrac{1}{2}\sigma^2(x)R''(x) + \mu(x)R'(x) - k(x)R(x) + g(x) = 0 \tag{3.40}$$

coupled to the boundary conditions $R(a) = R(b) = 0$.

For later purposes, we need to evaluate

$$P(x) = E_x\left[\exp\left\{-\int_0^T k(X(\tau))\,d\tau\right\}\right]. \tag{3.41}$$

Applying (3.37) with the specification $f(x) = e^{-x}$ and $g(x) = k(x)$ and noting that $f'(x) = -f(x)$ in the case at hand shows that $P(x)$ solves

$$\tfrac{1}{2}\sigma^2(x)P''(x) + \mu(x)P'(x) - k(x)P(x) = 0 \tag{3.42}$$

and obeys the boundary conditions $P(a) = P(b) = 1$.

An extension of (3.41) concerns the evaluation of the functional

$$Q(x) = E_x\left[\exp\left\{-\int_0^T k(X(\tau))\,d\tau\right\}\int_0^T f(X(\tau))\,d\tau\right] \tag{3.43}$$

with the usual smoothness assumptions stipulating that f and k are continuous at x. The relevant differential equation becomes

$$\tfrac{1}{2}\sigma^2(x)Q''(x) + \mu(x)Q'(x) - k(x)Q(x) + f(x)P(x) = 0 \tag{3.44}$$

with the function $P(x)$ above being that of (3.41) with the necessary boundary conditions $Q(a) = Q(b) = 0$.

4: Some Concrete Cases of the Functional Calculations

With the solutions (3.10)-(3.12) at hand we expound a number of cases of interest in applications.

A. STANDARD BROWNIAN MOTION

Let $\{X(t), t \geq 0\}$ be standard Brownian motion whose infinitesimal parameters are $\mu(x) \equiv 0$, $\sigma^2(x) \equiv 1$. Obviously we may take

$$s(x) = \exp\left\{-2\int^x [\mu(\xi)/\sigma^2(\xi)]\, d\xi\right\} = 1,$$

and for the scale measure $S(x) = x$, so that $u(x)$, the probability of reaching b prior to a with initial state x, is

$$u(x) = \frac{x - a}{b - a}, \qquad a \leq x \leq b, \tag{4.1}$$

verifying the same result obtained by other means in Chapter 7. The speed density (cf. (3.7)) is

$$m(\xi) = \frac{1}{s(\xi)} = 1,$$

and the Green function (3.15) for the interval $[a, b]$ is

$$G(x, \xi) = \begin{cases} \dfrac{2(x - a)(b - \xi)}{(b - a)}, & a \leq x \leq \xi \leq b, \\[2mm] \dfrac{2(\xi - a)(b - x)}{(b - a)}, & a \leq \xi \leq x \leq b. \end{cases} \tag{4.2}$$

Then a direct calculation from (3.12) gives

$$v(x) = E[T_{a,b} | X(0) = x] = (x - a)(b - x), \qquad a \leq x \leq b. \tag{4.3}$$

B. BROWNIAN MOTION WITH DRIFT

If $\{X(t), t \geq 0\}$ is Brownian motion with nonzero drift $\mu(x) \equiv \mu$ and variance σ^2, then

$$s(x) = \exp(-2\mu x/\sigma^2),$$

$$S(x) = A \exp(-2\mu x/\sigma^2) + B \qquad (A \text{ and } B \text{ constants}),$$

and

$$u(x) = \frac{e^{-2\mu x/\sigma^2} - e^{-2\mu a/\sigma^2}}{e^{-2\mu b/\sigma^2} - e^{-2\mu a/\sigma^2}},$$

which also was obtained in Chapter 7.

C. THE WRIGHT–FISHER DIFFUSION MODEL FOR GENE FREQUENCY

Wright's model for the fluctuation of gene frequency under random reproduction was described in Example G of Section 2, Chapter 2. A more elaborate
version involving mutation and selection pressures was set forth in Section 2 of
this chapter, where a variety of approximating diffusion processes were
highlighted.

The simplest Wright–Fisher diffusion (a model for depicting fluctuations of
gene frequency of the A-type within a population having both A- and a-types
subject to selection influences) has the state space $[0, 1]$ and diffusion
coefficients

$$\mu(x) = \sigma x(1 - x), \qquad \sigma^2(x) = x(1 - x). \tag{4.4}$$

Recall that the state variable $X(\tau)$ is the fraction of the A-type in a population
of N individuals, where $\tau = 1$ corresponds to about N generations.

(a) $\sigma = 0$ (No Selection Differences between the A-type and the a-type)

The boundary states 0 and 1 are absorbing points and signify that the
population exhibits no A-genes or all A-genes. Since $\mu(x) = 0$, then $s(x) = 1$ for
$0 < x < 1$ and $S(x) = x$. Hence, for any a, b, where $0 < a < b < 1$,

$$u(x) = \frac{x - a}{b - a} \qquad \text{for} \quad 0 < a \le x \le b < 1, \tag{4.5}$$

gives the probability of reaching a fraction b of A-types before reaching a
fraction a when the initial A frequency is x. It is intuitive and readily justified
that this formula holds in the limit as $a \downarrow 0$ and $b \uparrow 1$. Therefore, if $X(0) = x$, the
probability of ultimate fixation into a population including only A-types is x,
while fixation for the alternative type occurs with probability $1 - x$.

In the same way, *in this particular example* we shall obtain a valid result for
the mean time to fixation if we compute $v(x)$ using $a = 0$ and $b = 1$ provided
we use the form in Eq. (3.12). We have

$$G(x, \xi) = \begin{cases} \dfrac{2x(1 - \xi)}{\xi(1 - \xi)} & \text{for} \quad 0 < x < \xi < 1, \\[2ex] \dfrac{2(1 - x)\xi}{\xi(1 - \xi)} & \text{for} \quad 0 < \xi < x < 1, \end{cases}$$

and

$$m(\xi) = \frac{1}{\sigma^2(\xi)s(\xi)} = \frac{1}{\xi(1 - \xi)}, \qquad 0 < \xi < 1.$$

Hence

$$v(x) = \int_0^1 G(x, \xi)\, d\xi$$

$$= 2\left[\int_0^x (1 - x)\frac{1}{1 - \xi}\, d\xi + \int_x^1 x\frac{1}{\xi}\, d\xi\right]$$

$$= -2[(1 - x)\log(1 - x) + x \log x]. \tag{4.6}$$

The maximum occurs at $x = \frac{1}{2}$, and we have then $v(\frac{1}{2}) = 2\ln 2$. As mentioned earlier, since one unit of time corresponds to N generations, the value of (4.6) in terms of generations should be multiplied by N.

Expected local time. It is of interest to calculate the expected time, before fixation, that the A frequency has values in (x_1, x_2) where the initial A frequency was x. This is equivalent to evaluating

$$w(x) = E\left[\int_0^T g(X(\tau))\, d\tau \,|\, X(0) = x\right] \tag{4.7}$$

for the function

$$g(\xi) = \begin{cases} 1 & \text{for} \quad x_1 < \xi < x_2, \\ 0 & \text{otherwise,} \end{cases}$$

and T represents the time to some fixation. Following the procedure used in Remark 3.3, we have

$$w(x) = \int_{x_1}^{x_2} G(x, \xi)\, d\xi,$$

which reduces to

$$w(x) = \begin{cases} 2x \ln\dfrac{x_2}{x_1} & \text{for} \quad x < x_1, \\[2mm] 2\left[(1 - x)\ln\left(\dfrac{1 - x_1}{1 - x}\right) + x \ln\left(\dfrac{x_2}{x}\right)\right] & \text{for } x_1 < x < x_2 \\[2mm] 2(1 - x)\ln\dfrac{1 - x_1}{1 - x_2} & \text{for} \quad x > x_2. \end{cases}$$

The measure function $G(x, \xi)\, d\xi$ can thereby be interpreted as the expected time the process spends in $(\xi, \xi + d\xi)$ prior to fixation starting from $X(0) = x$. (See Remark 3.3.)

(b) $\sigma > 0$ (*Selection Favors the A-type over the a-type*)

Recall from (4.4) that in this case

$$\mu(\xi) = \sigma\xi(1 - \xi) \qquad \text{and} \qquad \sigma^2(\xi) = \xi(1 - \xi).$$

It follows, referring to (3.5), that

$$s(x) = \exp\left[-\int^x \frac{2\sigma\xi(1-\xi)}{\xi(1-\xi)}\,d\xi\right] = e^{-2\sigma x},$$

and therefore

$$S(x) = \frac{1 - e^{-2\sigma x}}{1 - e^{-2\sigma}} \tag{4.8}$$

is the probability of fixation of the population in the A-type where $X(0) = x$ is the initial fraction of A-individuals. In the usual population genetics literature on the subject, the scaling of the selection parameter σ of (4.4) is taken as $\sigma = N\gamma$, and then (4.8) becomes

$$S(x) = \frac{1 - e^{-2N\gamma x}}{1 - e^{-2N\gamma}}$$

which for $x = 1/N$ (that is, where the initial population consists of a single A-type and all others a-types) is approximately $S(1/N) = 2\gamma/(1 - e^{-2N\gamma})$ ($\sim 2\gamma$ if $N\gamma$ is large).

D. WRIGHT–FISHER MODEL WITH ONE-WAY MUTATION

Suppose no mutations of a to A are allowed, while mutations of A to a occur at a rate $\alpha = \gamma/N$. The relevant approximating diffusion is that of Case (a), Example F of Section 2 with diffusion coefficients

$$\mu(x) = -\gamma x \qquad \text{and} \qquad \sigma^2(x) = x(1 - x) \qquad \text{for} \quad 0 \le x \le 1.$$

In this case we obtain

$$s(x) = \exp\left(+\int^x \frac{2\gamma}{1-\xi}\,d\xi\right) = \exp[-2\gamma \ln(1-x)] = \frac{1}{(1-x)^{2\gamma}}. \tag{4.9}$$

The scale function for the interval $(0, 1)$ can be taken to be

$$S(x) = \frac{1 - (1-x)^{-2\gamma+1}}{1 - 2\gamma}, \qquad 2\gamma \ne 1, \tag{4.10}$$

so that $S(x)$ is increasing. Notice in all circumstances that

$$u(x) = \frac{S(x) - S(a)}{S(b) - S(a)} = \frac{(1-x)^{-2\gamma+1} - (1-a)^{-2\gamma+1}}{(1-b)^{-2\gamma+1} - (1-a)^{-2\gamma+1}} \tag{4.11}$$

is the probability of reaching b before a. For $b \to 1$, then $u(x) \to 0$ when $2\gamma > 1$, while if $2\gamma < 1$ the limit is positive. Thus, only if the mutation rate is sufficiently high is the endpoint 1 (signifying a state of only A-types) unattainable. Of course, fixation of the A-gene is never possible, but where $2\gamma < 1$ the analysis at the boundary point 1 is more subtle. We shall deal with these kinds of problems in the discussion and classification of boundary behavior for diffusion processes in Sections 6–8.

From the biology of the model we should expect ultimate fixation of the a-type, and this indeed occurs. The time of fixation is then T_0, the hitting time of state 0. As just seen, when $2\gamma > 1$, state 1 is unattainable and thus $T_0 = T_{0,1}$. We may compute the expected time to fixation using

$$E[T_0 | X(0) = x] = \lim_{a\downarrow 0, b\uparrow 1} E[T_{a,b} | X(0) = x],$$

where $T_{a,b}$ is the first time the level a or b is reached, $0 < a < b < 1$. Following (3.7), we set $m(\xi) = 1/\xi(1 - \xi)s(\xi) = (1 - \xi)^{2\gamma-1}/\xi$. Now, according to (3.12), when $2\gamma > 1$ we obtain

$$E[T_{a,b} | X(0) = x] = \int_a^b G_{a,b}(x, \xi)\, d\xi$$

$$= 2u(x)\int_x^b [S(b) - S(\xi)]m(\xi)\, d\xi$$

$$+ 2[1 - u(x)]\int_a^x [S(\xi) - S(a)]m(\xi)\, d\xi,$$

where $u(x)$ is that appearing in (4.11),

$$= 2\left[\frac{(1 - x)^{-2\gamma+1} - (1 - a)^{-2\gamma+1}}{(1 - b)^{-2\gamma+1} - (1 - a)^{-2\gamma+1}}\right]$$

$$\times \int_x^b \left(\frac{(1 - b)^{-2\gamma+1} - (1 - \xi)^{-2\gamma+1}}{2\gamma - 1}\right)\frac{(1 - \xi)^{2\gamma-1}}{\xi}\, d\xi$$

$$+ 2\left[\frac{(1 - b)^{-2\gamma+1} - (1 - x)^{-2\gamma+1}}{(1 - b)^{-2\gamma+1} - (1 - a)^{-2\gamma+1}}\right]$$

$$\times \int_a^x \left(\frac{(1 - \xi)^{-2\gamma+1} - (1 - a)^{-2\gamma+1}}{2\gamma - 1}\right)\frac{(1 - \xi)^{2\gamma-1}}{\xi}\, d\xi. \tag{4.12}$$

Letting a decrease to 0 and b increase to 1 and noting that, when $2\gamma > 1$, then $(1 - b)^{-2\gamma+1} \to \infty$ and correspondingly $u(x) \to 0$, we obtain

$$E[T_0 | X(0) = x] = \lim_{a\downarrow 0, b\uparrow 1} E[T_{a,b} | X(0) = x]$$

$$= 2\left[\frac{(1 - x)^{-2\gamma+1} - 1}{2\gamma - 1}\right]\int_x^1 \frac{(1 - \xi)^{2\gamma-1}}{\xi}\, d\xi$$

$$+ 2\int_0^x \frac{(1 - \xi)^{2\gamma-1}}{\xi}\left(\frac{(1 - \xi)^{-2\gamma+1} - 1}{2\gamma - 1}\right)d\xi$$

$$= 2\int_0^x \frac{1}{(1 - \xi)^{2\gamma}}\int_\xi^1 \frac{(1 - \eta)^{2\gamma-1}}{\eta}\, d\eta\, d\xi, \tag{4.13}$$

the last equation resulting by integration by parts of the second integral, where one of the boundary terms and the first integral cancel. The above formula also works for $2\gamma = 1$.

In the case $2\gamma < 1$, the state 1 can be attained prior to fixation, and so $T_0 \neq T_{0,1}$ with positive probability. The limit procedure of (4.13) yields the mean of $T_{0,1}$ and not the mean of T_0. The mean fixation time

$$v_0(x) = E[T_0 | X(0) = x]$$

satisfies the same differential equation $\frac{1}{2}\sigma^2(x)v_0''(x) + \mu(x)v_0'(x) = -1$ for $0 < x < 1$, and the boundary condition $v_0(0) = 0$ maintains. However, the boundary condition $v_0(1) = 0$ is no longer correct and must be replaced by an appropriate alternative that correctly models the phenomenon being studied. This example highlights the importance of understanding boundary behavior, the subject of Sections 6-8. The analysis will be done in another manner. We seek a solution of

$$\frac{x(1-x)}{2} v_0''(x) - \gamma x v_0'(x) + 1 = 0 \tag{4.14}$$

obviously obeying the boundary condition $v_0(0) = 0$. This requirement alone does not uniquely determine the solution. On intuitive grounds we should expect $v_0(x)$ to be monotone increasing, and so we add this condition. The general solution of (4.14) has the form

$$v_0(x) = \int_0^x \frac{2}{(1-\xi)^{2\gamma}} \left[\int_\xi^1 \frac{(1-\eta)^{2\gamma-1}}{\eta} \, d\eta \right] d\xi + A(1-x)^{1-2\gamma} + B,$$

where A and B are arbitrary constants $[A(1-x)^{1-2\gamma} + B$ is the general solution of the homogeneous equation $\frac{1}{2}\xi(1-\xi)y''(\xi) - \gamma\xi y'(\xi) = 0]$. The condition $v_0(0) = 0$ implies $A = -B$. The fact that $v_0(x)$ is increasing near 1 means that $A < 0$. With these conditions fulfilled we write

$$v_0(x) = \int_0^x \frac{2}{(1-\xi)^{2\gamma}} \int_\xi^1 \frac{(1-\eta)^{2\gamma-1}}{\eta} \, d\eta \, d\xi + B[1 - (1-x)^{1-2\gamma}], \quad B \geq 0.$$

The constant B is forced to be zero by imposing either of the following two constraints:

    (i) $v_0(x)$ is the smallest positive monotone solution satisfying $v_0(0) = 0$, or
    (ii) $v_0(0) = 0$ and $v_0'(1) < \infty$.

An intuitive argument shows that $v_0'(1) = 1/\gamma$. Because $\mu(1) = -\gamma$ while $\sigma^2(1) = 0$, the motion at $x = 1$ is essentially deterministic, and given $X(t) = 1$, then $X(t + h) \simeq 1 - \gamma h$. Then

$$v_0(1) = h + E[v_0(X(t+h)) | X(t) = 1] = h + v_0(1 - \gamma h) + o(h)$$
$$= h + v_0(1) - \gamma h v_0'(1) + o(h).$$

Subtracting $v_0(1)$, dividing by h, and sending h to zero shows $v_0'(1) \sim 1/\gamma$.

The nub of the above discussion is that where $2\gamma < 1$ some further stipulations are necessary to extract the proper solution of (4.14) corresponding to $E[T_0 | X(0) = x] = v_0(x)$. We emphasize again that the last formula is not uniquely defined unless the behavior of the diffusion at the state point $x = 1$ is made more explicit.

E. A CASH INVENTORY MODEL

Let $Z(t)$ be the amount of cash an organization has on hand at time t. We suppose that in the absence of intervention $\{Z(t), t \geq 0\}$ behaves as a Brownian motion process with zero drift and variance parameter $\sigma^2 = 1$.

Holding cash involves an opportunity cost since this cash could be invested. We therefore suppose that holding cash at level z incurs costs at the rate cz. Since transactions into and out of cash are also costly, we include a cost K for each transaction.

Consider the following (s, S)-type policy for controlling the cash level: "If the cash reaches level S, invest $S - s$ and reduce the cash level to s. This transaction incurs a cost of K. If the cash ever dips to zero, sell investments, and bring the cash level up to s. This again costs K."

Consider a cycle to be from one intervention returning the cash level to s to the next such intervention. The long-run cost per unit time will be the expected cost per cycle divided by the expected cycle time, or

$$(K + A)/B,$$

where

$$A = E\left[\int_0^T cZ(\tau)\, d\tau \,\Big|\, Z(0) = s \right]$$

and

$$B = E[T | Z(0) = s].$$

[The above formulas can be derived by direct renewal arguments (cf. Chapter 5).] Here

$$T = \min\{t > 0 : Z(t) = S \text{ or } Z(t) = 0\}.$$

From Eq. (4.3) in the Brownian motion example we have (using $a = 0$ and $b = S$)

$$B = v(s) = s(S - s).$$

Applying (3.14) with $g(z) = cz$, and taking $G(x, \xi)$ appropriate for the interval $(a, b) = (0, S)$ (as given by (4.2)), we obtain

$$E\left[\int_0^T g(Z(t))\, dt \,\Big|\, Z(0) = x \right] = w(x) = \int_0^S G(x, \xi) g(\xi)\, d\xi = c(\tfrac{1}{3}xS^2 - \tfrac{1}{3}x^3),$$

and in particular

$$A = w(s) = \tfrac{1}{3}cs(S^2 - s^2).$$

The average cost is

$$\frac{K + A}{B} = \frac{K + \frac{1}{3}cs(S^2 - s^2)}{s(S - s)} = \frac{K}{s(S - s)} + \frac{c(S + s)}{3}.$$

To minimize, change variables, letting $\bar{S} = S - s$. The average cost is

$$C(s, \bar{S}) = \frac{K}{s\bar{S}} + \frac{c(\bar{S} + 2s)}{3},$$

which we differentiate to get

$$\frac{\partial C}{\partial s} = -\frac{K}{\bar{S}s^2} + \frac{2c}{3} \quad \text{and} \quad \frac{\partial C}{\partial \bar{S}} = -\frac{K}{s\bar{S}^2} + \frac{c}{3}.$$

We equate the derivatives to zero to obtain the cost minimizing $\bar{S} = \bar{S}^*$ and $s = s^*$:

$$(s^*)^2 \bar{S}^* = \frac{3K}{2c}, \qquad s^*(\bar{S}^*)^2 = \frac{3K}{c}, \tag{4.15}$$

whence

$$\frac{s^*}{\bar{S}^*} = \frac{1}{2}, \tag{4.16}$$

or

$$s^* = \frac{1}{2}(S^* - s^*) \qquad \text{or} \qquad s^* = \frac{1}{3}S^*. \tag{4.17}$$

We substitute (4.16) in (4.15) to get

$$(s^*)^3 = \frac{3K}{4c}.$$

The optimal control parameters are

$$s^* = \left(\frac{3K}{4c}\right)^{1/3} \qquad \text{and} \qquad S^* = 3s^*.$$

F. A BROWNIAN MOTION CONTROL PROBLEM

Let $X(t)$ be the state of some system which evolves stochastically from an initial position $X(0) = 0$. While the system is evolving, costs are accrued at a rate proportional to the square of the state with proportionality constant c. To avoid these costs, the observer may, at his will, pay a fixed cost K and then restart the process at zero. Thus, if T is the time of restart, then the total operating and restart costs up to time T are $W(T) = K + \int_0^T c[X(s)]^2 \, ds$.

We are going to assume that $\{X(t), t \geq 0\}$ is a Brownian motion process, at least until restarted, that the drift parameter is zero, and that the variance parameter is $\sigma^2 = 1$. We shall restart at times

$$T = \min\{t \geq 0 : |X(t)| \geq \lambda\}$$

and try to choose an optimal λ. Our long-run cost per unit time is the expected cost over a cycle divided by the expected cycle time. Thus we need to compute $A = E[W(T)|X(0) = 0]$ and $B = E[T|X(0) = 0]$. From Eq. (4.3) in the Brownian motion example we have, using $a = -\lambda$ and $b = \lambda$,

$$B = v(0) = \lambda^2.$$

Applying (3.11) with $g(x) = cx^2$, we obtain

$$w(x) = 2\left[u(x) \int_x^\lambda (\lambda - \xi)g(\xi)\,d\xi + [1 - u(x)] \int_{-\lambda}^x (\xi + \lambda)g(\xi)\,d\xi \right]$$

with

$$u(x) = \frac{x + \lambda}{2\lambda} \qquad \text{and} \qquad g(\xi) = c\xi^2.$$

Since $u(0) = \frac{1}{2}$, by straightforward integration and simplification we get

$$A = w(0) = \tfrac{1}{6}c\lambda^4.$$

The average cost is

$$C(\lambda) = \frac{K + A}{B} = \frac{K}{\lambda^2} + \frac{c\lambda^2}{6}.$$

We differentiate with respect to λ and equate to zero to obtain the minimizing cost for $\lambda = \lambda^*$ satisfying

$$0 = -\frac{2K}{(\lambda^*)^3} + \frac{c\lambda^*}{3}$$

or

$$\lambda^* = (6K/c)^{1/4}.$$

5: The Nature of Backward and Forward Equations and Calculation of Stationary Measures

Let $\{X(t), t \geq 0\}$ be a regular time homogeneous diffusion process on the open interval $I = (l, r)$. We designate by $P(t, x, y) = \Pr\{X(t) \leq y | X(0) = x\}$ the transition distribution function of $X(t)$ subject to the initial distribution

$$P(0, x, y) = \begin{cases} 1, & \text{if } x \leq y, \\ 0, & \text{if } x > y, \end{cases} \tag{5.1}$$

i.e., a point distribution concentrating at x. We shall assume throughout this chapter that $P(t, x, y)$ derives from a continuous density on (l, r), namely,

$$\frac{dP(t, x, y)}{dy} = p(t, x, y) \qquad \text{for } t > 0. \tag{5.2}$$

(The existence of a continuous density in (5.2) is not an assumption for a regular diffusion which has smooth infinitesimal coefficients. The analytic validation of this property is beyond the level of this text.)

KOLMOGOROV BACKWARD DIFFERENTIAL EQUATION

Our next objective in the spirit of Section 3 is to derive a partial differential equation for the function

$$u(t, x) = E[g(X(t))| X(0) = x], \tag{5.3}$$

where $g(x)$ is bounded and piecewise continuous on I.

Under mild conditions (those of Section 1 suffice) we will ascertain that $u(t, x)$ satisfies the partial differential equation

$$\frac{\partial u}{\partial t} = \frac{1}{2}\sigma^2(x)\frac{\partial^2 u}{\partial x^2} + \mu(x)\frac{\partial u}{\partial x} \tag{5.4}$$

with the initial condition $u(0+, x) = g(x)$, where $u(0+, x) = \lim_{h \downarrow 0} u(h, x)$.

The specification $g(\eta) = 1$ for $\eta \leq y$ and 0 for $\eta > y$, produces the transition distribution function

$$u(t, x) = P(t, x, y).$$

Equation (5.4) in this case is referred to as the *Kolmogorov backward equation*, that is,

$$\frac{\partial P(t, x, y)}{\partial t} = \frac{1}{2}\sigma^2(x)\frac{\partial^2 P(t, x, y)}{\partial^2 x} + \mu(x)\frac{\partial P(t, x, y)}{\partial x}, \tag{5.5}$$

applicable for $t > 0$ and $l < x, y < r$. The initial condition attendant to (5.5) is

$$P(0+, x, y) = \begin{cases} 1 & \text{if} \quad x \leq y, \\ 0 & \text{if} \quad x > y. \end{cases} \tag{5.6}$$

The transition density $p(t, x, y)$ also satisfies the Kolmogorov backward equation:

$$\frac{\partial p}{\partial t} = \frac{1}{2}\sigma^2(x)\frac{\partial^2 p}{\partial x^2} + \mu(x)\frac{\partial p}{\partial x} \tag{5.7}$$

for $t > 0$ and x, y in (l, r).

Remark. It is *not* true that (5.5) and (5.7) always admit unique solutions, even if we require that $P(t, x, y)$ be a distribution function in y for each t and x. As an example, the transition distribution functions for absorbing and reflecting Brownian motion both satisfy (5.5) but are manifestly unequal. The problem is that (5.5) makes no mention of the behavior of the process at the boundaries of the state space, exactly where absorbing and reflecting Brownian motion differ.

The assertions of (5.4)–(5.7) are surprisingly difficult to prove. In particular, it is rather formidable to prove that the functions P, p, and u are differentiable in t and twice differentiable in x. However, beginning from this point it is not hard to show that $u(t, x)$ satisfies (5.4).

We proceed to the proof of (5.4) assuming that $u(t, x)$ is differentiable in t and possesses two continuous derivatives in x. Begin with

$$E[g(X(h + t))|X(h) = x] = u(t, x)]$$

or

$$E[g(X(h + t))|X(h)] = u(t, X(h))$$

so that

$$u(t + h, x) = E[g(X(h + t))|X(0) = x] = E[E[g(X(h + t))|X(h)]|X(0) = x]$$

(by the law of total probability) $= E[u(t, X(h))|X(0) = x]$. Subtracting $u(t, x)$ from both sides and dividing by h yields

$$\frac{u(t + h, x) - u(t, x)}{h} = \frac{1}{h} E[u(t, X(h)) - u(t, x)|X(0) = x]. \tag{5.8}$$

Under our stipulation we know that $u(t, x)$ is differentiable in t and possesses two continuous derivatives in x. (In most applications this assumption would be satisfied.) As h approaches 0 the left-hand side of (5.8) passes into $\partial u/\partial t$. The right-hand side is evaluated by implementing a Taylor series expansion. First, under the condition $X(0) = x$, let $\varepsilon > 0$ be an arbitrary positive number, to be determined later, and let $\Delta X = X(h) - x$ and

$$\Delta X_\varepsilon = \begin{cases} \Delta X & \text{if } |\Delta X| \le \varepsilon, \\ 0 & \text{if } |\Delta X| > \varepsilon. \end{cases}$$

Since g is a bounded function, say, $|g(x)| \le A$ for all x, it follows that $|u(t, x)| \le A$ for all t and x and

$$E[|u(t, x + \Delta X) - u(t, x + \Delta X_\varepsilon)|] \le 2A \Pr\{|\Delta X| > \varepsilon|X(0) = x\}.$$

The right-hand side is of order less than h as $h \to 0$ (cf. (1.1) of Section 1). Thus, the right-hand side of (5.8) may be evaluated through

$$\frac{1}{h} E[u(t, X(h)) - u(t, x)|X(0) = x] = \frac{1}{h} E[u(t, x + \Delta X_\varepsilon)$$
$$- u(t, x)|X(0) = x] + \frac{o(h)}{h}.$$

Expanding in a Taylor series, we get

$$u(t, x + \Delta X_\varepsilon) - u(t, x) = \Delta X_\varepsilon \frac{\partial u}{\partial x}(t, x) + \tfrac{1}{2}(\Delta X_\varepsilon)^2 \frac{\partial^2 u}{\partial x^2}(t, x + Z), \tag{5.9}$$

where Z is a random variable satisfying $|Z| \le |\Delta X_\varepsilon| \le \varepsilon$. Let $\delta > 0$ be given. Since $\partial^2 u/\partial x^2$ is continuous in x we may determine ε sufficiently small to ensure that

$$\left| \frac{\partial^2 u}{\partial x^2}(t, x + Z) - \frac{\partial^2 u}{\partial x^2}(t, x) \right| \le \delta \qquad (5.10)$$

in (5.9). Divide by h, take expected values, let $h \to 0$, and use (5.10) to conclude

$$\overline{\lim_{h \to 0}} \left| \frac{E[u(t, x + \Delta X_\varepsilon) - u(t, x) | X(0) = x]}{h} - \mu(x) \frac{\partial u}{\partial x} - \frac{1}{2}\sigma^2(x) \frac{\partial^2 u}{\partial x^2} \right| \le \tfrac{1}{2}\sigma^2(x)\delta.$$

$$(5.11)$$

Since $\delta > 0$ is arbitrary, it follows that the limit in (5.11) is 0. Thus, if $h \to 0$ in (5.8) we obtain

$$\frac{\partial u}{\partial t} = \frac{1}{2} \sigma^2(x) \frac{\partial^2 u}{\partial x^2} + \mu(x) \frac{\partial u}{\partial x} \qquad (5.12)$$

for $t > 0$, $l < x < r$. The appropriate initial condition is $\lim_{t \downarrow 0} u(t, x) = g(x)$.

Now suppose g is the function

$$g(\xi) = \begin{cases} 1 & \text{if } \xi \le y, \\ 0 & \text{if } \xi > y. \end{cases}$$

Then $u(t, x) = E[g(X(t))|X(0) = x] = \Pr\{X(t) \le y | X(0) = x\} = P(t, x, y)$ so that (5.12) would become the backward equation (5.5). However, the function $g(\cdot)$ just specified is not continuous as required in the preceding development. But g can be suitably approximated by such smooth functions, and through such approximations the argument leading to the backward equation can be justified with some effort. Equation (5.7) is obtained formally by differentiating (5.5) with respect to y.

If the diffusion process $\{X(t), t \ge 0\}$ is not assumed time homogeneous then (5.7) is modified as follows: Consider, for $t > s$ and $l < x < r$,

$$u(t, s, x) = E[g(X(t))|X(s) = x], \qquad (5.13)$$

displaying its dependence on the initial time s and state x and current time t. The corresponding backward differential equation for $u(t, s, x)$ is

$$-\frac{\partial u(t, s, x)}{\partial s} = \frac{1}{2}\sigma^2(x, t) \frac{\partial^2 u(t, s, x)}{\partial x^2} + \mu(x, t) \frac{\partial u(t, s, x)}{\partial x} \qquad (5.14)$$

coupled to the initial condition $\lim_{t \downarrow s} u(t, s, x) = g(x)$.

Example 1. *Standard Brownian motion B(t).* The associated backward differential equation is the so-called heat equation

$$\frac{\partial p}{\partial t} = \frac{1}{2}\frac{\partial^2 p}{\partial x^2} \qquad \text{for} \quad t > 0, \quad -\infty < x < \infty. \tag{5.15}$$

The unique transition probability density function satisfying (5.15) together with the appropriate initial condition at $t = 0$ is the *Gauss kernel*

$$\varphi(t, x, y) = \frac{1}{\sqrt{2\pi t}} \exp\left\{-\frac{(y-x)^2}{2t}\right\} \qquad \text{for} \quad t > 0, \quad -\infty < x, \quad y < \infty. \tag{5.16}$$

The straightforward task of verifying that (5.16) satisfies (5.15) is left to the reader.

Example 2. *Brownian motion with drift $W(t) = \sigma B(t) + \mu t$.* Associated with the infinitesimal parameters $\mu(x) = \mu$ and $\sigma^2(x) = \sigma^2$ for $-\infty < x < \infty$ is the backward equation

$$\frac{\partial p}{\partial t} = \frac{1}{2}\sigma^2 \frac{\partial^2 p}{\partial x^2} + \mu \frac{\partial p}{\partial x} \qquad \text{for} \quad t > 0, \quad -\infty < x, \quad y < \infty. \tag{5.17}$$

The unique transition probability density function satisfying (5.17) together with the appropriate initial condition at $t = 0$ is

$$p(t, x, y) = \varphi(\sigma^2 t, x + \mu t, y), \tag{5.18}$$

where $\varphi(t, x, y)$ is the Gauss kernel in (5.16).
We will verify (5.18) satisfies (5.17) by using the chain rule for differentiation. Introduce the notation

$$\varphi_x = \frac{\partial \varphi}{\partial x}, \qquad \varphi_{xx} = \frac{\partial^2 \varphi}{\partial x^2}, \qquad \text{and} \qquad \varphi_t = \frac{\partial \varphi}{\partial t}. \tag{5.19}$$

Then (5.15) becomes

$$\varphi_t = \tfrac{1}{2}\varphi_{xx}. \tag{5.20}$$

Differentiation of (5.18) then gives

$$\frac{\partial p}{\partial t} = \sigma^2 \varphi_t + \mu \varphi_x,$$

$$\mu \frac{\partial p}{\partial x} + \frac{1}{2}\sigma^2 \frac{\partial^2 p}{\partial x^2} = \mu \varphi_x + \frac{1}{2}\sigma^2 \varphi_{xx}.$$

In conjunction with (5.20), then (5.17) is verified.

Example 3. *The Ornstein–Uhlenbeck process $V(t)$.* The backward equation

$$\frac{\partial p}{\partial t} = \frac{1}{2}\sigma^2 \frac{\partial^2 p}{\partial x^2} - \alpha x \frac{\partial p}{\partial x} \qquad \text{for} \quad t > 0, \quad -\infty < x, \quad y < \infty \qquad (5.21)$$

corresponds to the coefficients $\mu(x) = -\alpha x$ and $\sigma^2(x) = \sigma^2$ for $-\infty < x < \infty$. The unique solution as a transition probability density function is

$$p(t, x, y) = \varphi\left(\frac{\sigma^2}{2\alpha}(1 - e^{-2\alpha t}), xe^{-\alpha t}, y\right), \qquad (5.22)$$

where $\varphi(t, x, y)$ is the Gauss kernel given in (5.16). The verification that (5.22) satisfies (5.21) uses the chain rule for differentiation in a manner paralleling but more arduous than that used in Example 2.

The Ornstein–Uhlenbeck process $\{V(t), t \geq 0\}$ can be realized from standard Brownian motion $B(t)$ through the succinct representation

$$V(t) = e^{-\alpha t} B\left[\frac{\sigma^2(e^{2\alpha t} - 1)}{2\alpha}\right] \qquad (5.23)$$

requiring a deterministic change of the time clock and a rescaling of the state variable.

It is manifest that the process $V(t)$ in (5.23) inherits only continuous sample paths from Brownian motion, and that $V(t)$ is a Markov process. It remains to identify the infinitesimal parameters of $V(t)$ as those of (5.21). To this end it is convenient to abbreviate $\tau = \sigma^2(e^{2\alpha t} - 1)/2\alpha$, and then

$$E[V(t + h) - V(t) | V(t) = x]$$

$$= e^{-\alpha t}\left\{E\left[e^{-\alpha h}B\left(\sigma^2 \frac{e^{2\alpha t}e^{2\alpha h} - 1}{2\alpha}\right) - xe^{\alpha t}\bigg| B(\tau) = xe^{\alpha t}\right]\right\}.$$

$$= e^{-\alpha t}(e^{-\alpha h} - 1)xe^{\alpha t} = -\alpha x h + o(h). \qquad (5.24)$$

It is clear from (5.24) that the drift coefficient of $V(t)$ is $\mu_V(x) = -\alpha x$.

The calculation $\lim_{h \downarrow 0}(1/h)E[\{V(t + h) - V(t)\}^2 | V(t) = x] = \sigma^2$ is done similarly.

The representation (5.23) shows that the Ornstein–Uhlenbeck process is a Gaussian process; i.e., the finite-dimensional distributions are multivariate normal with mean zero and covariance kernel

$$E[V(t)V(s)] = \sigma^2 e^{-\alpha(t - s)}\frac{(e^{2\alpha s} - 1)}{2\alpha} \qquad \text{for} \quad s < t.$$

The existence of a limiting distribution for $V(\infty) = \lim_{t \to \infty} V(t)$ (referring here to convergence in law) is clear from the representation (5.23) by noting that $V(t)$ is normally distributed with mean 0 and variance $\sigma^2(1 - e^{-2\alpha t})/2\alpha$, which converges to $\sigma^2/2\alpha$ as t increases to ∞. We highlight this fact.

Proposition 5.1.   For the Ornstein–Uhlenbeck process $\{V(t); t \geq 0\}$ with drift coefficient $\mu(x) = -\alpha x$, and diffusion coefficient $\sigma^2(x) = \sigma^2$,

$$\lim_{t \to \infty} \Pr\{V(t) \leq y\} = \Pr\{V(\infty) \leq y\},$$

where $V(\infty)$ is normally distributed with mean zero and variance $\sigma^2/2\alpha$.

THE FORWARD EQUATION

We shall now derive a second partial differential equation satisfied by the transition density $p(t, x, y) = dP(t, x, y)/dy$. The equation can be regarded as dual to (5.70) and the pertinent variables are t and y (the state variable at time t rather than the initial state x). For this reason, among others, this equation is commonly called the *forward* or *evolution equation*. For versions of this equation with a discrete sample space we refer the reader to Chapter 14. The derivation of the forward equation is considerably more complex than that of the backward equation and often requires modifications in its statement.

For the "formal adjoint" (dual) differential equation to (5.17) we write

$$\frac{\partial p}{\partial t}(t, x, y) = \frac{1}{2}\frac{\partial^2}{\partial y^2}[\sigma^2(y)p(t, x, y)] - \frac{\partial}{\partial y}[\mu(y)p(t, x, y)]. \tag{5.25}$$

We proceed with a heuristic derivation and subsequently we shall indicate some of the formidable problems encountered in making the analysis rigorous. Let $\varphi(t, y)$ be an arbitrary smooth function satisfying the identity

$$\varphi(t + s, y) = \int \varphi(t, \xi)p(s, \xi, y)\, d\xi \qquad \text{for all} \quad t, s > 0. \tag{5.26}$$

Differentiate both sides of (5.26) with respect to s and utilize the backward equation satisfied by $p(s, \xi, y)$. This gives

$$\frac{\partial \varphi}{\partial t}(t + s, y) = \frac{\partial \varphi}{\partial s}(t + s, y) = \int \varphi(t, \xi)\frac{\partial p}{\partial s}(s, \xi, y)\, d\xi$$

$$= \int \varphi(t, \xi)\left[\frac{1}{2}\sigma^2(\xi)\frac{\partial^2 p}{\partial \xi^2} + \mu(\xi)\frac{\partial p}{\partial \xi}\right] d\xi. \tag{5.27}$$

Next, integration by parts, assuming that the contributions from the boundaries vanish, transforms (5.27) into

$$\frac{\partial \varphi}{\partial t}(t + s, y) = \int \left\{\frac{\partial^2}{\partial \xi^2}\left[\frac{\sigma^2(\xi)}{2}\varphi(t, \xi)\right] - \frac{\partial}{\partial \xi}[\mu(\xi)\varphi(t, \xi)]\right\}p(s, \xi, y)\, d\xi. \tag{5.28}$$

Since $p(s, \xi, y)$ approaches the delta (degenerate) measure concentrating at y, then as $s \to 0$ relation (5.28) passes into

$$\frac{\partial \varphi}{\partial t}(t, y) = \frac{1}{2} \frac{\partial^2}{\partial y^2} [\sigma^2(y)\varphi(t, y)] - \frac{\partial}{\partial y} [\mu(y)\varphi(t, y)], \tag{5.29}$$

which equals (5.25). In particular, the choice $\varphi(t, y) = p(t, x, y)$ certainly obeys (5.26) (it merely expresses the Markov property of the diffusion process) and consequently $p(t, x, y)$ "satisfies" the forward equation (5.25).

The analysis was quite loose and to justify the steps requires very stringent assumptions. Actually, in some cases $p(t, x, y)$ *does not* satisfy (5.25), but further terms have to be added partly reflecting boundary behavior of the process. The full analysis of the forward equation is well beyond the scope of this introductory text.

STATIONARY DISTRIBUTIONS

If it exists, a stationary density $\psi(x)$ necessarily satisfies

$$\psi(y) = \int \psi(x)p(t, x, y) \, dx \qquad \text{for all} \quad t > 0. \tag{5.30}$$

Mimicking the derivation of (5.29) we can deduce that $\psi(y)$ satisfies

$$0 = \frac{1}{2} \frac{\partial^2}{\partial y^2} [\sigma^2(y)\psi(y)] - \frac{\partial}{\partial y} [\mu(y)\psi(y)]. \tag{5.31}$$

We should also expect by analogy with the fundamental limit theorem of Markov chains that the stationary density is approached to the extent that

$$\lim_{t \to \infty} p(t, x, y) = \psi(y) \tag{5.32}$$

holds in some appropriate sense. Under strong conditions on the process, relation (5.32) obtains, but the development of such facts requires a more sophisticated context. If, indeed, (5.32) holds, then we can most likely pass to a limit in (5.29), with $p(t, x, y)$ substituted for $\varphi(t, y)$, and since

$$\lim_{t \to \infty} \partial p(t, x, y)/\partial t = 0$$

is expected, we arrive once again at (5.31). If the convergence in (5.32) occurs boundedly then we can attempt to interchange the limit $s \to \infty$ with the integral in the Chapman–Kolmogorov relation $p(t + s, x, y) = \int p(t, x, z)p(t, z, y) \, dz$ to obtain (5.30). In all circumstances we achieve (as $s \to \infty$) the inequality (by Fatou's lemma of real analysis)

$$\psi(y) \geq \int \psi(z)p(t, z, y) \, dz.$$

The existence of the limit in (5.32) with $\psi(y)$ representing a bona fide probability density on the state space (l, r) implies in particular that the process is strongly recurrent (positive ergodic) such that probability mass cannot escape to the boundaries. Actually, all the possibilities that arise in Markov chains, and even more pathologies, can occur in the case of a diffusion process.

CALCULATION OF THE STATIONARY DISTRIBUTION

Integrating Eq. (5.31) gives

$$\frac{d}{dy}\left[\frac{\sigma^2(y)}{2}\psi(y)\right] - \mu(y)\psi(y) = \tfrac{1}{2}C_1, \tag{5.33}$$

where C_1 is a constant. Multiplying by the integrating factor

$$s(y) = \exp\left\{-\int^y\left[\frac{2\mu(\xi)}{\sigma^2(\xi)}\right]d\xi\right\},$$

we can write (5.33) in the compact form $d[s(y)\sigma^2(y)\psi(y)]/dy = C_1 s(y)$. Another integration, with $S(x) = \int^x s(y)\,dy$, gives

$$\psi(x) = C_1\frac{S(x)}{s(x)\sigma^2(x)} + C_2\frac{1}{s(x)\sigma^2(x)}.$$

$$= m(x)[C_1 S(x) + C_2]. \tag{5.34}$$

The constants are determined to guarantee the constraints $\psi(x) \geq 0$ on (l, r) and $\int_l^r \psi(\xi)\,d\xi = 1$. If this is possible then a stationary density exists and otherwise not.

Example 4. For the Ornstein–Uhlenbeck process (5.23), Eq. (5.34) with $\sigma^2(x) = \sigma^2$, $s(x) = e^{\gamma x^2}$ $(\gamma = \alpha/\sigma^2)$ becomes

$$\psi(x) = C_1\left(\int_0^x e^{\gamma\xi^2}\,d\xi\right)e^{-\gamma x^2} + C_2 e^{-\gamma x^2}. \tag{5.35}$$

To insure that $\psi(x)$ is positive for $|x|$ large entails $C_1 = 0$. It follows that the unique stationary measure based on (5.34) is the normal density $\psi(x) = ce^{-\gamma x^2}$, in agreement with Proposition 5.1.

Example 5. *Wright–Fisher frequency model with mutation* [cf. Case (a), Example F of Section 2]. Let $X(t)$ be the fraction of the A-type at time t in a population of N individuals comprised of A- and a-types. Suppose the rates of mutation are $A \xrightarrow{\alpha} a$, $a \xrightarrow{\beta} A$ where $N\alpha \to \gamma_1$, $N\beta \to \gamma_2$. The approximating diffusion has infinitesimal parameters $\sigma^2(x) = x(1 - x)$, $\mu(x) = -\gamma_1 x + \gamma_2(1 - x)$, and therefore

$$s(x) = \exp\left[-\int^x \frac{2\mu(\xi)}{\sigma(\xi)}\,d\xi\right] = \exp\left\{\int^x\left(+\frac{2\gamma_1}{1-\xi} - \frac{2\gamma_2}{\xi}\right)d\xi\right\} = \frac{1}{(1-x)^{2\gamma_1}x^{2\gamma_2}}.$$

By (5.34) a set of possible stationary measures is determined from the formula

$$\psi(y) = C_1\left[\int_a^y \frac{dx}{(1-x)^{2\gamma_1}x^{2\gamma_2}}\right](1-y)^{2\gamma_1-1}y^{2\gamma_2-1} + C_2(1-y)^{2\gamma_1-1}y^{2\gamma_2-1}.$$

$$(5.36)$$

The possibilities depend now on the magnitudes of γ_1 and γ_2. For either $2\gamma_1$ or $2\gamma_2$ exceeding or equal to 1, the integral of the first term in (5.36) is not finite. The condition $\psi(y) \geq 0$ requires that $a = 0$ or 1, and then integrability of $\psi(y)$ on $I = (0, 1)$ compels that $C_1 = 0$. Thus, the stationary distribution is, with the correct normalizing constant,

$$\psi(y) = \frac{\Gamma(2\gamma_1)\Gamma(2\gamma_2)}{\Gamma(2\gamma_1 + 2\gamma_2)}(1-y)^{2\gamma_1-1}y^{2\gamma_2-1}.$$

In the circumstance

$$2\gamma_1 < 1 \quad \text{and} \quad 2\gamma_2 < 1, \qquad (5.37)$$

formula (5.36) provides a two-parameter family of stationary distributions and which distribution is appropriate (if any) requires further assumptions or information on the real process. Because the boundaries are both "attainable" it is essential to spell out the influence of the boundaries on the sample realizations. This involves extra specifications beyond the effects of the basic infinitesimal parameters. With this in mind, in Sections 6–8 we elaborate on several important facets of diffusion theory including classifications and characterizations of boundary behavior.

Example 6. *Wright–Fisher diffusion with mutation and selection.* [Compare to the conjunction of Cases (a) and (b), Example F of Section 2.] The infinitesimal parameters are $\mu(x) = -\gamma_1 x + \gamma_2(1-x) + sx(1-x)$, $\sigma^2(x) = x(1-x)$. The stationary distribution derived from (5.34) for $2\gamma_1 \geq 1$ and $2\gamma_2 \geq 1$ now takes the form

$$\psi(y) = \frac{(1-y)^{2\gamma_1-1}y^{2\gamma_2-1}e^{2sy}}{\int_0^1 (1-\xi)^{2\gamma_1-1}\xi^{2\gamma_2-1}e^{2s\xi}\,d\xi}.$$

THE BACKWARD EQUATION FOR THE KAC FUNCTIONAL

Let $\{X(t), t \geq 0\}$ be a diffusion and $k(x)$ a nonnegative function defined on the state space $I = (l, r)$. Consider the function

$$w(x, t) = E_x\left[\exp\left\{-\int_0^t k[X(\tau)]\,d\tau\right\}g(X(t))\right], \qquad (5.38)$$

defined for $g(x)$ bounded and continuous on I.

We shall develop the analog of the backward differential equation for the function $w(x, t)$. To this end, we start with some useful rearrangements of

$$\exp\left[-\int_0^t k(X(\tau))\,d\tau\right]g(X(t))$$

$$= \exp\left[-\int_0^h k(X(\tau))\,d\tau\right]\exp\left[-\int_h^t k(X(\tau))\,d\tau\right]g(X(t))$$

$$= \left\{\exp\left[-\int_0^h k(X(\tau))\,d\tau\right] - 1\right\}\exp\left[-\int_h^t k(X(\tau))\,d\tau\right]g(X(t))$$

$$+ \exp\left[-\int_h^t k(X(\tau))\,d\tau\right]g(X(t)).$$

We next implement a Taylor expansion and take expectations to get

$$w(x, t) = E_x\left[-hk(x)E_{X(h)}\left[\exp\left\{-\int_0^{t-h} k(X(\tau))\,d\tau\right\}g(X(t-h))\right]\right]$$

$$+ E_x E_{X(h)}\left[\exp\left\{-\int_0^{t-h} k(X(\tau))\,d\tau\right\}g(X(t-h))\right] + o(h).$$

Using the time homogeneity of the diffusion process, we have

$$w(x, t) = E_x[\{1 - hk(x)\}w(X(h), t-h)] + o(h).$$

We next apply a Taylor expansion to $w(\cdot, t-h)$ about its state variable, yielding

$$w(x, t) = E_x\left[\{1 - hk(x)\}\left(w(x, t-h) + \{X(h) - x\}\frac{\partial w}{\partial x}(x, t-h)\right.\right.$$

$$\left.\left. + \tfrac{1}{2}\{X(h) - x\}^2 \frac{\partial^2 w}{\partial x^2}(x, t-h)\right)\right] + o(h)$$

and then take expectations to get

$$w(x, t) = [1 - hk(x)]\left[w(x, t-h) + \mu(x)h\frac{\partial w}{\partial x}(x, t-h)\right.$$

$$\left. + \tfrac{1}{2}\sigma^2(x)h\frac{\partial^2 w}{\partial x^2}(x, t-h)\right] + o(h)$$

or

$$w(x, t) - w(x, t-h) = -hk(x)w(x, t-h) + h\mu(x)\frac{\partial w}{\partial x}(x, t-h)$$

$$+ h\tfrac{1}{2}\sigma^2(x)\frac{\partial^2 w}{\partial x^2}(x, t-h) + o(h).$$

Dividing by h and passing to the limit yields

$$\frac{\partial w}{\partial t}(x, t) = -k(x)w(x, t) + \mu(x)\frac{\partial w}{\partial x}(x, t) + \frac{1}{2}\sigma^2(x)\frac{\partial^2 w}{\partial x^2}(x, t). \qquad (5.39)$$

The necessary initial condition is $w(x, 0) = g(x)$.

P. LEVY'S ARCSINE LAW

For a standard Brownian motion we construct the occupation time process

$$\Phi(t) = \text{the time spent in the positive half line up to time } t. \qquad (5.40)$$

We claim that

$$\Pr\{\Phi(t) < \tau | X(0) = 0\} = \frac{2}{\pi}\arcsin\sqrt{\frac{\tau}{t}}, \qquad 0 \le \tau \le t \qquad \text{(arcsine law)}$$

$$(5.41)$$

whose density is $f_\Phi(s) = 1/\pi\sqrt{s(t-s)}$ for $0 < s < t$.
   In order to validate (5.41) we define

$$k(x) = \begin{cases} \beta & \text{if} \quad x > 0, \\ 0 & \text{if} \quad x \le 0, \end{cases} \qquad (5.42)$$

where $\beta > 0$ is a free parameter at our disposal. Obviously

$$\beta\Phi(t) = \int_0^t k(X(s))\, ds.$$

Consider

$$v(x, t) = E_x[e^{-\beta\Phi(t)}] = E_x\left[\exp\left\{-\int_0^t k(X(s))\, ds\right\}\right]. \qquad (5.43)$$

This is exactly (5.38) with $g(x) \equiv 1$. Accordingly, from the development of (5.39) we should expect that $v(x, t)$ uniquely solves

$$\frac{\partial v(x, t)}{\partial t} = \begin{cases} -\beta v(x, t) + \frac{1}{2}\frac{\partial^2 v(x, t)}{\partial x^2}, & x > 0, \quad t > 0, \\[2mm] \frac{1}{2}\frac{\partial^2 v(x, t)}{\partial x^2}, & x \le 0, \quad t > 0, \end{cases} \qquad (5.44)$$

subject to the initial condition $v(x, 0+) \equiv 1\ (=g(x))$ and the continuity relations

$$v(0+, t) = v(0-, t) \qquad \text{and} \qquad \frac{\partial v}{\partial x}(0+, t) = \frac{\partial v}{\partial x}(0-, t), \qquad t > 0. \qquad (5.45)$$

Observe that we expect continuity for $v(x, t)$ and its derivative despite the discontinuity in the second derivative manifested in the differential equation (5.44).

It is difficult to handle (5.44) directly and more convenient to work with the (Laplace transform) function

$$V(x, \lambda) = \int_0^\infty e^{-\lambda t} v(x, t)\, dt \qquad \text{defined for} \quad \lambda > 0. \tag{5.46}$$

The corresponding differential equation for $V(x, \lambda)$ (obtained directly from (5.44) involving also an integration by parts) takes the form

$$(\lambda + \beta)V(x, \lambda) - \frac{1}{2}\frac{\partial^2 V(x, \lambda)}{\partial x^2} - 1 = 0 \qquad \text{for all} \quad x > 0,$$

$$\lambda V(x, \lambda) - \frac{1}{2}\frac{\partial^2 V(x, \lambda)}{\partial x^2} - 1 = 0 \qquad \text{for all} \quad x \le 0, \tag{5.47}$$

subject to the obvious conditions that $V(x, \lambda)$ is bounded for all real x, $\lambda > 0$, and, owing to (5.45),

$$V(0+, \lambda) = V(0-, \lambda), \qquad V'(0+, \lambda) = V'(0-, \lambda) \tag{5.48}$$

(the prime indicates differentiation with respect to x). Relations (5.47) on the positive and negative real line present ordinary differential equations which can be readily solved with insistence on a *bounded* solution, thereby yielding

$$V(x, \lambda) = \begin{cases} \dfrac{1}{\lambda + \beta} + Ae^{-\sqrt{2(\lambda+\beta)}\,x}, & x > 0, \\[2ex] \dfrac{1}{\lambda} + Be^{\sqrt{2\lambda}\,x}, & x \le 0. \end{cases} \tag{5.49}$$

The continuity conditions (5.48) reduce to $A + 1/(\lambda + \beta) = B + 1/\lambda$ and $A\sqrt{2(\lambda + \beta)} = -B\sqrt{2\lambda}$. Their solution is

$$A = \frac{1}{\sqrt{\lambda}(\lambda + \beta)}(\sqrt{\lambda + \beta} - \sqrt{\lambda})$$

and

$$B = -\frac{1}{\lambda\sqrt{\lambda + \beta}}(\sqrt{\lambda + \beta} - \sqrt{\lambda}), \tag{5.50}$$

and

$$\int_0^\infty e^{-\lambda t} v(0, t)\, dt = V(0, \lambda) = A + \frac{1}{\lambda + \beta} = \frac{1}{\sqrt{\lambda}\sqrt{\lambda + \beta}} \qquad \text{for} \quad \lambda > 0. \tag{5.51}$$

Being a Laplace transform, the solution $v(0, t)$ to (5.51) is unique, and thus, to establish that

$$v(0, t) = \int_0^t \frac{1}{\pi \sqrt{s} \sqrt{t - s}} e^{-\beta s} \, ds \quad \text{for} \quad \beta > 0 \tag{5.52}$$

we need only to verify that (5.52) satisfies (5.51). To this end,

$$\int_0^\infty e^{-\lambda t} \left\{ \int_0^t \frac{1}{\pi \sqrt{s} \sqrt{t - s}} e^{-\beta s} \, ds \right\} dt = \frac{1}{\pi} \int_0^\infty \frac{1}{\sqrt{s}} e^{-\beta s} \left\{ \int_s^\infty e^{-\lambda t} \frac{1}{\sqrt{t - s}} \, dt \right\} ds$$

$$= \frac{1}{\pi} \left\{ \int_0^\infty s^{-1/2} e^{-(\lambda + \beta)s} \, ds \right\} \left\{ \int_0^\infty v^{-1/2} e^{-\lambda v} \, dv \right\}$$

$$= \frac{1}{\sqrt{\lambda} \sqrt{\lambda + \beta}}.$$

Finally, (5.52) and the definition $v(0, t) = E[e^{-\beta \Phi} | X(0) = x] = \int_0^t e^{-\beta s} f_\Phi(s) \, ds$ shows that the density for Φ is

$$f_\Phi(s) = \frac{1}{\pi \sqrt{s} \sqrt{t - s}} \quad \text{for} \quad 0 < s < t,$$

thereby completing the derivation of (5.41).

6: Boundary Classification for Regular Diffusion Processes

Let $\{X(t), t \geq 0\}$ be a regular diffusion process on an interval I having left boundary l and right boundary r. For x in (l, r) we postulate the continuous infinitesimal drift and variance coefficients $\mu(x)$ and $\sigma^2(x) > 0$, respectively.

In this section we develop the modern classifications of possible behavior near the boundaries l and r. We concentrate on the left boundary l, the right being entirely similar. The approach is to let a decrease to l in the quantities

$$u(x) = u_{a,b}(x) = \Pr\{T_b < T_a | X(0) = x\}, \qquad l < a < x < b < r, \tag{6.1}$$

and

$$v(x) = v_{a,b}(x) = E[T_{a,b} | X(0) = x], \qquad l < a < x < b < r, \tag{6.2}$$

where T_z is the hitting time to z and $T_{a,b} = T_a \wedge T_b = \min\{T_a, T_b\}$.

Recall the scale function $S(x)$ with explicit expression

$$S(x) = \int_{x_0}^x s(\xi) \, d\xi; \qquad s(\xi) = \exp\left\{ -\int_{\xi_0}^\xi [2\mu(\eta)/\sigma^2(\eta)] \, d\eta \right\}, \tag{6.3}$$

where x_0 and ξ_0 are arbitrary fixed points inside (l, r). As indicated in Section 3, the particular choice of x_0 and ξ_0 is of no relevance.

It simplifies the exposition if we introduce the *scale measure*, the function $S[J]$ of closed intervals $J = [c, d] \subset (l, r)$ defined by

$$S[J] = S[c, d] = S(d) - S(c).$$

We denote both the scale function and scale measure by the same symbol S; no confusion results.

We freely use the scale measure $dS(x) = S[dx]$ of an infinitesimal interval $[x, x + dx]$ with $S[dx] = S(x + dx) - S(x) = dS(x) = s(x)\, dx$. For example, we evaluate $\int_c^d f(x)\, dS(x)$ using the usual integral $\int_c^d f(x)s(x)\, dx$, for, say, piecewise continuous functions $f(x)$.

The reader should check that $0 < S[c, d] < \infty$ for $l < c < d < r$, and that

$$S[c, d] = S[c, x] + S[x, d] \qquad \text{for} \quad l < c < x < d < r. \tag{6.4}$$

Similarly we introduce the speed measure M induced by the speed density $m(x) = 1/[\sigma^2(x)s(x)]$, where

$$M[J] = M[c, d] = \int_c^d m(x)\, dx, \qquad J = [c, d] \subset (l, r).$$

Again, $M[J]$ is positive and finite for $J = [c, d] \subset (l, r)$.

In terms of the scale and speed measures, (6.1) and (6.2) are written

$$u(x) = u_{a,b}(x) = S[a, x]/S[a, b], \qquad l < a < x < b < r, \tag{6.5}$$

and

$$v(x) = v_{a,b}(x) = 2\left\{ u(x) \int_x^b S[\eta, b]\, dM(\eta) + [1 - u(x)] \int_a^x S[a, \eta]\, dM(\eta) \right\}. \tag{6.6}$$

It follows from the nonnegativity of the measure S and (6.4) that $S[a, b]$ is monotonic in a for fixed b and that therefore we may define $S(l, b] \leq \infty$ by

$$S(l, b] = \lim_{a \downarrow l} S[a, b] \leq \infty, \qquad l < b < r. \tag{6.7}$$

If $[a, b] \subset (l, r)$, then $0 \leq S[a, b] < \infty$. As an easy consequence of this and (6.4) we have

$$S(l, b] = \infty \qquad \text{for some } b \text{ in } (l, r)$$

if and only if

$$S(l, b] = \infty \qquad \text{for all } b \text{ in } (l, r). \tag{6.8}$$

Please observe the judicious use of parentheses and brackets. We write $S(l, b]$ (and not $S[l, b]$) to emphasize the definition as a limit. The value of $S(l, b]$ depends only on the process parameters in the interior of the state space and, indeed, whether or not the boundary point l is included among the possible states is immaterial.

We turn to the hitting time of l. For each sample path starting at $X(0) = x$ in (a, b) the hitting time T_a is a monotonically nonincreasing function of a. It follows that we may define the random time $T_{l+} = \lim_{a \downarrow l} T_a \leq \infty$. We now show that $T_{l+} = T_l$, the hitting time to l. When $X(0) = x$ in (a, b), certainly $T_a \leq T_l$, whence $T_{l+} = \lim_{a \downarrow l} T_a \leq T_l$. If $T_{l+} = \infty$, then $T_l = T_{l+} = \infty$, so suppose $T_{l+} < \infty$. Because the paths are continuous,

$$X(T_{l+}) = \lim_{a \downarrow l} X(T_a) = \lim_{a \downarrow l} a = l > -\infty$$

when $T_{l+} < \infty$, and thus $T_{l+} \geq T_l = \inf\{t \geq 0, X(t) = l\}$. We have shown both $T_{l+} \leq T_l$ and $T_{l+} \geq T_l$. Thus $T_{l+} = T_l \leq \infty$.

Note that T_l is defined even when l is not a possible state (and then $T_l = \infty$).

The following result ensues directly from (6.5) and the preceding discussion.

Lemma 6.1.   (i)   *Suppose* $S(l, x_0] < \infty$ *for some* x_0 *in* (l, r). *Then*

$$\Pr\{T_{l+} \leq T_b | X(0) = x\} > 0 \text{ for all } l < x < b < r.$$

(ii)   *Suppose* $S(l, x_0] = \infty$ *for some* x_0 *in* (l, r). *Then*

$$\Pr\{T_{l+} < T_b | X(0) = x\} = 0 \text{ for all } l < x < b < r.$$

Remark 6.1.   For distinct points a, b in (l, r), the equality $T_a = T_b$ cannot occur, since a continuous path cannot be simultaneously in two distinct places. But $T_{l+} = T_b$ can occur, provided both are infinite. Brownian motion with negative drift to $l = -\infty$ provides a simple example. For those paths starting at $x < b$ which never reach b, then $T_b = \infty$ by convention, while the drift to $l = -\infty$ implies $T_{l+} = \infty$ in a possibly different sense.

In view of Lemma 6.1 we have the following definition.

Definition 6.1   The boundary l is *attracting* if $S(l, x] < \infty$ and this criterion applies independently of x in (l, r).

Example 1.   For a standard Brownian motion, $l = -\infty$ is not attracting since $S[a, x] = x - a$ (up to a fixed multiplicative constant) and $S(-\infty, x] = \infty$. On the other hand, suppose the process has drift $\mu = -\alpha < 0$. Then $S[a, x] = e^{2\alpha x} - e^{2\alpha a} \to e^{2\alpha x} < \infty$ as $a \to -\infty$. Thus $l = -\infty$ is attracting for a Brownian motion with negative drift.

We learn from this example that an attracting boundary need not be in the state space of the process. The example when $\mu = -\alpha < 0$ also shows that the probability of reaching an attracting boundary *in finite time* may be zero. (Lemma 6.2 will refine this conclusion.)

While $l = -\infty$ is attracting for a Brownian motion with negative drift, it is not attainable in finite expected time. This leads to the next refinement in the study of boundary behavior.

WHEN IS A BOUNDARY ATTAINABLE IN FINITE EXPECTED TIME?

Consider an attracting boundary l wherein $S(l, x] < \infty$ for all x in (l, r). As affirmed by Lemma 6.1, the boundary l can be reached prior to reaching an arbitrary state b with positive probability from any interior starting point $x < b$, although not necessarily in finite time.

We next evaluate, with b fixed, $b < r$,

$$\lim_{a \downarrow l} E_x[T_a \wedge T_b], \qquad x < b < r,$$

the expected first passage time for the attainment of the boundary l or the level b commencing from x. (We use freely, as in Section 5, the notation E_x to indicate the expectation corresponding to the initial state x.) Recall that $T_a \wedge T_b$ denotes the time of first hitting the level a or b.

Referring to (6.5) and (6.6), we see that

$$\lim_{a \downarrow l} E_x[T_a \wedge T_b] = \lim_{a \downarrow l} \frac{2S[a, x]}{S[a, b]} \int_x^b S[\xi, b] \, dM(\xi)$$

$$+ \lim_{a \downarrow l} \frac{2S[x, b]}{S[a, b]} \int_a^x S[a, \xi] \, dM(\xi).$$

Since l is attracting by assumption, it is obvious that $\lim_{a \downarrow l}(S[a, x]/S[a, b])$ is finite and positive, and accordingly the first right-hand term above is finite. Similarly, $\lim_{a \downarrow l}(S[x, b]/S[a, b])$ is finite and positive. It follows that

$$\lim_{a \downarrow l} E_x[T_a \wedge T_b] < \infty \qquad \text{if and only if} \qquad \lim_{a \downarrow l} \int_a^x S[a, \xi] \, dM(\xi) < \infty.$$
$$\tag{6.9}$$

Thusly motivated we define (employing some obvious interchanges of order of integration)

$$\Sigma(l) = \lim_{a \downarrow l} \int_a^x S[a, \xi] \, dM(\xi) = \int_l^x S(l, \xi] \, dM(\xi) = \int_l^x \left\{ \int_l^\xi s(\eta) \, d\eta \right\} m(\xi) \, d\xi$$

$$= \int_l^x \left\{ \int_\eta^x m(\xi) \, d\xi \right\} s(\eta) \, d\eta = \int_l^x M[\eta, x] \, dS(\eta). \tag{6.10}$$

Notice that we have introduced the notation $\Sigma(l, x) = \Sigma(l)$ to represent the above double integral. It depends on x but in later considerations only whether its value is finite or infinite is relevant and we can therefore suppress the dependence on x without ambiguity. (See the proof of Lemma 6.2, for example.)

Expressed in terms of $\Sigma(l)$ we have the following dichotomy.

Definition 6.2.   The boundary l is said to be

(i)   *attainable* if $\Sigma(l) < \infty$,
(ii)   *unattainable* if $\Sigma(l) = \infty$.

An elementary argument shows that $S(l, x] < \infty$ whenever $\Sigma(l) < \infty$, and hence, if l is attainable, then l is attracting. An unattainable boundary may or may not be attracting.

Example 2.   Consider a Brownian motion with negative drift. Then $l = -\infty$ is attracting but unattainable since the mean time to reach l or b from any state $x < b$ is infinite.

Example 3.   Consider an absorbing Brownian motion on the state space $[0, \infty)$ (Example B of Section 2) where $l = 0$ is an absorbing state. Then l is attracting and attainable since $E[T_0 \wedge T_b | X(0) = x] = x(b - x) < \infty$ for $0 < x < b$. However, $E[T_0 | X(0) = x] = \infty$. Thus the mean time to reach an attainable boundary may be infinite.

When a boundary l is attainable, then those realizations for which it is reached have finite expected time when limited by the alternative that a prescribed state b is reached first. Lemma 6.2, which follows, shows that an attainable boundary can be reached in finite time with positive probability. (Compare with Example 1.) The expected time to reach an unattainable boundary is always infinite.

Lemma 6.2.   *Let l be an attracting boundary and suppose $l < x < b < r$. Then the following are equivalent:*

(i) $\Pr\{T_l < \infty | X(0) = x\} > 0$;
(ii) $E[T_l \wedge T_b | X(0) = x] < \infty$;
(iii) $\Sigma(l) = \int_l^x S(l, \eta]\, dM(\eta) < \infty$.

Proof.   That (ii) and (iii) are equivalent was shown in (6.9). To show that (ii) implies (i), assume that $E[T_l \wedge T_b | X(0) = x] < \infty$. Then $T_l \wedge T_b < \infty$ (with probability 1) and, according to Remark 6.1, $T_l \neq T_b$. Since l is attainable, then $\Pr\{T_l \leq T_b | X(0) = x\} > 0$, and thus $\Pr\{T_l < T_b | X(0) = x\} > 0$. Finally,

$$\Pr\{T_l < \infty | X(0) = x\} \geq \Pr\{T_l < T_b \leq \infty | X(0) = x\} > 0.$$

The last part of the proof is to show that (i) implies (iii). Suppose that $\Pr\{T_l < \infty | X(0) = x\} > 0$. Then there exists $t > 0$ for which

$$\Pr\{T_l < t | X(0) = x\} = \alpha > 0.$$

Every path starting at x and reaching l prior to time t visits every intervening state ξ in $(l, x]$, and then travels from ξ to l. We formalize this observation by writing $T_l = T_\xi + T_l^+(\xi)$, where $T_l^+(\xi)$ is the duration from the first visit to ξ to the first visit to l. Then $T_l \geq T_l^+(\xi)$, whence

$$0 < \alpha \leq \Pr\{T_l^+(\xi) \leq t\} = \Pr\{T_l \leq t \mid X(0) = \xi\}$$
$$\leq \Pr\{T_l \wedge T_x \leq t \mid X(0) = \xi\}, \qquad \text{for any } \xi \text{ in } (l, x].$$

It follows that

$$\sup_{\xi \text{ in } (l,x]} \Pr\{T_l \wedge T_x > t \mid X(0) = \xi\} \leq 1 - \alpha < 1,$$

and, then by induction using the Markov property, that

$$\sup_{\xi \text{ in } (l,x]} \Pr\{T_l \wedge T_x > nt \mid X(0) = \xi\} \leq (1 - \alpha)^n \qquad \text{for} \quad n \geq 1.$$

From the latter expression, we derive $E_\xi[T_l \wedge T_x] \leq t/\alpha < \infty$. But, because (ii) and (iii) are equivalent, then $\int_l^\xi S(l, \eta] \, dM(\eta) < \infty$.

Finally, because $\int_\xi^x S(l, \eta] \, dM(\eta) < \infty$, then

$$\int_l^x S(l, \eta] \, dM(\eta) = \int_l^\xi S(l, \eta] \, dM(\eta) + \int_\xi^x S(l, \eta] \, dM(\eta) < \infty. \quad \blacksquare$$

Roughly speaking, $\Sigma(l)$ measures the time it takes to reach the boundary l or an alternative interior state b starting from an interior point $x < b$. We next introduce the quantities

$$M(l, x] = \lim_{a \downarrow l} M[a, x] \tag{6.11}$$

and

$$N(l) = \int_l^x S[\eta, x] \, dM(\eta) = \int_l^x M(l, \xi] \, dS(\xi). \tag{6.12}$$

Then $M(l, x]$ measures the speed of the process near l and $N(l)$ roughly measures the time it takes to reach an interior point x in (l, r) starting at the boundary l.

The modern classification of boundary behavior is based on the values (actually on whether they are finite or infinite) of the four functionals $S(l, x]$, $\Sigma(l)$, $N(l)$, and $M(l, x]$. These criteria are not independent. Lemma 6.3 delineates their relationships.

Lemma 6.3. *The following relations hold between $S(l, x]$, $\Sigma(l)$, $N(l)$, and $M(l, x]$:*

   (i) $S(l, x] = \infty$ *implies* $\Sigma(l) = \infty$,

  (ii) $\Sigma(l) < \infty$ *implies* $S(l, x] < \infty$,

 (iii) $M(l, x] = \infty$ *implies* $N(l) = \infty$,

 (iv) $N(l) < \infty$ *implies* $M(l, x] < \infty$,

  (v) $\Sigma(l) + N(l) = S(l, x] M(l, x]$.

Proof. (i) $S(l, x] = \infty$ implies $S(l, \xi] = \infty$ for all ξ in (l, r) and hence

$$\Sigma(l) = \int_l^x S(l, \xi] \, dM(\xi) = \infty$$

since M is a strictly positive measure. Statement (ii) is the contrapositive to (i).
(iii) $M(l, x] = \infty$ implies $M(l, \xi] = \infty$ for all ξ whence $N(l) = \int_l^x M(l, \xi] \, dS(\xi)$
$= \infty$ since S is a strictly positive measure. Statement (iv) is the contrapositive to
(iii). (v) Upon summing

$$\Sigma(l) = \int_l^x S(l, \xi] \, dM(\xi)$$

and

$$N(l) = \int_l^x S[\xi, x] \, dM(\xi),$$

we obtain

$$\Sigma(l) + N(l) = \int_l^x \{S(l, \xi] + S[\xi, x]\} \, dM(\xi)$$

$$= S(l, x] \int_l^x dM(\xi) \qquad [\text{by (6.4)}]$$

$$= S(l, x] M(l, x]. \qquad \blacksquare$$

Table 6.1 lists the 16 combinations of assignments of finite or infinite
values to the four quantities $S(l, x]$, $\Sigma(l)$, $N(l)$, and $M(l, x]$. Those that are ruled
out by Lemma 6.3 are indicated by X with the appropriate reason noted.

The six realizable combinations of boundary criteria values have been
grouped and labeled differently by different authors. William Feller introduced
the original classification, adhered to by most American probabilists. The
Russian school uses a slightly different formalization. Both groupings have
their merits and are now juxtaposed.

Table 6.2 lists again the six possible combinations, rows (1), (6) (8), (11),
(12), and (16) in Table 6.1 along with the terminology of both the Feller and
the Russian classification schemes.

(a) *Regular Boundary*

A diffusion process can both enter and leave from a regular boundary. For the
full characterization of the process, the behavior at the boundary must be
specified, and this can be done by assigning a speed $M[\{l\}]$ to the boundary l
itself. The behavior can range from absorption ($M[\{l\}] = \infty$) to reflection
($M[\{l\}] = 0$) as epitomized by absorbing and reflecting Brownian motion.

TABLE 6.1 All combinations of assignments of finite or infinite values to the four quantities $S(l, x]$, $\Sigma(l)$, $M(l, x]$, and $N(l)$ defined in (6.7) and (6.10)–(6.12), respectively

	$S(l, x]$	$M(l, x]$	$\Sigma(l)$	$N(l)$	Possible?	Reason
(1)	$< \infty$	$< \infty$	$< \infty$	$< \infty$	Yes	
(2)	$< \infty$	$< \infty$	$< \infty$	$= \infty$	X	$\Sigma(l) + N(l) = S(l, x] \times M(l, x]$
(3)	$< \infty$	$< \infty$	$= \infty$	$< \infty$	X	$\Sigma(l) + N(l) = S(l, x] \times M(l, x]$
(4)	$< \infty$	$< \infty$	$= \infty$	$= \infty$	X	$\Sigma(l) + N(l) = S(l, x] \times M(l, x]$
(5)	$< \infty$	$= \infty$	$< \infty$	$< \infty$	X	$N(l) < \infty \Rightarrow M(l, x] < \infty$
(6)	$< \infty$	$= \infty$	$< \infty$	$= \infty$	Yes	
(7)	$< \infty$	$= \infty$	$= \infty$	$< \infty$	X	$N(l) < \infty \Rightarrow M(l, x] < \infty$
(8)	$< \infty$	$= \infty$	$= \infty$	$= \infty$	Yes	
(9)	$= \infty$	$< \infty$	$< \infty$	$< \infty$	X	$\Sigma(l) < \infty \Rightarrow S(l, x] < \infty$
(10)	$= \infty$	$< \infty$	$< \infty$	$= \infty$	X	$\Sigma(l) < \infty \Rightarrow S(l, x] < \infty$
(11)	$= \infty$	$< \infty$	$= \infty$	$< \infty$	Yes	
(12)	$= \infty$	$< \infty$	$= \infty$	$= \infty$	Yes	
(13)	$= \infty$	$= \infty$	$< \infty$	$< \infty$	X	$\Sigma(l) < \infty \Rightarrow S(l, x] < \infty$
(14)	$= \infty$	$= \infty$	$< \infty$	$= \infty$	X	$\Sigma(l) < \infty \Rightarrow S(l, x] < \infty$
(15)	$= \infty$	$= \infty$	$= \infty$	$< \infty$	X	$N(l) < \infty \Rightarrow M(l, x] < \infty$
(16)	$= \infty$	$= \infty$	$= \infty$	$= \infty$	Yes	

Behavior somewhat between absorption and reflection $(0 < M[\{l\}] < \infty)$ is the sticky barrier phenomenon, where a strictly positive duration is spent at the boundary. This duration at the boundary contains no interval, however, and is not describable in an elementary way.†

Another possibility is to restart the process in the interior of the state space upon first attaining the regular boundary point. While such a process may be Markov it is not a diffusion in our strict sense because the paths are not continuous everywhere. However, the transition density of such a process will satisfy the backward differential equation, and indeed, while it is evolving in the interior of the state space such a process is indistinguishable from a diffusion and thus diffusion process techniques can be brought to bear in its study.

To establish that a boundary point l is regular, it suffices to check that $S(l, x] < \infty$ and $M(l, x] < \infty$.

(b) *Exit Boundary*

In Section 7 we shall show that at an exit boundary l we must have

$$\lim_{b \downarrow l} \lim_{x \downarrow l} \Pr\{T_b < t \,|\, X(0) = x\} = 0 \qquad \text{for all } t > 0.$$

Thus, starting at l (i.e., where the initial point x approaches l) it is impossible to reach any interior state b no matter how near b is to l. This is tantamount to the assertion that once the boundary l is attained, no continuous sample path can be extricated from l.

† The time at the boundary is similar to a Cantor set of positive Lebesgue measure.

TABLE 6.2 The terminology of the Feller and the Russian boundary classification schemes[a]

Row in table 6.1	Criteria $S(l,x]$	$M(l,x]$	$\Sigma(l)$	$N(l)$	Terminology Feller	Russian		
(1)	$<\infty^*$	$<\infty^*$	$<\infty$	$<\infty$	Regular	Regular	Attracting	Attainable
(6)	$<\infty$	$=\infty^*$	$<\infty^*$	$=\infty$	Exit—Trap—Absorbing		Attracting	Attainable
(8)	$<\infty^*$	$=\infty^*$	$=\infty^*$	$=\infty$		Attracting, unattainable	Attracting	Unattainable
(12)	$=\infty^*$	$<\infty^*$	$=\infty$	$=\infty^*$	Natural $(\Sigma(l)=\infty^*, N(l)=\infty^*)$	Natural $(S(l,x]=\infty^*)$	Nonattracting	Unattainable
(16)	$=\infty^*$	$=\infty^*$	$=\infty$	$=\infty$			Nonattracting	Unattainable
(11)	$=\infty^*$	$<\infty$	$=\infty$	$<\infty^*$	Entrance		Nonattracting	Unattainable

[a] The criteria of finite or infinite values for $S(l,x]$, $\Sigma(l)$, $M(l,x]$, and $N(l)$ are defined in (6.7) and (6.10)–(6.12), respectively. Minimal sufficient conditions for establishing each row are indicated by an asterisk. For example, to establish that a boundary is regular, it suffices to verify that $S(l,x] < \infty$ and $M(l,x] < \infty$.

It is possible to preserve the Markov property by having the process sojourn at l for a random duration, necessarily following an exponential distribution if the Markov property is to be maintained, followed by a jump into the interior (l, r) of the state space according to a prescribed probability law. If such discontinuous trajectories are precluded, that is, if the Markov process is a diffusion, then the exit boundary l must be a trap state, i.e., an absorbing point.

If l is an exit boundary, then T_l, the exit time at l, is the limit $T = \lim_{a \downarrow l} T_a$ for processes starting at $X(0)$ in (l, r). Thus, for example, $E[T_l \wedge T_b | X(0) = x]$ $= v_{l,b}(x) = \lim_{a \downarrow l} v_{a,b}(x) = \lim_{a \downarrow l} E[T_a \wedge T_b | X(0) = x]$, and the types of calculations of Section 4 find relevance here as well.

To establish that a boundary l is exit, it suffices to show that $\Sigma(l) < \infty$ but $M(l, x] = \infty$.

(c) Entrance Boundary

An entrance boundary cannot be reached from the interior of the state space, but it is possible, and in many applications quite natural, to consider processes that begin there. Such processes quickly move to the interior never to return to the entrance boundary.

Consider, for the moment, a diffusion process on $[l, r)$ for which l is an entrance boundary, and look at the problem of evaluating, for instance, $w(l) = E[\int_0^{T(b)} g(X(s)) \, ds | X(0) = l]$, where $T(b)$ is the hitting time for b. Because l is inaccessible,

$$T(b) = \lim_{a \downarrow l} T(a) \wedge T(b),$$

$$w(l) = \lim_{x \downarrow l} \lim_{a \downarrow l} w_{a,b}(x)$$

$$= \lim_{x \downarrow l} \lim_{a \downarrow l} E\left[\int_0^{T(a) \wedge T(b)} g(X(s)) \, ds | X(0) = x \right].$$

Thus $w(x)$ may be evaluated by taking the appropriate limits in formula (3.11) of Section 3.

To show that a boundary l is entrance it suffices to establish that $S(l, x] = \infty$ while $N(l) < \infty$.

(d) Natural (Feller) Boundary

Such a diffusion process can neither reach in finite mean time nor be started from a natural boundary. Natural boundaries are omitted from the state space, so that, for example, if l is a natural boundary then the state space may be taken of the form (l, r) or $(l, r]$.

To establish that a boundary l is natural in the Feller sense one needs to show that both $\Sigma(l) = \infty$ and $N(l) = \infty$.

The analogous criteria for the upper boundary r are

$$S[x, r] = \lim_{b \uparrow r} S[x, b], \qquad M[x, r] = \lim_{b \uparrow r} M[x, b],$$

$$\Sigma(r) = \int_x^r M[x, \xi] \, dS(\xi) = \int_x^r S[\eta, r] \, dM(\eta),$$

and

$$N(r) = \int_x^r S[x, \xi] \, dM(\xi) = \int_x^r M[\eta, r] \, dS(\eta).$$

It would be desirable to relate the boundary classification stipulated above with the usual notions of ergodicity, null recurrence, and the transientness of the process. When both boundary points represent natural boundaries, then although l and r are unattainable neither of the possibilities

$$\Pr\left\{ \lim_{t \to \infty} X(t) = l \,|\, X(0) = x_0 \right\} = 1,$$

$$\Pr\left\{ \underline{\lim_{t \to \infty}} \, X(t) = l, \, \overline{\lim_{t \to \infty}} \, X(t) = r \,|\, X(0) = x_0 \right\} = 1$$

is precluded. There may or may not exist a stationary measure approached by the probability distribution of $X(t)$ as $t \to \infty$. Later examples will verify that all possibilities indeed arise.

Where both boundaries are exit, the process is transient and absorption into one of the two boundaries occurs quite rapidly. It will be established later that a diffusion process displaying entrance boundaries at both ends l and r has a limiting stationary distribution.

Prior to developing several equivalent characterizations of the boundary classification we examine the boundary behavior in the above formulation for a number of concrete diffusion models important in the physical and biological sciences.

Example 4. Standard Brownian motion $[(l, r) = (-\infty, \infty)]$. Recall that $dS(\xi) = s(\xi) \, d\xi$ has $s(\xi) \equiv 1$ and $dM(\xi) = m(\xi) \, d\xi$ has $m(\xi) \equiv 1$. An elementary computation reveals that

$$\Sigma(l) = N(l) = \infty,$$

so that $l = -\infty$ is a *natural boundary* and the same is true for the right boundary $r = \infty$.

The detailed discussion of the one-dimensional Brownian motion process covered in Chapter 7 indicates that the $\pm \infty$ points are never attained although every finite point can be reached with probability one in finite time. Thus the process is recurrent. The first passage probability distribution associated with an arbitrary level a commencing from the origin, displayed explicitly in formula (3.4) of Chapter 7, has an infinite first moment.

Furthermore, in the development of the sample path behavior (consult especially Chapter 12) we found

$$\Pr\left\{\varlimsup_{t \to \infty} X(t) = +\infty, \varliminf_{t \to \infty} X(t) = -\infty\right\} = 1$$

i.e., the trajectories of the process oscillate continuously between $+\infty$ and $-\infty$ infinitely often so that the process is actually null recurrent.

Example 5. *Ornstein–Uhlenbeck (O.U.) process* $[(l, r) = (-\infty, \infty), \mu(\xi) = -\xi,$ $\sigma^2(\xi) \equiv 1]$. We obtain

$$S'(\xi) = s(\xi) = e^{\xi^2}, \qquad m(\xi) = e^{-\xi^2}.$$

In order to ascertain the character of the boundary $+\infty$ we need to estimate

$$\Sigma(\infty) = \int_x^\infty \left[\int_x^\xi dM(\eta)\right] dS(\xi) \qquad \text{and} \qquad N(\infty) = \int_x^\infty \left[\int_x^\eta dS(\xi)\right] dM(\eta).$$

For the case at hand, we have

$$N(\infty) = \int_x^\infty \left(\int_\eta^\infty e^{-\xi^2}\, d\xi\right) e^{\eta^2}\, d\eta$$

where x is for convenience taken large but fixed. Observe through exercising twice integration by parts

$$\int_\eta^\infty e^{-\xi^2}\, d\xi = \tfrac{1}{2}\int_\eta^\infty \frac{1}{\xi} 2\xi e^{-\xi^2}\, d\xi = \frac{1}{2\eta} e^{-\eta^2} - \frac{1}{2}\int_\eta^\infty \frac{e^{-\xi^2}}{\xi^2}\, d\xi$$

$$= \frac{1}{2\eta} e^{-\eta^2} - \frac{1}{4\eta^3} e^{-\eta^2} + \frac{3}{4}\int_\eta^\infty \frac{e^{-\xi^2}}{\xi^4}\, d\xi$$

and therefore

$$\int_\eta^\infty e^{-\xi^2}\, d\xi \simeq \frac{1}{2\eta} e^{-\eta^2}, \qquad \eta \to \infty$$

By choosing x_0 large enough and employing the above asymptotic relation we deduce that $N(\infty) = \infty$. It is simpler to verify that $\Sigma(\infty) = \infty$.

The boundary classification asserts that both $-\infty$ and $+\infty$ are natural boundaries. However, recall (cf. Proposition 5.1) that the Ornstein–Uhlenbeck process is strongly ergodic, meaning there exists a unique normal stationary distribution to which the process converges as $t \to \infty$ (cf. Section 5). Brownian motion (B.M.) and the Ornstein–Uhlenbeck (O.U.) process share the common state space, the real line with $\pm\infty$ being natural boundaries in both cases. It is worthwhile to highlight one of the principal differences in the two processes. The former process (B.M.) depicts a null recurrent sample path behavior while the latter process (O.U.) is positively ergodic, due to the restoring force directed toward the origin.

Example 6. *Bessel Process.* The Bessel process $\{Y(t)\}$ of parameter $\alpha \geq 0$ is the one-dimensional diffusion process on $[0, \infty)$ having the infinitesimal coefficients

$$\mu(x) = \frac{\alpha - 1}{2x}, \qquad \sigma^2(x) \equiv 1. \tag{6.13}$$

The transition density is explicitly given by

$$p(t, x, y) = \frac{\exp[-(x^2 + y^2)/2t]}{t(xy)^{(\alpha - 2)/2}} y^{\alpha - 1} I_{(\alpha - 2)/2}\left(\frac{xy}{t}\right), \qquad t > 0, \quad x, y > 0, \tag{6.14}$$

where $I_\nu(z)$ is the modified Bessel function

$$I_\nu(z) = \sum_{k=0}^{\infty} \frac{(z/2)^{2k+\nu}}{k!\,\Gamma(k + \nu + 1)}. \tag{6.15}$$

When $\alpha = n$, a positive integer, then $Y(t)$ is the radial part of an n-dimensional Brownian motion (see Example E of Section 2 and Chapter 7, Section 6).

The scale and speed functions are readily computed to be

$$s(\eta) = \exp\left[-\int^\eta (\alpha - 1)\frac{dz}{z}\right] = \eta^{1-\alpha},$$

$$S(\xi) = \int_1^\xi \eta^{1-\alpha}\, d\eta \tag{6.16}$$

$$= \begin{cases} \dfrac{1}{2 - \alpha}\xi^{2-\alpha} & \text{if } \alpha \neq 2, \\[2mm] \ln \xi & \text{if } \alpha = 2, \end{cases}$$

$$m(\eta) = \eta^{\alpha - 1}.$$

We examine the boundary point 0. Accordingly,

$$\Sigma(0) = \int_0^1 \left[\int_\xi^1 m(\eta)\, d\eta\right] s(\xi)\, d\xi = \int_0^1 \frac{1}{\alpha}(1 - \xi^\alpha)\xi^{1-\alpha}\, d\xi$$

$$= \frac{1}{\alpha}\left(\int_0^1 \xi^{1-\alpha}\, d\xi - 1\right)\begin{cases} < \infty & \text{if } \alpha < 2, \\ = \infty & \text{if } \alpha \geq 2. \end{cases}$$

$$N(0) = \int_0^1 \left[\int_\eta^1 s(\xi)\, d\xi\right] m(\eta)\, d\eta = \frac{1}{2 - \alpha}\int_0^1 (1 - \eta^{2-\alpha})\eta^{\alpha - 1}\, d\eta$$

$$= \frac{1}{2 - \alpha}\left(\int_0^1 \eta^{\alpha - 1}\, d\eta - \tfrac{1}{2}\right)\begin{cases} < \infty & \text{if } \alpha > 0, \\ = \infty & \text{if } \alpha = 0. \end{cases}$$

To sum up,

$$\text{the boundary 0 is} \begin{cases} \text{an entrance boundary} & \text{for} \quad \alpha \geq 2, \\ \text{a regular boundary} & \text{for} \quad 0 < \alpha < 2, \quad (6.17) \\ \text{an exit boundary} & \text{for} \quad \alpha = 0. \end{cases}$$

The infinite point $+\infty$ is a natural boundary for all specifications of α. (Check this.) The interpretation ascribed to (6.17) is the usual one. With $\alpha \geq 2$, and commencing the Bessel process at a point arbitrarily near 0, almost all the trajectories move onto the positive axis and ultimately drift to $+\infty$. However, where the constant drift rate in the positive direction is not of sufficient strength to the extent that $0 \leq \alpha < 2$, then 0 can be attained in finite expected time. To delimit the process completely in this case it is essential to describe the probabilistic laws governing the path behavior whenever the boundary point 0 is reached.

For a regular boundary a variety of boundary behavior can be prescribed in a consistent way, including the contingencies of complete absorption or reflecting, elastic or sticky barrier phenomena, and even the possibility of the particle (path), when attaining the boundary point, waiting there for an exponentially distributed duration followed by a jump into the interior of the state space according to a specified probability distribution function. In the latter event, the process only exhibits continuous sample paths over the interior of the state space.

For $\alpha = 0$ the 0 point behaves as an exit boundary with resulting total absorption in finite time.

Example 7. *Boundary classification of a population growth model.* The diffusion process described by

$$I = [l, r) = [0, \infty), \quad \mu(\xi) = \beta\xi, \quad \text{and} \quad \sigma^2(\xi) = \alpha\xi \quad (\alpha > 0, \beta \geq 0).$$

is a continuous state analog of a branching process. We find

$$dS(x) = e^{-2\beta x/\alpha}\,dx, \qquad dM(x) = \frac{1}{\alpha x}e^{2\beta x/\alpha}\,dx.$$

At the boundary 0 we get

$$\Sigma(0) < \infty, \qquad N(0) = \infty$$

and so 0 is an exit boundary. The infinite point is a natural boundary. Consult Elementary Problem 10 and Problem 9 for further insights into this model.

Example 8. *Wright–Fisher gene frequency model involving mutation pressures.* The model in Case (a), Example F of Section 2 has parameters

$$I = [l, r] = [0, 1], \quad \mu(\xi) = \gamma_2(1 - \xi) - \gamma_1\xi, \quad \text{and} \quad \sigma^2(\xi) = \xi(1 - \xi)$$
$$(\gamma_1, \gamma_2 > 0).$$

A direct calculation gives

$$-\log s(x) = \int^x \frac{2\mu(\xi)}{\sigma^2(\xi)}\,d\xi = \int^x \left(\frac{2\gamma_2}{\xi} - \frac{2\gamma_1}{1-\xi}\right) d\xi = \log[x^{2\gamma_2}(1-x)^{2\gamma_1}],$$

$$S(x) = \int^x s(\xi)\,d\xi = \int^x \frac{1}{\xi^{2\gamma_2}(1-\xi)^{2\gamma_1}}\,d\xi.$$

The speed measure is

$$dM(x) = x^{2\gamma_2-1}(1-x)^{2\gamma_1-1}\,dx$$

The description of the boundaries 0 and 1 depends on the range of γ_1, γ_2 determined by combination of 16 cases delimited by the inequalities

$$\gamma_1 = 0, \qquad 0 < \gamma_1 < \tfrac{1}{2}, \quad \gamma_1 = \tfrac{1}{2}, \quad \gamma_1 > \tfrac{1}{2},$$

$$\gamma_2 = 0, \qquad 0 < \gamma_2 < \tfrac{1}{2}, \quad \gamma_2 = \tfrac{1}{2}, \quad \gamma_2 > \tfrac{1}{2}.$$

We examine the character of the boundary 0. (The endpoint 1 is treated by similar means.)

$$\Sigma(0) = \int_0^{1/2} \left(\int_\xi^{1/2} dM\right) dS(\xi) = \int_0^{1/2} \left[\int_\xi^{1/2} x^{2\gamma_2-1}(1-x)^{2\gamma_1-1}\,dx\right] dS(\xi)$$

$$\approx \int_0^{1/2} (c_1\xi^{2\gamma_2} + c_2)\xi^{-2\gamma_2}\,d\xi$$

($\approx$ signifies "is of the order") where c_2 is a positive constant.

The above quantity is finite iff $2\gamma_2 < 1$. We deduce straightforwardly $N(0) = \int_0^{1/2} (\int_\xi^{1/2} dS)\,dM(\xi)$ is finite provided $\gamma_2 > 0$ and infinite for $\gamma_2 = 0$. It follows that

$$0 \begin{cases} \text{is an exit boundary} & \text{for} \quad \gamma_2 = 0, \\ \text{is a regular boundary} & \text{for} \quad 0 < \gamma_2 < \tfrac{1}{2}, \\ \text{is an entrance boundary} & \text{for} \quad \gamma_2 \geq \tfrac{1}{2}. \end{cases} \qquad (6.18)$$

In a similar manner, we find that

$$1 \begin{cases} \text{is an exit boundary} & \text{for} \quad \gamma_1 = 0, \\ \text{is a regular boundary} & \text{for} \quad 0 < \gamma_1 < \tfrac{1}{2}, \\ \text{is an entrance boundary} & \text{for} \quad \gamma_1 \geq \tfrac{1}{2}. \end{cases} \qquad (6.19)$$

The interpretation of the facts of (6.18) and (6.19) is as follows. When γ_1 and γ_2 exceed or equal $\tfrac{1}{2}$ (implying that 0 and 1 are entrance boundaries) then the process attains a unique stationary distribution (the process is strongly ergodic) whose explicit density is

$$\frac{\Gamma(2\gamma_1)\Gamma(2\gamma_2)}{\Gamma(2\gamma_1 + 2\gamma_2)} x^{2\gamma_2 - 1}(1 - x)^{2\gamma_1 - 1}. \tag{6.20}$$

This stationary density function is also meaningful for the range $0 < \gamma_1 < \frac{1}{2}$ and $0 < \gamma_2 < \frac{1}{2}$ but the boundary points of the process are now regular and it is essential to spell out the behavior of the process whenever 0 and/or 1 is attained. In most of the genetic applications of this mutation model the "appropriate" boundary condition to be imposed is that of a reflecting barrier. The rationale for this statement derives by examining more carefully the boundary behavior for the finite Wright–Fisher Markov chain approximating process to the above diffusion process. (On this point, see Example F of Section 2.)

In the regular case it is sometimes meaningful to impose an exponentially distributed waiting time on the path behavior at the boundaries and after each sojourn at the boundary have the process continue from a random point of $(0, 1)$ whose distribution is (6.20). The correct stationary measure of such a modified process involves mass jumps at the boundaries 0 and 1 plus a density portion of the form (6.20).

The case $\gamma_1 = \gamma_2 = 0$ is of course the model with absorbing boundaries depicting fluctuation of gene frequency in the presence of no mutation forces.

ENTRANCE BOUNDARIES AND STATIONARY MEASURES

For a one-dimensional regular diffusion where both ends constitute entrance boundaries we should expect that the process evolves $(t \to \infty)$ to a unique limiting stationary distribution. The fact that the process $X(t)$ is strongly ergodic (i.e., settles as $t \to \infty$ to its stationary distribution) requires a more recondite analysis which we will not enter into.

We pointed out that a stationary density can be formally calculated from (5.31) to yield

$$\psi(x) = m(x)[C_1 S(x) + C_2]. \tag{6.21}$$

Because l and r are entrance boundaries, then $S(x)$, which is monotonic throughout (l, r), must increase to ∞ as $x \uparrow r$, and must decrease to $-\infty$ as $x \downarrow l$. (Consult Table 6.2.) To maintain $\psi(x)$ positive throughout (l, r), we must have $C_1 = 0$. Then C_2 is chosen to ensure that $\int_l^r \psi(x) \, dx = 1$. Thus, the unique stationary density is

$$\psi(x) = \frac{m(x)}{\int_l^r m(\xi) \, d\xi} = \frac{1}{\sigma^2(x)s(x) \int_l^r [\sigma^2(\xi)m(\xi)]^{-1} \, d\xi}. \tag{6.22}$$

The denominator in (6.22) is finite because both boundaries are entrance. Indeed, the argument just cited requires only that $S(l, x] = S[x, r) = \infty$ while $M(l, x] < \infty$ and $M[x, r) < \infty$, and this holds for certain natural boundaries in addition to entrance boundaries. (Consult Table 6.2.)

7: Some Further Characterizations of Boundary Behavior

For the sake of variety, we focus in this section on the right boundary r, rather than on the left boundary point l, on which the developments of Section 6 are based. The relevant integrals of the boundary classification are

$$\Sigma(r) = \lim_{b \uparrow r} \int_x^b S[\eta, b] \, dM(\eta) = \int_x^r S[\eta, r) \, dM(\eta) = \int_x^r M[x, \xi] \, dS(\xi), \qquad (7.1)$$

$$N(r) = \lim_{b \uparrow r} \int_x^b S[x, \eta] \, dM(\eta) = \int_x^r S[x, \eta] \, dM(\eta) = \int_x^r M[\xi, r) \, dS(\xi), \qquad (7.2)$$

$$S[x, r) = \lim_{b \uparrow r} S[x, b], \qquad (7.3)$$

and

$$M[x, r) = \lim_{b \uparrow r} M[x, b]. \qquad (7.4)$$

We recall a number of facts pertaining to first passage probabilities. Let T_b denote the first hitting time of the level b. The random variable T_b is well defined in an extended sense that allows infinite values by the prescription

$$T_b = \infty \qquad\qquad\qquad \text{if} \quad X(t) \neq b \quad \text{for all} \quad t \geq 0,$$

$$= \inf \{ t \geq 0 : X(t) = b \} \qquad \text{otherwise.}$$

In some circumstances, for instance where certain boundaries are attainable and absorbing, T_b can take the value ∞ with positive probability.

Remark 7.1. For $x < c$, the random variables T_c increase as $c \uparrow r$, and consequently we may define the limit random variable

$$T_{r-} = \lim_{c \uparrow r} T_c \leq \infty.$$

We claim $T_{r-} = T_r$, the hitting time to r. Certainly $T_{r-} \leq T_r$ since $T_c \leq T_r$ for all c. Thus, if $T_{r-} = \infty$, then $T_r = \infty$ and $T_{r-} = T_r$. It remains to consider the case $T_{r-} < \infty$. Then $X(T_{r-}) = \lim_{c \uparrow r} X(T_c) = \lim_{c \uparrow r} c = r$, so that

$$T_{r-} \geq T_r = \inf \{ t \geq 0 : X(t) = r \}.$$

We have shown both $T_{r-} \geq T_r$ and $T_{r-} \leq T_r$. Thus $T_{r-} = T_r$.

An important functional of these hitting time random variables is the expectation

$$u(x) = E_x[e^{-\alpha T_b}], \qquad b < x < r,$$

$$= E_x\left[\exp\left\{-\alpha \int_0^T g(X(s))\,ds\right\}\right] \tag{7.5}$$

with $g(u) \equiv 1$ in the form as appears in Eq. (3.41). Of course, when $T_b = \infty$, then $e^{-\alpha T_b} = 0$ and therefore $u(x) = E_x[e^{-\alpha T_b}, T_b < \infty]$, the qualification under the expectation signifying that the contributions to the expectation occur only for the sample paths over which T_b is finite. The arguments of Section 3 indicate that $u(x)$ satisfies the differential equation [cf. (3.42)]

$$Lu(x) = \tfrac{1}{2}\sigma^2(x)u''(x) + \mu(x)u'(x) = \alpha u(x) \tag{7.6}$$

with the boundary condition $u(b) = 1$.

It is convenient to write (7.6) in its canonical form [see (3.9)]:

$$Lu(x) = \frac{1}{2}\frac{d}{dM}\frac{d}{dS}u(x) = \frac{1}{2}\frac{1}{m(x)}\frac{d}{dx}\left[\frac{1}{s(x)}\frac{d}{dx}u\right](x) = \alpha u(x). \tag{7.7}$$

A one-dimensional diffusion process cannot go from state b to state d without visiting all intermediate points. Coupled with the strong Markov property, this fact leads to an important identity.

Lemma 7.1. *Consider points $l < b < c < d < r$. Then*

(i)
$$E_d[e^{-T_b}] = E_d[e^{-T_c}]E_c[e^{-T_b}] \tag{7.8}$$

and

(ii)
$$E_b[e^{-T_d}] = E_b[e^{-T_c}]E_c[e^{-T_d}]. \tag{7.9}$$

Proof. Clearly, any sample path commencing at d and reaching b must pass through c first. This is expressed by the relationship $T_b = T_c + T_b^+(c)$, according to which a sample path starting at d and reaching the level b first achieves c at time T_c, and from c, first reaches b in the time length $T_b^+(c)$. The superscript $+$ refers to the sample path *after* time T_c.

Invoking the law of total probabilities, conditioning on the bivariate random variable $(X(T_c), T_c)$, we have

$$E_d[e^{-T_b}] = E_d[e^{-\{T_c + T_b^+(c)\}}] = E_d[E[e^{-\{T_c + T_b^+(c)\}}|(X(T_c), T_c)]],$$

and since T_c is trivially determined conditioned on $(X(T_c), T_c)$,

$$= E_d[e^{-T_c}E_c[e^{-T_b^+(c)}|T_c]].$$

Appeal to the strong Markov property now allows us to discard the conditioning clause and to reduce the inner expression to $E_c[e^{-T_b}]$. That the process starts afresh at c conditioned on T_c is exactly what the strong Markov property affirms. The combined result is (7.8).

The verification of (7.9) is entirely similar. ■

For $b < c$ the hitting time random variables T_c increase as c increases. Therefore, $E_b[e^{-T_c}]$, which is certainly bounded between 0 and 1, is decreasing as c increases to r and increasing as b increases. It follows that $\lim_{c \uparrow r} E_b[e^{-\alpha T_c}]$ exists and is an increasing function of b, and consequently the double limit $\lim_{b \uparrow r} \lim_{c \uparrow r} E_b[e^{-\alpha T_c}]$ exists. We now proceed to construct two relevant functionals of the boundary. We again take $b < c < r$ and define, for some $\alpha > 0$,

$$\varphi(r) = \lim_{b \uparrow r} \lim_{c \uparrow r} E_b[e^{-\alpha T_c}] = \lim_{b \uparrow r} E_b[e^{-\alpha T_r}], \tag{7.10}$$

$$\psi(r) = \lim_{b \uparrow r} \lim_{c \uparrow r} E_c[e^{-\alpha T_b}]. \tag{7.11}$$

Their interpretations and relevance for characterizing boundary behavior will be elaborated below.

Lemma 7.2. *Each of the quantities $\varphi(r)$ and $\psi(r)$ is either 0 or 1.*

Proof. Consider ψ first and for definiteness take $\alpha = 1$. Let $b < c < d < r$.

In the identity (7.8) we first let $d \uparrow r$, then $c \uparrow r$, and finally $b \uparrow r$, in that order. With these operations (7.8) yields $\psi(r) = [\psi(r)]^2$ and this equation is only feasible provided $\psi(r)$ equals zero or unity.

In order to prove the result for $\varphi(r)$ we commence from identity (7.9). Letting $d \uparrow r$, $c \uparrow r$, and then $b \uparrow r$, in that order, yields $\varphi(r) = [\varphi(r)]^2$, implying $\varphi(r) = 0$ or 1. This concludes the proof of the lemma. ■

We have assumed that $\alpha = 1$ in the above; it is obvious that the same proof works for any $\alpha > 0$.

Remark 7.2. We pointed out earlier that if $X(0) = b < c$, then the sequence of random variables T_c increases as $c \uparrow r$ and $T_r = \lim_{c \uparrow r} T_c$. We assert that

$$\varphi(r) = 1 \qquad \text{if and only if} \quad \Pr\{T_r < \infty \mid X(0) = b\} > 0. \tag{7.12}$$

When r is an attracting boundary, then $S[b, r) < \infty$ and

$$\Pr\{T_r < T_a \mid X(0) = b\} > 0 \text{ for } a < b < r.$$

The statement that $\Pr\{T_r < \infty \mid X(0) = b\} > 0$, now being affirmed as equivalent to $\varphi(r) = 1$, is a stronger statement. It corresponds to r being an attainable boundary. (See Examples 1–3 of Section 6 and Lemma 6.2.) Equivalent to

(7.12) is the assertion that $\varphi(r) = 0$ if and only if $\Pr\{T_r < \infty \,|\, X(0) = b\} = 0$. To show this, suppose $b < c < r$. Then

$$\lim_{c \uparrow r} E_b[e^{-\alpha T_c}] = \lim_{c \uparrow r} (E_b[e^{-\alpha T_c}; \, T_r = \infty] + E_b[e^{-\alpha T_c}; \, T_r < \infty]). \qquad (7.13)$$

If $\Pr\{T_{r-} < \infty \,|\, X(0) = b\} > 0$, then the right-hand side of (7.13) is strictly positive. Since $E_b[e^{-\alpha T_r}] \leq E_c[e^{-\alpha T_r}]$ for $b < c < r$, it follows that $\varphi(r) > 0$, and hence $\varphi(r) = 1$. If $\Pr\{T_r < \infty \,|\, X(0) = b\} = 0$, then

$$\Pr\{T_r = \infty \,|\, X(0) = b\} = 1,$$

and the right-hand side of (7.13) vanishes. Hence $\varphi(r) = 0$.

Remark 7.3. $\quad \psi(r) = 1$ if and only if $\lim_{c \uparrow r} E_c[e^{-\alpha T_b}] > 0$. This follows directly from its definition (7.11), and the relation $E_r-[e^{-\alpha T_b}] \leq E_r-[e^{-\alpha T_c}]$ for $b < c < r$, in conjunction with Lemma 7.2. Note that in this case there exists $M < \infty$ such that $\lim_{c \uparrow r} \Pr\{T_b \leq M \,|\, X(0) = c\} > 0$. In this sense, the process can escape from r into the interior.

We already introduced the function (7.5)

$$u(x) = E_x[e^{-\alpha T_b}], \qquad b < x < r, \qquad (7.14)$$

which satisfies the differential equation (7.6). We shall also need the function

$$v(x) = \frac{1}{E_b[e^{-\alpha T_x}]}, \qquad b < x < r. \qquad (7.15)$$

It is also correct that $v(x)$ satisfies the differential equation

$$\tfrac{1}{2}\sigma^2(x)v''(x) + \mu(x)v'(x) = \alpha v(x), \qquad b < x < r, \qquad (7.16)$$

together with the boundary condition $v(b) = 1$. Indeed, let $b < x < c < r$. Then, referring to (7.9) in Lemma 7.1, we have

$$E_b[e^{-\alpha T_c}] = E_b[e^{-\alpha T_x}]E_x[e^{-\alpha T_c}].$$

Hence

$$v(x) = \frac{1}{E_b[e^{-\alpha T_x}]} = \frac{E_x[e^{-\alpha T_c}]}{E_b[e^{-\alpha T_c}]} = \frac{u(x)}{\text{const}}.$$

Since v differs from u by a constant, (7.16) follows from (7.6).

The next three theorems relate the nature of the process behavior at the boundaries to the evaluations of the functionals (7.10) and (7.11).

Theorem 7.1. $\quad \varphi(r) = 1$ if and only if $\Sigma(r) < \infty$.

Note. $\quad \Sigma(r) < \infty$ implies that r must be a regular or exit boundary.

Proof. Consider $b < x < r$. Assume $\Sigma(r) < \infty$, which by Lemma 6.3(ii) implies that $S[x, r] < \infty$. Now consider the differential equation for v defined in (7.16), which we write in the canonical form

$$\tfrac{1}{2} d\left[\frac{d}{dS} v(x)\right] = \alpha v(x)\, dM(x). \tag{7.17}$$

Integrating (7.17) once yields $dv(\eta)/dS - dv(b)/dS = 2 \int_b^\eta \alpha v(\xi)\, dM(\xi)$. Integrating again gives

$$v(x) - v(b) - \frac{dv(b)}{dS}\,[S(x) - S(b)] = 2\int_b^x\left[\int_b^\eta \alpha v(\xi)\, dM(\xi)\right] dS(\eta). \tag{7.18}$$

Noting that $v(b) = 1$ and that $v(x)$ is increasing from its definition, we find that

$$v(x) \le 1 + \frac{dv(b)}{dS}\,[S(x) - S(b)] + 2\alpha v(x)\Sigma_b(r), \tag{7.19}$$

where the subscript b on Σ denotes the lower limit of the integration domain in (7.1). Now choose b close to r such that $2\alpha\Sigma_b(r) < \tfrac{1}{2}$, which is possible since we have assumed $\Sigma(r) < \infty$. With this choice of b, (7.19) reduces to

$$\tfrac{1}{2} v(x) \le 1 + \frac{dv(b)}{dS}\,[S(x) - S(b)]. \tag{7.20}$$

Letting $x \uparrow r$ in Eq. (7.20), we deduce $v(r) < \infty$ since $S(r) < \infty$. Comparing to (7.15) we have $\lim_{x\uparrow r} E_b[e^{-\alpha T_x}] > 0$. From this fact, paraphrasing the argument of Remark 7.1, it follows that $\varphi(r) = 1$.

On the other hand, suppose $\Sigma(r) = \infty$. From the identity (7.18), since $v(\xi)$ is increasing, we obtain the lower estimate

$$v(x) = 1 + \frac{dv(b)}{dS}\,[S(x) - S(b)] + 2\alpha\int_b^x\left[\int_b^\eta v(\xi)\, dM(\xi)\right] dS(\eta)$$

$$\ge 1 + \frac{dv(b)}{dS}\,[S(x) - S(b)] + 2\alpha v(b)\Sigma_b(x). \tag{7.21}$$

Letting $x \to r$ implies $v(r) = \infty$. Hence, $\lim_{x\uparrow r} E_b[e^{-\alpha T_x}] = 0$, or $\varphi(r) = 0$. ∎

Theorem 7.2. (i) *If r is an entrance boundary, then $\psi(r) = 1$.*

(ii) *If r is an exit boundary or natural boundary, then $\psi(r) = 0$.*

(iii) *If r is a regular boundary, then $\psi(r) = 0$ or 1.*

Proof. (i) Suppose r is an entrance boundary. In particular, we have that $N(r) < \infty$. Let $a < c < r$, and recall $N_x(r) = \int_x^r S[x, \eta]\, dM(\eta) = \int_x^r M[\xi, r]\, dS(\xi)$.

Now we use the property that $u(x)$ of (7.5) satisfies $Lu(x) = \alpha u(x)$ together with $u(b) = 1$, which again for convenience is displayed in the form

$$d\left[\frac{d}{dS}u(x)\right] = 2\alpha u(x)\, dM(x).$$

Its first integral is

$$\frac{du(c)}{dS} - \frac{du(a)}{dS} = \int_a^c 2\alpha u(\eta)\, dM(\eta). \tag{7.22}$$

In order to proceed we need to establish the following:

Claim. $\lim_{c\uparrow r} du(c)/dS = 0$. To argue for this assertion note that

$$u(x) = E_x[e^{-\alpha T_b}],\ b < x < r,$$

is monotone decreasing and $u(x) \geq 0$. Hence $[1/s(x)]\, du/dx = du/dS$ is negative, while $Lu = \alpha u$ implies

$$\frac{d}{dx}\left(\frac{du}{dS}\right) = \frac{d}{dx}\left[\frac{1}{s(x)}\frac{du}{dx}\right] = 2m(x)[\alpha u(x)] \geq 0.$$

The last inequality implies that $du(x)/dS$ is increasing. Therefore,

$$-1 \leq u(x) - u(b) = \int_b^x \frac{du}{dS}\, dS$$

$$\leq \frac{du(x)}{dS}[S(x) - S(b)] \leq 0. \tag{7.23}$$

The assumption that r is an entrance boundary entails

$$S[b, r) = S(r) - S(b) = \infty.$$

(See Table 6.2.) Finally, sending $x \uparrow r$ in (7.23), we see that $du(x)/dS \to 0$ as $x \uparrow r$. This validates the claim

Resuming the proof of the theorem, we let $c \uparrow r$ and use the just demonstrated fact that $\lim_{c\uparrow r} du(c)/dS = 0$ to reduce (7.22) to

$$-\frac{du(a)}{dS} = 2\int_a^r \alpha u(x)\, dM(x).$$

Integrating this equation with respect to dS yields

$$-u(y) + u(c) = \int_c^y \left[\int_a^r 2\alpha u(x)\, dM(x)\right] dS(a). \tag{7.24}$$

Now let $y \uparrow r$ in (7.24). Since $u(x)$ is decreasing, this gives

$$-u(r) + u(c) \leq 2\alpha u(c)N_c(r) < \infty. \tag{7.25}$$

Choose c close to r, fulfilling $\alpha N_c(r) < \frac{1}{2}$. Then $u(r) \geq \frac{1}{2}u(c) > 0$ and so $u(r) > 0$. Since $u(r) = \lim_{x \uparrow r} E_x[e^{-\alpha T_b}]$, it follows that $\psi(r) = 1$.

(ii) Suppose r is an exit or natural boundary, so that $N(r) = \infty$. Now integrating $d[du(\xi)/dS] = 2\alpha u(\xi)\, dM(\xi)$ from x to a ($x < a$) produces

$$\frac{du(a)}{dS} - \frac{du(x)}{dS} = 2\alpha \int_x^a u(\xi)\, dM(\xi). \tag{7.26}$$

Integrating (7.26) from y to a with respect to $dS(x)$ ($y < a$) yields

$$\frac{du(a)}{dS}\left[S(a) - S(y)\right] - \left[u(a) - u(y)\right] = 2\alpha \int_y^a \left[\int_x^a u(\xi)\, dM(\xi)\right] dS(x). \tag{7.27}$$

Since it is clear that $0 \leq u(a),\, u(y) \leq 1$, we infer

$$1 \geq \frac{du(a)}{dS}\left[S(y) - S(a)\right] + 2\alpha \int_y^a \left[\int_x^a u(\xi)\, dM(\xi)\right] dS(x). \tag{7.28}$$

Since $S(y)$ is increasing, and $y < a$, $S(y) - S(a) < 0$. Also, we already noted that $du(a)/dS \leq 0$. Therefore

$$1 \geq 2\alpha \int_y^a \left[\int_x^a u(\xi)\, dM(\xi)\right] dS(x).$$

By virtue of the property that u is a decreasing function, we deduce

$$1 \geq 2\alpha u(a) \int_y^a \int_x^a dM(\xi)\, dS(x) = 2\alpha u(a) N(a). \tag{7.29}$$

Now let $a \uparrow r$. Then $N(a) \uparrow N(r) = \infty$. In view of inequality (7.29) we must have $\lim_{a \uparrow r} u(a) = u(r-) = 0$, i.e., $\lim_{a \uparrow r} E_a[e^{-\alpha T_b}] = 0$.

Now let $b \uparrow r$, implying that $\psi(r) = 0$.

(iii) This is simply a restatement of the indeterminancy of $\psi(r)$ for a regular boundary. By Lemma 7.2, $\psi(r)$ is either 0 or 1.

The proof of Theorem 7.2 is complete. ∎

For our final theorem supplementing the interpretation of the boundary classifications, we focus on the scale measure in a neighborhood of r. Recall that $S(r) < \infty$ is equivalent to the statement $\lim_{c \uparrow r} \Pr\{T_c < T_b \mid X(0) = a\} > 0$, where $b < a < c < r$.

Assume that r is entrance or natural, which are the only cases in which the diffusion cannot approach r in finite time having started at a point $a < r$. Now we shall inquire whether this diffusion can approach r in infinite time or not; namely, when does $\Pr\{\lim_{t \uparrow \infty} X(t) = r \mid X(0) = a\} > 0$ hold? The next theorem shows that $S(r) < \infty$ is a necessary and sufficient condition for this property.

Remark. The positive probability of the event that $\{\lim_{t \uparrow \infty} X(t) = r\}$ and the continuity of paths well justifies the term attracting associated with $S(r) < \infty$.

Theorem 7.3. *If r is not regular then $S(r) < \infty$ if and only if $\Pr\{\lim_{t \to \infty} X(t) = r \mid X(0) = a\} > 0$.*

Proof. Suppose $S(r) < \infty$. Let $a = a_0 < a_1 < a_2 < \cdots$ be a sequence $a_n \to r$. The event $\{\lim_{t \to \infty} X(t) = r\}$ certainly contains the event $\{T_{a_2} < \infty$ and $T_{a_{n+1}}^{(n)} < T_{a_{n-1}}^{(n)}$ for all $n \geq 2\}$ where $T_x^{(n)}$ means the first passage time to x starting from a_n. We can use the strong Markov property of the process to split this event into a sequence of independent events. In fact,

$$\Pr\left\{\lim_{t \to \infty} X(t) = r \mid X(0) = a\right\} \geq \Pr\{T_{a_2} < \infty \mid X(0) = a\}$$

$$\times \prod_{n=2}^{\infty} \Pr\{T_{a_{n+1}} < T_{a_{n-1}} \mid X(0) = a_n\}$$

$$= \Pr\{T_{a_2} < \infty \mid X(0) = a\} \prod_{n=2}^{\infty} \frac{S(a_n) - S(a_{n-1})}{S(a_{n+1}) - S(a_{n-1})}.$$

$$(7.30)$$

Since $S(r) < \infty$, we can determine the a_n recursively close enough to r so that

$$[S(a_n) - S(a_{n-1})]/[S(r) - S(a_{n-1})] \geq 1 - n^{-2}.$$

Since S is an increasing function, this implies that

$$[S(a_n) - S(a_{n-1})]/[S(a_{n+1}) - S(a_{n-1})] \geq 1 - n^{-2}.$$

Therefore

$$\Pr\left\{\lim_{t \to \infty} X(t) = r \mid X(0) = a\right\} \geq \Pr\{T_{a_2} < \infty \mid X(0) = a\} \prod_{n=2}^{\infty} \left(1 - \frac{1}{n^2}\right) > 0.$$

Suppose $S(r) = \infty$. (This happens only when r is an entrance or natural boundary.) We shall prove that $\Pr\{\lim_{t \uparrow \infty} X(t) = r \mid X(0) = a\} = 0$. We fix a point c, with $c < a < r$. If $c < x < r$, then

$$\Pr\{T_c < \infty \mid X(0) = x\} = \Pr\{\text{there exists a level } b \text{ with } T_c < T_b \mid X(0) = x\}$$

$$= \lim_{b \uparrow r} \Pr\{T_c < T_b \mid X(0) = x\} = \lim_{b \uparrow r} \frac{S(b) - S(x)}{S(b) - S(c)} = 1,$$

$$(7.31)$$

and therefore,

$$\Pr\{T_c = \infty \mid X(0) = x\} = 0. \qquad (7.32)$$

Consider $\Pr\{\lim_{t \uparrow \infty} X(t) = r \mid X(0) = a\}$. A sample path with the property $\lim_{t \uparrow \infty} X(t) = r$ requires for any $c < r$ that there exists a t_0 depending on the path such that for all $s > 0$, $X(t_0 + s) > c$. Thus, the quantity (7.32) can be estimated above by

$$\Pr\left\{\lim_{t \uparrow \infty} X(t) = r \,|\, X(0) = a\right\} \leq \Pr\{\text{there is some } t \text{ such that } X(t + s) > c$$

$$\text{for all } s \geq 0 \,|\, X(0) = a\}$$

$$= \lim_{t \uparrow \infty} \Pr\{X(t + s) > c \text{ for all } s \geq 0 \,|\, X(0) = a\}$$

$$= \lim_{t \uparrow \infty} \Pr\{T_c^+(t) = \infty, c < X(t) < r \,|\, X(0) = a\}$$

(+ refers to the sample path after time t), and then the Markov property applies to show this equals

$$\lim_{t \uparrow \infty} E_a[\Pr\{T_c = \infty \,|\, X(t)\}; c < X(t) < r] = 0 \qquad \text{by (7.31)}.$$

Thus, we have proved that $S(r) = \infty$ implies

$$\Pr\left\{\lim_{t \to \infty} X(t) = r \,|\, X(0) = a\right\} = 0.$$

This completes the proof of Theorem 7.3.

The findings of Theorems 7.1–7.3 are highlighted in tabular form (Table 7.1) to enhance and solidify the boundary classifications.

Another approach to characterizing and discriminating boundary behavior is in terms of the semigroup of operators (11.3) and its associated infinitesimal operator with its domain of definition (see Section 11). We shall also develop there the nature of the infinitesimal operator as influenced by the boundary.

TABLE 7.1

	Exit	Entrance	Natural (Feller)	Regular
Σ	$< \infty$	∞	∞	$< \infty$
N	∞	$< \infty$	∞	$< \infty$
$S = \displaystyle\int^r dS$	$< \infty$	∞	$M, S, \leq \infty;$ and	$< \infty$
$M = \displaystyle\int^r dM$	∞	$< \infty$	at least one is ∞	$< \infty$
φ	1	0	0	1
ψ	0	1	0	0 or 1
$\Pr\left\{\lim\limits_{t \to \infty} X(t) = r\right\}$	> 0	0	0 or > 0	> 0
	Attain r in finite expected time and get stuck there	Cannot reach r from interior of (l, r). If at r, enter (l, r) rapidly		

8: Some Constructions of Boundary Behavior of Diffusion Processes

There are two principal tactics in dealing with the behavior of diffusion processes at boundary points. An analytic delineation uses an appropriate second-order differential operator (the basic infinitesimal operator of the process; see Section 11) coupled with boundary conditions. The nature of these boundary constraints delimits the boundary classification. A second procedure operates by examining the *local time process* at the boundary. We emphasize several constructions founded on the second approach next.

Boundary behaviors are combinations of five basic types:

- (i)   absorbing barrier phenomenon,
- (ii) reflecting barrier action,
- (iii) elastic boundary structure,
- (iv) sticky boundary complex,
- (v)   jump boundary behavior, and instantaneous return processes.

A. TWO CLASSICAL EXAMPLES

The prototypic process with an absorbing boundary is *absorbing Brownian motion*, where the origin of a standard Brownian motion is converted into an absorbing or trap state. More specifically, consider Brownian motion, $\{Z(t), t \geq 0\}$, where the state space is confined to the positive axis subject to the prescription that the sample paths fix at the zero position once it is attained. The explicit transition probability density for absorbing Brownian motion was displayed in Section 4 of Chapter 7. The infinitesimal parameters of the process, pertinent only to the restricted state space $(0, \infty)$, naturally coincide with those of standard Brownian motion.

Reflecting Brownian motion $\{Y(t), t \geq 0\}$ behaves as standard Brownian motion in the interior of its domain $(0, \infty)$. However, when it reaches its zero boundary, then the sample path returns to the interior in a manner reminiscent of that of a light wave reflecting off a mirror. Thus, if Brownian motion is defined on $[0, \infty)$ with a reflecting barrier, it can be viewed as the absolute value process $Y(t) = |W(t)|$ of a standard Brownian motion $W(t)$. The state space again consists of $[l, r) = [0, \infty)$, with 0 now acting as a regular boundary point. The sample paths are manifestly continuous [since those of $W(t)$ are] and at the times when the origin is hit the observed motion displays immediate reflection to the positive axis in a continuous manner.

Local Time†

In order to construct processes depicting the boundary behavior of (iii) and (iv) it is useful to introduce the concept of local time. To this end, we first consider

---

† Some aspects of local time and inverse local time, with applications to generalized Bessel diffusion processes, appear in Section 12.

a simpler notion. For any given subset A (say an open interval) contained in the state space (l, r) let

$$I_A(\zeta) = \begin{cases} 1 & \text{for} \quad \zeta \in A \\ 0 & \text{for} \quad \zeta \notin A \end{cases}$$

be the indicator function of the set A. The *occupation time* of the set A up to time t is defined by

$$L_A(t) = \int_0^t I_A(X(\tau)) \, d\tau. \tag{8.1}$$

Because the sample paths of a diffusion process are continuous, this random variable is well defined as an ordinary integral along each process realization. It is an easy matter to calculate its successive moments. Specifically, for an initial state $X(0) = x$, we have $\Pr\{X(\tau) \in B \mid X(0) = x\} = \int_B p(\tau, x, y) \, dy$, where $p(\tau, x, y)$ is the transition density function of the process. We obtain

$$
\begin{aligned}
E_x[L_A(t)] &= E_x\left[\int_0^t I_A(X(\tau)) \, d\tau \right] \\
&= \int_0^t E_x[I_A(X(\tau))] \, d\tau \\
&= \int_0^t \left[\int_A p(\tau, x, y) \, dy \right] d\tau.
\end{aligned}
\tag{8.2}
$$

The corresponding second-moment calculation produces

$$
\begin{aligned}
E_x[L_A(t)^2] &= E_x\left[\left\{ \int_0^t I_A((X(\tau)) \, d\tau \right\} \left\{ \int_0^t I_A(X(u)) du \right\} \right] \\
&= \int_0^t \int_0^t E_x[I_A(X(\tau)) I_A(X(u))] \, d\tau \, du \\
&= 2 \int_0^t \int_0^\tau \left[\int_A \int_A p(u, x, \zeta) p(\tau - u, \zeta, y) \, d\zeta \, dy \right] du \, d\tau.
\end{aligned}
$$

(An appeal to the Markov property justifies the last equation.)

Returning to the concept of occupation time we specialize $A = A(a, \varepsilon)$ to be $(a - \varepsilon, a + \varepsilon)$, an interval of length 2ε centered at a. Consider the limit random variable

$$\varphi(t, a) = \lim_{\varepsilon \downarrow 0} \frac{1}{2\varepsilon} L_{A(a, \varepsilon)}(t). \tag{8.3}$$

A deep and remarkable result is that

$$\varphi(t, a) \text{ defines a family of random variables,} \tag{8.4}$$

called the *local time* process, valid for $t > 0$, where a is an interior state point. In the case of Brownian motion it can be further established *that $\varphi(t, a)$ is jointly continuous and nondecreasing* in t for almost all sample paths. The modern development of these facts proceeds by the theory of stochastic integrals. We discuss a bit more the nature of $\varphi(t, a)$ and some facts on additive and multiplicative stochastic functionals in Section 12.

The random function $\varphi(t, a)$ can be construed as the density of $L_A(t)$ in the sense that

$$L_A(t) = \int_A \varphi(t, a)\, da. \tag{8.5}$$

The representation (8.5) is commonly used and furnishes a remarkable formula of substantial utility in many advanced theoretical deliberations on stochastic processes.

We should like to emphasize a recondite but important fact. The local time $\varphi(t, a)$ is a density and is not the same quantity as the occupation time of a point,

$$L_{\{a\}}(t) = \int_0^t I_{\{a\}}(X(\tau))\, d\tau \tag{8.6}$$

where

$$I_{\{a\}}(u) = \begin{cases} 1, & u = a, \\ 0, & u \neq a. \end{cases}$$

The right-hand integral in (8.6) is usually identically zero. This happens, in particular, for the case of Brownian motion where the integral vanishes but the local time process $\varphi(t, a)$ can be defined such that $\varphi(t, a)$ is strictly increasing in the neighborhood of $t = 0$ for almost all sample paths starting from $X(0) = a$.

Random Time Changes

We wish to develop the concept of a random time change in order to perform constructions of diffusion processes exhibiting special boundary behavior. The concept of time change has wide ramifications far beyond the applications reviewed here.

Let $A(t)$ be a continuous increasing random function with $A(0) = 0$. We sometimes indicate the dependence on the sample path ω by writing $A(t, \omega)$ rather than $A(t)$. Corresponding to each trajectory ω let $A^{-1}(s)$, $s \geq 0$, denote the inverse function of $A(t)$, defined formally by

$$A^{-1}(s) = \inf\{\zeta, A(\zeta) > s\}.$$

As an example consult Fig. 1. A constant piece for $A(t)$ induces a jump discontinuity in $A^{-1}(s)$. It can be checked routinely that $A^{-1}(s)$ is increasing and right continuous (not necessarily continuous). It suffices to assume that

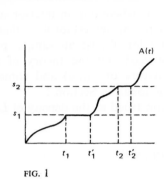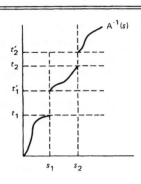

$A(t)$ is right continuous and increasing and then $A^{-1}(s)$ retains the same properties. In this case a jump discontinuity of $A(t)$ gives rise to a constant portion for $A^{-1}(s)$.

Let $X(t)$ be a strong Markov process. A process $A(t, \omega)$ is called an *additive functional associated with* $X(t)$ if

$$A(t + \tau, \omega) = A(t, \omega) + A(\tau, \omega_t^+), \tag{8.7}$$

where ω_t^+ denotes the sample path translated t time units. (Specifically, if ω is a sample path, then ω_t^+ refers to the same sample path but viewed over the time duration subsequent to t.) The function (8.1) provides an example of an additive increasing functional. We validate this assertion now.

Consider

$$L_A(t + \tau, \omega) = \int_0^{t+\tau} I_A(X(u, \omega))\, du = \int_0^t I_A(X(u, \omega))\, du + \int_t^{t+\tau} I_A(X(u, \omega))\, du.$$

$$\tag{8.8}$$

Implementing an obvious change of variables gives

$$\int_t^{t+\tau} I_A(X(u))\, du = \int_0^\tau I_A(X(u + t))\, du. \tag{8.9}$$

But the clear interpretations show that

$$X(u + t) = X(u + t, \omega) = X(u, \omega_t^+). \tag{8.10}$$

Inserting this relationship into (8.9) and then (8.8) we verify instantly that $L_A(t, \omega)$ satisfies (8.7).

Another example of a continuous additive functional is

$$G(t, \omega) = G(t) = g(X(t)) - g(X(0))$$

where g is a continuous function defined on the state space. Further applications of additive functionals and associated martingales are developed in Section 12.

With the availability of an increasing continuous additive functional with respect to $\{X(t)\}$, designated by $A(t) = A(t, \omega)$, we construct the new process

$$\overline{Y}(t) = X(A^{-1}(t)) \tag{8.11}$$

It is a nontrivial fact that $\overline{Y}(t)$ determines a strong Markov process with right continuous sample paths. The proof of this assertion is intricate and well beyond the scope of this book. The above result will be used without further validation to enable us to construct diffusions displaying special boundary behavior.

B. A DIFFUSION WITH AN ELASTIC BOUNDARY

An elastic boundary manifests the properties of both a reflecting and an absorbing boundary. For example, the left endpoint l is an elastic boundary if it is a regular boundary which behaves as a reflecting boundary for a certain random duration after which the process is killed.

We provide a concrete construction of an elastic boundary by confining attention to certain special paths of the standard Brownian motion process. Consider standard Brownian motion $\{B(t), t \geq 0\}$ and fix a level $-c < 0$. Let T_{-c} denote the first passage time to the point $-c$. That is, for a given sample path ω, $T_{-c}(\omega)$ indicates the time when the value $-c$ is first attained. Formally,

$$T_{-c} = \inf\{t \geq 0, B(t) = -c\}.$$

The fact that T_{-c} is finite with probability 1 was established in Chapter 7. A realization ω of a Brownian path until time T_{-c} is depicted in Fig. 2.

Now let $W_{-c}(t) = B(t)$ for $t < T_{-c}$, and undefined for $t \geq T_{-c}$. Thus $W_{-c}(t)$ is the Brownian motion killed upon first reaching $-c$. It remains a diffusion process.

Construct now the additive increasing functional

$$C(t) = C(t, \omega) = \int_0^t I_{[0, \infty]}(B(u)) \, du, \tag{8.12}$$

which is exactly the occupation time of the positive axis. For any sample path ω, $C(t)$ is certainly continuous and increasing but obviously strictly increasing only over the time durations where $B(t) \geq 0$, i.e., at the times when the process occupies the positive axis.

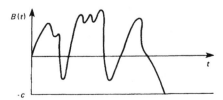

FIG. 2

Manifestly, $C(t)$ is constant during the excursions of $B(t)$ to the negative axis. Consider now the inverse $C^{-1}(s)$. Notice that this function jumps upwards (displays a discontinuity) for the time periods when $B(u)$ visits the negative axis. Now form the process

$$Z(s) = W_{-c}(C^{-1}(s)), \qquad 0 \le s < C(T_{-c}). \tag{8.13}$$

where $W_{-c}(t)$ is the Brownian motion process run only until the stopping time T_{-c}. Under this convention the $Z(s)$ process retains the strong Markov property. (It is tempting to define $Z(s) = 0$ for $s > C(T_{-c})$, but such a designation confounds this event with the events of contact with the zero point prior to time $C(T_{-c})$, and the process no longer possesses the strong Markov property.)

As stated immediately following (8.11), the $Z(s)$ process is strong Markov. Moreover, as the time $s > 0$ proceeds, a jump will occur in the value of $C^{-1}(s)$ exclusively where the $B(u)$ trajectory sojourns on the negative axis. For the obvious reasons the process $C^{-1}(s)$ is referred to as a *random time change*. In more picturesque language, the clock of $C^{-1}(s)$ runs only during the excursion periods of $B(u)$ on the positive axis and effectively ignores the time periods where $B(u)$ traverses the negative axis. It is wise to consult the schematization shown in Fig. 3. Manifestly the $Z(s)$ process terminates (is killed) at the moment ζ equivalent to when the $W(t)$ path crosses zero for the last time before reaching the level $-c$. The thin curve represents a trajectory of the Brownian process stopped at the hitting time of the level $-c$. The arcs of the trajectory where $W(u) \ge 0$ but prior to T_{-c} remain intact since $C^{-1}(s)$ strictly increases during these time periods. The corresponding trajectory of $Z(s)$ is precisely the thick portion of the curve. The Z process is killed after the time $\zeta \equiv C^{-1}(T_c)$ in view of the fact that the Brownian path W then moves on the negative axis until it crosses the level $-c$. It is important to observe that at a jump point of $C^{-1}(s)$ the initial and final value of the Brownian path are both zero, i.e., $W(C^{-1}(s)) = 0$ at these points. It follows in view of this comment that every path of $Z(s) = W(C^{-1}(s))$ is continuous. Indeed, where a path ω traverses the positive axis the Z trajectory coincides with its antecedent Brownian path, and where ω describes motion over the negative axis but prior to time T_{-c} these time durations produce jumps in $C^{-1}(s)$ with the consistent value 0 for Z. The

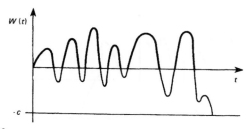

FIG. 3

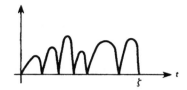

FIG. 4

specific path of $Z(s)$ emanating from the example of Fig. 3 is described in Fig. 4. Thus the boundary zero acts as a reflecting barrier point part of the time such that, upon contact, instantaneous continuous return to the positive axis transpires. This possibility occurs a random number of times, but at some attainment of zero the process is killed.

To sum up, a typical path of $Z(s)$ ranges on the positive axis with a number of continuous reflections from the zero boundary until finally culminating with killing at the origin.

C. A STICKY BOUNDARY

Consider a diffusion process $\{X(t), t \geq 0\}$ on the state space $[l, r)$. A regular boundary l is said to be *sticky* if the occupation time of the process at l

$$L_{\{l\}}(t) = \int_0^t I_{\{l\}}(X(s)) \, ds \qquad (8.14)$$

is positive for all $t > 0$, where $X(0) = l$. As pointed out earlier, the occupation time at a point is not the same as the local time.

To set forth a concrete case of a sticky boundary, take $Y(t)$ as *reflecting Brownian motion*, specifically $Y(t) = |B(t)|$ where $B(t)$ is standard Brownian motion. We take $\varphi(t) = \varphi_{\{0\}}(t)$ to be the local time at the origin. [Sec. (8.4).] Of course, as indicated earlier, $\varphi(t)$ determines an additive increasing functional. We now form the new additive increasing process with respect to $Y(t)$.

$$U(t) = t + \kappa\varphi(t) \qquad (8.15)$$

where κ is a fixed positive constant. Observe that $U(t)$ increases continuously for each sample path from 0 to ∞ [since the $\varphi(t)$ process is continuous nondecreasing]. We also see that $dU(t)/dt$ can differ from 1 only at the time points where a path contacts the origin.

Using $U^{-1}(s)$ as a random time change superimposed on $Y(t)$ we construct the new process

$$\Gamma(s) = Y(U^{-1}(s)) \qquad (8.16)$$

Obviously, $\Gamma(s)$ inherits the state space $[l, r) = [0, \infty)$ of reflecting Brownian motion. Because 0 is a regular boundary point of $Y(t) = |W(t)|$, it follows (why?) that 0 is a regular boundary of the $\Gamma(s)$ process. We now characterize its behavior at the origin by evaluating the occupation time process for $\Gamma(s)$ at $\{0\}$. (Some of the manipulations will be done without full justifications.)

Consider

$$L_{\{0\}}(t;\Gamma) = \int_0^t I_{\{0\}}(\Gamma(s))\,ds = \int_0^t I_{\{0\}}(Y(U^{-1}(s)))\,ds \tag{8.17}$$

Executing the change of variable (for each sample path) $s = U(\tau)$ in (8.17) yields

$$\begin{aligned} L_{\{0\}}(t) &= \int_0^{U^{-1}(t)} I_{\{0\}}[Y(\tau)]\,dU(\tau) \\ &= \int_0^{U^{-1}(t)} I_{\{0\}}(Y(\tau))\,d\tau + \kappa \int_0^{U^{-1}(t)} I_{\{0\}}(Y(\tau))\,d\varphi(\tau). \end{aligned} \tag{8.18}$$

The first integral on the right, corresponding to the occupation time of the set $\{0\}$ for the Brownian motion process evaluated at time $U^{-1}(t)$, is identically zero. In essence, the actual occupation time as measured by the first integral (caution, this is not the local time functional) of a specific state point for a Brownian path over a given time epoch is zero.

Next we examine the second integral. By its very definition, $\varphi(t)$ concentrates all its increase only at the time points where the $Y(t)$ process contacts the origin. In other words $d\varphi(\tau) > 0$ only where $I_{\{0\}}(Y(\tau)) = 1$. Accordingly the final integral in (8.18) reduces to

$$\int_0^{U^{-1}(t)} I_{\{0\}}(Y(\tau))\,d\varphi(\tau) = \int_0^{U^{-1}(t)} d\varphi(\tau) = \varphi(U^{-1}(t)) \tag{8.19}$$

It was pointed out following Eq. (8.6) that for a Brownian path emanating from the origin, the local time process $\varphi(t, 0)$ is strictly positive for all $t > 0$. Therefore, the quantity in (8.19) is positive, and since $\kappa > 0$ by stipulation, we find that the occupation time functional of (8.17) is positive for all $t > 0$. Thus, the barrier $\{0\}$ in the $\Gamma(s)$ process fulfills the properties of the definition of a sticky boundary.

D. JUMP DISCONTINUITY AT A BOUNDARY

Consider a regular diffusion process on (l, r) and suppose l is a regular boundary point. As has often been stated, the infinitesimal rates governing the sample path realizations apply only at the interior points of (l, r). The behavior at a boundary point l requires a separate specification. A feasible pattern is the following.

When l is reached, the process waits there for a random duration (following an exponential distribution) and leaves l by a jump into (l, r) according to a given distribution. The probabilistic transitions for the movement inside (l, r) are thereafter guided by the infinitesimal parameters until the next attainment of the boundary. The waiting and jump effects at successive visits to l are independent. To sum up, the process behavior consists of two types of epochs: In the first, continuous movement occurs until the boundary is attained, and in the second, at each boundary visit, an exponentially distributed sojourn transpires, culminating with a jump return into (l, r).

It is of interest to contrast the nature of the sojourn and jump phenomena possible from a boundary point and to ask whether such behavior can be manifested at interior points of the state space. Certainly, a diffusion process, which by definition necessarily has continuous sample paths cannot exhibit jump discontinuities. The question remains whether there can exist a state c in (l, r) such that, at each attainment of c, a random independently distributed sojourn time is registered at this point before moving on in a continuous manner. Of course, in order to preserve the Markov character of the process, the sojourn time at any given state is necessarily exponentially distributed (cf. Chapter 14, Section 3).

Consider the process beginning at c and suppose it waits a random period at c and moves off continuously. We shall establish presently that the *strong* Markov property coupled with the fact that $X(t)$ is a diffusion precludes the existence of a state c at which such a random waiting period can appear. Let η_c denote the first departure time from the position c, where $X(0) = c$. We claim that η_c is a Markov time. In order to validate this assertion it is necessary to show that $\{\omega, \eta_c(\omega) < t\} \in \mathscr{F}_t$ for each t; in other words we must show that to decide whether the event described by the inequality $\eta_c < t$ holds or not it is enough to know the process values $X(s)$ for all $s \leq t$. But, in view of the continuous paths, $\eta_c < t$ means there exists some rational $s < t$ where $X(s) \neq c$. Obviously the event $X(s) \neq c$ is certainly measurable with respect to $\mathscr{F}_s \subset \mathscr{F}_t$. It follows that η_c is a stopping time as stated.

Now let A be a Borel subset of (l, r). We shall now show that

$$\Pr\{X(\eta_c + t) \in A \,|\, X(\eta_c), X(0) = c\} \neq \Pr\{X(t) \in A \,|\, X(\eta_c)\}, \qquad (8.20)$$

violating the strong Markov property. In fact, let $p(t, a, x)$ denote the transition density function of the $X(t)$ process. That is,

$$p(t, c, x)\, dx = \Pr\{x \leq X(t) \leq x + dx \,|\, X(0) = c\}.$$

Since the *path is continuous*, we infer that $X(\eta_c) = c$. By the law of total probabilities and the strong Markov property, the left-hand side of (8.20) becomes

$$\int_0^t \left\{ \int_A p(t - \tau, c, y)\, dy \right\} \lambda e^{-\lambda \tau}\, d\tau. \qquad (8.21)$$

On the other hand, the right-hand side of (8.20) equals

$$\int_A p(t, c, y)\, dy \qquad (8.22)$$

and the equality of (8.21) and (8.22) is only possible if $\lambda = \infty$, meaning that movement through c is instantaneous. The contradiction in (8.20) can be avoided only if $E[\eta_c] = 0$ for all c in (l, r), or alternatively the sample paths are allowed to be discontinuous. In particular, for a diffusion process nonzero sojourn epochs cannot transpire at interior points of the state space.

E. INSTANTANEOUS RETURN PROCESS

Consider a standard Brownian motion $\{B(t), t \geq 0\}$ on the real line, with $B(0) = 0$. Fix the interval $(-1, 1)$. Construct the return process $Z(t)$ agreeing with $X(t)$ until the point 1 or -1 is reached. At that time, the particle is instantaneously returned to the origin, starting the Brownian motion afresh. This procedure is repeated at each attainment of 1 or -1, so that $Z(t)$ is a stochastic process on the state space $I = (-1, 1)$. Intuitively, it is expected that a limiting and stationary distribution of the return process exists. In this example, we find the form of this limiting distribution.

More generally, consider a regular diffusion $\{X(t), t \geq 0\}$ on the interval $I = (l, r)$, and let $l < a < b < r$. We construct the return process $Z(t)$ relative to $[a, b]$ as follows. Starting at a point x_0 in (a, b), the particle is returned instantaneously to x_0 whenever a or b is reached. After such a return, the subsequent motion of the process behaves just like $X(t)$; this process is repeated at each attainment of level a or b. The resulting process $Z(t)$ consists of recurrent cycles of random time duration $C_1, C_2, C_3, \ldots$, where the C_i are independently and identically distributed, with the same distribution as $T_{a,b} = \min\{T_a, T_b\}$, the first exit time from the interval (a, b), starting from x_0. It follows that

$$E[C_i | X(0) = x_0] = \int_a^b G(x_0, \xi) \, d\xi, \tag{8.23}$$

where $G(x_0, \xi) = G_{[a,b]}(x_0, \xi)$ is the Green function of the process $X(t)$ relative to the interval (a, b); this is displayed explicitly in (3.15).

Let $q(t, y)$ be the density function of $Z(t)$. That is,

$$q(t, y) \, dy = \Pr\{y \leq Z(t) < y + dy | Z(0) = x_0\}.$$

The objective is to evaluate $\alpha(y) = \lim_{t \to \infty} q(t, y)$, the limiting density of $Z(t)$.

To do this, we fix an interval $[y_1, y_2]$ with $a < y_1 < y_2 < b$ and define the process $\{I(t), t \geq 0\}$ by

$$I(t) = \begin{cases} 1 & \text{if } y_1 \leq Z(t) < y_2, \\ 0 & \text{otherwise.} \end{cases} \tag{8.24}$$

In the language of renewal theory, $I(t)$ is on if $y_1 \leq Z(t) < y_2$, and otherwise $I(t)$ is off. We can now split a typical cycle (of length C_i, say) into two parts: in one part, $I(t)$ is on, in the other $I(t)$ is off (see Fig. 5). We have shown that

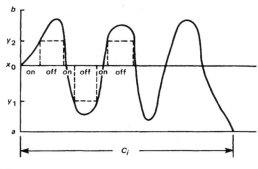

FIG. 5

$\Pr\{I(t) = 1\} = E[I(t)] = \int_{y_1}^{y_2} q(t, y)\,dy$, and we can now appeal to the renewal theorem as used in Section 7.C of Chapter 5 to deduce that

$$\lim_{t \to \infty} \Pr\{I(t) = 1\} = \frac{E[\text{time spent in } (y_1, y_2) \text{ in a cycle} \,|\, Z(0) = x_0)]}{E[\text{time duration of a cycle} \,|\, Z(0) = x_0]}. \quad (8.25)$$

But we showed in Section 3 that the numerator is $\int_{y_1}^{y_2} G(x_0, \xi)\,d\xi$, while, from (8.23), the denominator is just $\int_a^b G(x_0, \xi)\,d\xi$. Hence

$$\lim_{t \to \infty} \int_{y_1}^{y_2} q(t, y)\,dy = \frac{\int_{y_1}^{y_2} G(x_0, \xi)\,d\xi}{\int_a^b G(x_0, \xi)\,d\xi}.$$

Since this holds for every y_1, y_2 in (a, b), we may deduce that

$$\alpha(y) = \lim_{t \to \infty} q(t, y) = G(x_0, y)\bigg/\int_a^b G(x_0, \xi)\,d\xi. \quad (8.26)$$

Example. Take $\{X(t)\}$ to be a standard Brownian motion with $X(0) = 0$, $a = -1$, $b = +1$. Then the limiting density is $\alpha(y) = 1 - |y|$, $-1 < y < 1$.

9: Conditioned Diffusion Processes

Consider a regular diffusion process $\{X(t), t \geq 0\}$ with state space $[0, 1]$ and infinitesimal mean and variance $\mu(x)$ and $\sigma^2(x)$, respectively. Assume that 0 and 1 are exit boundaries, and let $p(t, x, y)$ be the transition density of $X(t)$. It is of interest, especially in certain biological contexts, to concentrate only on the realizations of the process that lead to absorption at 1. Accordingly, $\{X^*(t), t \geq 0\}$ is prescribed to be the process confined to the sample paths involving ultimate absorption at 1. That $X^*(t)$ exhibits only continuous paths is clear, since $X^*(t)$ is merely a restriction of $X(t)$ to part of its original sample space. Moreover, $X^*(t)$ is a Markov process since $X(t)$ is, and past history (beyond the current state) cannot affect where absorption occurs (the reader should apply a formal argument). It follows that $X^*(t)$ is a diffusion process.

For the $X^*(t)$ process it is intuitive that the boundary point 1 remains an exit boundary, while the boundary point 0 is converted into an entrance boundary. We will justify these conclusions later.

We suppose hereafter that the diffusion $X(t)$ satisfies the moment conditions of (1.4), and we shall refer to $X^*(t)$ as an associated *conditioned* diffusion process. The one constructed above is a special case of a large class of such processes.

Our immediate task will be to determine the infinitesimal parameters of the process $\{X^*(t), t \geq 0\}$, and subsequently we shall highlight a number of examples and applications.

To proceed more formally, we assume that the boundaries are attainable in finite expected time, i.e., $E_x[T_0 \wedge T_1] < \infty$ (As usual, T_z denotes the first hitting time of the point z).

We define

$$s(\xi) = \exp\left[-\int_0^\xi \frac{2\mu(\eta)}{\sigma^2(\eta)} d\eta\right] \quad \text{and} \quad S(x) = \int_0^x s(\xi) d\xi.$$

Because both boundaries are exit by assumption, $S(x)$ and even $S(1)$ are finite and $\Pr\{T_1 < T_0 | X(0) = x\} = S(x)/S(1)$, for $0 \leq x \leq 1$.

To simplify the development we assume that

$$S''(x) = s'(x) = -[2\mu(x)/\sigma^2(x)]s(x) \text{ is bounded for } 0 \leq x \leq 1.$$

Let $p^*(t, x, y)$ be the transition density function of the conditioned process $X^*(t)$, or equivalently, the density of the $X(t)$ process, conditioned on the event $T_1 < T_0$. Since the conditioning event has positive probability, we obtain from Bayes' rule

$$p^*(t, x, y) \, dy = \Pr\{y \leq X(t) < y + dy | X(0) = x, T_1 < T_0\}$$

$$= \frac{\Pr\{y \leq X(t) < y + dy | X(0) = x\} \Pr\{T_1 < T_0 | X(0) = x, X(t) = y\}}{\Pr\{T_1 < T_0 | X(0) = x\}}$$

By the Markov property

$$\Pr\{T_1 < T_0 | X(0) = x, X(t) = y\} = \Pr\{T_1 < T_0 | X(0) = y\} = S(y)/S(1).$$

Combining, we have

$$p^*(t, x, y) = \frac{p(t, x, y)S(y)}{S(x)}. \tag{9.1}$$

Consider now

$$\mu^*(\xi) = \lim_{h \downarrow 0} \frac{E[\{X^*(h) - X^*(0)\} | X^*(0) = \xi]}{h} = \lim_{h \downarrow 0} \frac{1}{h} \int p^*(h, \xi, y)(y - \xi) \, dy$$

$$= \lim_{h \downarrow 0} \frac{1}{h} \int \frac{p(h, \xi, y)S(y)(y - \xi)}{S(\xi)} \, dy. \tag{9.2}$$

A Taylor expansion of $S(y)$ about $y = \xi$ (S is differentiable at least twice by definition) leads to

$$S(y) = S(\xi) + (y - \xi)s(\xi) + \frac{(y - \xi)^2}{2} s'(z), \qquad (9.3)$$

where z is in the interval (ξ, y) and depends, of course, on ξ and y.

Substituting from (9.3) into (9.2) yields

$$\mu^*(\xi) = \lim_{h \downarrow 0} \frac{1}{h} \left[\int (y - \xi)p(h, \xi, y) \, dy + \frac{s(\xi)}{S(\xi)} \int (y - \xi)^2 p(h, \xi, y) \, dy \right.$$

$$\left. + \frac{1}{2S(\xi)} \int (y - \xi)^3 p(h, \xi, y)s'(z) \, dy \right]. \qquad (9.4)$$

Using the assumptions that $\lim_{h \to 0}(1/h)E[|X(h) - X(0)|^3 \,|\, X(0) = \xi] = 0$ and that $S''(z) = s'(z)$ is bounded, (9.4) reduces to

$$\mu^*(\xi) = \mu(\xi) + \frac{s(\xi)}{S(\xi)} \sigma^2(\xi). \qquad (9.5a)$$

Consider next

$$\sigma^{2*}(\xi) = \lim_{h \to 0} \frac{1}{h} \int (y - \xi)^2 p^*(h, \xi, y) \, dy$$

$$= \lim_{h \to 0} \frac{1}{h} \left\{ \int (y - \xi)^2 p(h, \xi, y) \left[1 + (y - \xi) \frac{s(z)}{S(\xi)} \right] dy \right\} \qquad (9.5b)$$

$$= \sigma^2(\xi) \qquad \text{since } s \text{ is bounded.}$$

We are now in a position to be able to classify the boundary behavior of the conditioned process $X^*(t)$. It follows from the definitions ((3.5)–(3.7)) that the scale functions $S^*(x)$ and $s^*(x)$, and the speed density $m^*(x)$ of the conditioned process are given by

$$s^*(x) = [S(x)]^{-2}s(x), \qquad S^*(x) = -1/S(x), \qquad (9.6)$$

$$m^*(x) = [S(x)]^2 m(x). \qquad (9.7)$$

We shall now show that 0 is an entrance boundary for $X^*(t)$. To do this, we need to show that

$$S^*(0, \eta) = \int_0^\eta s^*(\xi) \, d\xi = \infty \quad \text{and} \quad N^*(0) = \int_0^b \int_\eta^b dS^*(\xi) \, dM^*(\eta) < \infty.$$

But, because $S(0) = 0$, it follows that

$$\int_0^\eta dS^*(\xi) = \int_0^\eta \frac{s(\xi)}{S^2(\xi)} \, d\xi = \frac{1}{S(0)} - \frac{1}{S(\eta)} = \infty.$$

Next,

$$N^*(0) = \int_0^b \int_\eta^b dS^*(\xi)\, dM^*(\eta)$$

$$= \int_0^b \left[\frac{1}{S(\eta)} - \frac{1}{S(b)} \right] S^2(\eta) m(\eta)\, d\eta$$

$$< \int_0^b S(\eta) m(\eta)\, d\eta = \int_0^b \int_0^\eta dS(\xi) m(\eta)\, d\eta = \Sigma(0) < \infty$$

since 0 is an exit boundary for $X(t)$. We have now established that 0 is an entrance boundary for $X^*(t)$. The proof that state 1 remains an exit boundary is similar and is left as an exercise.

GREEN FUNCTIONS FOR THE CONDITIONED PROCESS

We can define the Green function on the interval $(0, 1]$ for the conditioned process using the results of (9.6) and (9.7). Direct substitution into (3.15) using the fact that $S^*(x) = -[S(x)]^{-1}$ leads to

$$G_*(x, \xi) = \begin{cases} \dfrac{2[S(x)][S(1) - S(\xi)]}{S(1)} \dfrac{S(\xi)}{\sigma^2(\xi)s(\xi)S(x)} & \text{for} \quad 0 \le x \le \xi \le 1 \\[2em] \dfrac{2[S(1) - S(x)][S(\xi)]}{S(1)} \dfrac{S(\xi)}{\sigma^2(\xi)s(\xi)S(x)} & \text{for} \quad 0 \le \xi \le x \le 1. \end{cases}$$

$$(9.8)$$

Here, we have used the subscript $*$ to indicate conditional Green functions, as opposed to the functions $G^*(x, \xi)$ of (3.30). Then $G_*(x, \xi)\, d\xi$ is interpreted as the mean time the conditioned process spends in the interval $[\xi, \xi + d\xi)$ prior to the absorption time T_1. A formal derivation proceeds by first considering the Green function on a subinterval $[a, b]$ of $(0, 1)$ and letting $a \downarrow 0$ and $b \uparrow 1$. Because 0 is an entrance boundary for X^*, we have

$$\lim_{a \downarrow 0, b \uparrow 1} T_a^* \wedge T_b^* = T_1^*.$$

(Compare to Example C of Section 4. See also Section 6.)

We can use the above result to compute, for example, the mean of the absorption time T_1^* when the exit boundary is reached. We proceed as follows.

$$E_x[T_1^*] = \int_0^1 G_*(x, \xi)\, d\xi$$

$$= \frac{2[S(1) - S(x)]}{S(1)S(x)} \int_0^x \frac{S^2(\xi)}{\sigma^2(\xi)s(\xi)}\, d\xi + 2 \int_x^1 \frac{S(\xi)[S(1) - S(\xi)]\, d\xi}{\sigma^2(\xi)s(\xi)S(1)},$$

$$(9.9)$$

using (9.8). This should be compared to (3.12).

Example 1. *Wright–Fisher Diffusion with No Mutation or Selection.* Recall that $X(t)$ is a diffusion on $[0, 1]$ with infinitesimal parameters $\mu(x) = 0$, $\sigma^2(x) = x(1 - x)$. We verified in Section 4, Example C, that $S(x) = x$, so that for the $X^*(t)$ process we have

$$\mu^*(x) = 1 - x, \qquad \sigma^{*2}(x) = x(1 - x). \tag{9.10}$$

The expected time $E_x[T_1^*]$ to reach 1 for the conditioned process starting at x is given, using (9.10), by

$$E_x[T_1^*] = \frac{2(1 - x)}{x} \int_0^x \frac{y}{1 - y} \, dy + 2 \int_x^1 dy = -\frac{2(1 - x)}{x} \ln(1 - x). \tag{9.11}$$

Interpreting the units of time in the standard way such that one unit of the diffusion reflects about N generations of the discrete approximating model, we infer that the expected time until fixation under the initial condition $x = 1/N$ (i.e., a single mutant type) is approximately

$$N \left\{ \frac{-2[1 - (1/N)]}{1/N} \log[1 - (1/N)] \right\} \sim 2N. \tag{9.12}$$

Among population genetics investigators, (9.12) is commonly replaced by $4N$, the extra factor of 2 is attributable to the fact that population size consists of $2N$ genes instead of N individuals.

Example 2. *An Infinite Allele Neutral Mutation Genetic Model.* We assume that in each generation an existing type may mutate to a new type with probability μ. We focus attention on a specific allelic type A currently represented in the population and consider a population of N individuals composed of i A-types, and $N - i$ non–A-types. Postulating that only mutation and random sampling effects operate, the mean proportion of A-types after mutation is $\pi_i = (i/N)(1 - \mu)$.

Let X_n be the number of A individuals in generation n. In line with the Wright–Fisher formulation of Section 2, Example F, the next generation is determined by N binomial trials, such that

$$\Pr\{X_{n+1} = j \mid X_n = i\} = \binom{N}{j} \pi_i^j (1 - \pi_i)^{N-j}, \qquad 0 \le i, j \le N.$$

Clearly, state 0 is absorbing, whereas state N is not. This follows because mutation will ultimately cause the specific A-type to disappear, so that a population configuration consisting exclusively of A-types can last only temporarily. We say that a *quasi-fixation* occurs at some time n if $X_n = N$, signifying that the population comprises only A-types. It is of some interest in biological applications to evaluate (at least to a good approximation for large N) the probability of quasi-fixation, and the expected time to quasi-fixation, conditional on this event occurring. We can do this by calculating the corresponding quantities for appropriate diffusion approximations.

In the spirit of the derivations of the diffusion approximations of Section 2, we obtain a limiting diffusion process $\{X(t), t \geq 0\}$ on the interval $I = [0, 1]$, with infinitesimal mean and variance

$$\mu(x) = -\gamma x, \qquad \sigma^2(x) = x(1 - x), \tag{9.13}$$

respectively. (We have assumed that $\gamma = \lim_{N \to \infty} N\mu$.) Using (9.13), it is easy to check that the scale density $s(x)$ and scale measure $S(x)$ of $X(t)$ are given by

$$s(x) = (1 - x)^{-2\gamma}, \quad S(x) = \frac{1 - (1 - x)^{1 - 2\gamma}}{1 - 2\gamma}, \qquad 2\gamma \neq 1. \tag{9.14}$$

It follows from the results of Section 6, Table 6.2, that if $\gamma \geq \frac{1}{2}$ then the boundary point 1 is nonattracting and hence unattainable, and thus $\pi(x)$, the probability of a quasi-fixation starting from $X(0) = x$, is identically zero. When $0 < \gamma < \frac{1}{2}$, then $\pi(0) = S(x)/S(1)$ is given explicitly by

$$\pi(x) = 1 - (1 - x)^{1 - 2\gamma}, \qquad 0 < \gamma < \frac{1}{2}. \tag{9.15}$$

We now compute the expected time to quasi-fixation, conditional on this event happening. From the above discussion, we require $0 < \gamma < \frac{1}{2}$, and then we consider the diffusion process $\{X^*(t), t \geq 0\}$ restricted only to those realizations leading to quasi-fixation. It is clear that the conditioning results developed in this section apply. In particular, the mean time $E_x[T_1^*]$ to reach state 1 starting from $X^*(0) = x$ is given by the formula (9.9), with $S(x)$ as in (9.14). Of particular interest is the case where $x = 1/N$, corresponding to the introduction of a single copy of the mutant A-type. Explicitly, we have

$$E_{1/N}[T_1^*] = \frac{2(1 - 1/N)^{1 - 2\gamma}}{1 - (1 - 1/N)^{1 - 2\gamma}} \int_0^{1/N} \frac{[1 - (1 - \xi)^{1 - 2\gamma}]^2}{(1 - 2\gamma)\xi(1 - \xi)^{1 - 2\gamma}} \, d\xi$$

$$+ 2 \int_{1/N}^1 \frac{1 - (1 - \xi)^{1 - 2\gamma}}{(1 - 2\gamma)\xi} \, d\xi$$

$$= I_1(N) + I_2(N),$$

say. But

$$I_1(N) \approx \left(\frac{2}{1 - 2\gamma} \right) \left(\frac{(1 - 1/N)^{1 - 2\gamma}}{1 - (1 - 1/N)^{1 - 2\gamma}} \right) \left(\frac{[1 - (1 - 1/N)^{1 - 2\gamma}]^2}{(1/N)(1 - 1/N)^{1 - 2\gamma}} \frac{1}{N} \right)$$

$$= \frac{2}{1 - 2\gamma} \left[1 - \left(1 - \frac{1}{N} \right)^{1 - 2\gamma} \right] \approx \frac{2}{N},$$

so that $NI_1(N) \to 2$ as $N \to \infty$. Also, $NI_2(N) \sim Nc$, where c is a finite constant,

$$c = \frac{2}{1 - 2\gamma} \sum_{k=1}^{\infty} \left(\frac{1 - 2\gamma}{k} \right) \frac{(-1)^{k+1}}{k}.$$

It follows that the conditional mean quasi-fixation time starting from $x = 1/N$ is, in time units of the original model, $NI_1(N) + NI_2(N) \approx N(2 + c)$ as $N \to \infty$.

We construct a number of processes extracted from the Brownian motion process $\{B(t), t \geq 0\}$ by imposing appropriate restrictions on the sample space.

Let α and β be fixed real numbers. Let $W^*(t)$ for $0 \leq t \leq 1$ denote the constrained Brownian motion conditioned that

$$\alpha < B(1) < \beta. \tag{9.16}$$

Intuitively, $W^*(t)$ is a diffusion process with time parameter confined to the interval $0 \leq t \leq 1$. We shall now determine the infinitesimal parameters of the process, which can be time dependent because of the requirement of (9.16). We follow the recipe of the calculations underlying (9.5a) and (9.5b) with one essential modification. Let $\pi(x, t)$ be the probability that from the state value x at time t the sample path of $B(t)$ satisfies (9.16) at time 1.

If $p(t, x, y)$ denotes the transition density of the unconditioned process at time t then the transition density of W^* is given by

$$p^*(t, x; s, y)\, dy = \Pr\{y < W^*(s) \leq y + dy \,|\, W^*(t) = x\}$$

$$= \frac{p(s - t, x, y)\pi(y, s)\, dy}{\pi(x, t)}, \qquad 0 < t < s < 1. \tag{9.17}$$

The justification of (9.17) follows that of (9.1) mutatis mutandis. We are now prepared to compute

$$\mu^*(x, t) = \lim_{h \downarrow 0} \frac{1}{h} \int (y - x)p^*(t, x; t + h, y)\, dy. \tag{9.18}$$

All our calculations are done in a formal setting in the spirit of the arguments of Section 3. A rigorous validation is substantially more technical. We postulate sufficient regularity for $\pi(x, t)$ to permit the use of the Taylor expansion,

$$\pi(y, t + h) = \pi(x, t) + (y - x)\frac{\partial \pi}{\partial x}(x, t) + h\frac{\partial \pi}{\partial t}(x, t) + o(y - x) + o(h).$$

This is a mild assumption readily verified [see (9.22)] in the case at hand. Substituting from (9.17) into (9.18) gives

$$\mu^*(x, t) = \lim_{h \downarrow 0} \frac{1}{h} \int (y - x)p(h, x, y)\, dy$$

$$+ \frac{\partial \pi(x, t)/\partial x}{\pi(x, t)} \lim_{h \downarrow 0} \frac{1}{h} \int (y - x)^2 p(h, x, y)\, dy, \tag{9.19}$$

again stipulating a strong form of the infinitesimal relations for the $B(t)$ process. In the case of the Brownian motion process no obstacles are encountered in these presumptions. By virtue of (9.19) we obtain

$$\mu^*(x, t) = \mu(x) + \frac{\partial \pi(x, t)/\partial x}{\pi(x, t)} \sigma^2(x). \tag{9.20}$$

By similar means we secure

$$\sigma^*(x, t) = \sigma(x). \tag{9.21}$$

For the specific case under consideration we have $\mu(x) \equiv 0$, $\sigma^2(x) \equiv 1$, and therefore

$$\mu^*(x, t) = \frac{\partial \pi(x, t)/\partial x}{\pi(x, t)}, \qquad \sigma^*(x, t) = 1,$$

$$\pi(x, t) = \int_\alpha^\beta \frac{1}{\sqrt{2\pi(1 - t)}} e^{-(y - x)^2/2(1 - t)} \, dy = \frac{1}{\sqrt{2\pi}} \int_{(\alpha - x)/\sqrt{1 - t}}^{(\beta - x)/\sqrt{1 - t}} e^{-\xi^2/2} \, d\xi$$

$$\tag{9.22}$$

It is sometimes convenient to use the notation $\Pi_{\alpha, \beta}(x, t)$ for expression (9.22).

Of special interest is the specification $\alpha = -\varepsilon$, $\beta = +\varepsilon$, and then the ascertainment of the limiting conditioned diffusion is determined after sending ε to zero. The process obtained is called the *Brownian bridge*, for reasons indicated later. A simple direct computation shows that for $\Pi_{-\varepsilon, \varepsilon}(x, t) = \Pi_\varepsilon(x, t)$ we get

$$\lim_{\varepsilon \to 0} \frac{1}{\Pi_\varepsilon(x, t)} \frac{\partial \Pi_\varepsilon(x, t)}{\partial x} = -\frac{x}{1 - t}. \tag{9.23}$$

Thus Brownian motion conditioned on $B(1) = 0$, i.e., the Brownian bridge process, corresponds to a diffusion process $\{\tilde{W}(t), 0 \le t < 1\}$ with infinitesimal coefficients

$$\tilde{\mu}(x, t) = -\frac{x}{1 - t}, \qquad \tilde{\sigma}^2(x, t) \equiv 1. \tag{9.24}$$

Sharply contrasting with the conditioned diffusions associated with (9.16), the diffusion of (9.24) is constructed by conditioning on the realization from a collection of sample paths having probability 0. Despite the complication of dealing with constraints inducing events of probability 0, the diffusion process determined by the parameters of (9.24) is well defined. Later in this section we shall achieve a representation of this process from which a description of its properties is readily forthcoming.

More generally, Brownian motion conditioned on $B(1) = a$ gives rise to the diffusion process $\{\tilde{W}_a(t), 0 \le t < 1\}$ with infinitesimal parameters

$$\mu_a(x, t) = \frac{(a - x)}{1 - t}, \qquad \sigma_a^2(x, t) \equiv 1. \tag{9.25}$$

We shall provide another approach and representation for the Brownian bridge: Consider Brownian motion subject to a deterministic change of time variable and a scaling of the state variable of the explicit form

$$\Gamma(t) = (1 - t)B\left(\frac{t}{1 - t}\right), \qquad 0 < t < 1. \tag{9.26}$$

The process $\Gamma(t)$ is obviously a diffusion, and it remains to ascertain its infinitesimal mean and variance. This will be done by direct evaluation of

$$\lim_{h \downarrow 0} \frac{1}{h} E[\Gamma(t + h) - x \mid \Gamma(t) = x] = \mu_\Gamma(x, t), \tag{9.27}$$

$$\lim_{h \downarrow 0} \frac{1}{h} E[\{\Gamma(t + h) - x\}^2 \mid \Gamma(t) = x] = \sigma_\Gamma^2(x, t)$$

by reducing the calculation to a familiar one involving Brownian motion. Note that $\Gamma(t) = x$ corresponds to $B(t/(1 - t)) = x/(1 - t)$, and because Brownian motion has zero drift, that

$$E\left[B\left(\frac{t + h}{1 - t - h}\right) \middle| B\left(\frac{t}{1 - t}\right) = y\right] = y \qquad \text{for} \quad t > 0, \quad h > 0.$$

Therefore

$$E[\Gamma(t + h) - x \mid \Gamma(t) = x] = E\left[(1 - t - h)B\left(\frac{t + h}{1 - t - h}\right)\right.$$

$$\left. - x \mid B\left(\frac{t}{1 - t}\right) = \frac{x}{1 - t}\right]$$

$$= (1 - t - h)\frac{x}{1 - t} - x$$

$$= x\left(\frac{1 - t - h}{1 - t} - 1\right)$$

$$= -\frac{xh}{1 - t}. \tag{9.28}$$

It follows when dividing by h and sending h to 0 that (9.28) converges to $-x/(1 - t)$. A similar calculation verifies that $\sigma_\Gamma^2(x, t) \equiv 1$.

The above analysis establishes that the Brownian bridge identified as Brownian motion conditioned such that $B(1) = 0$ can be realized by the transformed process

$$\Gamma(t) = (1 - t)B\left(\frac{t}{1 - t}\right), \qquad 0 < t < 1, \tag{9.29}$$

where $B(s)$ is standard Brownian motion. Every sample path of $B(s)$ induces a sample path of the Brownian bridge process through the representation (9.28).

Notice that $\Gamma(1)$ vanishes since $B(u)/u \to 0$ as $u \to \infty$ with probability 1 (this fact is established in Chapter 7). Observe the further property that $\Gamma(t)$ is a Gaussian process [since $B(s)$ is Gaussian] and we now compute its covariance function. Clearly $E[\Gamma(t)] = (1 - t)E[B(t/(1 - t))] = 0$. Consider now

$$
E[\Gamma(s_1)\Gamma(s_2)] = (1 - s_1)(1 - s_2)E\left[B\left(\frac{s_1}{1 - s_1} \right) B\left(\frac{s_2}{1 - s_2} \right) \right]
$$

$$
= (1 - s_1)(1 - s_2) \min\left(\frac{s_1}{1 - s_1}, \frac{s_2}{1 - s_2} \right)
$$

$$
= \begin{cases} s_1(1 - s_2), & s_1 < s_2 \\ s_2(1 - s_1), & s_1 > s_2. \end{cases} \tag{9.30}
$$

A third version of the Brownian bridge process is

$$
\hat{B}(t) = B(t) - tB(1), \qquad 0 \le t < 1. \tag{9.31}
$$

To validate this statement we need merely verify three properties:

   (i) $\hat{B}(t)$ is Markov. (Why?)
   (ii) $\hat{B}(t)$ is a Gaussian process. (This is obvious since $B(t)$ is Gaussian and $\hat{B}(t)$ is constructed as a linear operation on a Gaussian process.)
   (iii) $E[\hat{B}(t)] = 0$ and

$$
E[\hat{B}(s_1)\hat{B}(s_2)] = \begin{cases} s_1(1 - s_2) & \text{for} \quad s_1 < s_2, \\ s_2(1 - s_1) & \text{for} \quad s_2 < s_1. \end{cases}
$$

Indeed,

$$
E[\hat{B}(s_1)\hat{B}(s_2)] = E[\{B(s_1) - s_1 B(1)\}\{B(s_2) - s_2 B(1)\}]
$$

$$
= E[B(s_1)B(s_2)] - s_1 E[B(1)B(s_2)] - s_2 E[B(s_1)B(1)]
$$

$$
+ s_1 s_2 E[B(1)]^2
$$

$$
= \min(s_1, s_2) - s_1 s_2,
$$

and this formula is identical to (9.30).

Two Gaussian processes with the identical covariance function share the same finite-dimensional distribution functions. Thus $\hat{B}(t)$ and $\Gamma(t)$ identify the same diffusion, characterized by the infinitesimal parameters (9.24).

A final direct approach to the acquisition of the Brownian bridge is to exploit the interpolation formulas of Theorem 2.1 of Chapter 7. The joint distribution of

$$
B(t_1), \ldots, B(t_r), \qquad 0 < t_1 < \cdots t_r < 1,
$$

conditioned on $B(0) = 0$ and $B(1) = 0$ is multivariate normal having a covariance matrix of elements

$$\sigma_{ij} = E[B(t_i)B(t_j)] = t_i(1 - t_j), \qquad i < j.$$

This is a direct extension of Theorem 2.1 of Chapter 7.

The name Brownian bridge derives from the constraints $B(0) = B(1) = 0$ and the feature that if a path $\{B(t), 0 \le t \le 1\}$ is also nonnegative its picture looks like a bridge (see Fig. 6). A synonym is "tied down Brownian motion."

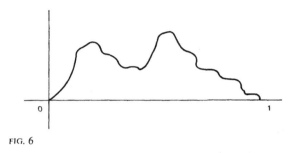

FIG. 6

The Brownian bridge process is intimately associated with certain random functionals of empirical distribution functions based on independent observations (cf. Chapter 13).

AN EXAMPLE OF BROWNIAN MOTION CONDITIONED ON ITS GROWTH BEHAVIOR

Finally, we shall construct a conditioned diffusion process obtained by imposing a growth bound on the sample trajectory. The potential usefulness of such constructions is clearly shown by the following concrete case. Let $\{B(t), t \ge 0\}$ represent standard Brownian motion and define the process

$$Z^*(t) = B(t) \qquad \text{constrained such that} \qquad B(t) \le \alpha t + \beta \quad \text{for all } t > 0. \tag{9.32}$$

The $Z^*(t)$ process is a diffusion (why?), and to ascertain the associated infinitesimal parameters it is essential to evaluate

$$\pi(x) = \Pr\{B(t) \le \alpha t + \beta \text{ for all } t \,|\, B(0) = x\}.$$

Consulting Elementary Problem 7, Chapter 7, we find that

$$\pi(x) = \begin{cases} 1 - e^{-2\alpha(\beta - x)}, & x < \beta, \\ 0, & x > \beta. \end{cases}$$

We can now compute the infinitesimal parameters of the $Z^*(t)$ process following the recipe of $(9.5a,b)$. In this vein, we obtain

$$\sigma^{*2}(x) = \sigma^2(x) \equiv 1$$

$$\mu^*(x) = \begin{cases} \mu(x) + \dfrac{\pi'(x)}{\pi(x)} = 0 + \dfrac{e^{-2\alpha\beta}e^{2\alpha x}2\alpha}{1 - e^{-2\alpha(\beta - x)}} = \dfrac{2\alpha}{e^{2\alpha(\beta - x)} - 1} & \text{for} \quad x < \beta \\ 0 & \text{for} \quad x > \beta. \end{cases}$$

Note that $\mu^*(x)$ is discontinuous in this example.

10: Some Natural Diffusion Models with Killing

A. A DIFFUSION APPROXIMATION TO A PROBLEM IN THE GENETICS OF RECOMBINATION

Consider a finite diploid population in which two alleles (types) are segregating at each of two linked loci. The problem is to determine the probability that a recombination event will occur before the population fixes on one chromosome type.

Specifically, suppose there is a population of N individuals. Each individual has a number of chromosomes which occur in homologous pairs. We are concerned with the genetic constitution at just two loci, which are "linked" so that the genes occurring at the two loci are not independent. Suppose that at locus (position) 1 there are two possible alleles, A and a, and at locus 2 there are possible alleles, B and b. An individual can then express any of ten possible genotypes

$$\begin{array}{cc} \underline{\underline{\text{A} \quad \text{B}}} & \leftarrow \text{chromosome} \\ \text{a} \quad \text{B} & \leftarrow \text{the homologous chromosome} \\ \uparrow \quad \uparrow & \\ \text{locus 1} \quad \text{locus 2} & \end{array}$$

The alternative chromosome pairs are

$$\frac{\overline{\overline{\text{AB}}}}{\text{AB}}, \frac{\overline{\overline{\text{AB}}}}{\text{Ab}}, \frac{\overline{\overline{\text{Ab}}}}{\text{Ab}}, \frac{\overline{\overline{\text{AB}}}}{\text{aB}}, \frac{\overline{\overline{\text{AB}}}}{\text{ab}}, \frac{\overline{\overline{\text{Ab}}}}{\text{aB}}, \frac{\overline{\overline{\text{Ab}}}}{\text{ab}}, \frac{\overline{\overline{\text{aB}}}}{\text{aB}}, \frac{\overline{\overline{\text{aB}}}}{\text{ab}}, \frac{\overline{\overline{\text{ab}}}}{\text{ab}}$$

[Note that the chromosome ordering is irrelevant: e.g., $\dfrac{\overline{\overline{\text{AB}}}}{\text{aB}}$ and $\dfrac{\overline{\overline{\text{aB}}}}{\text{AB}}$ are indistinguishable. However the phase is important, i.e., $\dfrac{\overline{\overline{\text{Ab}}}}{\text{aB}}$ and $\dfrac{\overline{\overline{\text{AB}}}}{\text{ab}}$ are different genotypes.]

In the formation of the next generation, individuals produce (segregate) gametes, the equivalent of one chromosome from each homologous pair. Two gametes in the population will then join to form an individual of the next generation. It is in the segregation of the gametes that recombination may occur. Thus an $\frac{Ab}{aB}$ individual may produce an Ab-gamete, or an aB-gamete: but also, by recombination, it may produce AB- and ab-gametes. When two loci are linked these recombinant gametes are less likely to be produced than the other (parental) gametes. The recombination parameter r quantifies the linkage thus by:

$$r = \text{probability of a recombination.}$$

For example, the gametic output of an $\frac{Ab}{aB}$ individual contains the four chromosomal types Ab, aB, AB, ab in the proportions $\frac{1}{2}(1 - r)$, $\frac{1}{2}(1 - r)$, $\frac{1}{2}r$, $\frac{1}{2}r$, respectively. (Note that for many genotypes recombination is irrelevant to the production of gametes: e.g., $\frac{AB}{Ab}$ segregates $\frac{1}{2}AB$ and $\frac{1}{2}Ab$ irrespective of recombination.)

Suppose that the population at hand initially contains only the gamete types Ab and aB, where all individuals are of genotypes $\frac{Ab}{Ab}$, $\frac{Ab}{aB}$ or $\frac{aB}{aB}$.

If r is small, then it is possible that many generations will pass by without the appearance of an individual with a recombinant chromosome (AB or ab) produced from the genotype $\frac{Ab}{aB}$. During that time, it could happen that the population reaches a state in which everyone is of the type $\frac{Ab}{Ab}$: the population can never leave that state—for only Ab-gametes can possibly be produced— the population is "fixed" on Ab. Similarly the population could fix on aB.

The problem is then to determine the probability, given the initial makeup, that a recombinant gamete appears in the population before fixation occurs and prevents that possibility.

We now set up a model for this problem, in the course of which the mechanics of the formation of one generation from the preceding one will be clarified. Suppose that no recombination has yet occurred. There are $2N$ gametes of N individuals in the population: suppose there are i of type Ab, and hence $2N - i$ of type aB. Of course, to fully describe the population we need not only the gamete frequencies but also the genotype frequencies—that is, how the gametes are paired in individuals (are they mostly $\frac{Ab}{Ab}$ and $\frac{aB}{aB}$ with just a few $\frac{Ab}{aB}$ or vice versa?) However, if N is not too small, the following approximation gives sufficient accuracy, and greatly simplifies the analysis: namely, that the (assumed) random mating of individuals in the previous generation, which is equivalent to the random union of their gametes, entails the genotype frequencies

$$\frac{Ab}{Ab} \qquad \left(\frac{i}{2N}\right)^2$$

$$\frac{Ab}{aB} \qquad 2\left(\frac{i}{2N}\right)\left(1 - \frac{i}{2N}\right)$$

$$\frac{aB}{aB} \qquad \left(1 - \frac{i}{2N}\right)^2.$$

From this population, an Ab-gamete could be produced either

$$\text{by an } \frac{Ab}{Ab} \text{ with probability 1}$$

$$\text{or by an } \frac{Ab}{aB} \text{ with probability } \tfrac{1}{2}(1 - r).$$

So, when the number of Ab-types is i, then

$$p_i = \Pr\{\text{a gamete produced is Ab}\}$$

$$= \left(\frac{i}{2N}\right)^2 + 2\left(\frac{i}{2N}\right)\left(1 - \frac{i}{2N}\right)\tfrac{1}{2}(1 - r)$$

$$= \frac{i}{2N} - r\left(\frac{i}{2N}\right)\left(1 - \frac{i}{2N}\right).$$

Similarly,

$$q_i = \Pr\{\text{a gamete produced is aB}\}$$

$$= \left(1 - \frac{i}{2N}\right)^2 + 2\left(\frac{i}{2N}\right)\left(1 - \frac{i}{2N}\right)\tfrac{1}{2}(1 - r)$$

$$= 1 - \frac{i}{2N} - r\left(\frac{i}{2N}\right)\left(1 - \frac{i}{2N}\right).$$

Finally, the probability that a gamete produced is of a recombinant type is

$$1 - (p_i + q_i) = 2\left(\frac{i}{2N}\right)\left(1 - \frac{i}{2N}\right)r = \frac{i}{N}\left(1 - \frac{i}{2N}\right)r.$$

The $2N$ gametes which will make up the individuals of the next generation are chosen by random binomial sampling from this pool of gametes.

Thus the transition probability, P_{ij}, that there are j gametes of type Ab, and $2N - j$ of type aB is

$$P_{ij} = \binom{2N}{j} p_i^j q_i^{2N-j}, \qquad j = 0, 1, \ldots, 2N.$$

The probability that no recombinant gamete makes it into the next generation is $(p_i + q_i)^{2N}$, so that the probability that one or more recombinants do appear is

$$1 - (p_i + q_i)^{2N}.$$

Notice that it is assumed that the population size N is the same for each generation, and that there are no selection pressures—every individual survives to breeding age and has an equal chance to contribute to the next generation.

The situation is now that of a Markov chain with $2N + 2$ states, $2N + 1$ of which represent the state in which no recombinant gamete has yet appeared, and the population now contains j Ab- and $2N - j$ aB-gametes $(j = 0, 1, \ldots, 2N)$. The other state is that a recombinant gamete has appeared in an individual. For our purposes this is a killing state. Admittedly an AB- or ab-gamete could appear and then not make it into the next generation, and never appear again, but what matters is that a recombination *has* occurred. Strictly, this state is that AB or ab has been present in some individual, past or present.

As noted earlier, the states $j = 0$ and $2N$ are absorbing.

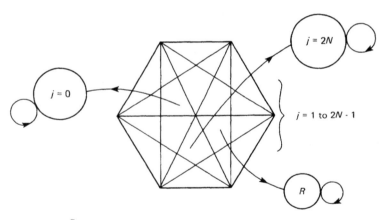

FIG. 7

Starting from the initial state i, a population will almost surely reach one of the absorbing states or be killed. What is the probability R_i that it eventually reaches the recombinant state R?

From the theory of Markov chains this can be solved in the following way: if $u_i = 1 - R_i =$ probability that the population eventually fixes, then $(u_0, u_1, \ldots, u_{2N})$ satisfy the equations

$$u_i = \sum_{j=0}^{2N} P_{ij} u_j, \qquad i = 0, 1, \ldots, 2N$$

with the boundary conditions

$$u_0 = u_{2N} = 1.$$

For large N the solution of these equations is, in practice, impossible, analytically or numerically. We proceed to approximate the Markov chain by a diffusion process. Numerical computation indicates that for $N > 10$, this approximation gives satisfactory accuracy. Strictly speaking, the relevant approximating process is a "diffusion process with killing": we follow the progress of the frequency of the Ab gamete, until the process is "killed" by the appearance of a recombinant gamete or fixation occurs.

At the nth generation, the state variable $X(n)$ can take the values $X(n) = j/2N, j = 0, 1, ..., 2N$. *Given* that no recombinant then appears, $X(n + 1)$ is then distributed as $1/2N$ of a binomial variable with parameters $\{2N, p_j/(p_j + q_j)\}$. The probability that no recombinant does appear is, as we have noted before, $(p_j + q_j)^{2N}$. Then

$$E\left[X(n + 1) - X(n)\,|\,X(n) = \frac{j}{2N} = x\right] = \left(\frac{p_j}{p_j + q_j} - x\right)(p_j + q_j)^{2N}$$

$$= [p_j - x(p_j + q_j)](p_j + q_j)^{2N - 1}$$

and

$$E\left[(\Delta X)^2\,|\,X(n) = \frac{j}{2N} = x\right] = \left[\frac{1}{2N}\left(\frac{p_j}{p_j + q_j}\right)\left(\frac{q_j}{p_j + q_j}\right)\right.$$

$$\left. + \left(\frac{p_j}{p_j + q_j} - x\right)^2\right](p_j + q_j)^{2N}$$

$$= \left[\frac{p_j q_j}{2N} + (p_j - x(p_j + q_j))^2\right](p_j + q_j)^{2N - 2}$$

and $\Pr\{\text{killed}\,|\,X(n) = j/2N = x\} = 1 - (p_j + q_j)^{2N}$.

Introducing a new recombination parameter ρ defined by

$$r = \rho/N^2,$$

and regarding ρ as a constant, we see that

$$E[\Delta X\,|\,X(n) = x] = O(1/N^2)$$

$$E[(\Delta X)^2\,|\,X(n) = x] = x(1 - x)/2N + O(1/N^2)$$

$$\Pr\{\text{killed}\,|\,X(n) = x\} = 4\rho x(1 - x)/N + O(1/N^2).$$

Now we rescale time so that one unit of time corresponds to the passage of N generations, i.e., let

$$Y_N(\tau) = X([N\tau]).$$

Finally, let $N \to \infty$. Then the processes $Y_N(\tau)$ will tend in the limit to a diffusion process $Y(\tau)$ with killing. The infinitesimal parameters of the Y-process are to be found in the limit from the X parameters, remembering that ΔX is over a rescaled time period of $1/N$ units: thus,

$$\mu(x) = \lim_{N \to \infty} \left(\frac{1}{1/N} \right) E[\Delta X \,|\, X(n) = x] = 0$$

$$\sigma^2(x) = \lim_{N \to \infty} \frac{1}{(1/N)} E[(\Delta X)^2 \,|\, X(n) = x] = \frac{x(1-x)}{2}$$

$$k(x) = \lim_{N \to \infty} \frac{1}{(1/N)} \Pr\{\text{killed} \,|\, X(n) = x\} = 4\rho x(1-x).$$

Y is then a diffusion process with killing on the interval $[0, 1]$: the problem is to determine the probability that the process reaches the absorbing states 0 or 1 (where the killing rate is zero) before it gets killed.

There are two ways of approaching this question. We may treat this as a modified type of diffusion process, with three relevant parameters (μ, σ^2, k) and obtain the solution essentially from first principles. Alternatively, we can approach it via the Kac functional (Section 5), treating it as an ordinary (μ, σ^2) diffusion process, but modifying the expectations and probabilities by the killing factor $\exp[-\int_0^t k(Y(\tau))\, d\tau]$. These two methods are, of course, equivalent, and they both lead to the following result:

Let $u(x, t) = \Pr\{\text{starting at } x, \text{ the } Y \text{ process has not been killed by time } t\}$. Then u satisfies the differential equation [cf. (5.39)]

$$\frac{\partial u}{\partial t} = \frac{1}{2}\sigma^2(x)\frac{\partial^2 u}{\partial x^2} + \mu(x)\frac{\partial u}{\partial x} - k(x)u$$

together with

$$u(0, t) \equiv u(1, t) \equiv 1, \qquad u(x, 0) \equiv 1.$$

Clearly, $u(x, t)$ is decreasing in t and is bounded below by 0. So, as $t \to \infty$, $u(x, t)$ tends to a limiting value $u(x, \infty) = u(x)$, and $\partial u/\partial t \to 0$.

Therefore, $u(x)$ satisfies

$$0 = \frac{1}{2}\sigma^2(x)\frac{d^2 u}{dx^2} + \mu(x)\frac{du}{dx} - k(x)u, \qquad u(0) = u(1) = 1,$$

and $u(x)$ has the interpretation,

$$u(x) = \Pr\{\text{process is never killed} \,|\, X(x) = x\}$$

$$= \Pr\{\text{process reaches 0 or 1} \,|\, X(n) = x\}.$$

The probability we require is $Z(x) = 1 - u(x)$. This satisfies

$$\tfrac{1}{2}\sigma^2(x)Z'' + \mu(x)Z' - k(x)Z = -k(x), \qquad Z(0) = Z(1) = 0.$$

Substituting for μ, σ^2, k and canceling common factors, this simplifies to

$$\tfrac{1}{4}Z'' - 4\rho Z = -4\rho, \qquad Z(0) = Z(1) = 0.$$

Noting the trivial particular solution, $Z(x) \equiv 1$ to this inhomogeneous differential equation, we try to fit the general solution

$$Z(x) = 1 + A \sinh(4\sqrt{\rho}x) + B \cosh(4\sqrt{\rho}x)$$

to the boundary conditions. This is straightforward and yields

$$Z(x) = 1 - \frac{\sinh(4\sqrt{\rho}x) + \sinh(4\sqrt{\rho}(1-x))}{\sinh(4\sqrt{\rho})}.$$

Of greater interest is the following modification: suppose that only the appearance of an AB-recombinant gamete is of concern and kills the process, and that if an ab-gamete appears in an individual it is recognized, discarded, and replaced by a new, randomly chosen gamete. Obviously, this simply has the effect of halving the rate of killing, so that $R(x)$, the probability of eventually obtaining an AB-gamete, is

$$R(x) = 1 - \frac{\sinh(\sqrt{8\rho}x) + \sinh[\sqrt{8\rho}(1-x)]}{\sinh\sqrt{8\rho}}.$$

As would be expected, we can see immediately that if ρ is large (so that $r \gg 1/N^2$) then $R \sim 1$, and recombination before fixation is virtually certain. On the other hand, if ρ is small, then $R \sim 0$, and fixation almost certainly occurs first.

The maximum probability for the appearance of an AB-recombinant occurs when the initial state is $x = \frac{1}{2}$:

$$R(\tfrac{1}{2}) = 1 - (2 \sinh \sqrt{2\rho}/\sinh 2\sqrt{2\rho}) = 1 - \operatorname{sech}\sqrt{2\rho}$$
$$= 1 - \operatorname{sech} N\sqrt{2r} \approx 1 - 2e^{-N\sqrt{2r}}$$

the last approximation applying if $N\sqrt{r}$ is large.

The formula $N = (1/\sqrt{2r}) \operatorname{sech}^{-1}(1 - R(\tfrac{1}{2}))$ enables one to calculate the population size to obtain an AB-recombinant with a specified confidence $R(\tfrac{1}{2})$, starting from the initial state $x = \frac{1}{2}$. Thus, to achieve $R(\tfrac{1}{2}) = 0.99$, we need $N \sim (1/2r)\log 200$.

B. THE DETECTION OF A RECESSIVE VISIBLE GENE IN A FINITE POPULATION

Another interesting application of a diffusion with killing is provided by the following problem from population genetics. Consider a finite population in which two alleles, denoted by A and a, are segregating at a particular locus. What is the probability that an aa-genotype will be formed before the population fixes on the AA-genotype?

We can formulate the problem in the following way. Suppose the population comprises N individuals, and that the aa-genotype is visible as soon as it is formed. This might correspond, for example, to the a-allele being lethal in homozygous (aa) condition. Let X_n denote the number of heterozygotes (Aa) at time n. Assuming that we have not yet detected an aa-genotype, we can suppose that the population comprises i Aa-individuals and $N - i$ AA-individuals. To find out the population composition at the next time point, we will assume there is random mating in the population, and so the possible mating types have the following relative frequencies:

$$\begin{array}{ccc} \text{AA} \times \text{AA} & \text{Aa} \times \text{Aa} & \text{Aa} \times \text{AA} \\ (1 - i/N)^2 & (i/N)^2 & 2(i/N)(1 - i/N) \end{array}$$

From matings of this type, an AA-individual can be produced with probability 1 from an AA $\times$ AA mating, with probability $\frac{1}{4}$ from an Aa $\times$ Aa mating, and with probability $\frac{1}{2}$ from the final mating type. Hence,

$$p_i = \Pr\{\text{produce an AA-type}\}$$

$$= \left(1 - \frac{i}{N}\right)^2 + \frac{1}{4}\left(\frac{i}{N}\right)^2 + \frac{1}{2} \cdot \frac{2i}{N}\left(1 - \frac{i}{N}\right) = \left(1 - \frac{i}{2N}\right)^2.$$

Similarly,

$$q_i = \Pr\{\text{produce an Aa-type}\}$$

$$= \frac{1}{2}\left(\frac{i}{N}\right)^2 + \frac{1}{2}\frac{2i}{N}\left(1 - \frac{i}{N}\right) = \frac{i}{N}\left(1 - \frac{i}{2N}\right),$$

while

$$r_i = \Pr\{\text{produce an aa-type}\} = \frac{i^2}{4N^2}.$$

To form the next generation of N individuals, we take a (multinomial) sample of size N according to the probabilities p_i, q_i, r_i. Recall that if we sample any aa-individuals, then the process stops and so if $X_n = i$, then the process continues to $X_{n+1} = j$ with probability

$$\tilde{P}_{ij} = \binom{N}{j}(p_i)^{N-j}(q_i)^j, \qquad 0 \le i, \ j \le N.$$

This transition probability matrix is substochastic, that is, the rows no longer all sum to one. This corresponds to the fact that the chain can be *killed* by detection of an aa-individual. One way to rectify this defect is to add on an extra state H (for detected homozygote), and then

$$P_{iH} = 1 - \sum_{j=0}^{N} P_{ij} = 1 - (1 - r_i)^N, \qquad 0 \le i \le N$$

and

$$P_{HH} = 1, \qquad P_{Hi} = 0, \qquad 0 \le i \le N.$$

Notice that in this model we have assumed that there are no mutation or selection pressures acting on the population, and that the population size is fixed. State 0 is an absorbing state, corresponding to fixation of the A-allele, while state H is the killing state, and corresponds to formation of an aa-individual.

The first problem we analyse involves finding the probability u_i that fixation occurs before detection when the population starts from $X_0 = i$. As in the previous example in this section, this probability can in principle be found by solving the system of equations

$$u_i = P_{i0} + \sum_{j=1}^{N} P_{ij} u_j,$$

subject to $u_0 = 1$. In practice, the system cannot be solved exactly, and even numerical computation is difficult for large values of N. In the spirit of the previous example, we will resort to diffusion approximation to determine the detection probability $v_i = 1 - u_i$. Once again, the relevant approximating process is a diffusion with killing.

Using the form of the transition matrix given above, we find that

$$E[X_{n+1} - i \,|\, X_n = i] = [Nq_i - i(q_i + p_i)][1 - r_i]^{N-1},$$
$$E[(X_{n+1} - i)^2 \,|\, X_n = i] = [N(N-1)q_i^2 - (2i-1)Nq_i(1 - r_i) + i^2(1 - r_i)^2]$$
$$\times [1 - r_i]^{N-2},$$

and

$$\Pr\{X \text{ killed at time } n + 1 \,|\, X_n = i\} = 1 - (1 - r_i)^N.$$

We define the process $X_N(n) = X_n/(2N)^{1/3}$. (The factor 2 is a convenience that simplifies later formulas.) Then

$$E[X_N(n + 1) - x \,|\, X_N(n) = x = i/(2N)^{1/3}] = O(N^{-2/3}),$$

$$E[(X_N(n + 1) - x)^2 \,|\, X_N(n) = x] = \frac{x}{(2N)^{1/3}} + O(N^{-2/3}),$$

and

$$\Pr\{X_N(n + 1) \text{ killed} \,|\, X_N(n) = x\} = \frac{x^2}{(2N)^{1/3}2} + O(N^{-2/3}).$$

We now rescale time so that one unit of time in the new process corresponds to the passage of $(2N)^{1/3}$ generations. That is, we set

$$Y_N(t) = X_N([(2N)^{1/3}t]), \qquad t \ge 0.$$

If we let $N \to \infty$, the processes $Y_N(t)$ will tend to a diffusion $Y(t)$ with killing on the interval $[0, \infty)$, whose infinitesimal parameters are determined from those of X_N. Recalling that time is now measured over rescaled time units of $(2N)^{-1/3}$, we obtain

$$\mu(x) = \lim_{N \to \infty} (2N)^{1/3} E[\Delta X_N | X_N(n) = x] = 0$$

$$\sigma^2(x) = \lim_{N \to \infty} (2N)^{1/3} E[(\Delta X_N)^2 | X_N(n) = x] = x$$

$$k(x) = \lim_{N \to \infty} (2N)^{1/3} \Pr[\text{killed} | X_N(n) = x] = x^2/2.$$

Since state $\{0\}$ is an exit boundary for $Y(t)$, we will determine the probability $u(x)$ that fixation at 0 occurs before killing. As in the previous example, $u(x)$ satisfies the equation

$$\frac{\sigma^2(x)}{2} u''(x) + \mu(x)u'(x) - k(x)u(x) = 0,$$

with boundary conditions $u(0) = 1$ and $u(\infty) = 0$. Substituting for σ^2, μ and k, and canceling common factors leads to

$$u''(x) - xu(x) = 0, \qquad u(0) = 1, \quad u(\infty) = 0. \tag{10.1}$$

This differential equation is known as Airy's equation, and arises in the study of radio waves, light spectra and the theory of differential equations. The two standard solutions are known as the Airy functions, $A(x)$ and $B(x)$, and are defined by

$$A(x) = \frac{x^{1/2}}{3} \left[I_{-1/3}\left(\frac{2x^{3/2}}{3}\right) - I_{1/3}\left(\frac{2x^{3/2}}{3}\right) \right]$$

and

$$B(x) = \left(\frac{x}{3}\right)^{1/2} \left[I_{-1/3}\left(\frac{2x^{3/2}}{3}\right) + I_{1/3}\left(\frac{2x^{3/2}}{3}\right) \right],$$

where $I_v(x)$ is the modified Bessel function of order v. The function $A(x)$ can be characterized (up to a multiplicative constant) as the unique strictly decreasing solution of (10.1) vanishing at ∞. The solution we require is

$$u(x) = \frac{A(x)}{A(0)},$$

and so the probability of detection (before fixation) is given by

$$v(x) = 1 - \frac{A(x)}{A(0)}.$$

This function can be readily evaluated using tabulated values of the Airy function.

From a biological viewpoint, the most interesting part of the evolution of $Y(t)$ involves those sample paths that result in eventual detection, as opposed to absorption at state 0. In what follows, we derive some properties of the process $Y^*(t)$ obtained by conditioning on eventual detection (killing).

If $p(t, x, y)$ is the transition density function of $Y(t)$, then a simple conditioning argument (cf. (9.1)) shows that the transition density of the conditioned process is given by

$$p^*(t, x, y) = p(t, x, y) \frac{v(y)}{v(x)}.$$

Consider now

$$\mu^*(x) = \lim_{h \downarrow 0} \frac{E[Y^*(h) - x \,|\, Y^*(0) = x]}{h}$$

$$= \lim_{h \downarrow 0} \int \frac{(y - x)p^*(h, x, y)}{h} \, dy$$

$$= \lim_{h \downarrow 0} \frac{1}{v(x)h} \int (y - x)p(h, x, y)v(y) \, dy.$$

Expanding $v(y)$ about x gives

$$v(y) = v(x) + (y - x)v'(x) + \frac{(y - x)^2}{2} v''(z),$$

for some z between x and y. Then

$$\mu^*(x) = \lim_{h \downarrow 0} \frac{1}{h} \left\{ \int (y - x)p(h, x, y) \, dy + \frac{v'(x)}{v(x)} \int (y - x)^2 p(h, x, y) \, dy \right.$$

$$\left. + \frac{1}{2v(x)} \int (y - x)^3 v''(z)p(h, x, y) \, dy \right\}.$$

If we assume that $v''(x)$ is bounded, then we obtain

$$\mu^*(x) = \mu(x) + \sigma^2(x) \frac{v'(x)}{v(x)},$$

an expression formally the same as the result in (9.5a). In our case, $\mu(x) = 0$, $\sigma^2(x) = x$ and, since $v(x)$ satisfies the equation $v''(x) = -xu(x)$, $v''(x)$ is bounded, and so for our model

$$\mu^*(x) = \frac{xv'(x)}{v(x)}.$$

In a similar way, the infinitesimal variance $\sigma^{*2}(x)$ can be shown to be given by

$$\sigma^{*2}(x) = \sigma^2(x) = x.$$

It remains to compute the new killing rate $k^*(x)$. We have

$$\Pr\{Y^* \text{ killed in } (t, t + h) | Y^*(t) = x\}$$

$$= \Pr\{Y \text{ killed in } (t, t + h) | Y(t) = x, Y \text{ killed eventually}\}$$

$$= \frac{k(x)h + o(h)}{v(x)},$$

and hence $k^*(x) = k(x)/v(x) = x^2/2v(x)$. The conditioned process $Y^*(t)$ is now a diffusion process with killing on the interval $(0, \infty)$, with an entrance boundary at 0 (check this).

In order to evaluate functionals of the diffusions Y and Y^*, it is useful to compute the relevant Green functions (cf. Section 3).

Paraphrasing the constructions of Section 3 we determine a positive solution $p_1(x)$ of (10.1) vanishing at infinity and $p_2(x)$ a second positive solution vanishing at zero. Clearly, $p_1(x) = A(x)$ and $p_2(x) = B(x) - \sqrt{3}A(x)$ qualify. Moreover, $p_1(x)$ is decreasing, while $p_2(x)$ is increasing. Since $\sigma^2(x) = x$ for the unconditioned process $Y(t)$, we can expect that

$$G(x, y) = \begin{cases} 2\pi A(x) \dfrac{[B(y) - \sqrt{3}A(y)]}{y}, & 0 < y \le x \\[2ex] 2\pi[B(x) - \sqrt{3}A(x)] \dfrac{A(y)}{y}, & y \ge x. \end{cases} \tag{10.2}$$

The derivation of (10.2) can be justified by confining attention to the state space $[0, r]$, r large, and treating 0 and r as absorbing barriers. Let $G_r(x, y)$ be the associated Green function on $[0, r]$. Sending r to infinity we find that $G_r(x, y)$ converges to $G(x, y)$ as set forth in (10.2).

The Green functions for the conditioned process $Y^*(t)$ in the spirit of Section 9 is given by

$$G_*(x, y) = G(x, y) \frac{v(y)}{v(x)}.$$

The mean time to detection, conditional on detection occurring, is given by

$$M^*(x) = \int_0^\infty G_*(x, y)\, dy$$

$$= \frac{2\pi A(0)A(x)}{A(0) - A(x)} \int_0^x \frac{B(y) - \sqrt{3}A(y)}{y}\left[1 - \frac{A(y)}{A(0)}\right] dy$$

$$+ \frac{2\pi A(0)[B(x) - \sqrt{3}A(x)]}{A(0) - A(x)} \int_x^\infty \frac{A(y)}{y}\left[1 - \frac{A(y)}{A(0)}\right] dy,$$

where we have used the fact that $v(x) = 1 - A(x)/A(0)$.

For small values of x, $M^*(x)$ is approximately constant ($M^*(x) = C$), from which we deduce the following. In populations of size N, the mean time to detection starting from one heterozygote Aa in the initial generation (conditional on detection occurring) is of the order of $N^{1/3}$ generations. This result stands in marked contrast to the conditional result for the random drift model derived in (9.10) and (9.12).

Finally, we will derive the distribution of the maximum functional for the conditioned process. That is, we find the probability $w(x) = w(x; y)$ that starting from $Y^*(0) = x$, the maximum will exceed y ($y > x$) before detection. So

$$w(x) = w(x; y) = \Pr\left\{ \max_{0 \le s < \infty} Y^*(s) > y \,|\, Y^*(0) = x \right\}, \qquad y > x.$$

Thus $w(x)$ is just the probability that Y^* ever gets above y before detection, and is therefore the solution of the equation

$$\frac{x}{2} w''(x) + x \frac{v'(x)}{v(x)} w'(x) - \frac{x^2 w(x)}{2v(x)} = 0$$

satisfying $w(y) = 1$, $w(0) < \infty$. Canceling common factors and making the substitution $w(x) = \eta(x)/v(x)$ leads to the following equation for $\eta(x)$:

$$\eta''(x) - x\eta(x) = 0.$$

Two linearly independent solutions of this equation are $A(x)$ and $[B(x) - \sqrt{3}A(x)]$. Hence,

$$w(x) = C_0 \frac{A(x)}{v(x)} + C_1 \frac{[B(x) - \sqrt{3}A(x)]}{v(x)}.$$

Since $v(x) \to 0$ as $x \to 0$, and we require $w(0) < \infty$, it follows that $C_0 = 0$, and the appropriate solution is

$$w(x) = w(x; y) = \frac{B(x) - \sqrt{3}A(x)}{B(y) - \sqrt{3}A(y)} \cdot \frac{A(0) - A(y)}{A(0) - A(x)}.$$

The mean maximum number of heterozygotes $L^*(x)$ is then given by

$$L^*(x) = \int_0^\infty \Pr\left\{ \max_{0 \le s < \infty} Y^*(s) > y \,|\, Y^*(0) = x \right\} dy$$

$$= x + \int_y^\infty w(x; y)\, dy.$$

The examples given in this section highlight the usefulness of diffusion processes with killing in the analysis of Markov chains.

11: Semigroup Formulation of Continuous Time Markov Processes†

Let $\{X(t), t \geq 0\}$ be a regular time-homogeneous diffusion process on the open interval $I = (l, r)$. We designate by $P(t, x, y) = \Pr\{X(t) \leq y \mid X(0) = x\}$ the transition distribution function of $X(t)$ subject to the initial distribution

$$P(0, x, y) = \begin{cases} 1, & \text{if } x \leq y \\ 0, & \text{if } x > y, \end{cases} \tag{11.1}$$

i.e., a point distribution concentrating at x. We will assume for ease of exposition that $P(t, x, y)$ derives from a continuous density on (l, r), namely

$$\frac{dP(t, x, y)}{dy} = p(t, x, y) \qquad \text{for } t > 0. \tag{11.2}$$

We now consider the family of operators $\{U_t, t \geq 0\}$ transforming each bounded continuous function f on (l, r) into the function $U_t f$ by the formula

$$(U_t f)(x) = E_x[f(X(t))] = E[f(X(t)) \mid X(0) = x]. \tag{11.3}$$

In terms of (11.2), it is useful to display (11.3) explicitly as

$$(U_t f)(x) = \int_l^r p(t, x, \eta) f(\eta) \, d\eta. \tag{11.4}$$

Note that $U_t f$ is well defined for any bounded measurable function, We shall assume throughout this section that U_t preserves continuity, in that

$$(U_t f)(x) \text{ is jointly continuous with respect to } t > 0 \text{ and } x \text{ in } (l, r) \tag{11.5}$$

provided f is piecewise continuous and bounded on (l, r).

SEMIGROUP STRUCTURE

The operators U_t enjoy the semigroup property

$$U_{t+s} = U_t U_s \qquad \text{for all } t, s > 0, \tag{11.6}$$

in the sense that for every bounded piecewise-continuous f, we have

$$U_{t+s} f = U_t(U_s f) = U_s(U_t f). \tag{11.7}$$

Verification of (11.7) rests on the Markov nature of the process as now indicated. Consider

$$(U_{t+s} f)(x) = E[f(X(t + s)) \mid X(0) = x]$$
$$= E\{E[f(X(t + s)) \mid X(t)] \mid X(0) = x\}.$$

† Sections 11–13 that follow are more advanced and use several basic facts and devices of real variable and measure theory, e.g., standard properties of Lebesgue integrals. Sections 14 and 15 do not depend on these sections.

For the moment, let $g = U_s f$, so that

$$g(z) = E[f(X(s))|X(0) = z]$$
$$= E[f(X(t + s))|X(t) = z] \qquad \text{(by time homogeneity)},$$

and

$$g(X(t)) = E[f(X(t + s))|X(t)].$$

Then

$$(U_{t+s}f)(x) = E[g(X(t))|X(0) = x]$$
$$= U_t g(x)$$
$$= (U_t(U_s f))(x), \qquad \text{since} \quad g = U_s f.$$

The identity in (11.7) is variously referred to as the *Fokker–Planck equation* or the *Chapman–Kolmogorov equation*. In terms of the transition densities the Chapman–Kolmogorov equation becomes

$$p(t + s, x, y) = \int_l^r p(t, x, z)p(s, z, y)\,dz. \qquad (11.8)$$

The semigroup property (11.7) underlies a broad spectrum of theory and applications of dynamical systems, e.g., the reference Hille and Phillips† describes semigroup constructions pertinent to heat dissipation, wave propagation, biological population evolution, functional analysis for summability methods, approximation theory, and eigenfunction representations. For our purposes and to provide perspective, it is instructive to characterize the nature of a linear transformation semigroup in the finite-dimensional case, where the key concepts are more accessible. Later in this and the following section we shall elaborate on the semigroup approach to the study of general Markov processes.

Recall that every nonconstant real valued function $U(t)$ satisfying $U(t + s) = U(t)U(s)$ for $t, s \geq 0$, and $|U(t)| \leq 1$ over $t \geq 0$ is necessarily of the form $U(t) = e^{At}$ for some negative constant A. When $U(t)$ is continuous at $t = 0$, so that $\lim_{t \downarrow 0} U(t) = 1$, then A is finite, $U(t)$ is differentiable, and $A = dU(t)/dt|_{t=0}$. For obvious reasons, the coefficient A is called the *infinitesimal generator* of the semigroup. Where $t > 0$ we have

$$\frac{dU(t)}{dt} = AU(t) = U(t)A.$$

---

† E. Hille and R. S. Phillips, "Functional Analysis and Semi-Groups," Colloquium Publications Series, Vol. 31. Amer. Math. Soc., Providence, Rhode Island, 1974.

In a similar manner, a nonsingular continuous semigroup of finite matrices $\{U(t), t \geq 0\}$ ($U_{t+s} = U_t U_s$ here signifies matrix multiplication) can be represented in the form

$$U(t) = e^{At},$$

where the infinitesimal matrix A is obtained by differentiation:

$$A = \lim_{h \downarrow 0} \frac{U(h) - I}{h}. \tag{11.9}$$

(Compare to Theorems 2.1 and 2.2, Chapter 14.)

Laplace transforms provide another method to ascertain A. In the one-dimensional case we define

$$R_\lambda = \int_0^\infty e^{-\lambda t} U(t)\, dt, \qquad \lambda > 0.$$

Then

$$R_\lambda = \int_0^\infty e^{-\lambda t} e^{At}\, dt = \int_0^\infty e^{-(\lambda - A)t}\, dt = (\lambda - A)^{-1},$$

so that

$$\lambda - A = R_\lambda^{-1}, \qquad \text{and} \qquad A = \lambda - R_\lambda^{-1} \qquad \text{for any} \quad \lambda > 0.$$

The similar result $\lambda I - A = R_\lambda^{-1}$ holds in the matrix case, where R_λ^{-1} is the matrix inverse to R_λ and I is the identity matrix.

The theory of semigroups of operators takes these finite-dimensional results of real analysis and generalizes them to function valued operators of functions. The development is a powerful tool for the study of continuous time Markov processes, as will emerge later in this section.

Returning to the diffusion process with U_t given in (11.3), we define the associated resolvent operators R_λ of the process by

$$(R_\lambda f)(x) = \int_0^\infty e^{-\lambda t} (U_t f)(x)\, dt, \qquad \lambda > 0, \tag{11.10}$$

where, as before, f is a bounded piecewise continuous function on (l, r).

By interchanging the order of integration, we secure the resolvent as a kernel operator

$$(R_\lambda f)(x) = \int_l^r G_\lambda(x, y) f(y)\, dy, \tag{11.11}$$

where

$$G_\lambda(x, y) = \int_0^\infty e^{-\lambda t} p(t, x, y)\, dt. \tag{11.12}$$

It is often feasible to permit $\lambda = 0$ in (11.12). In this case the kernel $G_0(x, y)$ is intimately related to appropriate Green functions of Section 3.

By analogy with (11.9), the *infinitesimal operator* of $\{U_t, t \geq 0\}$ is defined as the derivative at the origin of the semigroup, defined in a suitable function space sense. Observe that $U_t \to I$, the identity operator, as $t \downarrow 0$, since the transition density concentrates at $X(0)$ when $t \downarrow 0$ (see (11.1)). We define A as the appropriate limit of the operator sequence $(1/h)(U_h - I)$:

$$\lim_{h \downarrow 0} \left[\frac{U_h - I}{h} \right] = A. \tag{11.13}$$

The characterization of the domain of A is a delicate matter, and the meaning ascribed to the operator representation $U_t = e^{At}$ requires care. These issues, and some applications are developed later.

Example 1. *Standard Brownian Motion.* For any $f(x)$ bounded and continuous on $(-\infty, \infty)$, we have

$$(U_t f)(x) = \frac{1}{\sqrt{2\pi t}} \int_{-\infty}^{\infty} e^{-(x-y)^2/2t} f(y) \, dy. \tag{11.14}$$

The image function $U_t f$ is also continuous on $(-\infty, \infty)$.

The corresponding resolvent kernel has the form

$$G_\lambda(x, y) = \frac{1}{\sqrt{2\lambda}} e^{-\sqrt{2\lambda}|x - y|}, \qquad \lambda > 0. \tag{11.15}$$

In order to verify (11.15) we need to establish the formula

$$J = \int_0^\infty e^{-\lambda t} \frac{1}{\sqrt{2\pi t}} e^{-x^2/2t} \, dt = \frac{1}{\sqrt{2\lambda}} e^{-\sqrt{2\lambda}|x|}.$$

The change of variable $t = s^2$ on the left gives

$$J = \sqrt{2/\pi} \int_0^\infty e^{-(\lambda s^2 + x^2/2s^2)} \, ds.$$

Next, the substitutions $c = |x|/\sqrt{2\lambda}$, $\beta = |x|\sqrt{\lambda/2}$, and $u = s/\sqrt{c}$ transform J into

$$J = \sqrt{\frac{2c}{\pi}} e^{-2\beta} \int_0^\infty e^{-\beta(u - 1/u)^2} \, du = \sqrt{\frac{2c}{\pi}} e^{-2\beta} \left\{ \int_0^1 + \int_1^x \right\}$$

$$= \sqrt{\frac{2c}{\pi}} e^{-2\beta} \int_0^1 e^{-\beta(u - 1/u)^2} \left(1 + \frac{1}{u^2} \right) du$$

$$= \sqrt{\frac{2c}{\pi}} e^{-2\beta} \int_0^\infty e^{-\beta v^2} \, dv, \qquad \text{where} \quad v = u - \frac{1}{u}$$

$$= \sqrt{\frac{2c}{\pi}} e^{-2\beta} \frac{1}{2} \sqrt{\frac{\pi}{\beta}} = \sqrt{\frac{2|x|}{\pi\sqrt{2\lambda}}} e^{-\sqrt{2\lambda}|x|} \frac{1}{2} \sqrt{\frac{\pi}{|x|\sqrt{\lambda/2}}}$$

$$= \frac{1}{\sqrt{2\lambda}} e^{-\sqrt{2\lambda}|x|}.$$

We will now show that for standard Brownian motion, *the domain of the infinitesimal* operator A includes the set of all bounded continuous functions f possessing *second derivatives* f'' *which are themselves bounded and continuous and vanishing as* $|x| \to \infty$. For such functions f, we also show that

$$(Af)(x) = \tfrac{1}{2} f''(x) \qquad \text{for} \quad -\infty < x < \infty. \tag{11.16}$$

Later we will provide an exact characterization of the infinitesimal operator for Brownian motion. To achieve the formula (11.16), we begin by changing variables according to $y = x + z\sqrt{t}$ in the formula (11.14) for $(U_t f)(x)$ to obtain

$$(U_t f)(x) = \int_{-\infty}^{\infty} \frac{1}{\sqrt{2\pi}} e^{-z^2/2} f(x + z\sqrt{t}) \, dz.$$

Next, expand $f(x + z\sqrt{t})$ in a Taylor series in $z\sqrt{t}$ about the point x. Then

$$f(x + z\sqrt{t}) = f(x) + z\sqrt{t} f'(x) + \tfrac{1}{2} z^2 t f''(x) + R_t(z),$$

where $R_t(z)$ is the remainder term and is of the order $O(t^{3/2})$ for t small. We need the facts that $\int_{-\infty}^{\infty} (1/\sqrt{2\pi}) z e^{-z^2/2} \, dz = 0$ and $\int_{-\infty}^{\infty} (z^2/\sqrt{2\pi}) e^{-z^2/2} = 1$. It is elementary to check that

$$r_t = \int_{-\infty}^{\infty} \frac{1}{\sqrt{2\pi}} e^{-z^2/2} R_t(z) \, dz$$

is order less than t; that is, $r(t)/t \to 0$ as $t \to 0$. These facts then entail

$$A_t f(x) \equiv \frac{1}{t} \{(U_t f)(x) - f(x)\} \qquad (A_t \text{ defined by this equation})$$

$$= \tfrac{1}{2} f''(x) + \frac{r_t}{t},$$

and hence

$$(Af)(x) = \lim_{t \downarrow 0} (A_t f)(x) = \tfrac{1}{2} f''(x),$$

with the convergence uniform with respect to x, $-\infty < x < \infty$. We have thus established (11.16) for the collection of f as prescribed.

THE GENERAL SEMIGROUP FORMULATION OF CONTINUOUS TIME MARKOV PROCESSES

Consider the state space as a locally compact metric space which we designate as S. The prototypic case takes S to be an open region such as a sphere in E^p (Euclidean p-space), all of E^p, or an orthant of E^p. In one dimension S is usually an interval such as (l, r) or $[l, r]$ where $-\infty \le l < r \le \infty$. When S is not compact we execute a standard one-point compactification by adding an infinite point ∞ whose neighborhoods are defined to be the complements of compact sets in S. The resulting compact space is denoted by S^*.

For $S = (l, r)$ on the real line we usually compactify S by adding l and r, thereby closing S to $S^* = [l, r]$.

Let $C(S^*)$ denote the set of all continuous functions on S^* where for $f(x)$ in $C(S^*)$ *the value at* ∞ *exists finitely*. It is convenient to also adjoin to S^* an isolated killing (cemetery) state "Δ" and prescribe $f(\Delta) = 0$ for all $f \in C(S^*)$.

In the specific case of $S = (l, r)$ and $S^* = [l, r]$, the function class $C(S^*)$ consists of all continuous bounded functions obeying the conditions that $\lim_{x \uparrow r} f(x)$ and $\lim_{x \downarrow l} f(x)$ exist, but are not necessarily the same. When r (or l) belong to S, then we assume f is continuous at these boundary points.

We denote by $C_0(S^*) = C_0(S)$ the subset in $C(S)$ of functions which tend to zero outside increasing compact sets. The collection of functions $C(S^*)$ is a linear space (i.e., if f and g are in $C(S^*)$, then also $f + g$ are in $C(S^*)$) endowed with the norm $\|f\| = \sup_{x \in S} |f(x)|$ which obviously obeys the triangle inequality $\|f + g\| \leq \|f\| + \|g\|$.

We consider a strong Markov process $\{X(t), t \geq 0\}$ (not necessarily a diffusion), with state space S^*. We posit the existence of a transition probability distribution

$$P(t, x, A) = \Pr\{X(t) \in A \mid X(0) = x\} \tag{11.17}$$

defined for $t \geq 0$, $x \in S^*$ and A traversing the usual subsets of S^* (i.e., Borel sets of S^* plus sets of probability zero). The possibility of a killing event passing the process to state Δ may also occur (cf. Section 1). The transition distribution is possibly only substochastic on S^*, i.e., $P(t, x, S^*) \leq 1$, but always stochastic on $S^* \cup \{\Delta\}$ signifying

$$P(t, x, S^* \cup \{\Delta\}) = 1. \tag{11.18}$$

The contingency of $P(t, x, S^*) < 1$ is present when a killing event occurs with positive probability.

To ease the exposition we shall stipulate the existence of a transition density $p(t, x, y)$ for $t > 0$ with

$$P(t, x, A) = \int_A p(t, x, y)\, dy \qquad \text{for} \quad A \subset S \quad \text{and} \quad x \in S.$$

In almost all applications, $p(t, x, y)$ is available. Unless stated otherwise we assume throughout that

$$p(t, x, y) \text{ is jointly continuous in } t > 0 \text{ and } x, y \text{ in } S. \tag{11.19}$$

THE SEMIGROUP OPERATORS

We define the family of semigroup operators over $C(S^*)$ by

$$U_t f(x) = E_x[f(X(t))] \qquad \text{for} \quad t > 0, \quad x \in S^*. \tag{11.20}$$

When f belongs to $C_0(S)$, also vanishing at Δ by convention, then (11.20) becomes

$$U_t f(x) = E_x[f(X(t))] = \int_S p(t, x, y) f(y) \, dy. \qquad (11.21)$$

We assume the following properties of the semigroup operators U_t, $t > 0$:

$$U_t: C(S^*) \to C(S^*) \qquad (11.22\text{a})$$

(i.e., $U_t f$ is also a function in $C(S^*)$ if f is in $C(S^*)$);

$$U_t f(x) \text{ is jointly continuous in } t > 0 \text{ and } x \text{ in } S^*; \qquad (11.22\text{b})$$

$$U_0 = I \qquad \text{(the identity operator);} \qquad (11.22\text{c})$$

(this is the natural requirement that $P(0+, x, A)$ is the point distribution concentrating at x)

$$U_{t+s} = U_t U_s, \qquad t, s > 0. \qquad (11.22\text{d})$$

This semigroup identity is essentially tantamount to the Markov property (cf. (11.7) and (11.8)).

$$\| U_t f - f \| = \sup_{x \in S^*} |U_t f(x) - f(x)| \to 0 \qquad \text{as} \quad t \downarrow 0. \qquad (11.22\text{e})$$

This expresses strong continuity at $t = 0$. Sometimes a weak continuity postulate suffices so that $U_t f(x) \to f(x)$ pointwise with $\| U_t f \|$ uniformly bounded for f in $C(S^*)$ already implies the strong continuity of (11.22e):

$$U_t 1 \leq 1 \qquad \text{(1 is the unit function on } S\text{).} \qquad (11.22\text{f})$$

This condition holds because P is a probability measure on $S^* \cup \{\Delta\}$.

$$U_t \text{ maps nonnegative functions into nonnegative functions} \qquad (11.22\text{g})$$

(because P is a nonnegative measure).

If we define a norm on the operator $\| U_t \| = \sup_{f \in C(S^*), \, \|f\| \leq 1} \| U_t f \|$ then (11.22f) and (11.22g) together imply $\| U_t \| \leq 1$, i.e., U_t is a contraction semigroup. In fact,

$$|U_t f(x)| \leq \int_{S^*} p(t, x, y) |f(y)| \, dy \leq \| f \| \int_{S^*} p(t, x, y) \, dy \leq \| f \|. \qquad (11.22\text{h})$$

We say that a Markov semigroup U_t has the (*strong*) *Feller property* if $U_t f(x)$ is continuous on S^* for any $f(x)$ bounded and measurable (e.g., piecewise continuous) on S.

Example 2. *Standard Brownian Motion in E_n, n Dimensions.* (Notation: for $\mathbf{x} = (x_1, \ldots, x_n) \in E_n$, let $\| \mathbf{x} \|^2 = \Sigma x_i^2$.) In this case it is customary to take the one-point compactification $S^* = E_n \cup \{\infty\}$ identifying $C(S^*) = C(E_n^*)$ comprised of all continuous functions on E_n where $\lim_{\|\mathbf{x}\| \to \infty} f(\mathbf{x})$ has a single limit independent of the direction in which $\|x\| \to \infty$. The transition density is

$$p(t, x, y) = \frac{1}{(\sqrt{2\pi t})^n} \exp\left[-\frac{1}{2t} \sum_{i=1}^{n} (x_i - y_i)^2\right] = \frac{1}{(\sqrt{2\pi t})^n} \exp\left(-\frac{1}{2t} \|x - y\|^2\right).$$

(11.23)

The corresponding semigroup operators for f in $C(S^*)$ are

$$U_t f(x) = \int_{E_n} \frac{1}{(\sqrt{2\pi t})^n} \exp\left(-\frac{1}{2t} \|x - y\|^2\right) f(y) \, dy.$$ (11.24)

Example 3. *An Ornstein–Uhlenbeck Process in n Dimensions.* One version of an Ornstein–Uhlenbeck process in n dimensions is to take the variance coefficient constant, the same as that of Brownian motion, but to take the mean infinitesimal displacement to be that of a restoring force directly proportional to the distance from the origin. In the notation of (1.2″) and (1.3″) of Section 1,

$$\mu_i(x) = -\|x\| = -\sqrt{\sum_{i=1}^{n} x_i^2}, \quad \text{and} \quad \sigma_{ij}(x) = \begin{cases} 1 & \text{if } i = j, \\ 0 & \text{if } i \neq j. \end{cases}$$

These parameters in one dimension reduce to the classical Ornstein–Uhlenbeck infinitesimal rates (cf. Example C, Section 2).

A transition density function of the corresponding radial process is

$$p(t, x, y) = \frac{y^{n-1} e^{-y^2}}{(1 - e^{2t})} \exp\left[\frac{-(x^2 + y^2)e^{-2t}}{1 - e^{-2t}}\right] (xye^{-t})^{-(n-2)/2} I_{(n-2)/2}\left(\frac{xye^{-t}}{1 - e^{-2t}}\right),$$

$$x = \|x\|, \quad y = \|y\|.$$

THE RESOLVENT OPERATOR

The resolvent is defined as the Laplace transform of $U_t f$, depending on the parameter $\lambda > 0$:

$$(R_\lambda f)(x) = \int_0^\infty e^{-\lambda t} (U_t f)(x) \, dt.$$ (11.25)

Elaborating the integrand where f vanishes at ∞ and Δ yields

$$(R_\lambda f)(x) = \int_0^\infty e^{-\lambda t} \int_S p(t, x, y) f(y) \, dy \, dt = \int_S f(y) G_\lambda(x, y) \, dy$$

involving the extended Green kernel $G_\lambda(x, y) = \int_0^\infty e^{-\lambda t} p(t, x, y) \, dt$. In view of $\|U_t\| \leq 1$, it is elementary to check that $\|R_\lambda f\| \leq (1/\lambda)\|f\|$.

We assume henceforth that $R_\lambda f(x)$ is in $C(S^*)$ for any f bounded and measurable. This property is satisfied in most practical cases.

The following *resolvent operator equation* identity holds

$$R_\lambda - R_\mu = (\mu - \lambda)R_\lambda R_\mu. \tag{11.26}$$

Proof. For any f in $C(S^*)$ we need to verify the equation

$$(\mu - \lambda)\int_0^\infty e^{-\lambda t}U_t\left(\int_0^\infty e^{-\mu s}U_s f(x)\,ds\right)dt = \int_0^\infty (e^{-\lambda t} - e^{-\mu t})U_t f(x)\,dt.$$

Because U_t is a bounded linear operator, we can move U_t to the inner integral yielding

$$\int_0^\infty e^{-\lambda t}U_t\left(\int_0^\infty e^{-\mu s}U_s f(x)\,ds\right)dt = \int_0^\infty e^{-\lambda t}\left(\int_0^\infty e^{-\mu s}U_t(U_s f(x))\,ds\right)dt$$

$$= \int_0^\infty e^{-\lambda t}\left(\int_0^\infty e^{-\mu s}U_{t+s} f(x)\,ds\right)dt,$$

the last equation resulting by the semigroup property. Now an obvious change of variable, $\tau = s + t$, gives

$$\int_0^\infty e^{-\lambda t}\left(\int_t^\infty e^{-\mu(\tau - t)}U_\tau f(x)\,d\tau\right)dt,$$

and by interchanging orders of integration

$$= \int_0^\infty e^{-\mu\tau}U_\tau f(x)\left(\int_0^\tau e^{-(\lambda - \mu)t}\,dt\right)d\tau$$

$$= \frac{1}{(\lambda - \mu)}\int_0^\infty e^{-\mu\tau}U_\tau f(x)[1 - e^{-(\lambda - \mu)\tau}]\,d\tau$$

$$= \frac{1}{\lambda - \mu}[R_\mu f(x) - R_\lambda f(x)].$$

Combining these equations, the verification of (11.26) is achieved. ∎

As a corollary of (11.26) we infer the important fact:

Proposition 11.1 *The range of the resolvent operator* (i.e., the collection of all $g(x) = R_\lambda f(x)$ for some $f \in C(S^*)$) *is independent of λ for $\lambda > 0$.*

Proof. Consider $g(x) = R_\lambda f(x)$ in the range of R_λ. Then with the aid of (11.26), we have

$$g(x) = R_\lambda f(x) = R_\mu[f(x) + (\mu - \lambda)R_\lambda f(x)] = R_\mu h(x)$$

with $h(x) = f(x) + (\mu - \lambda)R_\lambda f(x)$ so that $g(x)$ is also in the range of R_μ. The argument is plainly symmetric in λ and μ. ∎

Another basic corollary emanating from the resolvent equation follows.

Proposition 11.2. $R_\lambda f(x) \equiv 0$ *for some* $\lambda > 0$ *entails* $f(x) \equiv 0$.

Proof.   From the resolvent equation, we find that $R_\lambda f(x) \equiv 0$ for all $\lambda > 0$, i.e., $\int_0^\infty e^{-\lambda t}(U_t f)(x)\, dt = 0$ for all $\lambda > 0$ and x in S^*. The uniqueness of the Laplace transform implies that $U_t f(x) \equiv 0$ for all $t > 0$ and x in S^* since $U_t f(x)$ is continuous in t and x (See (11.22b)). Now let $t \downarrow 0$ and by property (11.22e), i.e., strong continuity at zero, the conclusion that $f(x) \equiv 0$ ensues.   ∎

THE INFINITESIMAL OPERATOR

We develop several equivalent definitions.

Definition 11.1.   *For every* $u(x)$ *in* $\mathcal{D}$, *the range of* R_α *for some* $\alpha > 0$ (*Proposition 11.1 tells us that* $\mathcal{D}$ *is independent of* $\alpha > 0$), *we define*

$$\tilde{A}u = \alpha u - f \quad \text{with} \quad R_\alpha f = u. \tag{11.27}$$

This is meaningful since $R_\lambda g \equiv 0$ entails $g \equiv 0$ so the f of (11.27) is uniquely determined, and by virtue of the resolvent equation, we have

$$\alpha u - f = \beta u - g \quad \text{for} \quad u = R_\alpha f = R_\beta g. \tag{11.28}$$

In fact, let $w = (\beta - \alpha)u - (g - f)$. Then the resolvent equation gives

$$\begin{aligned}
R_\beta w &= (\beta - \alpha) R_\beta (R_\alpha f) - R_\beta (g - f) \\
&= R_\alpha f - R_\beta f - R_\beta g + R_\beta f \\
&= R_\alpha f - R_\beta g = u - u = 0,
\end{aligned}$$

and so $w \equiv 0$, i.e., (11.28) follows. This implies that $\tilde{A}$ is uniquely defined.

The next definition is more intuitive and conforms more closely to the elementary treatment leading up to (11.9).

Definition 11.2.   *We define* $Au = v$ *provided that* $u(x) \in C(S^*)$ *and the limit relation*

$$\lim_{h \downarrow 0} \left\| \frac{U_h u - u}{h} - v \right\| = 0 \tag{11.29}$$

holds for some $v \in C(S^*)$.

Our next objective is to show that A and $\tilde{A}$ are equivalent operators, meaning that they have the same domains of definition and produce the same transformation. We write $\mathcal{D}(A) = \text{domain}(A)$ for the domain of an operator A.

Let $u = R_\lambda f$ in domain($\tilde{A}$). Consider

$$\frac{U_h(R_\lambda f) - R_\lambda f}{h} = \frac{1}{h}(U_h - I)\int_0^\infty e^{-\lambda t}U_t f(x)\,dt$$

$$= \frac{1}{h}\int_0^\infty e^{-\lambda t}[U_{t+h}f(x) - U_t f(x)]\,dt \qquad (11.30)$$

and by repartitioning the limits of integration

$$= \frac{1}{h}(e^{\lambda h} - 1)\int_0^\infty e^{-\lambda t}U_t f(x)\,dt - \frac{e^{\lambda h}}{h}\int_0^h e^{-\lambda t}U_t f(x)\,dt.$$

Now let h decrease to zero. The first term converges uniformly to $\lambda R_\lambda f$ and the second converges uniformly to f on account of the fundamental theorem of calculus and the fact that $\|U_\xi f - f\| \to 0$ as $h \downarrow 0$ where $0 < \xi < h$. Thus,

$$\lim_{h\downarrow 0}\left\|\frac{U_h(R_\lambda f) - R_\lambda f}{h} - (\lambda(R_\lambda f) - f)\right\| = 0.$$

In accordance with Definition 11.2, we have that $u = R_\lambda f$ is in domain(A) and $Au = \lambda u - f$, which shows that A is an extension of $\tilde{A}$ $(A \supset \tilde{A})$ as the two operators coincide on the domain of definition of $\tilde{A}$.

It remains to show that $\tilde{A}$ is an extension of A, $\tilde{A} \supset A$. To this end, let u be in domain(A) and set $f = \alpha u - Au$ (which is manifestly in $C(S^*)$). Put $v = R_\alpha f$ which we have just shown is in domain(A) (because $A \supset \tilde{A}$ is already established) and $Av = \alpha v - f = \alpha v - (\alpha u - Au)$. Therefore

$$A(v - u) = \alpha(v - u).$$

From the definition of A, we have

$$U_h(v - u) = v - u + h\alpha(v - u) + o(h) = (1 + h\alpha)(v - u) + o(h).$$

Since U_h is a contraction operator, then $\|v - u\| \geq \|U_h(v - u)\| \geq (1 + h\alpha)\|v - u\| + o(h)$ and canceling the common additive term, dividing by h, and then sending $h \downarrow 0$ establishes $0 \geq \|v - u\|$ and hence $v = u$. Therefore, u in $\mathscr{D}(A)$ implies v in $\mathscr{D}(\tilde{A})$ and therefore $\mathscr{D}(A) \subset \mathscr{D}(\tilde{A})$. Therefore, their domains coincide and $A = \tilde{A}$ as claimed. ∎

SOME PROPERTIES OF THE INFINITESIMAL OPERATOR A

(i) A is linear (this follows from the "derivative" definition (11.29) since the derivative is additive. $\qquad (11.31)$

(ii) $\mathscr{D}(A)$ is dense in $C(S^*)$. (The range of R_λ is dense. Check this.)

(iii) A is a closed operator. This means that when $u_n \in \mathscr{D}(A)$ and $Au_n = v_n$ converge, $\|u_n - u\| \to 0$, $\|v_n - v\| \to 0$ then $u \in \mathscr{D}(A)$ and $Au = v$. Equivalently the pairings $\{u, Au\}$ for u traversing $\mathscr{D}(A)$ is a closed set in the norm.

We prove this last statement. Since for $\alpha > 0$

$$v_n = Au_n = \alpha u_n - f_n, \qquad \text{where} \quad R_\alpha f_n = u_n, \tag{11.32}$$

and v_n and u_n separately converge in the norm of $C(S^*)$, we see from (11.32) that f_n also converges in the norm of $C(S^*)$, say to $f(x)$. It follows since R_α is bounded that $R_\alpha f_n = u_n$ passes into $R_\alpha f = u$ which means that u is in the range of R_α and consequently in the domain of A. The limit relation $v = \alpha u - f$ shows that $Au = v$.

(iv) The relation $Au = \alpha u - f$ can be expressed concisely as follows. The operator $\alpha I - A$ ($I =$ identity) is invertible as a bounded operator on domain(A) and $(\alpha I - A)^{-1} = R_\alpha$ such that

$$\|(\alpha I - A)^{-1}\| \leq \frac{1}{\alpha}. \tag{11.33}$$

(See p. 292.) There is an important converse to the above description of A endowed with the properties (i)–(iv), which we state without proof.

Theorem 11.1 (Hille-Yosida). *Let A be a linear operator with a closed dense domain in a Banach space, say $C(S^*)$. If for some $\alpha > 0$, $\|(\alpha I - A)^{-1}\| \leq 1/\alpha$, i.e., for each f in $C(S^*)$ there exists a unique u in $\mathscr{D}(A)$ such that $\alpha u - Au = f$ and $\|u\| \leq (1/\alpha)\|f\|$, then there exists a strongly continuous contraction semigroup U_t (property (11.22h)) with A as its infinitesimal operator. If $R_\alpha = (\alpha I - A)^{-1}$ preserves positivity, then U_t is a positivity preserving semigroup.*

Example. From (11.15) for Brownian motion on $(-\infty, \infty)$ recall that

$$R_\alpha f(x) = \int_{-\infty}^{\infty} f(y)\, \frac{1}{\sqrt{2\alpha}}\, e^{-\sqrt{2\alpha}|x-y|}\, dy \qquad \text{for} \quad f(x) \in C(S^*),$$

with $S^* = [-\infty, \infty]$ the closed real line and continuity at ∞ means that $\lim_{|x|\to\infty} f(x) = f(\infty)$ exists and exhibits the same value for $x \to +\infty$ or $-\infty$. Then

$$R_\alpha f(x) - \frac{f(\infty)}{\alpha} = \int_{-\infty}^{\infty} [f(y) - f(\infty)]\, \frac{1}{\sqrt{2\alpha}}\, e^{-\sqrt{2\alpha}|x-y|}\, dy$$

$$\to 0 \quad \text{as} \quad |x| \to \infty \qquad \text{(prove this)},$$

i.e., $R_\alpha f(\infty) = f(\infty)/\alpha$.

We next show that $u(x) = R_\alpha f(x)$ is twice continuously differentiable over the real line. To this end, we write

$$R_\alpha f(x) = e^{-\sqrt{2\alpha}x} \int_{-\infty}^{x} f(y)\, \frac{1}{\sqrt{2\alpha}}\, e^{+\sqrt{2\alpha}y}\, dy + e^{\sqrt{2\alpha}x} \int_{x}^{\infty} f(y)\, \frac{1}{\sqrt{2\alpha}}\, e^{-\sqrt{2\alpha}y}\, dy.$$

Then by the fundamental theorem of calculus,

$$\frac{dR_\alpha f(x)}{dx} = -e^{-\sqrt{2\alpha}x} \int_{-\infty}^{x} f(y)e^{\sqrt{2\alpha}y}\,dy + e^{\sqrt{2\alpha}x} \int_{x}^{\infty} f(y)e^{-\sqrt{2\alpha}y}\,dy$$

and

$$u''(x) = \frac{d^2 R_\alpha f(x)}{dx^2} = 2\alpha R_\alpha f(x) - 2f(x)$$

are clearly bounded and continuous on $[-\infty, \infty]$ because $f(x)$ and $R_\alpha f(x)$ are such functions. Moreover, as $x \to \pm\infty$, $u(x) = R_\alpha f(x) \to f(\infty)/\alpha$ as indicated earlier. The above considerations show that $Au(x) = \frac{1}{2}u''(x) = \alpha u(x) - f(x)$ and also

$$u''(x) \to 0 \qquad \text{as} \quad |x| \to \infty. \tag{11.34}$$

We can now succinctly characterize the domain of A for Brownian motion with the help of (11.34). Accordingly, we define the operator

$$\mathscr{A}u = \tfrac{1}{2}u''(x) \tag{11.35}$$

with domain consisting of all functions for which u, u' and u'' are continuous and bounded on $[-\infty, \infty]$ and such that $u''(x) \to 0$ as $|x| \to \infty$.

Claim. $\mathscr{A} = A$, the infinitesimal operator of Brownian motion.

We pointed out in (11.16) that if $u(x)$ and $u''(x)$ are continuous and bounded with u'' vanishing at ∞ then $u \in \mathscr{D}(A)$. Thus, $\mathscr{A}$ is an extension of A. Consider now $v \in \mathscr{D}(\mathscr{A})$. Define $f(x) = \alpha v(x) - \frac{1}{2}v''(x)$ with $\alpha > 0$. Then $f(\infty) = \alpha v(\infty)$ and $f \in C(S^*)$ ($S^* = [-\infty, \infty]$ ($+\infty$ and $-\infty$ identified). Let $w = R_\alpha f$ in $\mathscr{D}(A) \subset \mathscr{D}(\mathscr{A})$. Note that $\mathscr{A}(w - v) = A(w - v)$ and therefore $\alpha z(x) - \frac{1}{2}z''(x) = 0$ with $z = w - v$. The general solution of this differential equation is $z(x) = c_1 e^{-\sqrt{2\alpha}x} + c_2 e^{+\sqrt{2\alpha}x}$. This is supposed to be bounded for all real x which compels $c_1 = c_2 = 0$ and $z(x) \equiv 0$. Therefore, $w = v$ and v belongs to $\mathscr{D}(A)$. Thus, $\mathscr{D}(A) = \mathscr{D}(\mathscr{A})$ and $A = \mathscr{A}$ as asserted.

FIRST DYNKIN FORMULA

We present an ancillary formula before proceeding to the main result. Let $u = R_\alpha f$ and let σ be a Markov time. Then

$$u(x) = E_x\left[\int_0^\sigma e^{-\alpha t}f(X(t))\,dt\right] + E_x[e^{-\alpha\sigma}u(X(\sigma))]. \tag{11.36}$$

Proof of (11.36)

$$
u(x) = \int_0^\infty e^{-\alpha t}(U_t f)(x)\, dt = \int_0^\infty e^{-\alpha t} E_x[f(X(t))]\, dt
$$

$$
= E_x\left[\int_0^\sigma e^{-\alpha t} f(X(t))\, dt + \int_\sigma^\infty e^{-\alpha t} f(X(t))\, dt\right]
$$

$$
= E_x\left[\int_0^\sigma e^{-\alpha t} f(X(t))\, dt\right] + E_x\left[e^{-\alpha\sigma}\int_0^\infty e^{-\alpha t} f(X(\sigma + t))\, dt\right]
$$

and using the strong Markov property

$$
= E_x\left[\int_0^\sigma e^{-\alpha t} f(X(t))\, dt\right] + E_x[e^{-\alpha\sigma} u(X(\sigma))]. \quad\blacksquare
$$

DYNKIN FORMULA WITH INFINITESIMAL GENERATOR

We now present the main Dynkin formula which we shall refer to as the Dynkin formula to distinguish it from the first formula (11.36).

Theorem 11.2. *Assume σ is a Markov time with finite expectation and $u(x)$ is in $\mathcal{D}(A)$. Then*

$$
E_x\left[\int_0^\sigma Au(X(t))\, dt\right] = E_x[u(X(\sigma))] - u(x). \tag{11.37}
$$

Remark. We will present two proofs of (11.37). The immediate one falls back on the first Dynkin formula (11.36). The second proof presented in Section 12 involves the theory of additive processes.

Proof. Let $u = R_\alpha f$ for some $f \in C(S^*)$, which exists because by stipulation $u \in \mathcal{D}(A)$. Also, $Au = \alpha u - f$. The first Dynkin formula provides

$$
u(x) = E_x\left[\int_0^\sigma e^{-\alpha t} f(X(t))\, dt\right] + E_x[e^{-\alpha\sigma} u(X(\sigma))]
$$

$$
= E_x\left[\int_0^\sigma e^{-\alpha t}\{\alpha u(X(t)) - Au(X(t))\}\, dt\right] + E_x[e^{-\alpha\sigma} u(X(\sigma))]. \tag{11.38}
$$

Now

$$
\left|\alpha E_x\left[\int_0^\sigma e^{-\alpha t} u(X(t))\, dt\right]\right| \le \alpha\|u\| E_x[\sigma] \to 0 \qquad \text{as} \quad \alpha \to 0
$$

because $E_x[\sigma] < \infty$ by hypothesis. Also, $e^{-\alpha t}, t < \sigma$, goes to 1 as $\alpha \downarrow 0$ because σ is finite for almost every sample path. We may apply the bounded convergence theorem when $\alpha \to 0$ to get $E_x[e^{-\alpha\sigma}u(X(\sigma))] \to E_x[u(X(\sigma))]$. Therefore, the right side of (11.38) converges to

$$-E_x\left[\int_0^\sigma Au(X(t))\,dt\right] + E_x[u(X(\sigma))]. \tag{11.39}$$

Rearranging terms in (11.38) and (11.39) leads to (11.37). ∎

APPLICATIONS OF THE DYNKIN FORMULA

Consider standard Brownian motion $X(t) = (X_1(t), \ldots, X_n(t)), t > 0$ in n dimensions. Starting at x, and writing $\|x\| = \sum_{i=1}^n x_i^2 < 1$, let σ be the first time the process departs the unit sphere, i.e., σ is the first passage time to the set $\|x\| \geq 1$. This is a Markov time (why? cf. Chapter 6).

Our task here is to calculate $E_x[\sigma]$. [We shall later, in (12.25), determine the variance of σ.]

Paraphrasing the analysis of (11.16) we readily prove that the domain of the infinitesimal operator A of n-dimensional Brownian motion includes at least those functions $w(x)$ having continuous second derivatives for which $w(x)$ and $\Delta w(x) = \sum_{i=1}^n \partial^2 w/\partial x_i^2$ (Δ is the Laplacian differential operator) converge to zero as $\|x\| \to \infty$. Then

$$Aw = \tfrac{1}{2}\Delta w. \tag{11.40}$$

Now define

$$u(x) = 1 - \|x\|^2 \qquad \text{for } \|x\| \leq 1 \text{ and extended to } \|x\| > 1 \text{ to be twice} \\ \text{continuously differentiable and in } C_0(E_n). \tag{11.41}$$

Apply (11.40) to u observing by continuity of paths that $\|X(\sigma)\| = 1$ and thus $u(X(\sigma)) = 0$. Moreover, for $t < \sigma$, $u(X(t)) = 1 - \|X(t)\|^2$ and at these times $Au(X(t)) = -n$. We do not know a priori that $E_x[\sigma] < \infty$ so we cannot directly apply Dynkin's formula (11.37). The following procedure (via truncation) to overcome this problem is a common device. Specifically, we define a sequence of approximating Markov times. For each positive integer N, let $\sigma_N = \sigma \wedge N = \min(\sigma, N)$. Plainly σ_N is bounded and $E_x[\sigma_N] \leq N < \infty$. Since with probability one every sample path escapes $\|x\| \leq 1$, σ_N increases to σ as $N \uparrow \infty$. Applying the Dynkin formula (11.37) to σ_N and u of (11.41), we obtain

$$E_x[u(X(\sigma_N))] - u(x) = E_x\left[\int_0^{\sigma_N} Au(X(t))\,dt\right] = -nE_x[\sigma_N]. \tag{11.42}$$

So $E_x[\sigma_N] \leq (2/n)\|u\|$. By monotone convergence, σ_N increases, and we deduce that $E_x[\sigma] \leq (2/n)\|u\|$. Now we can apply (11.37) with σ as prescribed and since $u(X(\sigma)) = 0$, the Dynkin formula reduces to $u(x) = nE_x[\sigma]$ or

$$E_x[\sigma] = \frac{1 - \|x\|^2}{n}. \tag{11.43}$$

DYNKIN REPRESENTATION OF THE INFINITESIMAL OPERATOR OF A DIFFUSION

Consider a diffusion on S. A state a is called a *trap point* (or *absorbing point*) if

$$\Pr\{X(t) = a \,|\, X(0) = a\} = 1 \qquad \text{for all} \quad t > 0. \tag{11.44}$$

For any open subset V of the state space we let

$$\sigma(V^c) = \text{first passage time into the complement of } V. \tag{11.45}$$

We will establish at the close of this section that with x in V, then $E_x[\sigma(V^c)] < \infty$ for most diffusions when V is a bounded open sphere. In fact, the random variable $\sigma(V^c)$ has all moments finite, and its distribution function tails off exponentially fast.

With these preliminaries in mind, we have

Theorem 11.3. *Let $X(t)$ be a diffusion on S. If a is not a trap state, then for u in $\mathscr{D}(A)$*

$$Au(a) = \lim_{\substack{V \text{ nbd of } a \\ V \downarrow a \\ E_a[\sigma(V^c)] < \infty}} \left(\frac{E_a[u(X(\sigma(V^c)))] - u(a)}{E_a[\sigma(V^c)]} \right). \tag{11.46}$$

Here the designation "V nbd" is an abbreviation for a bounded open region containing a neighborhood of a. Usually, we will take V to be a small spherical neighborhood of a. The symbol $V \downarrow a$ refers to a collection of neighborhoods shrinking to a. For V a family of spheres with center at a then $V \downarrow a$ means that the radius of V reduces to zero.

The proof of (11.46) needs the following Lemma.

Lemma 11.1. *If a is not a trap state then there exists a bounded neighborhood V of a such that*

$$E_a[\sigma(V^c)] < \infty.$$

We defer the proof of this lemma until after the proof of the theorem.

Proof of Theorem 11.3. Let V be a neighborhood of a for which $E_a[\sigma(V^c)] < \infty$. Obviously, if $V \supset W \supset a$ then $\sigma(W^c) \leq \sigma(V^c)$ (in order to depart from V it is necessary to depart from W earlier) and therefore $E_a[\sigma(W^c)] \leq E_a[\sigma(V^c)]$. By Dynkin's formula

$$E_a[u(X(\sigma(V^c)))] - u(a) - Au(a)E_a[\sigma(V^c)] = E_a\left[\int_0^{\sigma(V^c)} \{Au(X(t)) - Au(a)\} \, dt \right].$$

$$\tag{11.47}$$

Because $Au(\cdot)$ in $C(S^*)$ is continuous at a in S, then for any prescribed $\varepsilon > 0$, we can choose a neighborhood W of a small enough and contained in V such that

$$|Au(x) - Au(a)| \le \varepsilon \qquad \text{for all} \quad x \in W. \tag{11.48}$$

This gives for W replacing V in (11.47)

$$E_a\left[\int_0^{\sigma(W^c)} |Au(X(t)) - Au(a)|\, dt\right] \le \varepsilon E_a[\sigma(W^c)]. \tag{11.49}$$

Combining (11.49) in (11.47) produces

$$\left| \frac{E_a[u(X(\sigma(W^c)))] - u(a)}{E_a[\sigma(W^c)]} - Au(a) \right| \le \varepsilon,$$

and this inequality persists for any contraction of W about a. Since ε is arbitrary, equality holds in the limit, i.e., (11.46) ensues. ∎

In order to prove Lemma 11.1 we develop some further characterizations of trap points.

Lemma 11.2. *If a is not a trap point, then there exists $u(x)$ in the domain of A such that $Au(a) \ne 0$.*

Proof. Consider the collection of functions $u = R_\alpha f$, which span the domain of A as f traverses $C(S^*)$, $\alpha > 0$. Suppose to the contrary that $Au(a) = 0$ for all such u. Then $Au(a) = \alpha u(a) - f(a) = 0$ for all $f \in C(S^*)$ with $u = R_\alpha f$ for all α. In particular

$$\int_0^\infty e^{-\alpha t}[U_t f(a) - f(a)]\, dt = 0 \qquad \text{for all} \quad \alpha > 0.$$

By the uniqueness of the Laplace transform, since $U_t f(a) - f(a)$ is continuous in t we deduce that

$$U_t f(a) - f(a) = 0 \qquad \text{for all} \quad t > 0 \quad \text{and} \quad f \in C(S^*), \quad a \in S,$$

that is, $E_a[f(X(t))] = f(a)$ for all $f \in C(S^*)$ and $t > 0$. Now a is not a trap point implies that $X(t)$ is not near a for some t. Take f to be smooth in $C(S^*)$) concentrated away from a, i.e., $f(x)$ is positive for some $x \ne a$ but $f(a) = 0$. Then for this f, we have $0 \ne E_a[f(X(t))] = f(a) = 0$ which is absurd. To avert this contradiction the statement of Lemma 11.2 necessarily prevails. ∎

Proof of Lemma 11.1. By Lemma 11.2, there exists $u \in \mathscr{D}(A)$ such that $Au(a) \ne 0$, say $Au(a) \ge \varepsilon > 0$. Let V be a neighborhood of a on which

$Au(x) \geq \varepsilon/2$. Define $\sigma(V^c)$ as in (11.45) and let $\sigma_N = \sigma(V^c) \wedge N$ be the corresponding truncated Markov time. We apply Dynkin's formula with σ_N yielding

$$E_a[u(\sigma_N)] - u(a) = E_a\left[\int_0^{\sigma_N} Au(X(t))\, dt\right]$$

$$\geq \frac{\varepsilon}{2} E_a[\sigma_N].$$

Therefore

$$E_a[\sigma_N] \leq \frac{2\|u\|}{\varepsilon},$$

where $\varepsilon > 0$ is independent of N. Letting $N \uparrow \infty$ entails $\sigma_N \uparrow \sigma(V^c)$ and the monotone convergence theorem implies

$$E_a[\sigma(V^c)] \leq \frac{2\|u\|}{\varepsilon} < \infty.$$

The proof of Lemma 11.1 is complete. ∎

BOUND ON EXPECTED TIME TO EXIT AN OPEN SET

If x_0 is not a trap state, there exists a neighborhood U_0 around x_0 and a time t_0 such that $P(t_0, x_0, U_0^c) = \beta > 0$ for some β. We may without loss of generality replace U_0 with a sphere U centered at x_0 contained in U_0 such that $\bar{U} \subset U_0$. (Specifically, take the radius of U equal to one-half the minimum distance from x_0 to the boundary of U.) Then

$$P(t_0, x_0, \bar{U}^c) \geq \beta.$$

Let r equal the radius of $\bar{U}$ and consider the sequence of functions

$$f_n(x) = \begin{cases} 0, & x \in \bar{U} \\ n[\|x - x_0\| - r], & r \leq \|x - x_0\| \leq r + 1/n, & x \notin U \\ 1, & \|x - x_0\| > r + 1/n, & x \notin U. \end{cases}$$

By assumption (11.22a), U_t maps continuous bounded functions into continuous bounded functions (for $t > 0$) so that

$$\int p(t_0, x, y) f_n(y)\, dy = \varphi_n(x)$$

is a continuous function of x and in particular the set $\{x \mid \varphi_n(x) > \alpha\}$ is an open set. Also $\bigcup_{n=1}^{\infty} \{x \mid \varphi_n(x) > \alpha\}$ is an open set as the union of open sets. But f_n converges to the indicator function f^* of $\bar{U}^c$ so $\varphi_n(x) \to P(t_0, x, \bar{U}^c)$ as a monotone increasing sequence of functions. Hence, for each $\alpha > 0$, $\{x \mid P(t_0, x, \bar{U}^c) > \alpha\}$ is an open set.

Consider $\tilde{V} = \{x \,|\, P(t_0, x, \overline{U}^c) > \beta/2\}$. This is an open set containing x_0. Let $V = \tilde{V} \cap U$ which is an open set containing x_0. Let $\sigma_{x_0}(V^c)$ designate the first time of attaining the boundary of V from x_0. Now $\overline{V} \subset \overline{U}$ entails $\overline{V}^c \supset \overline{U}^c$ so it follows that $P(t_0, x, \overline{U}^c) \leq P(t_0, x, \overline{V}^c)$ for all x in $\overline{V}$. Therefore, for all x in $\overline{V}$,

$$\Pr\{\sigma(\overline{V}) > t_0 \,|\, X(0) = x\} \leq P(t_0, x, \overline{V}) \leq P(t_0, x, \overline{U})$$

$$= 1 - P(t_0, x, \overline{U}^c) \leq 1 - \frac{\beta}{2}.$$

By the strong Markov property for $x \in \overline{V}$,

$$\Pr\{\sigma(\overline{V}^c) > 2t_0 \,|\, X(0) = x\} = \Pr\{X(s) \in \overline{V} \text{ for all } s \leq 2t_0\}$$

$$= \int_{z \in \overline{V}} \Pr\{\sigma(\overline{V}^c) > t_0 \,|\, X(t_0) = z\}$$

$$\times \Pr\{X(t_0) = z, X(s) \in \overline{V} \text{ for all } s \leq t_0 \,|\, X(0) = x\} \, dz$$

$$\leq \left(1 - \frac{\beta}{2}\right)^2$$

since both factors are less than $1 - \beta/2$ for any state $x \in \overline{V}$. By induction,

$$\Pr\{\sigma(\overline{V}^c) > nt_0 \,|\, X(0) = x\} \leq \left(1 - \frac{\beta}{2}\right)^n \qquad \text{for all} \quad x \in \overline{V}, \qquad (11.50)$$

and this inequality easily provides a bound on all moments of the exit time from V.

On account of (11.50) it is easy to deduce that the distribution function of $\sigma(\overline{V}^c)$ decreases exponentially fast. In particular, for a regular diffusion the first passage random variable $\sigma(\overline{V}^c)$ has all finite moments.

RELATIONSHIP OF THE INFINITESIMAL OPERATOR A WITH THE DIFFERENTIAL OPERATOR L OF (3.9)

In what follows, let $\{X(t), t \geq 0\}$ be a regular diffusion process in natural scale $(S(x) = x)$, with state space $I = [l, r]$, and speed measure density $m(x) = [\sigma^2(x)]^{-1}$ continuous and positive on $I^0 = (l, r)$. Let A be the (strong) infinitesimal generator of the process (See Definition 11.2) and let $u \in \mathscr{D}(A)$.

By Theorem 11.3, we know that for $a \in I^0$,

$$Au(a) = \lim_{\substack{V \downarrow a \\ V \text{ nbd of } a \\ E_a[\sigma(V^c)] < \infty}} \frac{E_a[u(X(\sigma(V^c)))] - u(a)}{E_a[\sigma(V^c)]} \qquad (11.51)$$

exists, and equals $Au(a)$. Scrutiny of the proof of (11.46) reveals that for each compact subinterval J of I^0 the convergence in (11.46) is uniform over $a \in J$, where we take in (11.51) $V = (a - \varepsilon, a + \varepsilon)$ contracting to a.

We will assume for the moment that $Au(a) > 0$ for all a; we treat the other cases later.

Now since $X(\sigma(V^c)) = a - \varepsilon$ or $a + \varepsilon$, each with probability $\frac{1}{2}$, and since the scale measure is x, we have $E_a[u(X(\sigma(V^c)))] = \frac{1}{2}u(a + \varepsilon) + \frac{1}{2}u(a - \varepsilon)$. Further, $E_a[\sigma(V^c)] = \int_{a-\varepsilon}^{a+\varepsilon} G^*(a, \xi)m(\xi)\,d\xi$, where $G^*(a, \xi)$ is given by (3.30) with $S(x) = x$. Hence we can write

$$Au(a) = \lim_{\varepsilon \downarrow 0} \frac{\dfrac{1}{2}\left[\dfrac{u(a + \varepsilon) - u(a)}{\varepsilon} - \dfrac{u(a) - u(a - \varepsilon)}{\varepsilon}\right]}{\dfrac{1}{\varepsilon^2}\displaystyle\int_{a-\varepsilon}^{a+\varepsilon} G^*(a, \xi)m(\xi)\,d\xi} \tag{11.52}$$

Recall from Remark 3.2, p. 197, that the denominator converges uniformly to $m(a) > 0$ as $\varepsilon \downarrow 0$ for a restricted to a compact subinterval of I^0.

Since the convergence in (11.52) is uniform over a compact region of I^0, we can choose δ so small that the numerator is positive for all $0 < \varepsilon \le \delta$. That is,

$$\tfrac{1}{2}u(a + \varepsilon) + \tfrac{1}{2}u(a - \varepsilon) > u(a) \qquad \text{for all} \quad 0 < \varepsilon \le \delta \quad \text{and} \quad a \text{ in } J \subset I^0.$$

This implies that $u(a)$ is locally convex and therefore convex. A convex continuous function possesses left continuous left-hand derivatives and right continuous right-hand derivatives.[†] But from (11.52), these left- and right-hand derivatives must be equal, and so $u'(x)$ is continuous. Further, the limit

$$v(a) = \lim_{\varepsilon \downarrow 0} \frac{1}{\varepsilon^2}[u(a + \varepsilon) - 2u(a) + u(a - \varepsilon)] \tag{11.53}$$

exists. Since $u(x)$ is convex, $u''(x)$ exists almost everywhere, and whereever $u''(x)$ exists the fact of (11.53) and a simple two-term Taylor expansion shows that $u''(x) = v(x)$.

Finally, $v(x)$ is continuous [since $v(x) = 2m(x)Au(x)$, and both m and Au are continuous] and so bounded on closed bounded intervals of I^0. So $u''(a)$ is bounded almost everywhere and hence $u'(a)$ is absolutely continuous. Accordingly, we can write

$$\int_x^y u''(\xi)\,d\xi = u'(y) - u'(x).$$

Hence,

$$\frac{1}{y - x}\int_x^y v(\xi)\,d\xi = \frac{1}{y - x}\int_x^y u''(\xi)\,d\xi = \frac{u'(y) - u'(x)}{y - x}.$$

As $y \to x$, the left-hand term goes to $v(x)$, entailing that

$$\lim_{y \to x}\left\{\frac{[u'(y) - u'(x)]}{(y - x)}\right\}$$

exists, and equals $v(x)$. That is $u''(x) = v(x)$ for all x.

[†] For properties of convex functions see G. H. Hardy, J. E. Littlewood, and G. J. Pólya, "Inequalities," 2nd ed., Cambridge Univ. Press, London and New York, 1952.

So we have shown that if u is in $\mathscr{D}(A)$ and $Au(a) > 0$ for all a, then

$$Au(a) = \tfrac{1}{2}\sigma^2(a)u''(a). \tag{11.54}$$

For any a such that $Au(a) = 0$, we know that since a is not a trap state, there exists u^* in $\mathscr{D}(A)$ such that $Au^*(a) > 0$. Take $\bar{u} = u + Cu^*$ so that $A\bar{u}(a) > 0$. Then $\bar{u}$ is twice differentiable at a, and since Cu^* is, so then is u. The case $Au(a) < 0$ follows in a similar way. We have now shown that $u \in \mathscr{D}(A)$ implies

$$Au(a) = Lu(a) = \tfrac{1}{2}\sigma^2(a)u''(a). \tag{11.55}$$

The result (11.55) extends easily to the case including an infinitesimal drift term $\mu(x)$ smooth over I. In fact, we can reduce considerations to a natural scale by passing to the process $S(X(t)) = Y(t)$ (cf. Remark 3.1). In this we achieve the identification: If $w \in \mathscr{D}(A)$ then $w''(x)$ is continuous on I^0 and

$$Aw(x) = \tfrac{1}{2}\sigma^2(x)w''(x) + \mu(x)w'(x). \tag{11.56}$$

12: Further Topics in the Semigroup Theory of Markov Processes and Applications to Diffusions

We deal with five topics in this section. (I) Some characterizations of boundary behavior for one dimensional diffusions in terms of the infinitesimal operator evaluated at the boundary. (II) Ramifications of the Dynkin formula (11.37) using the theory of additive processes and associated martingales. (III) The representation of processes with killing using multiplicative functionals. (IV) Some aspects of local time and inverse local time with applications to generalized Bessel diffusion processes. (V) The construction of some classes of space-time martingales and examples in diffusions.

A. SOME CHARACTERIZATIONS OF BOUNDARY BEHAVIOR FOR DIFFUSIONS BY THE NATURE OF THE DEFINITION OF THE INFINITESIMAL OPERATOR

The following discussion supplements the description of boundary behavior given in Sections 6 and 7. The emphasis here concerns the relationship of the Feller boundary classifications, Table 6.2, and the nature of the definition of the infinitesimal operator on or near the boundary.

Theorem 12.1. *Suppose the right end point r is an exit (or regular and absorbing) boundary, and $u(x)$ belongs to domain A. Then*

$$\lim_{x \to r} Au(x) = 0. \tag{12.1}$$

Proof. For $u \in \mathscr{D}(A) = $ domain A, $u = R_\alpha f$,

$$u(x) = E_x\left[\int_0^\infty e^{-\alpha t} f(X(t))\, dt\right] = E_x\left[\int_0^{T_{r-}} + \int_{T_{r-}}^\infty\right],$$

where T_{r-} is the first passage (hitting) time of the boundary r (cf. Remark 7.1). Consider first

$$\left| E_x\left[\int_0^{T_{r-}} e^{-\alpha t} f(X(t))\, dt\right] \right| \leq \|f\|\left[\frac{1 - E_x[e^{-\alpha T_{r-}}]}{\alpha}\right].$$

We have previously shown that $\lim_{x \uparrow r} E_x[e^{-\alpha T_{r-}}] = 1$ (see Theorem 7.2) at an exit or regular absorbing boundary. Since an exit boundary is absorbing it follows that

$$\lim_{x \uparrow r} u(x) = \lim_{x \uparrow r} E_x\left[\int_{T_{r-}}^\infty e^{-\alpha t} f(X(t))\, dt\right] = f(r) \lim_{x \uparrow r} E_x\left[\frac{e^{-\alpha T_{r-}}}{\alpha}\right] = \frac{f(r)}{\alpha}.$$

Therefore $\lim_{x \uparrow r} Au(x) = \alpha u(r) - f(r) = \alpha[f(r)/\alpha] - f(r) = 0$ at an exit or regular absorbing boundary as asserted in (12.1).

Remark 12.1. Theorem 12.1 emphasizes the importance of selecting the correct underlying function space when performing a particular evaluation in connection with a diffusion process.

To illustrate, consider the problem of finding $v(x) = E[T \,|\, X(0) = x]$, where $T = T_a \wedge T_b$ and $\{X(t)\}$ is standard Brownian motion on $[a, b]$ with both a and b regular and absorbing. As indicated in Section 3, $v(x)$ solves the differential equation $\frac{1}{2}v''(x) = -1$, $a < x < b$; $v(a) = v(b) = 0$. However, v is not in the domain of the infinitesimal operator A of the process, since were it to be so, then we would have $\frac{1}{2}v''(x) = -1$ for all x in (a, b), contradicting Theorem 12.1.

The problem lies in the function space $C[a, b]$ being too restrictive for the computation at hand.

Theorem 12.2. *Suppose* r *is an entrance boundary and* $u \in \mathscr{D}(A)$. *Then* $\lim_{x \uparrow r} u(x) = u(r-)$ *exists and* $\lim_{x \uparrow r}[(du/dS)(x)] = (du/dS)(r-) = 0$ (S *is the scale measure*).

Proof. Let $u = R_\alpha f$. We know by Theorem 7.2 that for $b < c < r$, $\lim_{b \uparrow r} \lim_{c \uparrow r} E_c[e^{-\alpha T_b}] = 1$, where T_b is the first time for reaching b. Now apply the first Dynkin formula (11.36) for the Markov time T_b to get

$$u(x) = E_x\left[\int_0^{T_b} e^{-\alpha t} f(X(t))\, dt\right] + E_x[e^{-\alpha T_b}]u(b).$$

Note that

$$\left| E_x\left[\int_0^{T_b} e^{-\alpha t} f(X(t))\, dt \right] \right| \le \|f\| \left(\frac{1 - E_x[e^{-\alpha T_b}]}{\alpha} \right).$$

Hence,

$$\overline{\lim_{x\uparrow r}}\, u(x) \le \frac{\|f\|}{\alpha} \left(1 - \lim_{x\uparrow r} E_x[e^{-\alpha T_b}] \right) + \lim_{x\uparrow r}(E_x[e^{-\alpha T_b}])u(b).$$

Now, since the left side is independent of b in sending $b\uparrow r$, we obtain

$$\overline{\lim_{x\uparrow r}}\, u(x) \le \frac{\|f\|}{\alpha} \left[1 - \lim_{b\uparrow r} \lim_{x\uparrow r}(E_x[e^{-\alpha T_b}]) \right] + \overline{\lim_{b\uparrow r}} \left[\lim_{x\uparrow r}(E_x[e^{-\alpha T_b}]u(b)) \right].$$

$$(12.2)$$

Taking cognizance of the fact that $\lim_{b\uparrow r} \lim_{c\uparrow r} E_c[e^{-\alpha T_b}] = 1$ (Theorem 7.2) we have on the basis of (12.2) that $\overline{\lim}_{x\uparrow r}\, u(x) \le \underline{\lim}_{b\uparrow r}\, u(b)$ and hence the limit exists.

Recall that at an entrance boundary $S[x, r) = \infty$ and $M[x, r) < \infty$. Since $Au = \frac{1}{2}(d/dM)(d/dS)u = \alpha u - f$ we have

$$\frac{1}{2} \lim_{a\uparrow r} \left[\frac{du}{dS}(a) - \frac{du}{dS}(b) \right] = \lim_{a\uparrow r} \left[\int_b^a (\alpha u - f)\, dM \right].$$

Because u, f, and M are bounded, the right-hand limit and hence the left-hand limit exist. But if $\lim_{x\uparrow r} du(x)/dS = \gamma > 0$; then $du(x)/dS > \gamma/2$ for x sufficiently near r. The property $S[x, r) = \infty$ is then incompatible with the fact that $u(x)$ is bounded since for x close to r, $u(r) - u(x) = \int_x^r (du/dS)\, dS \ge (\gamma/2) \int_x^r dS = \infty$. The only way to avert this contradiction is the conclusion $\lim_{x\uparrow r} du(x)/dS = 0$. The proof is complete. ∎

Characterizing Domains of the Infinitesimal Generator

In what follows, we take $\{X(t), t \ge 0\}$ to be a regular diffusion on $I = (l, r)$ with $\sigma^2(x) > 0$ and continuous on I. We first consider the case in which both boundaries are exit, so that $\bar{I} = [l, r]$. Let u be a twice continuously differentiable function with $Lu(x) = \frac{1}{2}\sigma^2(x)u''(x) + \mu(x)u'(x) \in C_0(I)$, that is continuous and vanishing at l and r. We define the function f by $-f = Lu - \lambda u$, for some $\lambda > 0$, and set $w = R_\lambda f$. Since $w \in \mathcal{D}(A)$, (11.56) shows in particular that w is twice continuously differentiable on I. Theorem 12.1 shows that $Aw \in C_0(I)$. Now consider $z = u - w$. Then

$$Lz = L(u - w) = \lambda u - f - Lw$$

$$= \lambda u - f - Aw = \lambda u - f - (\lambda w - f) \quad \text{(by 11.27)}$$

$$= \lambda z.$$

We would like to show that $z(x) \equiv 0$, so that $w = u$, and hence $u \in \mathcal{D}(A)$. Since $Lu \in C_0(I)$, and $Lw(a) = Aw(a) \to 0$ as $a \to l$ or r, we see that $Lz \in C_0(I)$, so that $z \in C_0(I)$. Now suppose that $z(x)$ has a local positive maximum at x_0 interior to I. Then $z'(x_0) = 0$ and $z''(x_0) \leq 0$, and so $Lz(x_0) = \frac{1}{2}\sigma^2(x_0)z''(x_0) \leq 0$, whereas $\lambda z(x_0) > 0$. This contradiction shows that z has no local positive maxima (and, similarly, no local negative minima). Hence, $z \equiv 0$. Thus, we have shown that if $\{X(t), t \geq 0\}$ has exit boundaries at l and r, and if $\widetilde{\mathcal{D}} = \{u: u \in C^2(I), Lu \in C_0(I)\}$, then $\widetilde{\mathcal{D}} \subseteq \mathcal{D}(A)$.

As a second example, we assume the process is in natural scale on $I = (l, r)$ but that both l and r are entrance boundaries. By Theorem 12.2, we know that $u \in \mathcal{D}(A)$ implies $du(x)/dS \in C_0(I)$, and that $u \in C(\bar{I})$.

Let $\hat{\mathcal{D}} = \{u: u \in C(\bar{I}), Lu \in C(I), du/dS \in C_0(I)\}$. As in the previous example, take $u \in \hat{\mathcal{D}}$ and define the function f by $-f = Lu - \lambda u$, $\lambda > 0$, and let $w = R_\lambda f$. Then, $w \in \mathcal{D}(A)$ and hence $w \in C(I)$. Then if $z = u - w$, we deduce as previously that $Lz = \lambda z$. Since both u and w are in $C(\bar{I})$, z must also be and dz/dS is in $C_0(I)$.

If z is of one sign, then integrating $Lz = (1/2m(x))(d/dx)(dz/dS(x)) = \lambda z(x)$ produces

$$0 \neq 2\lambda \int_{l+}^{r-} z(x)m(x)\,dx = \int_{l+}^{r-} \frac{d}{dx}\left(\frac{dz}{dS}\right) = \frac{dz}{dS}(r-) - \frac{dz}{dS}(l+) = 0$$

which is clearly absurd.

Suppose next that $z(l+) > 0 > z(r-)$. Let x_0 be the last zero of $z(x)$ in I. Then $z(x) < 0$ for $x_0 < x < r-$ and $[1/s(x_0)]\,dz(x_0)/dx = dz(x_0)/dS \leq 0$. Integrating over $(x_0, r-)$ gives

$$0 > 2\lambda \int_{x_0}^{r-} z(x)m(x)\,dx = \frac{dz}{dS}(r-) - \frac{dz}{dS}(x_0) = -\frac{dz}{dS}(x_0) \geq 0.$$

The foregoing contradictions imply $z(x) \equiv 0$ and so $u \in \mathcal{D}(A)$. So if both boundaries are entrance then $\hat{\mathcal{D}} \subseteq \mathcal{D}(A)$.

Further consider the case in which both boundaries are natural. Then if $\mathcal{D}^* = \{u: Lu \in C_0(I)\}$, similar methods show that $\mathcal{D}^* \subseteq \mathcal{D}(A)$.

For an interval $I = (l, r)$ where l is an entrance boundary and r a natural boundary the preceding analysis applies to show that the functions of $\mathcal{D} = \{u: Lu \in C(I), du(l+)/dS = 0$ and $Lu(r-) = 0\}$ belong to $\mathcal{D}(A)$.

B. ADDITIVE FUNCTIONALS

A more natural proof of the Dynkin formula (11.37) involves the concepts of additive functionals and associated martingales. We need the concept of a *shifted sample path*. Let ω_t^+ designate the sample path ω shifted by a time epoch t. In other words, for ω_t^+ we ignore the nature of ω up to time t and delimit ω_t^+ as the part of ω commencing at time t.

Definition 12.1. *A process $C(t) = C(t, \omega)$, $t > 0$ is said to constitute an additive functional adapted to an increasing family of fields $\{\mathscr{F}_t\}$ (see Section 8) if*

$$C(t, \omega) \text{ is measurable } \mathscr{F}_t \text{ for each } t \tag{12.3}$$

and with probability one,

$$C(t + s, \omega) = C(t, \omega) + C(s, \omega_t^+) \qquad \text{for all } t \text{ and } s. \tag{12.4}$$

We will operate in the setting of Brownian motion $B(t)$ as the underlying stochastic process on which all functionals are defined, although all analogous conclusions are also valid where $B(t)$ is replaced by any strong Markov process.

Examples of Additive Processes

(i) Let $f(x)$ be an integrable function taken over any finite segment of the real line. Then

$$C(t, \omega) = \int_0^t f(B(s, \omega))\, ds \quad \text{determines an additive functional.} \tag{12.5}$$

In fact

$$C(t + s, \omega) = \int_0^{t+s} f(B(\tau, \omega))\, d\tau = \int_0^t f(B(\tau, \omega))\, d\tau + \int_t^{t+s} f(B(\tau, \omega))\, d\tau.$$

By definition of ω and ω_t^+, $B(\tau, \omega) = B(\tau - t, \omega_t^+)$ for all $\tau > t$. The change of variable $\tau' = \tau - t$ in the last integral gives

$$\int_t^{t+s} f(B(\tau, \omega))\, d\tau = \int_t^{t+s} f(B(\tau - t, \omega_t^+))\, d\tau = \int_0^s f(B(\tau', \omega_t^+))\, d\tau'. \tag{12.6}$$

The relation (12.4) for the functional (12.5) ensues by virtue of (12.6).

(ii)
$$C(t) = f(B(t)) - f(B(0)) \tag{12.7}$$

is an additive functional adapted to $\{\mathscr{F}_t\}$. (Prove this.)

(iii) The sum of additive functionals is an additive functional.

The following results on additive functionals have wide applications.

Lemma 12.1. *A nonnegative additive functional $C(t, \omega)$ is increasing.*

Proof. Indeed, $C(t + s, \omega) = C(t, \omega) + C(s, \omega_t^+) \geq C(t, \omega)$.

Lemma 12.2. *An additive functional process $C(t, \omega)$ satisfying*

$$E_x[|C(t)|] < \infty \quad \text{and} \quad E_x[C(t)] = 0 \qquad \text{for all } x \text{ and } t \geq 0 \tag{12.8}$$

is a martingale with respect to $\{\mathscr{F}_t\}$.

Proof. We compute

$$E[C(t + s)|\mathscr{F}_s] = E[C(s)|\mathscr{F}_s] + E[C(t, \omega_s^+)|\mathscr{F}_s] \tag{12.9}$$

the equation resulting from the postulate (12.4).

By the Markov property

$$E[C(t, \omega_s^+)|\mathscr{F}_s] = E_{B(s)}[C(t)] = 0 \quad \text{with probability one,} \tag{12.10}$$

the last equation by virtue of the hypothesis (12.8). By definition, $C(s)$ is measurable $\mathscr{F}_s$ (property (12.3)) entailing

$$E[C(s)|\mathscr{F}_s] = C(s) \tag{12.11}$$

so that with the information of the foregoing discussion (12.9) reduces to

$$E[C(t + s)|\mathscr{F}_s] = C(s) \tag{12.12}$$

the desired martingale property. ∎

The Dynkin Formula (11.37) *and Additive Functionals*

Let $u(x)$ be in $\mathscr{D}(A)$. We form the process

$$Y(t) = u(X(t)) - u(X(0)) - \int_0^t Au(X(\tau))\, d\tau, \qquad t > 0. \tag{12.13}$$

We shall show that $Y(t)$ is a martingale by showing that $Y(t)$ is an additive functional and that $E_x[Y(t)] = 0$ so that Lemma 12.2 applies. To this end, observe first that $Y(t)$ is an additive functional by virtue of the facts of (12.5) and (12.7). The pertinent sigma fields are those generated by the process $X(t)$, i.e., $\mathscr{F}_t = \{$the Borel field induced by the sample values of $X(u)$, over the time span $0 \le u \le t\}$.

The next string of equalities follows by passing the expectation across the integral sign, using the definition of the semigroup operator U_t, employing the identity $dU_\tau/d\tau = U_\tau A$, and finally invoking the fundamental theorem of calculus. We compute

$$E_x[Y(t)] = E_x[u(X(t))] - u(x) - E_x\left[\int_0^t Au(X(\tau))\, d\tau\right]$$

$$= U_t u(x) - u(x) - \int_0^t U_\tau Au(x)\, d\tau$$

$$= U_t u(x) - u(x) - \int_0^t \left(\frac{d}{d\tau} U_\tau u\right)(x)\, d\tau = 0. \tag{12.14}$$

The Dynkin formula follows by applying the optional sampling theorem (see Theorem 3.2, Chapter 6) to the process $Y(t)$, that is, let σ be a stopping time with finite expectation. Then

$$E_x[Y(\sigma)] = E_x[u(X(\sigma))] - u(x) - E_x\left[\int_0^\sigma Au(X(\tau))\,d\tau\right]$$

$$= E_x[Y(t)] = 0 \qquad \text{for all } t \tag{12.15}$$

in agreement with the Dynkin formula (11.37).

An Extension

Let $f(x)$ be in the $\mathscr{D}(A)$ and let $g(t)$ and $g'(t)$ be bounded and continuous over $0 \le t < \infty$. Then

$$g(t + \tau)U_{t+\tau}f(x) - g(t)U_t f(x) = \int_t^{t+\tau} g'(s)U_s f(x)\,ds$$

$$+ \int_t^{t+\tau} g(s)U_s Af(x)\,ds. \tag{12.16}$$

The proof of (12.16) relies on the identity

$$\frac{d}{d\tau}U_\tau f = U_\tau Af = AU_\tau f$$

valid for any f in $\mathscr{D}(A)$ and straightforward integration by parts. The identity underlying the original Dynkin formula corresponds to the choice $g(s) \equiv 1$.

With (12.16) in hand and $v(x)$ in $\mathscr{D}(A)$ we form the process

$$Z(t) = g(t)v(X(t)) - g(0)v(X(0))$$

$$- \int_0^t g'(s)v(X(s))\,ds - \int_0^t g(s)Av(X(s))\,ds. \tag{12.17}$$

Claim. $Z(t)$ is a martingale with respect to the family of fields $\mathscr{F}_t$ determined by the process realizations of $\{X(s)\}$ up to time t.

Proof. Observe that

$$Z(t + \tau) - Z(t) = g(t + \tau)v(X(t + \tau)) - g(t)v(X(t))$$

$$+ \int_t^{t+\tau} g'(s)v(X(s))\,ds + \int_t^{t+\tau} g(s)Av(X(s))\,ds.$$

The expectation conditioned on knowing $X(\tau)$ up to time t ($\tau \le t$) gives

$$E[(Z(t + \tau) - Z(t))|\mathscr{F}_t] = 0 \tag{12.18}$$

since the right-hand side under the conditioning in view of the strong Markov property merely reduces to the identity (12.15) for $x = X(t)$.

The equation (12.18) affirms that $\{Z(t), t \geq 0\}$ constitutes a martingale adapted to $\{\mathscr{F}_t\}$. Since $E_x[Z(t)] = 0$, then for any Markov time σ of finite expectation satisfying

$$E_x\left[\int_0^\sigma |g'(X(s))|\, ds\right] < \infty \qquad \text{and} \qquad E_x\left[\int_0^\sigma |g(X(s))|\right] < \infty,$$

the optional sampling theorem (Theorem 3.2 of Chapter 6) implies

$$E_x[Z(\sigma)] = E_x[Z(t)] = 0. \tag{12.19}$$

Some Specializations

Suppose v, Av, A^2v, ..., A^kv belong to the domain of A and suppose σ is a Markov time such that $E_x[\sigma^k] < \infty$. Then,

$$E_x[v(X(\sigma))] = v(x) + \sum_{m=1}^{k} \frac{(-1)^{m-1}}{m!} E_x[\sigma^m A^m v(X(\sigma))]$$

$$+ \frac{(-1)^k}{k!} E_x\left[\int_0^\sigma s^k A^{k+1} v(X(s))\, ds\right]. \tag{12.20}$$

In fact, the result of (12.19), $E_x[Z(\sigma)] = 0$, for the case $g(s) = s^m$, $m \geq 1$, gives

$$E_x[\sigma^m v(X(\sigma))] = mE_x\left[\int_0^\sigma s^{m-1} v(X(s))\, ds\right] + E_x\left[\int_0^\sigma s^m Av(X(s))\, ds\right]. \tag{12.21}$$

We define

$$a_m = \frac{(-1)^m}{m!} E_x\left[\int_0^\sigma s^m A^{m+1} v(X(s))\, ds\right].$$

and substitute $A^m v$ for v in (12.21) to get

$$E_x[\sigma^m A^m v] = (-1)^m m! [a_m - a_{m-1}]. \tag{12.22}$$

This holds for $m = 1, 2, \ldots, k$. The case $k = 0$ is the original Dynkin formula, that is

$$E_x[v(X(\sigma))] = v(x) + E_x\left[\int_0^\sigma Av(X(s))\, ds\right]. \tag{12.23}$$

$$= v(x) + a_0.$$

If we add up (12.22) and (12.23) the identity (12.20) ensues.

Application

We will use these ideas to calculate $E_x[\sigma^2]$ where σ is the first departure time from the unit sphere $\|\mathbf{x}\|^2 \leq 1$ where the initial state is $X(0) = x$ and $X(t)$ is standard Brownian motion in E^n. Consult page 299 for the derivation of

$E_x[\sigma] = (1 - \|\mathbf{x}\|^2)/n$. A fully rigorous proof requires truncating σ by $\sigma_N = \sigma \wedge N$ and operating in terms of σ_N as we did in the analysis of (11.43). We leave this technical point to the reader and proceed as if $E_x[\sigma^2] < \infty$.

We apply (12.20) for $k = 1$ to the function

$$v(x) = \begin{cases} (1 - \|\mathbf{x}\|^2)^2 & \text{if} \quad \|\mathbf{x}\| \le 1 \\ \text{very smooth} & \text{if} \quad \|\mathbf{x}\| > 1 \end{cases} \tag{12.24}$$

with $v(\mathbf{x})$ vanishing rapidly as $\|\mathbf{x}\| \to \infty$. Recall from (11.40) that $Av = \frac{1}{2}\Delta v = \frac{1}{2}\sum_{i=1}^{n} \partial^2 v/\partial x_i^2$ so that

$$Av(\mathbf{x}) = -2n(1 - \|\mathbf{x}\|^2) + 4\|\mathbf{x}\|^2 \qquad \text{for} \quad \|\mathbf{x}\| \le 1$$

and

$$A^2 v(\mathbf{x}) = (2n + 4)n \qquad \text{for} \quad \|\mathbf{x}\| \le 1.$$

The equation (12.20) for $k = 1$ gives

$$0 = (1 - \|\mathbf{x}\|^2)^2 + E_x[4\sigma] - E_x\left[(4 + 2n)n \frac{\sigma^2}{2} \right].$$

We already determined $E_x[\sigma] = (1 - \|\mathbf{x}\|^2)/n$ in (11.43). Hence

$$E_x[\sigma^2] = \frac{4(1 - \|\mathbf{x}\|^2)}{(2 + n)n^2} + \frac{(1 - \|\mathbf{x}\|^2)^2}{n(2 + n)}. \tag{12.25}$$

C. MULTIPLICATIVE FUNCTIONALS

A positive multiplicative functional $Z(t)$ of the process $\{X(t)\}$ is such that $\log Z(t)$ is an additive process in the sense of Definition 12.1. Accordingly,

$$\exp\left[-\int_0^t k(X(\tau))\,d\tau \right] \tag{12.26}$$

with $k(x)$ bounded and continuous is a multiplicative functional (cf. (12.5)). In particular, the Kac functional (Section 5) reduces the contributions of $X(t)$ by a multiplicative functional factor since $k(x)$ is a nonnegative function and hence (12.26) operates as a killing rate.

For a regular diffusion $X(t)$ without killing the local behavior is governed by the drift and variance coefficients. The possibility of killing introduces a further infinitesimal rate

$$\lim_{h\downarrow 0} \frac{1}{h} \Pr\{\text{the process killed during the time interval}$$
$$(t, t + h)\,|\,X(t) = x\} = k(x). \tag{12.27}$$

Let the diffusion process with killing as in (12.27) with the same drift and variance coefficient as $X(t)$ be denoted by $Y(t)$. There is a simple procedure by which to realize the process $Y(t)$ in terms of the sample paths of $X(t)$ which we now describe. We will assume $k(x)$ positive and continuous defined on S^* such that

$$\int_0^\infty k(X(t))\, dt = \infty \tag{12.28}$$

for almost all sample paths of $X(t)$. Let R be a positive valued random variable following an exponential distribution with parameter 1. We sample a value of R independently of $X(t)$ and define for each sample path ω of $X(t)$ the positive random variable

$$\zeta = \inf\left\{ s: \int_0^s k(X(\tau))\, d\tau > R \right\}. \tag{12.29}$$

Now, set

$$Y(t) = \begin{cases} X(t) & \text{for } t < \zeta, \\ \Delta & \text{for } t \geq \zeta \end{cases} \tag{12.30}$$

(Δ = killing state). By reasoning paralleling the analysis in Section 3, it is easily shown that $Y(t)$ is a diffusion involving the same infinitesimal mean and variance parameters as $X(t)$ with killing rate $k(x)$ of (12.27).

We conclude this section with a more formal description of the Kac semigroup

$$U_t^{(H)} f(x) = H_t f(x) = E_x\left[\exp\left[\int_0^t q(X(s))\, ds \right] f(X(t)) \right] \tag{12.31}$$

incorporating the multiplicative functional $\exp[\int_0^t q(X(s))\, ds]$ with rate function $q(x)$ not necessarily negative on $C(S^*)$.

The resolvent operator for (12.31) analogous to (11.25),

$$R_\alpha^{(H)} f(x) = \int_0^\infty e^{-\alpha t} H_t f(x)\, dt \tag{12.32}$$

is well defined for $\alpha > \|q^+\| = \sup_x q(x)$.

Theorem 12.3. *Let A be the infinitesimal operator of U_t corresponding to the process $\{X(t)\}$. Then the infinitesimal generator for the semi-group $U_t^{(H)}$ is $A_H = A + q$ with domain $\mathcal{D}(A_H) = \mathcal{D}(A)$ and $A_H u(x) = (Au)(x) + q(x)u(x)$.*

Proof. We first verify the semigroup property of $U_t^{(H)}$. The additive process $Q_t = Q_t(\omega) = \int_0^t q(X(\tau, \omega))\, d\tau$ is certainly measurable with respect to the Borel fields $\mathcal{F}_t$ of the X process.

$$H_{t+s} f(x) = E_x[e^{Q_{t+s}} f(X(t+s))]$$
$$= E_x[e^{Q_t} e^{Q_s(\omega_t^+)} f(X(s, \omega_t^+))],$$

where ω_t^+ again refers to the sample path corresponding to ω subsequent to time t. By the strong Markov property

$$H_{t+s}f(x) = E_x[e^{Q_t}E_{X(t)}[e^{Q_s}f(X(s))]] = E_x[e^{Q_t}H_sf(X(t))]$$
$$= H_t(H_sf)(x).$$

Our next task concerns the determination of the infinitesimal operator of $U_t^{(H)}$. For $\alpha > \|q^+\|$ we claim that

$$R_\alpha^{(H)}f = R_\alpha f + R_\alpha^{(H)}(qR_\alpha f) = R_\alpha f + R_\alpha(qR_\alpha^{(H)}f). \tag{12.33}$$

To validate (12.33), set $v = R_\alpha^{(H)}f$ and $u = R_\alpha f$. Consider

$$v(x) - u(x) = E_x\left[\int_0^\infty e^{-\alpha t}f(X(t))(e^{Q_t} - 1)\,dt\right] \tag{12.34}$$

$$= E_x\left[\int_0^\infty e^{-\alpha t}f(X(t))\left(\int_0^t e^{Q_s}q(X(s))\,ds\right)dt\right] \quad \text{(by pathwise integration)}$$

$$= E_x\left[\int_0^\infty e^{Q_s}q(X(s))\left(\int_s^\infty e^{-\alpha t}f(X(t))\,dt\right)ds\right] \quad \begin{array}{l}\text{(interchanging}\\ \text{orders of integrals;}\\ \text{see below for}\\ \text{validation)}\end{array}$$

$$= E_x\left[\int_0^\infty e^{Q_s}q(X(s))e^{-\alpha s}\left(\int_0^\infty e^{-\alpha t}f(X(t, \omega_s^+))\,dt\right)\right] \quad \begin{array}{l}\text{(change of}\\ \text{variables)}\end{array}$$

$$= E_x\left[\int_0^\infty e^{Q_s}q(X(s))e^{-\alpha s}E_{X(s)}\left(\int_0^\infty e^{-\alpha t}f(X(t))\,dt\right)ds\right]$$
$$\text{(Markov property)}$$

$$= E_x\left[\int_0^\infty e^{Q_s}q(X(s))e^{-\alpha s}u(X(s))\,ds\right] = R_\alpha^{(H)}(qu)$$

yielding the first formula of (12.33).

The justification for interchanging orders of integration rests on the existence of

$$E_x\left[\int_0^\infty \int_0^t |e^{-\alpha t}f(X(t))e^{Q_s}q(X(s))|\,ds\,dt\right]$$

$$\leq \|f\|\|q\|E_x\left[\int_0^\infty \left(\int_0^t e^{-\alpha t}e^{\|q^+\|s}\,ds\right)dt\right] \quad \left(\|q^+\| = \sup_x q(x)\right)$$

$$\leq \frac{1}{\alpha}\|f\|\|q\|\int_0^\infty e^{-(\alpha - \|q^+\|)s}\,ds < \infty.$$

A similar analysis leads to

$$v(x) - u(x) = E_x\left[\int_0^\infty e^{-\alpha t}f(X(t))(e^{Q_t} - 1)\,dt\right]$$

$$= E_x\left[\int_0^\infty e^{-\alpha t}f(X(t))\left(\int_0^t e^{Q_t - Q_s}q(X(s))\,ds\right)dt\right]$$

$$= E_x\left[\int_0^\infty q(X(s))\left(\int_s^\infty e^{-\alpha t}f((X(t)))e^{Q_t - Q_s}\,dt\right)ds\right]$$

$$= E_x\left[\int_0^\infty e^{-\alpha s}q(X(s))\left(\int_0^\infty e^{-\alpha t}f(X(t+s))e^{Q_{t+s} - Q_s}\,dt\right)ds\right]$$

$$= E_x\left[\int_0^\infty e^{-\alpha s}q(X(s))\left(\int_0^\infty e^{-\alpha t}f(X(t,\omega_s^+)e^{Q_t(\omega_s^+)}\,dt\right)ds\right]$$

$$= E_x\left[\int_0^\infty e^{-\alpha s}q(X(s))v(X(s))\,ds\right] = R_\alpha(qv)$$

which is the second equation of (12.33).

From the resolvent equation

$$0 = R_\alpha^H - R_\beta^H + (\alpha - \beta)R_\alpha^H R_\beta^H, \qquad \alpha, \beta > \|q^+\|$$

we find that the domain of the generator of the Kac semigroup is the range

$$\mathcal{D}(A_{(H)}) = \text{range of } R_\alpha^{(H)} \text{ independent of } \alpha > \|q^+\|$$

(cf. Section 11, pages 293–295) and

$$A_{(H)} = \alpha u - f, \qquad \text{where} \quad u = R_\alpha^H f \qquad \text{for some } f \in C(S^*).$$

By virtue of (12.33), we have

$$R_\alpha f = R_\alpha^{(H)}(f - qR_\alpha f) \text{ in } \mathcal{D}(A^{(H)}), \qquad R_\alpha^{(H)}f = R_\alpha(f + qR_\alpha^{(H)}f) \text{ in } \mathcal{D}(A)$$

for $\alpha > \|q^+\|$, indicating that $\mathcal{D}(A_{(H)}) = \mathcal{D}(A)$. Suppose $u \in \mathcal{D}(A)$ and set $u = R_\alpha f$, then $u = R_\alpha^{(H)}(f - qu)$ from (12.33) and by its determination $A_{(H)}u = \alpha u - (f - qu)$ which reduces to $\alpha u - f + qu = Au + qu$. This completes the proof of Theorem 12.3. ■

D. LOCAL TIME PROCESS

We already encountered the existence of local time in Section 8. This is a powerful and important concept underlying the analysis of a broad spectrum of functionals of sample paths. Consider a regular diffusion on $I = (l, r)$ in natural scale (cf. Remark 3.1), i.e., scale density $s(\xi) \equiv 1$, and speed measure $m(x) > 0$ for x in I. The formal infinitesimal operator of the process is

$$Au = \frac{1}{2m(x)}\frac{d^2u}{dx^2}. \tag{12.35}$$

Under this normalization we determine the general *local time process* at the origin

$$\theta(t) = \theta(t, \omega) = \lim_{\varepsilon \downarrow 0} \frac{\int_0^t I_{(-\varepsilon, \varepsilon)}(X(s, \omega)) \, ds}{\int_{-\varepsilon}^{\varepsilon} m(\xi) \, d\xi}, \tag{12.36}$$

where

$$I_{(-\varepsilon, \varepsilon)}(\xi) = \begin{cases} 1, & |\xi| < \varepsilon, \\ 0, & |\xi| \geq \varepsilon \end{cases}$$

is the indicator function of the open interval $(-\varepsilon, \varepsilon)$. As usual, ω denotes a sample path of the $X(t)$ process.

The numerator of (12.36) measures the span of time from 0 to t that $X(s)$ resides in $I_{(-\varepsilon, \varepsilon)}$. The denominator is a measure of how fast the diffusion process clock runs when $X(s)$ is located in $(-\varepsilon, \varepsilon)$. *The existence of the limit $\theta(t, \omega)$ for almost every sample path ω is a remarkable and recondite fact.*

Since $\int_0^t f(X(s, \omega)) \, ds$ is an additive process for any nice function f as given in (12.5), the limit $\theta(t, \omega)$ is also an additive process so that

$$\theta(t + s, \omega) = \theta(t, \omega) + \theta(s, \omega_t^+)$$

where ω_t^+ denotes the sample path ω shifted by time t. Moreover, it is expected and correct that $\theta(t, \omega)$ is continuous in t for almost every ω. Plainly, $\theta(t, \omega)$ is nondecreasing in t. Derivation of the fact that $\theta(t)$ is strictly increasing at those times t where $X(t) = 0$ is subtle. A local time process can be defined with respect to each state point (not only the origin) where

$$\theta_a(t, \omega) = \lim_{\varepsilon \downarrow 0} \frac{\int_0^t I_{a - \varepsilon, a + \varepsilon}(X(s, \omega)) \, ds}{\int_{a + \varepsilon}^{a - \varepsilon} m(\xi) \, d\xi}.$$

We would expect, and it is correct but formidable to demonstrate rigorously, that

$$\int_0^t f(X(s, \omega)) \, ds = \int_{-\infty}^{\infty} f(a) \theta_a(t, \omega) \, da. \tag{12.37}$$

This formula arises formally by a change of variables. On the right the level a is occupied about $\theta_a(t, \omega)$ time units by the process excursion over the time period $(0, t)$ so that $f(a)\theta_a(t, \omega) \, da$ is the contribution to $\int_0^t f(X(s, \omega)) \, ds$ when $X(s, \omega)$ hovers at the level a.

When there are absorbing boundaries with T^* the absorption time, then

$$E_x[\theta_a(T^*, \omega)] = G(x, a),$$

where G is the appropriate Green function (cf. page 198).

Inverse Local Time Process

Since $\theta(t, \omega)$ is increasing and continuous, the inverse function is well defined (see the discussion of random time changes in Section 8). We define

$$\theta^{-1}(t, \omega) = \theta_0^{-1}(t, \omega) = \theta^{-1}(t), \tag{12.38}$$

suppressing ω when no ambiguity occurs. The process $\theta^{-1}(t, \omega)$ is referred to as *inverse local time at the origin*. By the nature of the construction of the inverse function, $\theta(v, \omega)$ is necessarily strictly increasing at $v = \theta^{-1}(s, \omega)$, meaning that $\theta(v + \varepsilon, \omega) > \theta(v, \omega)$ for every $\varepsilon > 0$ since $v = \theta^{-1}(s, \omega) = \inf\{v: \theta(v, \omega) > s\}$. Since $\theta(t, \omega)$ is continuous, we see that

$$\theta^{-1}(t, \omega) < s \quad \text{is equivalent to} \quad \theta\left(s - \frac{1}{n}, \omega\right) \geq t \quad \text{for some } n, \quad (12.39)$$

so that we can decide the validity of $\theta^{-1}(t, \omega) < s$ by knowing the process values $\{X(\tau)\}$ up to time s. *Therefore $\theta^{-1}(t, \omega)$ is a Markov time.*

We already remarked that the local time process (12.36) is defined as a limit of additive functionals so it is also an additive functional. This means that for $v > u$,

$$\theta(v, \omega) = \theta(u, \omega) + \theta(v - u, \omega_u^+), \quad (12.40)$$

or letting $\theta^{-1}(s) = v$ and $\theta^{-1}(t) = u$ for $t < s$, then $s = t + \theta(v - u, \omega_u^+)$. Now, $v - u = \theta^{-1}(s - t, \omega_u^+)$ because $\theta(\tau, \omega)$ is strictly increasing at $\tau = v$ so that $\theta(\tau, \omega_u^+)$ is also strictly increasing at $\tau = v - u$.

Thus, we have proved for $s > t$, that

$$\theta^{-1}(s, \omega) - \theta^{-1}(t, \omega) = \theta^{-1}(s - t, \omega_{\theta^{-1}(t)}^+). \quad (12.41)$$

Using the foregoing identity and the strong Markov property, we have for $0 = t_0 < t_1 < \cdots < t_n$, that

$$E_0\left[\prod_{i=1}^{n} \exp\{-\alpha_i[\theta^{-1}(t_i) - \theta^{-1}(t_{i-1})]\}\right]$$

$$= E_0\left[\prod_{i=1}^{n-1} \exp\{-\alpha_i(\theta^{-1}(t_i) - \theta^{-1}(t_{i-1}))\} \exp\{-\alpha_n\theta^{-1}(t_n - t_{n-1}, \omega_{\theta^{-1}(t_{n-1})}^+)\}\right]$$

and by conditional independence

$$= E_0\left[\left(\prod_{i=1}^{n-1} \exp\{-\alpha_i(\theta^{-1}(t_i) - \theta^{-1}(t_{i-1}))\}\right)\right]E_0[\exp\{-\alpha_n\theta^{-1}(t_n - t_{n-1})\}],$$

and inductively,

$$= \prod_{i=1}^{n} E_0[\exp\{-\alpha_i\theta^{-1}(t_i - t_{i-1})\}]. \quad (12.42)$$

Setting $\alpha_j = 0 \ (j \neq i)$, the above reduces to

$$E_0[\exp\{-\alpha_i[\theta^{-1}(t_i) - \theta^{-1}(t_{i-1})]\}] = E_0[\exp\{-\alpha_i\theta^{-1}(t_i - t_{i-1})\}].$$

Definition 12.2. *A process $\psi(t) = \{\psi(t, \omega), t \geq 0\}$ is said to be a time homogeneous Levy process if for every finite set of time points $0 = t_0 < t_1 < \cdots < t_n$, and all real $\alpha_1, \ldots, \alpha_n$, then*

$$E_0\left[\prod_{i=1}^{n} \exp\{-\alpha_i\psi(t_i - t_{i-1})\}\right] = E_0\left[\prod_{i=1}^{n} \exp\{-\alpha_i(\psi(t_i) - \psi(t_{i-1}))\}\right]$$

$$= \prod_{i=1}^{n} E_0[\exp\{-\alpha_i\psi(t_i - t_{i-1})\}]. \qquad (12.43)$$

Lévy processes have been characterized as limits of compositions of independent families of compound Poisson processes and Brownian motion. (This topic is an important part of a more advanced course in stochastic analysis; we refer to Chapter 16 for further elaborations.)

Combining (12.41) and (12.42), we have

$$E_0[\exp\{-\alpha\theta^{-1}(s + t, \omega)\}] = E_0[\exp\{-\alpha\theta^{-1}(s, \omega)\}]E_0[\exp\{-\alpha\theta^{-1}(t, \omega)\}]$$

Thus, the function $M(s, \alpha) = E_0[\exp\{-\alpha\theta^{-1}(s)\}]$ satisfies for each $\alpha > 0$ the familiar functional equation

$$M(s + t, \alpha) = M(s, \alpha)M(t, \alpha), \qquad M(0, \alpha) = 1. \qquad (12.44)$$

The evaluation at $t = 0$ is valid because $\theta(s, \omega)$ is strictly increasing at $s = 0$ provided $X(0) = 0$. We also know that $M(s, \alpha)$ is confined between 0 and 1. It follows by the characterization of the exponential function (cf. Theorem 2.2, Chapter 4) that

$$M(t, \alpha) = e^{-t\varphi(\alpha)} \qquad (12.45)$$

for some nonnegative increasing function $\varphi(\alpha)$ of $\alpha > 0$.

We concentrate henceforth in this section on the diffusion with state space $I = (-\infty, \infty)$ characterized by the speed measure m and scale density s

$$s(\xi) = 1, \qquad m(\xi) = |\xi|^\gamma \quad (\gamma > -1). \qquad (12.46)$$

The restriction $\gamma > -1$ assures that $m(\xi)$ is integrable on any bounded interval. The infinitesimal operator is $A = (1/2|x|^\gamma)(d^2/dx^2)$. Essentially equivalent to (12.46) we consider the diffusion $X_\gamma(= X_\gamma(t), t \geq 0)$ with

$$\mu(x) = 0 \qquad \text{and} \qquad \sigma^2(x) = \frac{1}{m(x)} = \frac{1}{|x|^\gamma}. \qquad (12.47)$$

Notice that $\sigma^2(x)$ is infinite (when $\gamma > 0$) at $x = 0$, but this causes no problem since m is locally integrable and, therefore, 0 is a regular point (why?). Of course, Brownian motion corresponds to $\gamma = 0$.

It is easy to check that the boundaries $-\infty$ and $+\infty$ are natural for each process X_y.

Invariance Properties of the Diffusion Process $X_y(t)$

For typographical convenience, we suppress the index y when no ambiguities occur. For each sample path $X(t)$ we construct a new path $\tilde{X}(t)$ by changing the time scale and multiplying the state variable by a factor in the manner

$$\tilde{X}(t) = \frac{1}{c} X(tc^{y+2}), \tag{12.48}$$

where $c > 0$ is fixed. The mapping (12.48) of the sample paths transforms the probability measure $P^{(y)}$ on Ω_y associated with X_y into a probability measure $\tilde{P}^{(y)}$ over the image paths $\tilde{X}_y$ in $\tilde{\Omega}_y$. The probability calculation of a collection $\mathscr{C}$ of sample paths in $\tilde{\Omega}_y$ is that of the collection of sample paths in Ω_y which yield $\mathscr{C}$ by the construction in (12.48). We claim that $\tilde{P}$ is the probability measure of a regular diffusion $\tilde{X}$ with scale and speed measure the same as those of X. In other words, P and $\tilde{P}$ are equivalent measures. To see this, let T_a ($\tilde{T}_a$) be the first passage time to the point a for the process X ($\tilde{X}$). Let $\tilde{S}_y$ be the scale measure of $\tilde{X}_y$, and observe that

$$\Pr\{\tilde{T}_b < \tilde{T}_a \mid \tilde{X}(0) = x\} = \frac{\tilde{S}_y(x) - \tilde{S}_y(a)}{\tilde{S}_y(b) - \tilde{S}_y(a)}. \tag{12.49}$$

It is clear from the definition (12.48) that sample paths of $\tilde{X}_y$ *starting from* $\tilde{X}_y(0) = x$ for which $\tilde{T}_b < \tilde{T}_a$ coincide with sample paths of X_y starting from $X(0) = cx$ for which $T_{cb} < T_{ca}$. Therefore, the right-hand side of (12.49) is exactly

$$= \Pr\{T_{bc} < T_{ac} \mid X(0) = cx\} = \frac{cx - ca}{cb - ca} = \frac{x - a}{b - a}$$

$$= \Pr\{T_b < T_a \mid X(0) = x\}.$$

These equations prove that $\tilde{S}(x) = S(x)$.

Examination of (12.48) also reveals that the time taken for $\tilde{X}$ to leave the interval (a, b) with $\tilde{X}(0) = x$ is the time taken for X to leave (ca, cb) starting from cx, stretched by a factor $c^{-(y+2)}$. Therefore

$$\tilde{E}_x[\tilde{T}_a \wedge \tilde{T}_b] = c^{-(y+2)} E_{cx}[T_{ca} \wedge T_{cb}]. \tag{12.50}$$

For the diffusion $\tilde{X}$, it follows from (12.49) that

$$\tilde{G}_{a,b}^*(x, \xi) = 2(x - a)(b - \xi)/(b - a), \quad x < \xi,$$

is the Green function (3.30) with absorbing boundaries at a and b since the process $\tilde{X}_y$ is on natural scale. We know by virtue of (3.13)–(3.15) that

$$\tilde{E}_x[\tilde{T}_a \wedge \tilde{T}_b] = \int_a^b \tilde{G}^*_{a,b}(x, y)\tilde{m}(y)\, dy, \qquad (12.51)$$

where $\tilde{m}(y)$ is the speed measure density of the $\tilde{X}$ process. We also know that

$$E_{cx}[T_{ca} \wedge T_{cb}] = \int_{ca}^{cb} G^*_{ca,cb}(cx, \xi)|\xi|^\gamma\, d\xi.$$

The change of variables ξ to $c\eta$ in (12.52) yields

$$E_{cx}[T_{ca} \wedge T_{cb}] = c^{\gamma+1} \int_a^b G^*_{ca,cb}(cx, c\eta)|\eta|^\gamma\, d\eta.$$

But since $G^*_{ca,cb}(cx, c\eta) = cG^*_{a,b}(x, \eta)$, the last expression gives

$$E_{cx}[T_{ca} \wedge T_{cb}] = c^{\gamma+2} \int_a^b G^*_{a,b}(x, \eta)|\eta|^\gamma\, d\eta. \qquad (12.52)$$

Now (12.50) in light of (12.51) and (12.52) gives

$$\begin{aligned}
\tilde{E}_x[\tilde{T}_a \wedge \tilde{T}_b] &= \int_a^b \tilde{G}^*_{a,b}(x, \xi)\tilde{m}(\xi)\, d\xi \\
&= c^{-(\gamma+2)} E_{cx}[T_{ca} \wedge T_{cb}] \\
&= c^{-(\gamma+2)} c^{\gamma+2} \int_a^b G^*_{a,b}(x, \eta)|\eta|^\gamma\, d\eta \\
&= \int_a^b G^*_{a,b}(x, \eta)|\eta|^\gamma\, d\eta. \qquad (12.53)
\end{aligned}$$

As noted earlier in (12.49), $\tilde{G}^*_{a,b}(x, \eta) = G^*_{a,b}(x, \eta)$ applies to any choice of $a < x < b$. The identity (12.53) then implies that $\tilde{m}(\eta) = |\eta|^\gamma = m(\eta)$. (Prove this.) Of course, if the scale and speed measures $S(x)$ and $M(x)$, respectively, of a regular diffusion are known, then the diffusion coefficients $\mu(x)$ and $\sigma^2(x)$ can be calculated from the formulas

$$m(x) = M'(x) = \frac{1}{s(x)\sigma^2(x)}$$

and

$$S'(x) = s(x) = \exp\left\{ -\int^x \frac{2\mu(\xi)}{\sigma^2(\xi)}\, d\xi \right\}.$$

See (3.6) and (3.7).

By the relationship (12.48), we see that the time spent in $(-\varepsilon, \varepsilon)$ for the X process up to time $tc^{\gamma+2}$ is the same as that spent for the corresponding process $\tilde{X}$ in the interval $(-\varepsilon/c, \varepsilon/c)$ up to time t. When $X(0) = \tilde{X}(0) = 0$, these facts imply that for the associated local time processes, $\tilde{\theta}(t)$ is equivalent to $\theta(tc^{\gamma+2})/c$. To establish this more formally, notice that

$$\theta(tc^{\gamma+2}) = \lim_{\varepsilon \downarrow 0} \frac{\int_0^{tc^{\gamma+2}} I_{(-c\varepsilon, c\varepsilon)}[X(s)]\, ds}{\int_{-c\varepsilon}^{c\varepsilon} |\xi|^\gamma \, d\xi}$$

$$= \lim_{\varepsilon \downarrow 0} \frac{c^{\gamma+2} \int_0^t I_{(-c\varepsilon, c\varepsilon)}[X(uc^{\gamma+2})]\, du}{\int_{-c\varepsilon}^{c\varepsilon} |\xi|^\gamma \, d\xi} \qquad \text{(substituting } s = uc^{(\gamma+2)})$$

$$= \lim_{\varepsilon \downarrow 0} \frac{c^{\gamma+2} \int_0^t I_{(-\varepsilon, \varepsilon)}[X(uc^{\gamma+2})/c]\, du}{\int_{-c\varepsilon}^{c\varepsilon} |\xi|^\gamma \, d\xi}$$

$$= \lim_{\varepsilon \downarrow 0} \frac{c^{\gamma+2} \int_0^t I_{(-\varepsilon, \varepsilon)}[X(uc^{\gamma+2})/c]\, du}{\int_{-\varepsilon}^{\varepsilon} c^{\gamma+1} |\eta|^\gamma \, d\eta} \qquad \text{(putting } \xi = c\eta)$$

$$= c \lim_{\varepsilon \downarrow 0} \frac{\int_0^t I_{(-\varepsilon, \varepsilon)}[\tilde{X}(u)]\, du}{\int_{-\varepsilon}^{\varepsilon} |\eta|^\gamma \, d\eta} \qquad \text{(by (12.48))}$$

$$= c\tilde{\theta}(t).$$

It follows that the inverse local time processes agree in the sense that

$$\theta^{-1}(s) \text{ is equivalent to } c^{\gamma+2}\tilde{\theta}^{-1}(s/c), \qquad s > 0, \quad c > 0. \tag{12.54}$$

Theorem 12.4.   *For the process X_γ of (12.48) we have*

$$E_0[e^{-\alpha\theta^{-1}(t)}] = e^{-t\varphi(\alpha)}, \tag{12.55}$$

where $\varphi(\alpha) = \alpha^{1/(\gamma+2)}\varphi(1)$, and $\varphi(1) > 0$.

Proof.   Since X_γ and $\tilde{X}_\gamma$ are equivalent processes, we infer that θ^{-1} and $\tilde{\theta}^{-1}$ are also equivalent processes. Referring to (12.54), we have

$$e^{-s\varphi(\alpha)} = E_0[\exp\{-\alpha\theta^{-1}(s)\}] = E_0[\exp\{-\alpha c^{\gamma+2}\tilde{\theta}^{-1}(s/c)\}]$$

$$= E_0[\exp\{-\alpha c^{\gamma+2}\theta^{-1}(s/c)\}] = \exp\{-(s/c)\varphi(\alpha c^{\gamma+2})\}$$

valid for all s, c, and α positive. It follows that $\varphi(\alpha) = (1/c)\varphi(\alpha c^{\gamma+2})$ for all α and c positive. Now determine c such that $\alpha c^{\gamma+2} = 1$ or $c = \alpha^{-1/(\gamma+2)}$. Then $\varphi(\alpha) = \alpha^{1/(\gamma+2)}\varphi(1)$ and the proof of Theorem 12.4 is complete.   ∎

Example.   For Brownian motion ($\gamma = 0$).

$$E_0[e^{-\alpha\theta^{-1}(t)}] = e^{-\sqrt{\alpha}Kt} \qquad \text{for all} \quad \alpha > 0 \tag{12.56}$$

for some constant K.

We recognize the right of (12.56) as the Laplace transform of the first passage time from 0 to a level t (see Chapter 7, page 362). Thus, the inverse local time sample paths at the origin for Brownian motion are realized as follows. For each level $a > 0$ on the positive axis, (a is to be viewed as time for the inverse local time process), we have $\theta^{-1}(a, \omega) = T_a$, where T_a is the first passage time from the origin to the point a. Clearly, from its meaning $T_{a+b} = T_a + T_b(\omega_a^+)$ which is the formula (12.40). Thus, the additive process with Laplace transform

$$E_0[e^{-\alpha\Psi(t)}] = e^{-K\sqrt{\alpha}t} = E_0[e^{-\alpha\theta^{-1}(t)}]$$

is that of a one-sided stable process of order $\frac{1}{2}$ (consult Chapter 7, page 362).

THE PROBABILITY DISTRIBUTION OF THE LOCAL TIME $\theta(t)$ AND THE OCCUPATION TIME OF THE POSITIVE AXIS FOR THE DIFFUSION X_γ OF ORDER γ.

Let $\theta(t)$ be the local time for the diffusion $X_\gamma(t)$ of index γ. Then,

Theorem 12.5

$$\Pr\left\{u \le \frac{\theta(t)}{ct^{1/(\gamma+2)}} \le u + du\right\} = f_\beta(u)\,du, \tag{12.57}$$

where

$$c = \frac{\Gamma(1/(\gamma+2))}{2^{(\gamma+3)/(\gamma+1)}\Gamma((\gamma+1)/(\gamma+2))}, \qquad \beta = \frac{1}{\gamma+2} \tag{12.58}$$

and $f_\beta(u)$ is the Mittag–Leffler density of index β, namely

$$f_\beta(u) = \frac{1}{\pi\beta}\sum_{k=1}^{\infty}\frac{(-1)^{k-1}}{k!}(\sin \pi\beta k)\Gamma(\beta k + 1)u^{k-1}. \tag{12.59}$$

The name "Mittag–Leffler distribution" comes from the fact that the Laplace transform, $M_\beta(-x)$ of $f_\beta(u)$ has $M_\beta(z) = \sum_{k=0}^{\infty} z^k/\Gamma(\beta k + 1)$, the Mittag–Leffler function of complex variables with index β.

Proof. Observe

$$\int_0^\infty e^{-\alpha t} E_0[e^{-\lambda\theta(t)}]\, dt$$

$$= E_0\left[\int_0^\infty e^{-\alpha t} e^{-\lambda\theta(t)}\, dt\right] \qquad \text{(this involves the expectation with initial position } X_\gamma(0) = 0\text{)}$$

$$= \frac{1}{\alpha} + \frac{1}{\alpha} E_0\left[\int_0^\infty e^{-\alpha t} d(e^{-\lambda\theta(t)})\right] \qquad \text{(integration by parts)}$$

$$= \frac{1}{\alpha} + \frac{1}{\alpha} E_0\int_0^\infty [e^{-\alpha\theta^{-1}(t)} d(e^{-\lambda t})] \qquad \text{(change of variable } \theta(t) \text{ to } t\text{)}$$

$$= \frac{1}{\alpha} - \frac{\lambda}{\alpha}\int_0^\infty E_0[e^{-\alpha\theta^{-1}(t)}] e^{-\lambda t}\, dt$$

$$= \frac{1}{\alpha} - \frac{\lambda}{\alpha}\int_0^\infty e^{-t\alpha^\beta/c} e^{-\lambda t}\, dt \qquad \text{(by Theorem 12.4)}$$

$$= \frac{1}{\alpha} - \frac{1}{\alpha}\left(\frac{1}{(\alpha^\beta/c) + \lambda}\right) = \frac{1}{\alpha}\frac{1}{(1 + \lambda c\alpha^{-\beta})} = \sum_{k=0}^\infty (-1)^k c^k \alpha^{-k\beta-1}\lambda^k.$$

Now take the inverse Laplace transform in α, i.e., observe

$$\alpha^{-(k\beta+1)} = \frac{1}{\Gamma(k\beta+1)}\int_0^\infty e^{-\alpha t} t^{k\beta}\, dt$$

yields

$$E_0[e^{-\lambda\theta(t)}] = \sum_{k=0}^\infty (-1)^k c^k \frac{t^{k\beta}}{\Gamma(k\beta+1)} \lambda^k = M_\beta(-ct^\beta\lambda) \qquad (12.60)$$

so that

$$E_0\left[\exp\left\{-\frac{\lambda}{ct^\beta}\theta(t)\right\}\right] = M_\beta(-\lambda),$$

i.e.,

$$\int_0^\infty e^{-u\lambda} \Pr\left\{u \le \frac{\theta(t)}{ct^\beta} \le u + du\right\} = \int_0^\infty e^{-u\lambda} f_\beta(u)\, du$$

which proves Theorem 12.5. ∎

E. SOME CONSTRUCTIONS OF SPACE-TIME MARTINGALES AND APPLICATIONS TO DIFFUSION PROCESSES

There is a simple way to derive a variety of martingales for diffusion processes. The result is a natural generalization of the Dynkin martingale of (12.13); the method arose initially in the study of first passage problems through moving boundaries.

Let $\{X(t), t \geq 0\}$ be a time homogeneous strong Markov process on the interval I, with a strong infinitesimal generator A, whose domain is $\mathscr{D}(A)$.

Theorem 12.6. *Let $v(t, x)$ be a continuous differentiable function of t, i.e., $v_t(t, x) = \partial v(t, x)/\partial t$ is jointly continuous in t and x and bounded for x in S, and suppose that for each fixed $t \geq 0$, the functions $v(t, x)$ and $v_t(t, x)$ are in $\mathscr{D}(A)$. Then the process $\{Z(t), t \geq 0\}$ defined by*

$$Z(t) = v(t, X(t)) - \int_0^t \{v_s(s, X(s)) + Av(s, X(s))\}\, ds \qquad (12.61)$$

is a martingale with respect to the sigma-fields $\mathscr{F}_t$ generated by the process values $X(u)$, $0 \leq u \leq t$.

Proof. It is convenient to split the proof into several parts. First, recall that if $w \in \mathscr{D}(A)$, the process $w(X(t)) - w(X(0)) - \int_0^t Aw(X(s))\, ds$ is a martingale (See 12.13). In particular, if we fix t_1, then the process $v(t_1, X(t)) - v(t_1, X(0)) - \int_0^t Av(t_1, X(s))\, ds$ is a martingale. Hence,

$$E\left[\int_{t_1}^{t_2} Av(t_1, X(s))\, ds \,\middle|\, \mathscr{F}_{t_1}\right] = E[v(t_1, X(t_2)) - v(t_1, X(t_1)) | \mathscr{F}_{t_1}]. \qquad (12.62)$$

Next, a direct integration gives

$$E\left[\int_{t_1}^{t_2} v_s(s, X(t_2))\, ds \,\middle|\, \mathscr{F}_{t_1}\right] = E[v(t_2, X(t_2)) - v(t_1, X(t_2)) | \mathscr{F}_{t_1}]. \qquad (12.63)$$

Adding (12.62) and (12.63) gives

$$E\left[\int_{t_1}^{t_2} [v_s(s, X(t_2)) + Av(t_1, X(s))]\, ds \,\middle|\, \mathscr{F}_{t_1}\right] = E[v(t_2, X(t_2)) - v(t_1, X(t_1)) | \mathscr{F}_{t_1}]. \qquad (12.64)$$

Now we evaluate

$$E\left[\int_{t_1}^{t_2}\int_s^{t_2} Av_s(s, X(\xi))\, d\xi\, ds \,\middle|\, \mathscr{F}_{t_1}\right] = \int_{t_1}^{t_2} E\left[\int_s^{t_2} Av_s(s, X(\xi))\, d\xi \,\middle|\, \mathscr{F}_{t_1}\right] ds. \qquad (12.65)$$

But for fixed s another application of the Dynkin martingale result conveys that

$$v_s(s, X(t)) - v_s(s, X(0)) - \int_0^t Av_s(s, X(\xi))\, d\xi$$

is a martingale. Hence, for $t_1 < s < t_2$ analogous to (12.62)

$$E\left[\int_s^{t_2} Av_s(s, X(\xi))\, d\xi \,|\, \mathscr{F}_{t_1}\right] = E[v_s(s, X(t_2)) - v_s(s, X(s)) \,|\, \mathscr{F}_{t_1}],$$

and so (12.65) becomes, say,

$$E\left[\int_{t_1}^{t_2} [v_s(s, X(t_2)) - v_s(s, X(s))]\, ds \,|\, \mathscr{F}_{t_1}\right] = I_1.$$

Evaluating the integral at (12.65) by switching the order of integration gives

$$I_1 = E\left[\int_{t_1}^{t_2}\int_s^{t_2} Av_s(s, X(\xi))\, d\xi\, ds \,|\, \mathscr{F}_{t_1}\right] = E\left[\int_{t_1}^{t_2}\int_{t_1}^{\xi} Av_s(s, X(\xi))\, ds\, d\xi \,|\, \mathscr{F}_{t_1}\right]$$

and since A is linear

$$I_1 = E\left[\int_{t_1}^{t_2} A\int_{t_1}^{\xi} v_s(s, X(\xi))\, ds\, d\xi \,|\, \mathscr{F}_{t_1}\right]$$

$$= E\left[\int_{t_1}^{t_2} [Av(\xi, X(\xi)) - Av(t_1, X(\xi))]\, d\xi \,|\, \mathscr{F}_{t_1}\right]$$

$$= I_2, \tag{12.66}$$

say.

With these preliminaries at hand, we can now show that $\{Z(t), t \geq 0\}$ is a martingale. We have to show that

$$E[Z(t_2) - Z(t_1) \,|\, \mathscr{F}_{t_1}] = 0, \qquad t_2 \geq t_1 \geq 0,$$

or, what is equivalent, that

$$E[v(t_2, X(t_2)) - v(t_1, X(t_1)) \,|\, \mathscr{F}_{t_1}] = E\left[\int_{t_1}^{t_2} [v_s(s, X(s)) + Av(s, X(s))]\, ds \,|\, \mathscr{F}_{t_1}\right]. \tag{12.67}$$

But the right-hand side of the above expression is

$$E\left[\int_{t_1}^{t_2} [v_s(s, X(t_2)) + Av(t_1, X(s))]\, ds \,|\, \mathscr{F}_{t_1}\right]$$

$$- E\left[\int_{t_1}^{t_2} [-v_s(s, X(s)) + v_s(s, X(t_2))]\, ds \,|\, \mathscr{F}_{t_1}\right]$$

$$+ E\left[\int_{t_1}^{t_2} [Av(s, X(s)) - Av(t_1, X(s))]\, ds \,|\, \mathscr{F}_{t_1}\right]$$

$$= E[v(t_2, X(t_2)) - v(t_1, X(t_1)) \,|\, \mathscr{F}_{t_1}] - I_1 + I_2 \qquad \text{[using (12.64)]}$$

$$= E[v(t_2, X(t_2)) - v(t_1, X(t_1)) \,|\, \mathscr{F}_{t_1}],$$

The last resulting since $I_1 = I_2$ as was shown in (12.66). The identity (12.67) is thereby established so that $\{Z(t), t \geq 0\}$ is a martingale, and the proof of Theorem 12.6 is complete. ∎

There are a number of special cases of this result worth mentioning. We begin with a simple example which illustrates that if $v(t, x) \notin \mathscr{D}(A)$, then $Z(t)$ need not be a martingale.

Example. Let $\{X(t), t \geq 0\}$ be Brownian motion reflected at the origin, and set $v(t, x) = x$. Clearly, $\{X(t), t \geq 0\}$ is not a martingale. The problem here is that $v(t, x) \in \mathscr{D}(A)$ requires $v_x(t, 0) = 0$ (see Problem 46).

Corollary 12.1. *If $v(t, x)$ satisfies the conditions of Theorem 12.6, and $v(t, x)$ satisfies $v_s(s, x) + Av(s, x) = 0$, then $v(t, X(t))$ is a martingale [cf. Chapter 7, (5.3)].*

Corollary 12.2. *If $f \in \mathscr{D}(A)$, then $v(t, x) = U_t f(x)$ satisfies the conditions of Theorem 12.6 and so $Z(t)$ is a martingale.*

More specifically, for $v(t, x) = U_t f(x)$,

$$Z(t) = v(t, X(t)) - \int_0^t [v_s(s, X(s)) + Av(s, X(s))]\, ds$$

$$= v(t, X(t)) - \int_0^t \left[\left(\frac{d}{ds} U_s \right) f(X(s)) + Av(s, X(s)) \right] ds$$

$$= v(t, X(t)) - 2 \int_0^t Av(s, X(s))\, ds$$

is a martingale.

In many examples, martingales can be constructed for functions not satisfying the conditions $v(t, x)$, $v_t(t, x) \in \mathscr{D}(A)$ by using the result of Theorem 12.6, and a truncation argument. We illustrate one such case now.

Example. Let $\{X(t), t \geq 0\}$ be a Bessel process on $(0, \infty)$ with parameter γ, $\gamma \geq \frac{1}{2}$. Then 0 is an entrance boundary, while ∞ is natural. The infinitesimal generator A is given by $Au = \frac{1}{2}u'' + (\gamma/x)u'$, with domain

$$\mathscr{D}(A) = \{u \in C^2(0, \infty): u, Au \to 0 \text{ as } x \to \infty; x^{2\gamma}u'(x) \to 0 \text{ as } x \to 0\}.$$

Proceeding formally for the moment, let $g(x, \lambda)$ be the unique (apart from a positive multiple) increasing solution with $\lim_{x \to 0+} x^{2\gamma}g(x, \lambda) = 0$ of $Ag = \lambda g$ ($\lambda > 0$), and set $v(t, x) = e^{-\lambda t}g(x, \lambda)$. Then $v_t(t, x) = -\lambda v(t, x)$, while $Av = \lambda v$ (by determination). Hence, $v_t(t, x) + Av(t, x) = 0$, and so by Theorem 12.6 $Y(t) = v(t, X(t))$ is a martingale. The only flaw in this argument is that $g \notin \mathscr{D}(A)$, since g is unbounded in x. To prove that $Y(t)$ is a martingale, we use a truncation argument.

First, we identify g. We have to solve the equation

$$\tfrac{1}{2}g'' + (\gamma/x)g' - \lambda g = 0. \tag{12.68}$$

The increasing solution of this equation is

$$g(x) \equiv g(x, \lambda) = x^{(1/2) - \gamma} I_{\gamma - 1/2}(\sqrt{2\lambda}x), \tag{12.69}$$

where $I_\delta(\cdot)$ is the modified Bessel function of order δ. [See, for example, (6.15)]. By examining the form of $g(x, \lambda)$ given in (12.69), we find that $x^{2\gamma}g'(x, \lambda) \to 0$ as $x \to 0$. The truncation argument goes as follows. Define functions $f_n(x)$ such that $f_n \in C_0^\infty(0, \infty)$ and $f_n(x) \equiv 1$, $0 \le x \le n$. Set $g_n(x, \lambda) = g(x, \lambda)f_n(x)$. Then $g_n \in \mathscr{D}(A)$, and by Theorem 12.6, we conclude that for $v^{(n)}(x, t) = e^{-\lambda t}g_n(x, \lambda)$,

$$v^{(n)}(t, X(t)) - \int_0^t [v_s^{(n)}(s, X(s)) + Av^{(n)}(s, X(s))]\, ds$$

is a martingale. Let σ_n be the first passage time of the process $X(t)$ to n. Since $v_s(s, x) + Av(s, x) = 0$ for $0 \le x \le n$, an application of the optional stopping theorem (Theorem 8.1, Chapter 6) shows that

$$v(t \wedge \sigma_n, X(t \wedge \sigma_n)) - \int_0^{t \wedge \sigma_n} [v_s(s, X(s)) + Av(s, X(s))]\, ds$$

$$= v(t \wedge \sigma_n, X(t \wedge \sigma_n)) = e^{-\lambda(t \wedge \sigma_n)}g_n(X(t \wedge \sigma_n), \lambda)$$

$$= e^{-\lambda(t \wedge \sigma_n)}g(X(t \wedge \sigma_n), \lambda)$$

is a martingale. It follows that

$$E_x[v((t \wedge \sigma_n), X(t \wedge \sigma_n))] = E_x[v(0, X(0))] = g(x, \lambda). \tag{12.70}$$

If we can justify the interchange of limit and expectation in (12.70), then we achieve

$$E_x[v(t, X(t))] = E_x[e^{-\lambda t}g(X(t), \lambda)] = g(x, \lambda), \tag{12.71}$$

which implies that $v(t, X(t))$ is a martingale. To see this, notice that since $g(x, \lambda)$ is monotone, $g(X(t), \lambda)$ is Markov. Hence, by (12.71),

$$E[v(t + s, X(t + s)) | \mathscr{F}_s] = E[v(t + s, X(t + s)) | X(s)]$$

$$= e^{-\lambda s}E[e^{-\lambda t}g(X(t + s), \lambda) | X(s)]$$

$$= e^{-\lambda s}g(X(s), \lambda) = v(s, X(s)).$$

To justify the interchange in (12.70), using the notation $I_{\{\sigma_n \le t\}}$ as the indicator function of the set $\{\omega: \sigma_n(\omega) \le t\}$, we write

$$E_x[v(t \wedge \sigma_n, X(t \wedge \sigma_n))] = E_x[v(t \wedge \sigma_n, X(t \wedge \sigma_n))I_{\{\sigma_n \le t\}}]$$

$$+ E_x[v(t \wedge \sigma_n, X(t \wedge \sigma_n))I_{\{\sigma_n > t\}}]$$

$$= E_x[v(\sigma_n, X(\sigma_n))I_{\{\sigma_n \le t\}}] + E_x[v(t, X(t))I_{\{\sigma_n > t\}}].$$

The second term converges to $E_x[v(t, X(t))]$, since the random variables $I_{\{\sigma_n > t\}}$ increase to the unit function as n increases, so the monotone convergence theorem applies. It remains to show that the first term goes to zero as $n \to \infty$. But

$$
\begin{aligned}
E_x[v(\sigma_n, X(\sigma_n))I_{\{\sigma_n \leq t\}}] &= E_x[e^{-\lambda \sigma_n} g(X(\sigma_n), \lambda) I_{\{\sigma_n \leq t\}}] \\
&\leq E_x[g(X(\sigma_n), \lambda) I_{\{\sigma_n \leq t\}}] = g(n, \lambda) E_x[I_{\{\sigma_n \leq t\}}] \\
&= g(n, \lambda) \Pr\{\sigma_n \leq t \mid X(0) = x\}.
\end{aligned}
\tag{12.72}
$$

From properties of the Bessel function, we know that $g(n, \lambda)$ increases like $n^{-\gamma} e^{\sqrt{2\lambda n}}$. We have then to show that $\Pr\{\sigma_n \leq t \mid X(0) = x\}$ decreases much faster. Now

$$
\Pr\{\sigma_n \leq t \mid X(0) = x\} = \Pr\left\{ \max_{0 \leq u \leq t} X(u) \geq n \mid X(0) = x \right\}
\tag{12.73}
$$

(and introducing the notation $\Pr_x) = \Pr_x\left\{ \max_{0 \leq u \leq t} X(u) \geq n \right\}$.

In what follows, for ease of exposition we will assume that $\gamma = (r - 1)/2$ for some integer $r \geq 2$. Then $X(t)$ is the radial process of r-dimensional Brownian motion $(Z_1(t), \ldots, Z_r(t))$, say. From (12.73),

$$
\begin{aligned}
\Pr_0&\left\{ \max_{0 \leq u \leq t} \sqrt{Z_1^2(u) + \cdots + Z_r^2(u)} \geq n \right\} \\
&= \Pr_0\left\{ \max_{0 \leq u \leq t} (Z_1^2(u) + \cdots + Z_r^2(u)) \geq n^2 \right\} \\
&\leq \Pr_0\left\{ \max_{1 \leq i \leq r} \max_{0 \leq u \leq t} |Z_i(u)| \geq \frac{n}{\sqrt{r}} \right\} \leq r \Pr_0\left\{ \max_{0 \leq u \leq t} |Z_1(u)| \geq \frac{n}{\sqrt{r}} \right\} \\
&\leq 2r \Pr_0\left\{ \max_{0 \leq u \leq t} Z_1(u) \geq \frac{n}{\sqrt{r}} \right\} = 2r \Pr_0\{T_{n/\sqrt{r}} \leq t\},
\end{aligned}
$$

where $T_a = \inf\{s > 0 : Z_1(s) > a\}$. But, using the result of Chapter 7, (3.4), this is just

$$
\begin{aligned}
2r &\int_0^t \frac{n s^{-3/2}}{\sqrt{r} \sqrt{2\pi}} \exp\left\{ -\frac{n^2}{2rs} \right\} ds \\
&= 2r \exp\left\{ -\frac{n^2}{2r(t+1)} \right\} \int_0^t \frac{n}{\sqrt{r} \sqrt{2\pi}} s^{-3/2} \exp\left\{ -\frac{n^2}{2r(t+1)s}(t + 1 - s) \right\} ds \\
&\leq 2r \exp\left\{ -\frac{n^2}{2r(t+1)} \right\} \int_0^t \frac{n}{\sqrt{r} \sqrt{2\pi}} s^{-3/2} \exp\left\{ -\frac{n^2}{2r(t+1)s} \right\} ds \\
&\leq 2r\sqrt{t+1} \exp\left\{ -\frac{n^2}{2r(t+1)} \right\}
\end{aligned}
\tag{12.74}
$$

since

$$\int_0^\infty \frac{ns^{-3/2}}{\sqrt{r(t+1)}\sqrt{2\pi}} \exp\left\{-\frac{n^2}{2r(t+1)s}\right\} ds = 1.$$

Hence, we can conclude from (12.73) and the inequality (12.74) that $g(n, \lambda) \operatorname{Pr}_x\{\sigma_n \le t\} \to 0$ as $n \to \infty$. Hence $v(t, X(t)) = e^{-\lambda t}g(X(t), \lambda)$ is a martingale whenever $\gamma = (r - 1)/2$. The more general cases $\frac{1}{2} \le \gamma$ can be proved with more arduous estimates of $\operatorname{Pr}_x\{\sigma_n \le t\}$ for Bessel processes.

13: The Spectral Representation of the Transition Density for a Diffusion

Consider at first a regular diffusion on a finite closed interval $I = [l, r]$, $\sigma^2(x)$ continuous and positive over I, with l, r separately exit and/or reflecting boundaries. With $f(x)$ bounded and continuous on (l, r) the function

$$u(t, x) = E_x[f(X(t))]$$

satisfies the backward differential equation [cf. (5.3)]

$$\frac{\partial u}{\partial t} = \frac{1}{2}\frac{1}{m(x)}\frac{d}{dx}\left(\frac{1}{s(x)}\frac{d}{dx}\right)u = Lu \tag{13.1}$$

with initial condition $u(0, x) = f(x)$ and boundary condition

$$\begin{aligned} u(t, l) &= 0 \qquad \text{if } l \text{ is an exit boundary,} \\ u'(t, l) &= 0 \qquad \text{if } l \text{ is reflecting,} \end{aligned} \tag{13.2}$$

and similarly at the right boundary r.

By the method of separation of variables we attempt a solution of (13.1) of the form

$$u(t, x) = c(t)\varphi(x). \tag{13.3}$$

Substituting into (13.1) and rearranging gives

$$\frac{c'(t)}{c(t)} = \frac{L\varphi(x)}{\varphi(x)}. \tag{13.4}$$

The left-hand side depends only on the variable t, while the right-hand side involves only the variable x. This is consistent only if (13.4) is a constant, yielding accordingly

$$c'(t) = -\lambda c(t) \tag{13.5}$$

and

$$L\varphi(x) = -\lambda\varphi(x), \tag{13.6}$$

where λ is a constant and $\varphi(x)$ obeys the proper boundary conditions, e.g., $\varphi(l) = \varphi(r) = 0$ if l and r are exit points.

For many specifications of boundary conditions, especially where $[l, r]$ is bounded, admissible solutions of (13.6) obeying the requisite boundary conditions can be realized only for a countable sequence $\{\lambda_n\}_0^\infty$, the "spectrum" (eigenvalues) of the operator L. In other words, there exists a unique infinite sequence

$$0 \le \lambda_0 < \lambda_1 < \cdots < \lambda_n < \cdots, \qquad \lambda_n \uparrow \infty \qquad (13.7a)$$

and associated eigenfunctions

$$\varphi_0(x), \varphi_1(x), \ldots, \varphi_n(x), \ldots \qquad (13.7b)$$

such that

$$L\varphi_n(x) = -\lambda_n \varphi_n(x) \qquad (13.8)$$

and then from (13.5), we have $c_n(t) = c_n e^{-\lambda_n t}$, where c_n is a free constant. Thus, this procedure produces the solution $c_n e^{-\lambda_n t} \varphi_n(x)$ of (13.1).

Integration by parts shows that $\{\varphi_n(x)\}$ constitutes an orthogonal sequence of functions with respect to the density $m(x)$, i.e.,

$$\int_l^r m(x)\varphi_i(x)\varphi_j(x)\,dx = 0, \qquad i \ne j. \qquad (13.9)$$

In fact,

$$-\lambda_j \int_l^r m(x)\varphi_i(x)\varphi_j(x)\,dx = \int_l^r m(x)\varphi_i(x)L\varphi_j(x)\,dx$$
$$= \frac{1}{2} \int_l^r \varphi_i(x)\left(\frac{d}{dx}\left(\frac{1}{s(x)}\frac{d}{dx}\varphi_j(x)\right)\right)dx.$$

Integration by parts twice, and use of the boundary conditions yields

$$-\int_l^r m(x)L\varphi_i(x)\varphi_j(x)\,dx = -\lambda_i \int_l^r m(x)\varphi_i(x)\varphi_j(x)\,dx.$$

If $i \ne j$, then $\lambda_i \ne \lambda_j$ by (13.7a), and relation (13.9) is valid.

As the differential equation is linear, superposition of the solutions $U_n(x, t) = c_n e^{-\lambda_n t}\varphi_n(x)$ produces a series solution

$$u(t, x) = \sum_{n=0}^\infty c_n e^{-\lambda_n t}\varphi_n(x), \qquad (13.10)$$

where the $\{c_n\}$ are to be determined to guarantee the initial condition $U(0, x) = f(x)$ at $t = 0$. On account of the orthogonality of $\{\varphi_n\}$ (13.9) and its "completeness" endowment, we may expect

$$c_n = \pi_n \int_l^r f(y)\varphi_n(y)m(y)\,dy,$$

where $1/\pi_n = \int_l^r \varphi_n^2(y)m(y)\,dy$.

For the specification

$$f(y) = \begin{cases} 1, & a < y < b \\ 0, & \text{elsewhere} \end{cases}$$

the above constructions suggest that

$$U(t, x) = \int_a^b p(t, x, y)\, dy = \sum_{n=0}^{\infty} e^{-\lambda_n t} \pi_n \varphi_n(x) \left(\int_a^b \varphi_n(y) m(y)\, dy \right).$$

Dividing by $(b - a)$ and letting the interval (a, b) shrink to y we obtain formally the spectral representation

$$p(t, x, y) = m(y) \sum_{n=0}^{\infty} e^{-\lambda_n t} \varphi_n(x) \varphi_n(y) \pi_n. \tag{13.11}$$

Such expansions are legitimate in many cases. We illustrate some classical examples where $\varphi_n(x)$ are identifiable. The appropriate boundary conditions entail:

(i) Vanishing at an exit boundary.
(ii) Derivative vanishing at a reflecting boundary.
(iii) $d/dS = 0$ at an entrance boundary.

Other kinds of lateral and/or boundary conditions relate to boundedness or integrability requirements.

It is useful to highlight the spectral representation for a variety of diffusion processes of classical types.

A. ORNSTEIN–UHLENBECK PROCESS

$$\mu(x) = -x, \qquad \sigma^2(x) = 1.$$

It is known (cf. Section 5) that the transition density is explicitly

$$p(t, x, y) = \frac{1}{\sqrt{\pi(1 - e^{-2t})}} \exp\left[-\frac{(xe^{-t} - y)^2}{1 - e^{-2t}} \right]$$

$$= \frac{e^{-y^2}}{\sqrt{\pi}\sqrt{1 - e^{-2t}}} \exp\left[-\frac{(x^2 + y^2)e^{-2t}}{1 - e^{-2t}} \right] \exp\left[\frac{2xye^{-t}}{1 - e^{-2t}} \right]. \tag{13.12a}$$

The equation (13.1) can be written as

$$\frac{\partial u}{\partial t} = \frac{1}{2} e^{x^2} \frac{\partial}{\partial x}\left(e^{-x^2} \frac{\partial u}{\partial x} \right) = \frac{1}{2}\left(\frac{\partial^2 u}{\partial x^2} - 2x \frac{\partial u}{\partial x} \right).$$

In the case at hand the eigenfunctions for (13.6) satisfy

$$L\varphi(x) = \frac{1}{2}\left(\frac{d^2\varphi}{dx^2} - 2x\frac{d\varphi}{dx}\right) = -\lambda\varphi(x), \qquad -\infty < x < \infty,$$

subject to the integrability conditions $\int_{-\infty}^{\infty} e^{-x^2}[\varphi(x)]^2\,dx < \infty$. These are the classical Hermite polynomials

$$H_n(x) = (-1)^n e^{x^2}\frac{d^n}{dx^n}(e^{-x^2})$$

with associated eigenvalues $\lambda_n = n$, respectively, and then we obtain the spectral representation

$$p(t, x, y) = e^{-y^2}\sum_{n=0}^{\infty} e^{-nt}H_n(x)H_n(y)\pi_n, \qquad (13.12b)$$

where $\pi_n^{-1} = (\sqrt{\pi}\,2^n n!)$. By virtue of a classical formula† for the sum of the series, we find that (13.12b) reduces to (13.12a).

B. RADIAL O.U. (ORNSTEIN–UHLENBECK) PROCESS IN N-DIMENSIONS

$$\mu(x) = \left(\frac{N-1}{2x} - x\right), \qquad \sigma^2(x) = 1, \qquad 0 < x < \infty. \qquad (13.13)$$

These constitute the infinitesimal parameters that arise when imposing on an N-dimensional Brownian motion a restoring force directly proportional to its distance from the origin.

The corresponding backward differential equation is

$$\frac{\partial u}{\partial t} = \frac{1}{2}\frac{1}{r^{N-1}}\frac{\partial}{\partial x}\left(x^{N-1}\frac{\partial u}{\partial x}\right) - x\frac{\partial u}{\partial x} = \frac{1}{2}\frac{\partial^2 u}{\partial x^2} + \left(\frac{N-1}{2x} - x\right)\frac{\partial u}{\partial x},$$
$$0 < x < \infty.$$

The infinite point is a natural boundary while the origin is an entrance boundary for $N > 1$ with the associated condition

$$\lim_{x\downarrow 0} x^{N-1}\frac{du}{dx} = \lim_{x\downarrow 0}\frac{du}{dS} = 0 \qquad \text{at} \quad x = 0.$$

Subject to this constraint, the solutions of

$$\frac{1}{2}\frac{d^2\varphi}{dx^2} + \left(\frac{N-1}{2x} - x\right)\frac{d\varphi}{dx} = -\lambda\varphi$$

† A. Erdélyi, ed., "Higher Transcendental Functions," Vol. II, p. 194. McGraw-Hill, New York, 1953.

are Laguerre polynomials. (The Laguerre polynomials $L_n^{(\alpha)}(\xi)$ with parameter α, usually normalized by the conditions $L_n^{(\alpha)}(0) > 0$ and

$$\int_0^\infty [L_n^{(\alpha)}(\xi)]^2 \frac{\xi^\alpha e^{-\xi}}{\Gamma(\alpha+1)} d\xi = \binom{n+\alpha}{n},$$

satisfy the differential equation $\xi L_n^{(\alpha)''}(\xi) + (\alpha + 1 - \xi)L_n^{(\alpha)'}(\xi) + nL_n^{(\alpha)}(\xi) = 0.$) The Laguerre system $L_n^{(\alpha)}(\xi)$ comprise the unique orthogonal polynomials with respect to the weight function $w(\xi) = \xi^\alpha e^{-\xi}/\Gamma(\alpha+1)$ for $\xi > 0$; $w(\xi) = 0$ for $\xi < 0$. When $\alpha = -\frac{1}{2}$, $L_n^{(-1/2)}(x^2)$ coincides with the Hermite polynomials, viz., $H_{2m}(x) = (-1)^m 2^{2m} m! L_m^{(-1/2)}(x^2)$.

The relevant spectral representation for (13.13) is

$$p(t, x, y) = y^{N-1}e^{-y^2} \sum_{n=0}^\infty e^{-2nt}L_n^{(\alpha)}(x^2)L_n^{(\alpha)}(y^2)w_n, \qquad (13.14)$$

where

$$w_n^{-1} = \tfrac{1}{2}\Gamma(\alpha+1)\binom{n+\alpha}{n} \quad \text{and} \quad \alpha = \frac{N-2}{2}.$$

The series (13.14) can be summed to give

$$p(t, x, y) = 2y^{N-1}e^{-y^2} \frac{1}{1-e^{-2t}} \exp\left[\frac{-(x^2+y^2)e^{-2t}}{1-e^{-2t}}\right](xye^{-t})^{-\alpha}I_\alpha\left(\frac{2xye^{-t}}{1-e^{-2t}}\right),$$

$$(13.14a)$$

where I_α is the modified Bessel function of order α.†

C. POPULATION GROWTH MODEL

$$\mu(x) = bx + c, \qquad \sigma^2(x) = 2ax, \qquad 0 < x < \infty.$$

a, b, c are constants, $a > 0$. A straightforward but arduous separation of variables provided $b < 0$ and $c > 0$ leads to

$$p(t, x, y) = \frac{(|b|/a)^{\alpha+1}y^\alpha e^{by/a}}{\Gamma(\alpha+1)} \sum_{n=0}^\infty e^{nbt}L_n^{(\alpha)}\left(\frac{-bx}{a}\right)L_n^{(\alpha)}\left(\frac{-by}{a}\right)\left(\frac{\Gamma(n+1)}{\Gamma(n+\alpha+1)}\right), \qquad (13.15)$$

where $\alpha = (c/a) - 1$, and so

$$p(t, x, y) = \frac{(|b|/a)^{\alpha+1}y^\alpha e^{by/a}}{\Gamma(\alpha+1)(1-e^{bt})} \exp\left\{\frac{e^{bt}(b/a)(x+y)}{1-e^{bt}}\right\}$$

$$\times \left(\frac{b^2}{a^2}e^{bt}xy\right)^{-\alpha/2}I_\alpha\left(\frac{2|b|\sqrt{xye^{bt}}}{a(1-e^{bt})}\right).$$

† See Erdélyi, "Higher Transcendental Functions," Vol. II, p. 189.

D. JACOBI DIFFUSION PROCESS

$$I = (-1, 1), \quad \mu(x) = \tfrac{1}{2}[(\beta + 1)(1 - x) - (\alpha + 1)(1 + x)], \quad \sigma^2(x) = (1 - x^2)$$

For α and β positive, the boundaries -1 and 1 are entrance boundaries. The corresponding differential equation (13.1) becomes

$$\frac{\partial u}{\partial t} = \frac{1}{2} \frac{1}{(1 - x)^\alpha (1 + x)^\beta} \frac{\partial}{\partial x}\left[(1 - x)^{\alpha + 1}(1 + x)^{\beta + 1} \frac{\partial u}{\partial x}\right].$$

The boundary conditions for entrance boundaries (see Theorem 12.2, page 306) are equivalent to

$$\tfrac{1}{2}(1 - x)^{\alpha + 1} \frac{\partial u}{\partial x}\bigg|_{x = 1^-} = 0, \qquad (1 + x)^{\beta + 1} \frac{\partial u}{\partial x}\bigg|_{x = -1^+} = 0. \tag{13.16}$$

The solution of

$$\frac{1}{(1 - x)^\alpha (1 + x)^\beta} \frac{d}{dx}\left[(1 - x)^{1 + \alpha}(1 + x)^{1 + \beta} \frac{d\varphi}{dx}\right] = -\lambda\varphi(x)$$

under the boundary conditions (13.16) singles out the Jacobi orthogonal polynomials.†

The spectral expansion becomes

$$p(t, x, y) = \frac{(1 - y)^\alpha (1 + y)^\beta \Gamma(\alpha + \beta + 2)}{2^{\alpha + \beta + 1}\Gamma(\alpha + 1)\Gamma(\beta + 1)} \sum_{n=0}^{\infty} e^{-n(n + \alpha + \beta + 1)t/2} P_n(x)P_n(y)\pi_n,$$

$$\tag{13.17}$$

where $P_n(x) = P_n^{(\alpha,\beta)}(x)$ are the Jacobi polynomials normalized to have value 1 at $x = 1$ and

$$\pi_n = \frac{\Gamma(\beta + 1)}{\Gamma(\alpha + 1)\Gamma(\alpha + \beta + 1)} \frac{\Gamma(n + \alpha + 1)\Gamma(n + \alpha + \beta + 1)(2n + \alpha + \beta + 1)}{\Gamma(n + 1)\Gamma(n + \beta + 1)(\alpha + \beta + 1)}.$$

E. RADIAL MOTION OF B.M. IN N-DIMENSION FOR A PARTICLE STARTED INSIDE THE UNIT SPHERE AND TERMINATED (FIXED) AT THE INSTANT IT ATTAINS THE SURFACE OF THE UNIT SPHERE

The process $Y(t)$ is a diffusion on $0 < x \leq 1$ for which the point 1 is a trap state. Its backward diffusion equation is

$$\frac{\partial u}{\partial t} = \frac{1}{2} \frac{1}{y^{N-1}} \frac{\partial}{\partial y}\left(y^{N-1} \frac{\partial u}{\partial y}\right)$$

with the boundary condition $u = 0$ at $y = 1$ at the trap boundary. The reflecting barrier (or entrance boundary) condition at $x = 0$ is $du/dS|_{x=0} = 0$. The diffusion coefficients are

$$\sigma^2(x) = 1, \qquad \mu(x) = \frac{\gamma}{x}, \qquad 0 < x < 1, \qquad \gamma = \frac{N - 1}{2}$$

† See Erdélyi, "Higher Transcendental Functions," Vol. II, p. 169.

(compare to the Bessel Process). The equation of (13.1) in the present context is

$$\frac{1}{2}\frac{1}{x^{N-1}}\frac{d}{dx}\left(x^{N-1}\frac{d\varphi}{dx}\right) = -\lambda\varphi(x) \tag{13.18}$$

coupled to the boundary conditions

$$x^{N-1}\frac{d\varphi}{dx}\bigg|_{x=0^{+}} = 0, \qquad \varphi(1) = 0. \tag{13.19}$$

The eigenfunctions of (13.18)–(13.19) can be identified to be

$$\varphi_n(x) = x^{-(N-2)/2}J_{(N-2)/2}(x\sqrt{\lambda_n}), \tag{13.20}$$

where $\sqrt{\lambda_n}$ is the sequence of positive zeros of the Bessel function $J_{(N-2)/2}(x)$. The spectral expansion of the transition density is

$$p(t, x, y) = y^{N-1}\sum_{n=0}^{\infty}e^{-2\lambda_n t}\varphi_n(x)\varphi_n(y)\pi_n, \tag{13.21}$$

where $\pi_n^{-1} = \int_0^1 \varphi_n^2(x)x^{N-1}\,dx$.

F. WRIGHT–FISHER GENE FREQUENCY DIFFUSION WITHOUT MUTATION (c.f. EXAMPLE F, SECTION 2)

$$\mu(x) = 0, \qquad \sigma^2(x) = 2\gamma x(1-x), \qquad I = [0, 1]. \tag{13.22}$$

The backward differential equation corresponding to (13.22) is

$$\frac{\partial u}{\partial t} = \gamma x(1-x)\frac{\partial^2 u}{\partial x^2} \tag{13.23}$$

with boundary conditions $u(t, 0) = u(t, 1) = 0$. The resulting eigenvalue differential equation problem (13.6) is

$$\gamma x(1-x)\frac{d^2\varphi(x)}{dx^2} = -\lambda\varphi(x), \tag{13.24}$$

subject to $\varphi(0) = \varphi(1) = 0$. Reference to texts on classical special functions shows that for $n \geq 1$,

$$\lambda_n = \gamma n(n+1)$$

and

$$\varphi_n(x) = x(1-x)P_{n-1}^{(1,1)}(1-2x), \tag{13.25}$$

where $P_{n-1}^{(1,1)}(\xi)$ are the Jacobi orthogonal polynomials of parameters $\alpha = 1$, $\beta = 1$ normalized so that $P_{n-1}^{(1,1)}(1) = 1$ [see after (13.17)].

These polynomials are orthogonal with respect to the weight function $6x(1-x)$ over the interval $[0, 1]$. The expansion of the transition density is

$$p(t, x, y) = \frac{1}{y(1-y)}\sum_{n=1}^{\infty}e^{-\gamma n(n+1)t}\varphi_{n-1}(x)\varphi_{n-1}(y)n(n+1)(2n+1). \tag{13.26}$$

Observe that $p(t, x, y)$ goes to zero at the rate $e^{-2\gamma t}$ as is anticipated because absorption at 0 and 1 are certain outcomes. The conditional limiting density of the process, given absorption has not occurred, is uniform:

$$\lim_{t \uparrow \infty} \Pr\{y < X(t) < y + dy \,|\, X(0) = x, X(t) \neq 0 \text{ or } 1\} = 1 \, dy.$$

Continuous Spectrum

There is not always available a series spectral representation of the transition density of a one dimensional diffusion. This situation occurs where the spectrum (set of eigenvalues of (13.6)) is not discrete but involves a continuous portion. The generalized spectral representation replaces the series by an integral of the form

$$p(t, x, y) = m(y) \int_{-\infty}^{\infty} e^{-\lambda t} \varphi(x, \lambda) \varphi(y, \lambda) \, d\psi(\lambda). \tag{13.27}$$

There are cases involving both discrete spectra (eigenvalues) and continuous spectral sets. The more intricate models will not be presented here. However, some of the standard diffusion examples have only continuous spectra.

G. BROWNIAN MOTION

There is no discrete spectrum in this model. In fact, the spectral density expansion of the density conforming to (13.27) becomes

$$p(t, x, y) = \frac{1}{\sqrt{2\pi t}} e^{-(x-y)^2/2t} = \frac{1}{\sqrt{2\pi}} \int_{-\infty}^{\infty} e^{-\lambda^2 t/2} e^{ix\lambda} e^{-iy\lambda} \, d\lambda \tag{13.28}$$

the Fourier transform of a Gaussian type kernel, actually the inverse Fourier transform of the Gaussian kernel. In this model $\varphi(x, \lambda) = e^{ix\lambda}$ which satisfies the equation $\varphi''(x, \lambda) = -\lambda^2 \varphi(x, \lambda)$ for each real λ where here the natural boundary conditions demand that $\varphi(x, \lambda)$ be bounded for all real x.

H. REFLECTING BROWNIAN MOTION ON $(0, \infty)$

The spectral expansion of the transition density for $x, y > 0$ is

$$p(t, x, y) = \frac{1}{\sqrt{2\pi t}} \left[e^{-(x-y)^2/2t} + e^{-(x+y)^2/2t} \right]$$

$$= \sqrt{\frac{2}{\pi}} \int_0^{\infty} e^{-\lambda^2 t/2} (\cos x\lambda)(\cos y\lambda) \, d\lambda, \tag{13.29}$$

the Fourier cosine transform of the Gaussian kernel.

The backward equation is

$$\frac{\partial u}{\partial t} = \frac{1}{2}\frac{\partial^2 u}{\partial x^2} + \frac{\gamma}{x}\frac{\partial u}{\partial x}, \qquad 0 < x < \infty, \quad \gamma = (N-1)/2.$$

The transition density has the form

$$p(t, x, y) = m(y)\int_0^\infty e^{-\lambda^2 t/2} T(\lambda x) T(\lambda y) m(\lambda)\, d\lambda$$

with

$$m(y) = \frac{2^{-(\gamma-1/2)}}{\Gamma(\gamma + \frac{1}{2})} y^{2\gamma}, \qquad T(x) = \Gamma(\gamma + 1)\left(\frac{x}{2}\right)^{1/2 - \gamma} J_{\gamma - 1/2}(x), \qquad (13.30)$$

where J_δ is the classical Bessel function of order δ.

Some Diffusions with Spherical Harmonics

In the remainder of this section we describe a geometrical construction which relates the transition probability (13.17) for the special case $\alpha = \beta = (N-1)/2$, $N = 0, 1, 2, \ldots$ with a Brownian motion on the surface of the unit sphere Ω in a Euclidean space of $N + 2$ dimensions. If $\xi_1, \ldots, \xi_{N+2}$ are Cartesian coordinates on the surface of the unit sphere $\Omega(\xi_1^2 + \cdots + \xi_{N+2}^2 = 1)$ then spherical coordinates $\theta_1, \ldots, \theta_N, \phi$ are defined by

$$\begin{aligned}
\xi_1 &= \cos\theta_1, \\
\xi_2 &= \sin\theta_1 \cos\theta_2, \\
&\vdots \\
\xi_N &= \sin\theta_1 \sin\theta_2 \cdots \sin\theta_{N-1}\cos\theta_N, \\
\xi_{N+1} &= \sin\theta_1 \cdots \sin\theta_N \cos\phi, \\
\xi_{N+2} &= \sin\theta_1 \cdots \sin\theta_N \sin\phi,
\end{aligned} \qquad (13.31)$$

where $0 \le \theta_i \le \pi$ and $0 \le \phi \le 2\pi$. The Laplace–Beltrami operator Δ (the analog of the Laplacian in N dimensions) referring to the unit sphere Ω is given by

$$\begin{aligned}
\Delta u = {}& (\sin\theta_1)^{-N}\frac{\partial}{\partial\theta_1}\left[(\sin\theta_1)^N\frac{\partial u}{\partial\theta_1}\right] \\
& + (\sin\theta_1)^{-2}(\sin\theta_2)^{1-N}\frac{\partial}{\partial\theta_2}\left[(\sin\theta_2)^{N-1}\frac{\partial u}{\partial\theta_2}\right] + \cdots \\
& + (\sin\theta_1 \cdots \sin\theta_{N-1})^{-2}(\sin\theta_N)^{-1}\frac{\partial}{\partial\theta_N}\left[\sin\theta_N\frac{\partial u}{\partial\theta_N}\right] \\
& + (\sin\theta_1 \cdots \sin\theta_N)^{-2}\frac{\partial^2 u}{\partial\phi^2}.
\end{aligned}$$

The diffusion equation $\partial u/\partial t = \Delta u$ has a unique fundamental solution on Ω, denoted by $p(t, \xi, \eta)$, and this solution is the transition probability density of a stationary Markov process on Ω (the term "density" means relative to the surface element $d\omega$ on Ω). We will assume that process $\xi(t)$, called Brownian motion on Ω, has continuous path functions. The transition density has the representation

$$p(t, \xi, \eta) = \sum_{n=0}^{\infty} e^{-\lambda_n t} \sum_{l=0}^{h(n)} S_n^{(l)}(\xi) S_n^{(l)}(\eta),$$

where   (13.32)

$$\lambda_n = n(n + \alpha + \beta + 1) = n(n + N),$$

$$h(n) = (n + N - 1)!(2n + N)/(n!N!)$$

and $S_n^{(l)}(\xi), l = 1, 2, \ldots, h(n)$ is a complete orthonormal set of surface harmonics of degree n, (e.g., see Erdélyi[†]). By virtue of the addition theorem[‡] for surface harmonics we have the alternative representation

$$p(t, \xi, \eta) = \sum_{n=0}^{\infty} e^{-\lambda_n t} \frac{h(n)}{w(N)} P_n(\langle\!\langle \xi, \eta \rangle\!\rangle),$$ (13.33)

where $\langle\!\langle \xi, \eta \rangle\!\rangle = \xi_1\eta_1 + \cdots + \xi_{N+2}\eta_{N+2}$ is the cosine of the angle between the unit vectors ξ and η, and $\omega(N)$ is the surface area of the unit sphere in $N + 2$ dimensions. P_n is an appropriate Jacobi polynomial.

Now we start a Brownian motion $\xi(t)$ on the sphere with an initial distribution which has the ξ_1 axis as an axis of symmetry, and which has no mass at the points $\xi_1 = \pm 1$. The distribution of $\xi(t)$ for any $t > 0$ is then also symmetric about the ξ_1 axis. The random variable $X(t) = \xi_1(t)$ which is the projection of $\xi(t)$ on the ξ_1-axis is evidently a Markov process with continuous path functions and with state space $-1 \leq x \leq 1$. To calculate the transition probability function of $X(t)$ we introduce polar coordinates

$$\xi \sim (\theta_1, \theta_2, \ldots, \theta_N, \phi), \qquad \eta \sim (\theta_1', \theta_2', \ldots, \theta_N', \phi')$$

according to (13.31). The transition probability is clearly given by

$$P(t, x, [-1, y]) = \Pr\{X(t) \leq y \mid X(0) = x\}$$

$$= \int_{-1 \leq \eta_1 \leq y} p(t, \xi, \eta) \, d\omega_\eta,$$ (13.34)

where ξ is any fixed unit vector with $\xi_1 = x$. By symmetry considerations we take without loss of generality, $\theta_1 = \cos^{-1} x, \theta_2 = \theta_3 = \cdots = \theta_N = 0$ so that $\langle\!\langle \xi, \eta \rangle\!\rangle = \cos \theta_1 \cos \theta_1' + \sin \theta_1 \sin \theta_1' \cos \theta_2'$. Using the area element

$$d\omega_\eta = (\sin \theta_1')^N (\sin \theta_2')^{N-1} \cdots (\sin \theta_N') \, d\theta_1' \, d\theta_2' \cdots d\theta_N' \, d\phi'$$ (13.35)

[†] Erdélyi, "Higher Transcendental Functions," Vol. II.
[‡] See Erdélyi, p. 243.

and the representation (13.33) together with (13.35) we obtain

$$P(t; x, [-1, y]) = \int_{\cos \theta_1' \leq y} (\sin \theta_1')^N \, d\theta_1' \sum_{n=0}^{\infty} e^{-\lambda_n t} \frac{\omega(N-2)h(n)}{\omega(N)} f_n(\theta_1, \theta_1'),$$

where

$$f_n(\theta_1, \theta_1') = \int_0^{\pi} P_n(\cos \theta_1 \cos \theta_1' + \sin \theta_1 \sin \theta_1' \cos \theta_2')(\sin \theta_2')^{N-1} \, d\theta_2'.$$

(This result has a slightly different form if $N = 0$.) Now by virtue of an identity of Gegenbauer† we have

$$f_n(\theta_1, \theta_1') = P_n(\cos \theta_1)P_n(\cos \theta_1') \int_0^{\pi} (\sin \theta_2)^{N-1} \, d\theta_2$$

and it follows that

$$P(t, x, [-1, y]) = c_{N-1}^{-1} \int_{-1}^{y} p(t; x, z)(1 - z^2)^{(N-1)/2} \, dz$$

with $p(t, x, z)$ as in (13.17), where $\alpha = \beta = (N-1)/2$ and $c_N = \int_{-1}^{1} (1 - \xi^2)^{N/2} \, d\xi$.

14: The Concept of Stochastic Differential Equations

The next three sections introduce the important subject of stochastic differential equations and stochastic integrals. The modern approach in dealing with diffusion processes is in these terms. Several chapters would be required to elaborate rigorously the concepts and techniques thereof. We discuss some of the principal ideas and methodology in an informal manner in Section 15, highlighting a variety of examples. In Section 16 several important results are stated that emphasize the martingale property and the power of the Ito transformation formula.

The motion of a particle suspended in a fluid is influenced by two principal forces. First, a nonrandom (deterministic) motion can be engendered by the nature of the underlying fluid flow or induced by some external force impressed on the system. Second, collisions and/or more general interaction relationships with other particles cause generally random movements which over short time durations are often well described by Brownian motion fluctuations. Thus, for a small duration from time t to $t + \Delta t$, the displacement of the particle (say along the x axis) is approximated by

$$X(t + \Delta t) - X(t) \approx \mu(x, t) \, \Delta t + \sigma(x, t) \, \Delta B(t), \tag{14.1}$$

† See G. Szegö, "Orthogonal Polynomials," p. 369. Colloquium Publications Series, Amer. Math. Soc., Providence, Rhode Island, 1959.

where $X(t) = X(t, \omega) = x$ is the location of the particle at time t. (we shall ordinarily suppress the variable ω identifying the sample path realization). Here $\mu(x, t)$ is the instantaneous velocity of the fluid at time t and position x while $\Delta B(t) = B(t + \Delta t) - B(t)$ is the incremental change associated with the standard Brownian motion process $B(t)$ and $\sigma^2(x, t) > 0$ measures the instantaneous variance associated with the collisions of the $X(t)$ process. The first term reflects the movement caused by the deterministic forces while the second term expresses the random component of the motion.

Since $B(t)$ involves continuous sample paths, we can infer that $X(t)$ is a continuous Markov process, provided $\mu(x, t)$ and $\sigma(x, t)$ are appropriately continuous deterministic functions. In fact the Markov property is implied by the independent increments of Brownian motion. From (14.1) we infer, heuristically, that $\{X(t), t \geq 0\}$ constitutes a diffusion process with infinitesimal drift coefficient $\mu(x, t)$ and infinitesimal variance (diffusion coefficient) $\sigma^2(x, t)$. Indeed, we have

$$\lim_{\Delta t \downarrow 0} \frac{E[\Delta X]}{\Delta t} = \lim_{\Delta t \downarrow 0} \frac{1}{\Delta t} E[\mu(x, t) \, \Delta t + \sigma(x, t) \, \Delta B(t)] = \mu(x, t) \quad (14.2)$$

(using the fact that $E[\Delta B] = 0$) and

$$\lim_{\Delta t \downarrow 0} \frac{1}{\Delta t} E[(\Delta X)^2] = \lim_{\Delta t \downarrow 0} \frac{1}{\Delta t} \{\text{var}(\Delta X) + (E[\Delta X])^2\}$$

$$= \lim_{\Delta t \downarrow 0} \frac{1}{\Delta t} \text{var}[\sigma(x, t) \, \Delta B] + \lim_{\Delta t \downarrow 0} \frac{1}{\Delta t} (E[\Delta X])^2$$

$$= \lim_{\Delta t \downarrow 0} \frac{1}{(\Delta t)} \sigma^2(x, t) \, \Delta t = \sigma^2(x, t) \quad (14.3)$$

since

$$(E[\Delta X])^2 = O(\Delta t)^2 \quad \text{and} \quad \text{var}[\Delta B] = \Delta t.$$

It is tempting to continue the foregoing reasoning and concomitantly evaluate the limit of $\Delta X/\Delta t$, but this random variable has variance of the order $1/\Delta t$. Notice that $\lim_{\Delta t \to 0} (1/\Delta t) \text{var}(\Delta X)$ exists with a finite limit while

$$\lim_{\Delta t \to 0} \text{var}\left(\frac{\Delta X}{\Delta t}\right)$$

does not exist and consequently convergence of $\Delta X/\Delta t$ is, a fortiori, precluded. To assign meaning to the limit relation attendant to (14.1), that is,

$$\frac{dX(t)}{dt} = \mu(x, t) + \sigma(x, t) \frac{dB}{dt}, \quad (14.4)$$

it is necessary to develop an extended version of stochastic differentials and integrals. Relation (14.4) is preferably written in the differential notation

$$dX = \mu(x, t)\, dt + \sigma(x, t)\, dB, \tag{14.5}$$

in which transformations on dX are more easily performed.

WHITE NOISE

The "process" $dB(t)/dt = W(t)$ (recall from Chapter 7 that the Brownian paths, although almost certainly continuous, are nowhere differentiable) is commonly referred to as the white noise "process" for a number of reasons, several of which we now elaborate.

Consider the correlation of the incremental displacements $\Delta_h B(t) = B(t + h) - B(t)$ and $\Delta_k B(s) = B(s + k) - B(s)$ corresponding to two non-overlapping epochs $(t, t + h)$ and $(s, s + k)$, where $t + h \leq s$ and $h, k > 0$.

The independent increment property of $B(t)$ guarantees

$$E\left[\frac{\Delta_h B(t)}{h} \frac{\Delta_k B(s)}{k}\right] = 0, \tag{14.6}$$

while

$$E\left[\left(\frac{\Delta_h B(t)}{h}\right)^2\right] = \frac{1}{h}.$$

Thus, we might expect the Dirac delta function

$$E\left[\frac{dB(t)}{dt} \frac{dB(s)}{ds}\right] = \begin{cases} 1/dt = \infty, & t = s, \\ 0, & t \neq s, \end{cases} \tag{14.7}$$

signifying that the continuous time stationary process dB/dt consists of a continuum of random variables, each random variable corresponding to a parameter value of t, which are all uncorrelated (and even mutually independent) and whose sample paths in some sense occur as the derivative of the realizations of Brownian motion. Of course, Brownian motion paths are not differentiable. There is no simple construction to realize such a process dB/dt. The analog of dB/dt in discrete time is simply a sequence of independent random variables, each normally distributed with zero means and unit variances. The spectral density of such a sequence (see Section 7, Chapter 9) is constant over the range $[0, 2\pi]$, signifying that all frequencies appear equally, and thus motivating the description "white noise." Difficulties arise in establishing a well-defined version of a process whose values at all points of a continuous time parameter are independent. Modern mathematics has developed a framework for differentiating all conceivable functions, even those usually considered nondifferentiable. These developments are encompassed by the theory of Schwartz distributions. We shall not follow this tack although it is somewhat pertinent to the present context.

In continuous time, white noise is not a physical process but is an abstraction wherein all frequencies over the range $(0, \infty)$ appear equally. What is germane for applications is to replace $W(t)$ by a smoother stationary Gaussian process with an essentially flat spectral density over a broad range of frequencies. For a Gaussian process $Y(t)$ having covariance function

$$R(t, s) = E[Y(t)Y(s)]$$

(to simplify the notation we have stipulated $E[Y(t)] \equiv 0$) for which $\rho(t, s) = \partial^2 R(t, s)/\partial t \, \partial s$ is continuous, the derivative process $\dot{Y}(t) = dY/dt$ is meaningful and constitutes a Gaussian process completely characterized by the covariance function

$$\rho(t, s) = \frac{\partial^2 R(t, s)}{\partial t \, \partial s} = E[\dot{Y}(t)\dot{Y}(s)] \tag{14.8}$$

In fact, expanding the expectation yields

$$E\left[\frac{Y(t + h) - Y(t)}{h} \frac{Y(s + k) - Y(s)}{k}\right]$$

$$= \frac{R(t + h, s + k) - R(t, s + k) - R(t + h, s) + R(t, s)}{hk}. \tag{14.9}$$

Letting h and k go to zero, the right-hand side converges to $\partial^2 R(t, s)/\partial t \, \partial s$. It follows that $[Y(t + h) - Y(t)]/h$ converges in mean square, and we then define $\dot{Y}(t)$ as the limit. It suffices to have $R(t, s)$ continuously differentiable on the diagonal $t = s$, since then $\dot{Y}(t)$ exists for all t. When the process $Y(t)$ is chosen appropriately, then the derivative process $\dot{Y}(t)$ will approximate white noise.

To compare with Brownian motion, recall that for it the covariance is

$$R_B(t, s) = \min(t, s), \tag{14.10}$$

which is not differentiable along the diagonal ($t = s$). The formal derivative is

$$\rho(t, s) = \frac{\partial_B^2 R(t, s)}{\partial t \, \partial s} = \begin{cases} \infty & \text{if} \quad t = s, \\ 0 & \text{if} \quad t \neq s. \end{cases} \qquad \text{(Compare with (14.7))}.$$

$$\tag{14.11}$$

The identification (14.11) compared to (14.7) suggests a consideration of the formal derivative $dB/dt = W(t)$ as a continuous time stationary Gaussian process with uncorrelated (and hence independent) points.

STOCHASTIC DYNAMICAL SYSTEMS

Extending (14.4), the solution of a differential equation of the form

$$\frac{dX(t)}{dt} = f(t, X(t), W(t)), \tag{14.12}$$

where f is a real-valued (generally smooth) function, is called a continuous *stochastic dynamical system*. The random disturbing function $W(t)$ is called a random forcing (driving, input) factor. The initial conditions in (14.12) may be fixed or random with known distribution function. Problems 6–11 deal with a number of discrete versions (difference equations) of (14.12).

A wide range of physical and engineering models expressed by the dynamic relation (14.12) can be found in numerous electrical engineering texts concerned with stochastic control, filtering, extrapolation, and prediction studies.

The analysis of (14.12) can be transferred to that of an equivalent integral equation,

$$X(t) - X(0) = \int_0^t f(\tau, X(\tau), W(\tau)) \, d\tau, \qquad (14.13)$$

provided one can appropriately define the limit involved in the integral.

We shall mostly concentrate on the special case of (14.12) of the type (14.4). In order to secure solutions of (14.12), or more specifically (14.4), we shall adapt to the stochastic context the classical method of successive approximations basic to solving differential equations. The simplest example of (14.12) has the right-hand side independent of $W(t)$ and randomness entering only in the initial conditions.

Another common physical model involves only a random nonhomogeneous component where (14.12) specifically takes the form

$$\dot{X}(t) = \frac{dX(t)}{dt} = f(X(t), t) + W(t). \qquad (14.14)$$

A well-studied physical dynamical system of the sort (14.14) is the mass spring linear oscillator driven by white noise. The displacement position corresponds to a solution of the second-order differential equation

$$\ddot{X}(t) + 2\beta \dot{X}(t) + \gamma^2 X(t) = \frac{dB}{dt} = W(t) \qquad (14.15)$$

where β (resistance parameter) and γ (inertial constant) are real constants. Introducing the vector

$$\mathbf{X}(t) = \begin{pmatrix} X_1(t) \\ X_2(t) \end{pmatrix}, \qquad X_2(t) = \dot{X}_1(t),$$

then (14.15) can be cast in the vector form

$$\dot{\mathbf{X}}(t) = \begin{pmatrix} 0 & 1 \\ -\gamma^2 & -2\beta \end{pmatrix} \mathbf{X}(t) + \mathbf{W}(t) \qquad (14.16)$$

with

$$\mathbf{W}(t) = \begin{pmatrix} 0 \\ W(t) \end{pmatrix},$$

and (14.16) can be regarded as a vector example of (14.14).

It is frequently more relevant to replace the right-hand side of (14.15) by a more general stationary Gaussian process $\dot{Y}(t)$ with a known covariance function, something other than white noise. For $\dot{Y}(t)$ as prescribed, the solution $X(t)$ is generally no longer a Markov process.

The Ornstein–Uhlenbeck Process as a Solution of a Dynamical Stochastic Differential Equation

We investigate the differential equation

$$\frac{dX(t)}{dt} = -\beta X(t) + W(t), \quad X(0) = 0 \qquad (\beta > 0), \qquad (14.17)$$

where the input process $\{W(t), t \geq 0\}$ is Gaussian white noise, $W(t) = dB(t)/dt$. This equation can be solved by direct methods. We rewrite (14.17), employing the obvious integrating factor, yielding

$$\frac{d}{dt}[e^{\beta t}X(t)] = e^{\beta t}\frac{dB(t)}{dt}.$$

Integrating both sides and invoking $X(0) = 0$ produces

$$e^{\beta t}X(t) = \int_0^t e^{\beta \tau}\frac{dB(\tau)}{d\tau}\, d\tau. \qquad (14.18)$$

Next we integrate by parts the right-hand side to get

$$e^{\beta t}X(t) = e^{\beta t}B(t) - \beta \int_0^t e^{\beta \tau}B(\tau)\, d\tau$$

or

$$X(t) = B(t) - \beta \int_0^t e^{-\beta(t-\tau)}B(\tau)\, d\tau. \qquad (14.19)$$

The representation of $X(t)$ in (14.19) as a linear operator on $B(\tau)$ implies that $X(t)$ is a Gaussian process (see Section 8 of Chapter 9). It remains to ascertain the mean and covariance of $X(t)$. Clearly,

$$E[X(t)] = E[B(t)] - \beta \int_0^t e^{-\beta(t-\tau)}E[B(\tau)]\, d\tau = 0.$$

Next we formally compute $E[X(t)X(s)]$, which is done most expeditiously proceeding from (14.18). We obtain for $t < s$

$$e^{\beta(t+s)}E[X(t)X(s)] = \int_0^t\int_0^s e^{\beta(\xi+\eta)}E[W(\xi)W(\eta)]\, d\xi\, d\eta$$

$$= \int_0^t e^{2\beta\xi}\, d\xi \quad \text{(since } E[W(\xi)W(\eta)]\, d\xi\, d\eta = \delta(\xi - \eta)\, d\xi\, d\eta)$$

$$= \frac{e^{2\beta t} - 1}{2\beta},$$

and then

$$E[X(t)^2] = \frac{1 - e^{-2\beta t}}{2\beta}.$$

With a general initial condition $X(0) = x$, we find that the solution $\{X(t), t \geq 0\}$ to (14.17) is a Gaussian process with moments

$$E[X(t)] = xe^{-\beta t} \quad \text{and} \quad \text{var}[X(t)] = \frac{1 - e^{-2\beta t}}{2\beta}. \qquad (14.20)$$

Comparing this with Example C of Section 2 of this chapter we recognize $X(t)$ as the Ornstein–Uhlenbeck process.

STOCHASTIC DIFFERENTIAL EQUATIONS

The modeling of physical and biological systems by stochastic differential equations of the form

$$dX(t) = \mu(X(t), t)\, dt + \sigma(X(t), t)\, dB(t) \qquad (14.21)$$

in terms of an input (driving) white noise process has two principal merits. First, it is usually amenable to mathematical analysis. An important fact is that the solution of (14.21) as implicit in the discussion of (14.4), determines a diffusion process with infinitesimal drift $= \mu(x, t)$ and infinitesimal variance $= \sigma^2(x, t)$. Second, although the white noise process is a mathematical abstraction, it well approximates a variety of noise or other background and environmental random input processes that occur naturally in physical and biological contexts. The resistance noise in certain electronic systems often yields a spectral density (see Chapter 9) for the voltage source which is flat for a wide range of frequencies.

The solution of the stochastic differential (14.21) is in terms of a stochastic integral $\int \sigma(X(\tau), \tau)\, dB(\tau)$. There are two prominent versions of stochastic integrals: one called the *Ito integral* (abbreviated the I-integral) and a second introduced by Stratonovich (called the S-integral). The construction and the most pertinent properties of the I-integral are set forth in Section 16. The concept and manipulations of the S-integral are reduced to calculations of related I-integrals. Apart from its value in solving stochastic differential equations, the *Ito indefinite integral generates an important class of martingales* as will be noted later (Section 16). However, the transformation properties of the Ito stochastic process are not concordant with the rules of ordinary calculus, which is perhaps slightly disconcerting. On the other hand, the solution of a stochastic differential equation in terms of the S-integral conforms more closely to the solution of the differential equation along each sample path induced by the random input process. The possibility of solving the stochastic differential equation pathwise is more satisfactory for many physical and biological systems and often more appropriate for the evaluation of data and the objectives of the model. The

S-integral generally differs from the I-integral by a corrective term. The precise relationships of the two integrals will be clarified later in this section.

Another important element in favor of the S-integral is as follows. Many natural discrete time, discrete state physical models are often approximated by stochastic differential systems or diffusion processes. In many cases the limiting process leads to a diffusion identified with the S-integral interpretation of the associated differential equation, rather than the I-integral. These contrasts are amply illustrated in the next section.

THE TRANSFORMATION LAW FOR THE ITO STOCHASTIC DIFFERENTIAL

We now proceed to determine the differential $dY(t)$ for $Y(t) = f(X(t), t)$ (f smooth), where $X(t)$ is a solution of

$$dX(t) = \mu(X(t), t)\, dt + \sigma(X(t), t)\, dB(t) \tag{14.22}$$

in the I sense. We carry out this task in a formal manner. The key to the analysis is the order relation

$$[dB(t)]^2 \approx dt \tag{14.23}$$

whose rationale stems from the formula

$$\lim_{n \to \infty} \sum_{i=0}^{n-1} [B(t_{i+1}^{(n)}) - B(t_i^{(n)})]^2 = t, \tag{14.24}$$

provided

$$0 = t_0^{(n)} < t_1^{(n)} < \cdots < t_n^{(n)} = t$$

and

$$\lim_{n \to \infty} \max_{0 \le i \le n-1} [t_{i+1}^{(n)} - t_i^{(n)}] = 0$$

hold. Relation (14.24) is always valid in the mean square sense. We derived (14.24) in Chapter 7 (Section 7) when $\{t_i^n\}$ was the partition of $[0, 1]$ delineated by the associated dyadic rationals. In that circumstance the convergence prevails for almost all sample path realizations as well as in the mean square sense.

The total differential of $f(X(t), t) = Y(t)$, where f is sufficiently smooth, is evaluated by applying Taylor's expansion:

$$\begin{aligned}
dY(t) = df(X(t), t) = {} & f_x(X(t), t)\, dX(t) + f_t(X(t), t)\, dt \\
& + \tfrac{1}{2} f_{xx}(X(t), t)[dX(t)]^2 + f_{x,t}(X(t), t)\, dX\, dt \\
& + \tfrac{1}{2} f_{tt}(X(t), t)(dt)^2 + \cdots
\end{aligned} \tag{14.25}$$

(abbreviating $f_x = \partial f/\partial x$, $f_t = \partial f/\partial t$, etc.).

Substituting for dX from (14.22) invoking (14.23), and neglecting higher order terms we obtain

$$\begin{aligned}
dY(t) = {} & [f_x(X(t), t)\mu(X(t), t) + f_t(X(t), t) + \tfrac{1}{2}f_{xx}(X(t), t)\sigma^2(X(t), t)]\, dt \\
& + f_x(X(t), t)\sigma(X(t), t)\, dB(t).
\end{aligned} \tag{14.26}$$

In integral form (14.26) is written

$$Y(\tau) - Y(0) = \int_0^\tau [f_x(X(t), t)\mu(X(t), t) + f_t(X(t), t)$$

$$+ \tfrac{1}{2}f_{xx}(X(t), t)\sigma^2(X(t), t)]dt + \int_0^\tau f_x(X(t), t)\sigma(X(t), t)\, dB(t),$$

$$(14.27)$$

referred to as the *Ito stochastic transformation formula*, where the final integral is interpreted as an I-integral.

The extra contribution to dt comes from the factor associated with $[dB(t)]^2$ $= dt$ owing to (14.23). A number of applications of (14.27) will be developed in Sections 15 and 16.

From many physical perspectives, since white noise in reality does not exist, it might be reasonable to suppose $[dB(t)]^2$ negligible compare to dt, and this property is inherent to the Stratonovich integral interpretation of S-$\int \sigma(X(t), t)\, dB(t)$. (The notation S- signifies that the integral is evaluated in the Stratonovich sense. Without S- the Ito definition applies.)

LIMITS OF INTEGRALS WITH RANDOM PROCESSES APPROXIMATING WHITE NOISE (WONG–ZAKAI LIMIT THEOREM)

In practical terms the white noise process is intended to represent a stationary Gaussian process with a spectral density that is flat over a wide range of frequencies. If we take $W(t)$ to be such a process, then there is no difficulty in treating (14.12) as an ordinary differential equation for each realization of the $W(t)$ process, provided the sample paths are well behaved. Consider a sequence of Gaussian processes $B_n(t)$ having piecewise differentiable sample paths. For each n and specified sample path, the equation

$$\frac{dX_n(t)}{dt} = \mu(X_n, t) + \sigma(X_n, t)\frac{dB_n(t)}{dt}, \qquad X_n(0) = x, \qquad (14.28)$$

can in principle be solved by standard means. Suppose now that B_n converges appropriately (as $n \to \infty$) to Brownian motion. If, independent of the approximation of B_n to B, the sequence $X_n(t)$, $0 \le t \le T$, also converges in a suitable sense to a process $X(t)$, $0 \le t \le T$, then we can propose $X(t)$ as a solution of

$$\frac{dX(t)}{dt} = \mu(X, t) + \sigma(X, t)W(t) \qquad \text{for} \qquad 0 \le t \le T,$$

where $W(t)$ is Gaussian white noise.

A natural choice is to take $B_n(t)$ as a polygonal approximation to $B(t)$ defined for each sample path by

$$B_n(t) = B(t_j^{(n)}) + [B(t_{j+1}^{(n)}) - B(t_j^{(n)})] \frac{(t - t_j^{(n)})}{t_{j+1}^{(n)} - t_j^{(n)}} \qquad \text{for} \quad t_j^{(n)} \leq t \leq t_{j+1}^{(n)},$$

(14.29)

where

$$0 = t_0^{(n)} < t_1^{(n)} < \cdots < t_n^{(n)} = T, \qquad \max_j [t_{j+1}^{(n)} - t_j^{(n)}] \to 0 \qquad \text{as} \quad n \to \infty.$$

The integral of (14.28) gives

$$X_n(t) - X_n(0) = \int_0^t \mu(X_n(\tau), \tau) \, d\tau + \int_0^t \sigma(X_n(\tau), \tau) \, dB_n(\tau).$$ (14.30)

Stipulating that $X_n(t)$ converges to $X(t)$ appropriately, and provided $\mu(x, t)$ is sufficiently smooth, the justification of the relation

$$\lim_{n \to \infty} \int_0^t \mu(X_n(\tau), \tau) \, d\tau = \int_0^t \mu(X(\tau), \tau) \, d\tau$$ (14.31)

is indeed forthcoming. On the other hand, the ascertainment of the limit of $\int_0^t \sigma(X_n(\tau), \tau) \, dB_n(\tau)$ poses some difficulties. We relate the limit of (14.30) in terms of appropriate Ito integrals.

We proceed in a heuristic fashion. Define

$$\psi(x, t) = \int_0^x \gamma(\xi, t) \, d\xi$$

($\gamma(\xi, t)$ to be determined explicitly later.) The total differential then becomes

$$\begin{aligned} d\psi(x, t) &= \psi_t(x, t) \, dt + \psi_x(x, t) \, dx \\ &= \psi_t(x, t) \, dt + \gamma(x, t) \, dx \end{aligned}$$ (14.32)

[$\gamma(x, t) = \psi_x(x, t)$ by the fundamental theorem of calculus].

Evaluating at $x = X_n(t)$ and substituting from (14.28) gives

$$\begin{aligned} d\psi(X_n(t), t) &= \psi_t(X_n(t), t) \, dt + \gamma(X_n(t), t) \, dX_n(t) \\ &= [\psi_t(X_n(t), t) + \gamma(X_n(t), t)\mu(X_n(t), t)] \, dt \\ &\quad + \gamma(X_n(t), t)\sigma(X_n(t), t) \, dB_n(t). \end{aligned}$$ (14.33)

Specifying $\gamma(\xi, t) = 1/\sigma(\xi, t)$, the coefficient of dB_n in (14.33) becomes identically 1. Now, integration produces

$$\psi(X_n(t), t) - \psi(X_n(a), a) = \int_a^t f(X_n(\tau), \tau) \, d\tau + B_n(t) - B_n(a)$$ (14.34)

where f is the coefficient of dt in (14.33).

If $X_n(t)$ converges to $X(t)$ and $B_n(t)$ to $B(t)$ appropriately then passing to the limit formally in (14.34) leads to

$$\psi(X(t), t) - \psi(X(a), a) = \int_a^t f(X(\tau), \tau) \, d\tau + B(t) - B(a) \qquad (14.35)$$

for all a and $t > a$. If we assume that $X(t)$ admits the representation

$$X(t) = X(a) + \int_a^t g(X(s), s) \, ds + \int_a^t h(X(s), s) \, dB(s), \qquad (14.36)$$

signifying that X is a solution of the Ito stochastic differential equation

$$dX = g(X(t), t) \, dt + h(X(t), t) \, dB,$$

and apply the stochastic transformation formula (14.27) to $\psi(X(t), t)$, we obtain

$$\psi(X(t), t) - \psi(X(a), a) = \int_a^t \{\psi_x(X(s), s)g(X(s), s) + \psi_t(X(s), s)$$

$$+ \tfrac{1}{2}\psi_{xx}(X(s), s)[h(X(s), s)]^2\} \, ds + \int_a^t \psi_x(X(s), s)h(X(s), s) \, dB(s). \quad (14.37)$$

Comparing (14.37) and (14.35), since $a < t$ is arbitrary we deduce that

$$\psi^x(X(s), s)h(X(s), s) \equiv 1$$

and

$$\psi_x(X(s), s)g(X(s), s) + \psi_t(X(s), s) + \tfrac{1}{2}\psi_{xx}(X(s), s)[h(X(s), s)]^2 = f(X(s), s)$$

$$= \psi_t(X(s), s) + \frac{\mu(X(s), s)}{\sigma(X(s), s)}. \qquad (14.38)$$

The first relation of (14.38) shows that

$$h(X(s), s) = \frac{1}{\psi_x(X(s), s)} = \sigma(X(s), s), \qquad (14.39)$$

the last equation resulting from the choice of $\psi_x(\xi, t) = \gamma(\xi, t) = 1/\sigma(\xi, t)$. The second equation of (14.38) reduces to

$$\psi_x(X(s), s)g(X(s), s) + \tfrac{1}{2}\psi_{xx}(X(s), s)[h(X(s), s)]^2$$

$$= \frac{\mu(X(s), s)}{\sigma(X(s), s)} = \mu(X(s), s)\psi_x(X(s), s).$$

Therefore

$$g(X(s), s) = \mu(X(s), s) - \tfrac{1}{2}\psi_{xx}(X(s), s)[h(X(s), s)]^3$$

$$= \mu(X(s), s) + \tfrac{1}{2}\sigma_x(X(s), s)\sigma(X(s), s) \qquad (14.40)$$

since

$$\psi_x(\xi, s) = \frac{1}{\sigma(\xi, s)} \quad \text{and} \quad \psi_{xx}(\xi, s) = -\frac{\sigma_x(\xi, s)}{[\sigma(\xi, s)]^2}.$$

Thus, by interpreting formally the solution of

$$dX(t) = \mu(X(t), t)\, dt + \sigma(X(t), t)\, dB(t) \tag{14.41}$$

as a limit of solutions of a sequence of equations of the type (14.28), we find that the solution agrees with the Ito solution of the modified stochastic differential equation

$$dX(t) = [\mu(X(t), t) + \tfrac{1}{2}\sigma_x(X(t), t)\sigma(X(t), t)]\, dt + \sigma(X(t), t)\, dB(t) \tag{14.42}$$

involving the *correction term* $\tfrac{1}{2}\sigma_x\sigma$ *contributing to the infinitesimal drift coefficient.* The solution obtained in this manner is referred to as the Wong–Zakai solution of (14.41). The Wong–Zakai solution of (14.41) also coincides with the S-solution; that is, with the process $X(t)$ satisfying

$$X(t) - X(0) = \int_0^t \mu(X(s), s)\, ds + \text{S-}\int_0^t \sigma(X(s), s)\, dB(s),$$

where the stochastic integral is the Stratonovich integral, as symbolized by writing S-$\int$. [See (14.50) which follows.] An appealing facet to the Wong–Zakai approach or the S-solution is that the transformation rule for differentials in (14.41) is now concordant with classical differential calculus.

Observe that where $\sigma(x, t)$ is independent of x and therefore $\sigma_x \equiv 0$ the S-solution and I-solution of (14.41) coincide.

A special case similar to (14.28) concerns the limit of X_n satisfying

$$dX_n = \sigma(B_n(t), t)\, dB_n,$$

where $B_n(t)$ is a sequence of smooth processes converging appropriately to $B(t)$. Equivalently, we would like to evaluate

$$\lim_{n \to \infty} \int_a^t \sigma(B_n(\tau), \tau)\, dB_n. \tag{14.43}$$

It suffices to assume that $B_n(t)$ has a continuous derivative in t.

The computation of (14.43) is analogous to that of (14.30), leading to

$$\lim_{n \to \infty} \int_a^t \sigma(B_n, \tau)\, dB_n(\tau) = \int_a^t \sigma(B(\tau), \tau)\, dB(\tau) + \tfrac{1}{2}\int_a^t \sigma_x(B(\tau), \tau)\, d\tau \tag{14.44}$$

involving a corrective term in the second integral on the right as in (14.42). Since the integral on the left in (14.44) coincides with the Stratonovich integral, we have

$$\text{S-}\int_a^t \sigma(B(\tau), \tau)\, dB(\tau) = \int_a^t \sigma(B(\tau), \tau)\, dB(\tau) + \tfrac{1}{2}\int_a^t \sigma_x(B(\tau), \tau)\, d\tau.$$

We summarize the analysis leading to (14.42) by highlighting the relationship of the S- and I-solutions of the stochastic differential equation

$$dX = \mu(X(t), t) \, dt + \sigma(X(t), t) \, dB. \tag{14.45}$$

Provided the coefficients μ and σ are sufficiently smooth, the Ito solution of (14.45) is a diffusion process having infinitesimal coefficients $\mu(x, t)$ and $\sigma^2(x, t)$. The S-solution of (14.45) can be obtained as the I-solution of the modified differential equation

$$dX = [\mu(X, t) + \tfrac{1}{2}\sigma_x(X, t)\sigma(X, t)] \, dt + \sigma(X, t) \, dB. \tag{14.46}$$

In view of the facts connected with (14.45) and (14.46), the S-solution is again a diffusion having the same variance coefficient while the drift coefficient is altered to

$$\mu(x, t) + \tfrac{1}{2}\sigma_x(x, t)\sigma(x, t).$$

The behavior of the two solutions can be markedly different. For much modeling of physical systems the S-solution interpretation appears to be more natural, while in biological contexts both perspectives can apply depending on the discrete approximation structure.

SOME FORMALITIES ON STOCHASTIC INTEGRALS

The solution of the stochastic differential equation (14.45) involves stochastic integrals of the form

$$\int \varphi(t, \omega) \, dB(t, \omega) = \int \varphi(t) \, dB(t), \tag{14.47}$$

where ω denotes a sample path and $\varphi(t, \omega) = \varphi(t)$ is a stochastic process with suitable smoothness properties, but $B(t)$ is not of bounded variation, and therefore the integral cannot be constructed in any standard sense. Nevertheless, it is possible to define (14.47) by the limiting process

$$\mathop{\text{l.i.m}}_{n \to \infty, \max_i |t_{i+1}^{(n)} - t_i^{(n)}| \to 0} \sum_{i=0}^{n-1} \varphi(t_i^{(n)})[B(t_{i+1}^{(n)}) - B(t_i^{(n)})] \tag{14.48}$$

(l.i.m signifies limit in mean square). In the sum (14.48), it is vital for the Ito definition of the integral to take the value of $\varphi(t)$ over $t_i^{(n)} \le t \le t_{i+1}^{(n)}$ at the smallest allowable t, namely, $t_i^{(n)}$.

The function $\varphi(t, \omega)$ is assumed to be measurable $\mathcal{F}_t = \sigma(B_\tau, \tau \le t)$ [that is, $\varphi(t, \omega)$ is determined completely by the realization of the process $B(\tau)$ up to time t] and provided $\int_0^\infty E[\varphi(t)^2] \, dt < \infty$ holds (a natural growth restriction), it can be proved that (14.48) exists. A remarkable result is that *the indefinite integral* $X(t) = X(t, \omega) = \int_0^t \varphi(\tau) \, dB(\tau)$ *determines a martingale adapted to the σ-fields* $\{\mathcal{F}_t\}$.

A symmetrical style integral has been proposed by Stratonovich in which the approximating sum is put in the symmetric form

$$\sum_{i=1}^{n-1} \tfrac{1}{2}[\varphi(t_i^{(n)}) + \varphi(t_{i+1}^{(n)})][B(t_{i+1}^{(n)}) - B(t_i^{(n)})]. \tag{14.49}$$

Under more restrictive conditions on $\varphi(t, \omega)$ than required in (14.48), the approximating sums of (14.49) converge to the Stratonovich integral

$$\text{S-}\int \phi(\tau) \, dB(\tau). \tag{14.50}$$

The indefinite integral based on (14.50) no longer generates a martingale but inherits more appealing transformation properties. The main one is that if $X(t)$ solves

$$dX(t) = \mu(X(t), t) \, dt + \sigma(X(t), t) \, dB(t)$$

entailing the S-integral, then $Y(t) = f(X(t), t)$ satisfies the standard differential relation

$$dY(t) = f_t(X(t), t) \, dt + f_x(X(t), t) \, dX(t), \tag{14.51}$$

where all differentials and integrals are now interpreted in the S sense.

It is illuminating to compare the evaluation of the I-integral and the S-integral for the function $\varphi(t) = B(t)$.

Observe for a partition $\{t_i^{(n)}\}$ of $[0, t]$

$$J_n = \sum_{i=0}^{n-1} \tfrac{1}{2}[B(t_{i+1}^{(n)}) + B(t_i^{(n)})][B(t_{i+1}^{(n)}) - B(t_i^{(n)})]$$

$$= \tfrac{1}{2} \sum_{i=0}^{n-1} \{[B(t_{i+1}^{(n)})]^2 - [B(t_i^{(n)})]^2\}$$

$$= \tfrac{1}{2}\{[B(t)]^2 - [B(0)]^2\} = \tfrac{1}{2}[B(t)]^2 \tag{14.52}$$

independent of the choice of $\{t_i^{(n)}\}$.

We wish next to evaluate

$$\lim_{n \to \infty} I_n = \lim_{n \to \infty} \sum_{i=0}^{n-1} B(t_i^{(n)})[B(t_{i+1}^{(n)}) - B(t_i^{(n)})].$$

To this end, we note the elementary identity

$$2J_n - \sum_{i=0}^{n-1} [B(t_{i+1}^{(n)}) - B(t_i^{(n)})]^2 = 2I_n. \tag{14.53}$$

We pointed out earlier, see (14.24), that

$$\sum_{i=0}^{n-1} [B(t_{i+1}^{(n)}) - B(t_i^{(n)})]^2 \to t \qquad \text{as} \qquad n \to \infty \tag{14.54}$$

(cf. Chapter 7). The facts of (14.52) and (14.54) together clearly establish by means of (14.53) that

$$I_n \quad \text{converges to} \quad \tfrac{1}{2}[B(t)]^2 - \tfrac{1}{2}t.$$

More generally we have, reminiscent of (14.44),

$$\text{S-} \int_0^t \varphi(B(\tau), \tau) \, dB(\tau) = \int_0^t \varphi(B(\tau), \tau) \, dB(\tau) + \tfrac{1}{2} \int_0^t \varphi_x(B(\tau), \tau) \, d\tau, \quad (14.55)$$

which we have just verified for $\varphi(t, \omega) = B(t)$.

Example. The Growth Equation. We shall solve the equation

$$dX = X \, dB \tag{14.56}$$

in the two senses of the S-integral and the I-integral. To this end we observe in accordance with the discussion of (14.46) that the S-solution of (14.56) is acquired as the Ito solution of

$$dX = \tfrac{1}{2}X \, dt + X \, dB. \tag{14.57}$$

In order to solve this equation we appeal to the transformation formula (14.27). Consider

$$f(x, t) = e^{x - t/2}. \tag{14.58}$$

By virtue of (14.26) and the trivial statement $dB = dB$ we find for this f that $f(B(t), t) = X(t)$ satisfies

$$dX(t) = (f_t + \tfrac{1}{2}f_{xx}) \, dt + f_x \, dB(t) = X(t) \, dB(t), \tag{14.59}$$

which is precisely (14.56). Because the solution of (14.56) [with initial condition $X(0) = 1$] is unique (Section 16), we can conclude that

$$X(t) = e^{B(t) - t/2}$$

is the required solution when interpreted as an Ito stochastic differential equation.

On the other hand

$$Y(t) = e^{B(t)} = g(B(t)) \tag{14.60}$$

satisfies (by (14.26)) the Ito equation

$$dY(t) = \tfrac{1}{2}Y(t) \, dt + Y(t) \, dB(t)$$

and also (compare to (14.57)) the S-equation

$$dY(t) = Y(t) \, dB(t) \quad \text{(S).} \tag{14.61}$$

The meaning of the S-differential equation in (14.61) is found in its integrated form

$$Y(t) - Y(0) = \text{S-}\int_0^t Y(\tau)\,dB(\tau).$$

Another approach to (14.56) proceeding by means of ordinary calculus yields

$$\frac{dY}{Y} = dB,$$

whose solution pathwise is

$$\log \frac{Y(t)}{Y(0)} = B(t)$$

or

$$Y(t) = e^{B(t)} \qquad \text{(for } Y(0) = 1),$$

which agrees with the S-solution of (14.56).

QUALITATIVE SUMMARY AND DISCUSSION OF DISCRETE AND CONTINUOUS STOCHASTIC MODELING

Many phenomena in control schemes, biology, economics, engineering systems, physics, and other areas can be modeled by differential equations with stochastic perturbation terms. (Section 15 elaborates a number of concrete examples.) The stochastic differential equation

$$\frac{dX(t)}{dt} = \varphi(X(t), t) + \psi(X(t), t)\frac{dB(t)}{dt}$$

$$\left(\frac{dB}{dt} = W(t), \text{ the white noise process}\right), \qquad (14.62)$$

or in differential notation

$$dX(t) = \varphi(X(t), t)\,dt + \psi(X(t), t)\,dB(t), \qquad (14.63)$$

can be interpreted in a number of ways which are *not* all equivalent.

A difficulty attendant to (14.62) already underscored in our earlier discussion is that the white noise process $dB(t)/dt = W(t)$ is not well defined. Three approaches have been principally set forth in seeking to resolve stochastic differential equations.

(I) Constructing a refined series of discretized versions of (14.63) and proposing the limit process where well defined to represent the solution of (14.63). There can be inconsistencies between diverse discretizations which lead to different limit processes. The interpretation may then depend on the relevance of a specific approximation procedure and the solution is accordingly model dependent. This consideration can be germane to the implications and interpretations of the analyses.

(II)   Method I suggests approximating the differential equations by discrete time systems, in this way circumventing the ambiguity in the meaning of $W(t)$. An alternative approach is to approximate $W(t)$ directly, replacing $dB/dt = W(t)$ by a sequence of processes $W_n(t)$ converging to $W(t)$ in some sense where all W_n are sufficiently smooth, permitting the solution of

$$\frac{dX_n(t)}{dt} = \varphi(X_n(t), t) + \psi(X_n(t), t)W_n(t) \tag{14.64}$$

along each sample path by the standard techniques of differential equations. When the processes $X_n(t)$ converge [say to $X(t)$], then $X(t)$ can be referred to as the solution of (14.63). Unfortunately in some cases, depending on the co-efficients $\varphi(t, x)$ and $\psi(t, x)$, the limit process $X(t)$ could again depend on the nature of the approximation $W_n(t)$.

(III)   A direct mathematical approach converts the differential equation (14.63) into the integral equation

$$X(t) = X(0) + \int_0^t \varphi(X(\tau), \tau) \, d\tau + \int_0^t \psi(X(\tau), \tau) \, dB(\tau), \tag{14.65}$$

where the last integral is interpreted in the Ito sense. Subject to reasonable growth restrictions and smoothness conditions (see Section 16) it can be established that (14.65) admits a unique solution on a prescribed time segment $0 \leq t \leq T$ where the initial condition $X(0)$ is specified. The solution of (14.63), or equivalently (14.65), obtained by these means is called the *Ito solution*.

(IV)   If the integral in (14.65) is taken in the Stratonovich sense the solution is achieved by incorporating into (14.63) an additional term in the dt part and then ascertaining its Ito solution. The modified equation often provides a process with behavior significantly different than the Ito solution of the original equation that lacks the corrective term.

A great advantage in the use of continuous stochastic differential equations versus discrete models in describing certain physical, engineering, biological, or economic processes is that explicit answers are frequently accessible in the continuous formulations. The dependence and sensitivity of the process on the parameters are therefore more easily discernible and interpretable. The process realizations (or expectation, variance, or distributional quantities) for the corresponding discrete models rarely admit explicit representations and so their qualitative discussion is more formidable. On the other hand, for compu-tational, numerical or simulation objectives, the benefits of a good discrete formulation are clear.

We now discuss each of these approaches more specifically. The technique extends the methods of standard differential equations to a framework entailing random coefficients and perturbation terms.

Consider first some aspects of approach I.

(i) A classical numerical technique used in dealing with ordinary differential equations is to construct the *Cauchy polygonal lines*. More specifically, consider a sequence of partitions $\Pi_n = \{t_i^{(n)}\}$,

$$0 = t_0^{(n)} < t_1^{(n)} < \cdots < t_n^{(n)} = T \tag{14.66}$$

with maximal mesh size

$$\delta_n = \max_{0 \le i \le n-1} (t_{i+1}^{(n)} - t_i^{(n)}).$$

In the case of (14.63) Khasminski proposed to define the discrete approximation at the grid points recursively by the prescription

$$
\begin{aligned}
X^{(n)}(t_{k+1}^{(n)}) &= X_{k+1}^{(n)} \\
&= X_k^{(n)} + \varphi(X_k^{(n)}, t_k^{(n)})(t_{k+1}^{(n)} - t_k^{(n)}) \\
&\quad + \psi(X_k^{(n)}, t_k^{(n)})[B(t_{k+1}^{(n)}) - B(t_k^{(n)})], \qquad k = 0, 1, 2, \ldots, n - 1,
\end{aligned}
\tag{14.67}
$$

with $X_0^{(n)} = X^{(n)}(0)$ fixed and $X^{(n)}(t)$ extended "linearly" between the partition points in the manner

$$
\begin{aligned}
X^{(n)}(t) &= X^{(n)}(t_k^{(n)}) + \varphi(X_k^{(n)}, t_k^{(n)})(t - t_k^{(n)}) \\
&\quad + \psi(X_k^{(n)}, t_k^{(n)})[B(t) - B(t_k^{(n)})], \quad t_k^{(n)} \le t < t_{k+1}^{(n)}.
\end{aligned}
\tag{14.68}
$$

This is not a linear extension since the adjunct $B(t) - B(t_k^{(n)})$ is obviously nonlinear in t over $t_k^{(n)} \le t < t_{k+1}^{(n)}$.

Subject to suitable smoothness requirements of $\varphi(x, t)$ and $\psi(x, t)$ it can be proved that as $\delta_n \to 0$, then

$$X^{(n)}(t) \text{ converges appropriately to the Ito solution } X(t) \text{ of (14.63).} \tag{14.69}$$

(ii) A natural smoothening of the white noise process is to modify the Brownian trajectories by polygonal approximations at the sequence of grids (14.66). More specifically, we define the polygonal sample path

$$B_n(t) = B(t_k^{(n)}) + \frac{B(t_{k+1}^{(n)}) - B(t_k^{(n)})}{t_{k+1}^{(n)} - t_k^{(n)}} (t - t_k^{(n)}), \qquad t_k^{(n)} \le t \le t_{k+1}^{(n)}, \tag{14.70}$$

or

$$W_n(t) = \frac{dB_n(t)}{dt} = \frac{B(t_{k+1}^{(n)}) - B(t_k^{(n)})}{t_{k+1}^{(n)} - t_k^{(n)}} \qquad \text{on} \qquad t_k^{(n)} < t < t_{k+1}^{(n)}. \tag{14.71}$$

The dynamic equation analog of (14.70) with $W_n(t)$ replacing dB/dt is

$$\frac{dX(t)}{dt} = \varphi(X(t), t) + \psi(X(t), t)W_n(t); \tag{14.72}$$

which is well defined except at the discrete set $\{t_k^{(n)}\}$ between which $W_n(t) = W_n(t, \omega)$ is a random piecewise constant trajectory. Subject to standard regularity stipulations on $\varphi(x, t)$ and $\psi(x, t)$ and treating (14.72) as an ordinary differential equation for each sample realization, we solve this equation uniquely producing the process $X_n(t) = X_n(t, \omega)$. In the presence of further smoothness restrictions on φ and ψ, the Wong–Zakai construction underlying (14.72) allows the existence of a limit process $\tilde{X}(t)$ such that $\lim_{n \to \infty} E[\{X_n(t) - \tilde{X}(t)\}^2] = 0$, and the probability limit

$$\Pr\left\{\tilde{X}(t) = \lim_{n \to \infty} X_n(t) \text{ uniformly } 0 \le t \le T\right\} = 1$$

holds along an appropriate subsequence.

The above limit process $\tilde{X}(t)$ coincides with the Stratonovich solution of (14.63). Generally but not always a limiting process associated with a smoothened process (where one shifts from white noise to a smooth Gaussian process) leads to a Stratonovich interpretation after evaluations through ordinary integrals.

The Stratonovich integral and the Stratonovich differential satisfy most of the conventional rules of the integral and differential calculus with respect to transformation formula and chain rule, the identity attendant to integration by parts, etc. This affords easy manipulation of the Stratonovich integral using familiar operations. A drawback of the Stratonovich integral contrasted to that of the Ito construct is that we lose the martingale property for the indefinite integral process.

Computations of expectations and moments are much easier with the Ito integral than with the Stratonovich integral. Generally, most theoretical work is more conveniently done in the Ito framework, exploiting heavily the martingale endowments of the Ito integral. Also, the Ito integral is defined for a significantly wider class of functions than that of the Stratonovich integral. In addition, the restrictions on the coefficients of the Stratonovich stochastic differential equation compared to the corresponding Ito stochastic differential equation are much more severe. In many natural cases (e.g., certain nonlinear filtering theory) a solution exists in the Ito setting but not for the associated Stratonovich stochastic differential equation.

We have described in (14.45) and (14.46) the transformations from Stratonovich to Ito for the stochastic integral and the conversions between the associated stochastic differential equations.

15: Some Stochastic Differential Equation Models

We shall describe two classes of examples, the first based on population ecological and genetic systems, the second involving economic structures and examples of stochastic control.

A. POPULATION GROWTH IN A RANDOM ENVIRONMENT

There is considerable interest in stochastic analogs of classical difference and differential equations describing phenomena in theoretical ecology and population genetics. In population genetics a number of recent studies have focused on models involving selection coefficients which vary systematically or randomly in time (cf. Case (g) of Section 2) in conjunction with random sampling effects that reflecting small population size fluctuations. Most of the ecological models incorporate randomness into their formulation by starting with the population growth model

$$\frac{dN}{dt} = f(N, t) + g(N, t)W(t), \qquad W(t) = \frac{dB(t)}{dt}, \tag{15.1}$$

where $N(t)$ is the population size at time t and $W(t)$ is the Gaussian white noise process. The simplest concrete case considers the "exponential" growth equation

$$\frac{dN}{dt} = r(t)N(t), \tag{15.2}$$

where $r(t)$ represents an instantaneous rate of growth at time t. To account for random environmental effects, it is stipulated that

$$r(t) = \alpha + \gamma W(t),$$

with $W(t)$ as in (15.1) and α and γ constants.

The stochastic differential equation (15.2) becomes

$$\frac{dN(t)}{dt} = N(t)\left(\alpha + \gamma \frac{dB(t)}{dt}\right)$$

$$\tag{15.3}$$

or in differential symbols,

$$dN(t) = N(t)(\alpha \, dt + \gamma \, dB(t)) = \alpha N(t) \, dt + \gamma N(t) \, dB(t).$$

The solution of (15.3) in terms of the I-integral produces a diffusion process on $(0, \infty)$ characterized by the drift and variance coefficients

$$\mu(x) = \alpha x \quad \text{and} \quad \sigma^2(x) = \gamma^2 x^2 \qquad \text{for} \quad 0 < x < \infty, \tag{15.4}$$

respectively, as can be seen formally from (14.21).

Recall from Example D, Section 2, that the process defined by $Y(t) = e^{\mu t + \sigma B(t)}$ is geometric Brownian motion, i.e., the diffusion process on $I = (0, \infty)$ with infinitesimal parameters $\mu(y) = (\mu + \frac{1}{2}\sigma^2)y$ and $\sigma^2(y) = \sigma^2 y^2$. Comparing this with (15.4) we see that the solution to $dN = \alpha N \, dt + \gamma N \, dB$ in the Ito sense is the geometric Brownian motion $N(t) = \exp[(\alpha - \frac{1}{2}\gamma^2)t + \gamma B(t)]$.

In another approach, because $N(t)$ factors on the right, we can solve along each sample path by writing (15.3) in the form

$$\frac{dN}{N} = \alpha \, dt + \gamma \, dB(t)$$

and then integrating both sides to get

$$\log \frac{N(t)}{N(0)} = \alpha t + \gamma B(t)$$

or

$$N(t) = N(0)e^{\alpha t + \gamma B(t)}. \tag{15.5}$$

This approach leads to a different geometric Brownian motion. The infinitesimal mean and variance, respectively, are computed as above to be

$$\mu(x) = (\alpha + \tfrac{1}{2}\gamma^2)x \quad \text{and} \quad \sigma^2(x) = \gamma^2 x^2, \qquad 0 < x < \infty. \tag{15.6}$$

In comparing (15.6) with (15.4) notice that the drift term entails a contribution from the white noise process. In fact, the coefficients are exactly those obtained by interpreting (15.3) in the Stratonovich sense.

The sample path behavior of the processes with diffusion coefficients (15.4) and (15.6) can differ significantly. In particular, if $0 < \alpha < \tfrac{1}{2}\gamma^2$, then following (15.4), $N(t) \to 0$ as $t \to \infty$, while following (15.6), $N(t) \to \infty$. The relevance of the method of solution of (15.3) is paramount in the interpretation.

The Stochastic Logistic Equation

One form of a stochastic logistic equation is

$$\frac{dN(t)}{dt} = N(t)[k(t) - N(t)] \quad \text{where} \quad k(t) = \alpha + \gamma W(t) = \alpha + \gamma \frac{dB(t)}{dt}. \tag{15.7}$$

Here we interpret $\alpha + \gamma W(t)$ as a stochastic analog of the carrying capacity of an ecological habitat, usually considered constant.

Another version sets

$$\frac{dN}{dt} = r(t)N - \beta(t)N^2$$

where $\beta(t)$ measures the individual effects on reproduction by survival of others. For $r(t) = k(t)$ and $\beta(t) = 1$, we recover the model of (15.7). In terms of the formulation (15.7) the intrinsic rate of increase r has been absorbed into the time scale.

The solution of (15.7) by means of the I-integral, is a diffusion on $(0, \infty)$ with infinitesimal coefficients

$$\mu(x) = x(\alpha - x) \quad \text{and} \quad \sigma^2(x) = \gamma^2 x^2. \tag{15.8}$$

The S-solution of (15.7) has the diffusion parameters

$$\mu(x) = x(\alpha + \tfrac{1}{2}\gamma^2 - x) \quad \text{and} \quad \sigma^2(x) = \gamma^2 x^2. \tag{15.9}$$

Differences in the behavior of the sample path of the two processes (15.8) as against (15.9) are significant. For example, in the case (15.9) by the criteria of Sections 5 and 6 we find a stationary probability density (a gamma distribution) always exists having expected population size $E[N(\infty)] = \alpha$ and $\text{var}[N(\infty)] = \frac{1}{2}\alpha\gamma^2$, while the existence of a stationary density for the process associated with (15.8) requires $2\alpha/\gamma^2 > 1$. The classification of the boundary point 0 (yielding criteria for and the nature of population extinction) also differs between the two models.

B. GENE FREQUENCY FLUCTUATION

Consider a large population composed of two types A_1 and A_2 and let $p(t)$ be the fraction of type A_1 at time t. If the relative intrinsic selective advantage of type A_1 over A_2 is the ratio $1 + s$ to 1, then a classical deterministic equation describing the changes in gene frequency has the form

$$\frac{dp}{dt} = sp(1 - p). \tag{15.10}$$

The rational for (15.10) is based on the following discrete time model. Let p_n be the frequency of A_1 in the nth generation, where generations are spaced Δt time units apart. As a result of natural selection forces, the relative proportions of A_1 to A_2 in the $(n + 1)$st generation are $(1 + s\,\Delta t)p_n$ to $1 - p_n$. Converting back to frequencies, we obtain

$$p_{n+1} = \frac{(1 + s\,\Delta t)p_n}{(1 + s\,\Delta t)p_n + (1 - p_n)} = \frac{p_n(1 + s\,\Delta t)}{1 + p_n s\,\Delta t}. \tag{15.11}$$

If Δt is small, allowing us to neglect orders of $(\Delta t)^2$, we obtain

$$\Delta p = p_{n+1} - p_n = \frac{(1 + s\,\Delta t)p_n}{(1 + s\,\Delta t)p_n + 1 - p_n} - p_n = \frac{p_n(1 - p_n)s\,\Delta t}{1 + p_n s\,\Delta t}$$

$$\simeq p_n(1 - p_n)s\,\Delta t. \tag{15.12}$$

Dividing by Δt and letting Δt shrink to zero leads to the differential equation of (15.10).

If the selection differential $s\,dt$ is assumed to be a random process, then (15.10) can be regarded as a stochastic differential equation

$$dp(t) = p(t)(1 - p(t))\,d\tilde{B}(t) \tag{15.13}$$

where $s\,dt$ is replaced by the white noise differential $d\tilde{B}(t)$ whose expected displacement and variance over the duration t to $t + dt$ are, respectively,

$$E[d\tilde{B}(t)] = a\,dt \quad \text{and} \quad E[(d\tilde{B}(t))^2] = v^2\,dt. \tag{15.14}$$

The solution $p(t)$ of (15.13), where $d\tilde{B}(t)/dt$ is white noise satisfying (15.14), conforms in the Ito sense to a diffusion process with coefficients

$$\mu(x) = ax(1 - x) \quad \text{and} \quad \sigma^2(x) = v^2 x^2 (1 - x)^2 \quad \text{for} \quad 0 < x < 1.$$
(15.15)

Consider the zero drift case wherein $a = 0$. Then the points 0 and 1 of the diffusion engendered by (15.15) are attracting but unattainable boundaries, and therefore cannot be reached in finite time. Moreover, the realizations of the frequency process exhibit persistent oscillations such that there is substantial time spent in the vicinity of 0 and 1. It should be emphasized that almost all sample paths of the process oscillate back and forth between 0 and 1 infinitely often.

The preceeding analysis derived a deterministic differential equation and then formed its stochastic analog. To consider another approach, we write (15.12) in the form

$$\Delta p(t) = \frac{p(t)(1 - p(t)) \Delta S(t)}{1 + p(t) \Delta S(t)} = p(t)(1 - p(t)) \Delta S(t)[1 - p(t) \Delta S(t) - \cdots],$$

where the selection process $\{S(t)\}$ is a diffusion process satisfying

$$E[\Delta S] = a \Delta t + o(\Delta t) \quad \text{and} \quad E[(\Delta S)^2] = v^2 \Delta t + o(\Delta t).$$

Because the term involving $(\Delta S)^2$ is not negligible, we calculate

$$E[\Delta p] = p(1 - p)(a - v^2 p) \Delta t \quad \text{and} \quad E[(\Delta p)^2] = p^2 (1 - p)^2 v^2 \Delta t.$$

The gene frequency process now is described by a diffusion process whose coefficients are

$$\mu(x) = x(1 - x)(a - v^2 x) \quad \text{and} \quad \sigma^2(x) = v^2 x^2 (1 - x)^2 \quad \text{for} \quad 0 < x < 1.$$
(15.16)

Notice that the variance coefficient v^2 now contributes to the drift term for the gene frequency process.

SOME STOCHASTIC DIFFERENTIAL MODELS OF ECONOMIC PROCESSES

During this past decade there has been increasing effort to describe various facets of dynamic economic interactions with the help of stochastic differential processes. There are two aspects to this approach. Traditional mathematical economics modeling focuses on transient and equilibrium interrelationships among production and consumption factors. Stochastic differential processes provide a mechanism to incorporate the influences associated with randomness, uncertainties, and risk factors operating with respect to various economic units (stock prices, labor force, technology variables, etc.). Secondly, modeling the stochastic fluctuations in terms of continuous stochastic differential equations (rather than in discrete time) often lends itself to explicit analysis with concomitantly easier interpretation and screening for sensitivity to the parameters.

Stochastic differential equation processes have been introduced in the study of three principal categories of economic phenomena: (a) *description of growth of certain factors under uncertainty*, (b) *the nature of option price variations preserving certain market conditions*, and (c) *stochastic dynamic programming and control objectives*.

We shall present one or two typical examples for each category.

C. PRODUCTION AND CONSUMPTION VARIATION UNDER UNCERTAINTY IN A ONE-SECTOR ECONOMY

The variables of the model are as follows.

(a) $K(t)$ is the capital assets at time t.

(b) $L(t)$ is the labor force available at time t and is assumed proportional to population size.

(c) $C(t)$ is the consumption rate at time t.

A production function $F(K, L)$ is postulated obeying appropriate conditions. In many situations a set of natural requirements is that F is concave and homogeneous with respect to K and L. The example $F(K, L) = K^\alpha L^{1-\alpha}$, $0 < \alpha < 1$, is a classical choice known as the Cobb–Douglas production function. The capital goods accumulation equation is expressed as

$$\frac{dK(t)}{dt} = F(K(t), L(t)) - \lambda K(t) - C(t), \qquad (15.17)$$

where λ is the rate of depreciation of capital (assumed to be nonnegative and constant) and $C(t)$ is the aggregate rate of consumption. Relation (15.17) is locally certain and dK/dt is instantaneously determined when K, L, and C are known.

In some situations it is natural to postulate that population size, or equivalently $L(t)$, is subject to random perturbations. Branching and certain related limiting diffusion processes are the main processes used to describe the changes in $L(t)$. The most common model of growth [cf. (15.3)] stipulates that L satisfies the stochastic differential equation

$$dL(t) = L(t)[\alpha \, dt + \sigma \, dB(t)] \qquad (15.18)$$

or, in accordance with (15.5),

$$L(t)/L(0) = e^{(\alpha - \sigma^2/2)t + \sigma B(t)} \qquad (15.19)$$

is geometric Brownian motion with an exponential drift. As developed in Chapter 7, Section 4, $L(t)/L(0)$ follows a log-normal distribution with moments

$$E\left[\log \frac{L(t)}{L(0)}\right] = \left(\alpha - \frac{\sigma^2}{2}\right)t \quad \text{and} \quad \text{Var}\left[\log \frac{L(t)}{L(0)}\right] = \sigma^2 t. \quad (15.20)$$

With the dynamics of labor size established according to (15.19) we can work out the dynamics for the capital accumulation. To this end, it is judicious to work in terms of the per capita variables. Accordingly, consider

$$k(t) = \frac{K(t)}{L(t)} = \text{capital–labor ratio,}$$

$$c(t) = \frac{C(t)}{L(t)} = \text{per capita consumption.}$$

(15.21)

Because F is homogeneous (of order 1), we have $F(\lambda K, \lambda L) = \lambda F(K, L)$, and we may introduce the per capita output function

$$f(k) = F(k, 1) = \frac{F(K, L)}{L}, \qquad k = K/L. \tag{15.22}$$

We further introduce the variable

$$s(t) = 1 - \frac{c(t)}{f(k(t))} \qquad \text{(savings per unit output)}$$

in which we allow $s(t)$ to be positive (real savings) or negative (debts). In order to achieve a one-dimensional diffusion we shall assume in this model that $s(t) \equiv s$ is a constant independent of the rate of output. Equivalently, the gross consumption rate per capita is a fixed fraction of the gross capital output.

Consider now the function of the two variables L and K

$$G = G(L, K) = \frac{K}{L} = k, \tag{15.23}$$

where $K(t)$ is the function obeying locally the deterministic equation (15.17) when L and C are known. Where L satisfies the stochastic differential equation (15.18), the Ito transformation formula (14.26) gives

$$dG(t) = (G_t + G_L L\alpha + \tfrac{1}{2}G_{LL} L^2\sigma^2) \, dt + G_L L\sigma \, dB(t). \tag{15.24}$$

(Of course, $G_t = \partial G/\partial t$, $G_L = \partial G/\partial L$, $G_{LL} = \partial^2 G/\partial L^2$.)

A direct calculation gives

$$G_t = \frac{dK(t)/dt}{L}$$

$$= \frac{F(K, L)}{L} - \lambda\frac{K}{L} - \frac{C}{L} \qquad \text{by (15.17)}$$

$$= f(k(t)) - \lambda k(t) - c(t) \qquad \text{[see (15.21)]}$$

$$= s(t)f(k(t)) - \lambda k(t)$$

$$= sf(k(t)) - \lambda k(t)$$

by the assumption that $s(t) = s$ is constant. From the definition

$$G_L L = -\frac{K}{L^2} L = -k.$$

Observe also that

$$\tfrac{1}{2}G_{LL} = \frac{K}{L^3} = \frac{k(t)}{L^2}.$$

Combining the preceding into (15.24) we get the diffusion

$$dk = [sf(k) - \lambda k]\, dt - \alpha k\, dt + k\sigma^2\, dt - \sigma k\, dB.$$

Thus, the diffusion governing $k(t)$ has infinitesimal parameters

$$\mu(k) = sf(k) - (\lambda + \alpha - \sigma^2)k \quad \text{and} \quad \sigma^2(k) = \sigma^2 k^2. \quad (15.25)$$

D. MODEL OF OPTION (WARRANT) PRICES

Assume the stock prices vary stochastically in accordance with the stochastic differential equation

$$dS = S[\mu\, dt + \sigma\, dB(t)]. \quad (15.26)$$

The warrant price (the price of the right to buy a stock over an appropriate time horizon) is assumed to be a function of the form

$$W = F(S, \tau), \quad (15.27)$$

where S is the current stock price and τ is the current expiration time in which to exercise the option. *The policy sought is to determine the function $F(S, \tau)$ in a manner so as to eliminate any uncertainty (risk) of the value of the option price.* The total investment I is assumed to be allocated among warrants and stocks as

$$I = F(S, \tau) + \alpha S, \quad (15.28)$$

where α units of stock is equivalent to 1 unit of a warrant. Applying the Ito transformation formula to (15.27), we have (since $d\tau = -dt$, as time is measured back from the instant that the option expires)

$$dW = (F_S \mu S - F_t + \tfrac{1}{2}\sigma^2 S^2 F_{SS})\, dt + F_S \sigma S\, dB.$$

From (15.26) we have

$$\alpha\, dS = \alpha S \mu\, dt + \alpha S \sigma\, dB. \quad (15.29)$$

The total rate of investment return (15.28) behaves according to

$$dI = dW + \alpha\, dS$$
$$= [(\alpha + F_S)\mu S - F_t + \tfrac{1}{2}\sigma^2 S^2 F_{SS}]\, dt + (\alpha + F_S)\sigma S\, dB \quad (15.30)$$

whose risk term (i.e., coefficient of dB) is $(\alpha + F_S)\sigma S$. We determine

$$\alpha = -F_S \tag{15.31}$$

to eliminate the risk. Then the expected income per unit time is the coefficient of dt, namely,

$$\tfrac{1}{2}F_{SS}S^2\sigma^2 - F_t,$$

which, in order to keep a stable market, should equal the total return rI from secure investments. Thus, the available income for investment would yield rI and by the portfolio policy of (15.28) should equal $r[F + \alpha S] = r[F - F_S S]$ since α is determined by (15.31) for reasons indicated before. Under these market conditions the resulting equation for F becomes

$$\tfrac{1}{2}F_{SS}S^2\sigma^2 - F_t = r[F - F_S S] \quad \text{or} \quad F_t = \tfrac{1}{2}S^2\sigma^2 F_{SS} + rSF_S - rF. \tag{15.32}$$

The initial condition having S stock units available is

$$F(S, 0) = \max(0, S - a)$$

where a is the fixed cost of exercising the purchase of the warrant.

   We recognize (15.32) as the backward equation for a diffusion with infinitesimal parameters

$$\mu(S) = rS, \qquad \sigma^2(S) = S^2\sigma^2,$$

and killing rate function $k(S) = r$. The diffusion on $0 < S < \infty$ with

$$\mu(S) = rS, \qquad \sigma^2(S) = \sigma^2 S^2 \tag{15.33}$$

is recognized as geometric Brownian motion (cf. Example D of Section 2). Accordingly, we can represent the solution of (15.32) as the probability expectation of the Kac functional [cf. (5.38) and (5.40)]

$$F(S, t) = E_S[e^{-rt}h(X(t))], \quad \text{where} \quad h(x) = \max(0, x - a)$$

with respect to $X(t)$, the diffusion process corresponding to (15.33). (Problem 32 asks for the explicit solution.)

E. AN OPTIMAL GROWTH MODEL

Let the output process be

$$X(t) = F(K(t), L(t), Z(t)) \tag{15.34}$$

where $K(t)$ is the capital, $L(t)$ is labor, and $Z(t)$ is the technology at time t. A classical specification takes $F(K, L, Z) = K^\beta L^{1-\beta}\varphi(Z)$ the so-called Cobb–Douglas form. The following relations are postulated:

$$\frac{dK(t)}{dt} = X(t) - C(t) - \delta K(t) \quad \text{(cumulation equation)}, \tag{15.35}$$

where the rate of change of capital is proportional to production output minus consumption minus depreciation of the capital goods at rate δ. If population growth is assumed constant then

$$\frac{dL(t)}{dt} = rL(t). \tag{15.36}$$

The change in technologies is assumed to vary stochastically according to the differential equation

$$dZ(t) = \pi(Z) \, dt + \sigma(Z) \, dB(t). \tag{15.37}$$

Let $U(C(t))$ be the utility derived per individual from a rate of consumption $C(t)$ at time t. The problem then is to determine $C(t)$ subject to (15.34)–(15.37) which maximizes the total, properly discounted, expected utility:

$$E\left[\int_0^\infty U(C(t))e^{-\rho t} \, dt\right]. \tag{15.38}$$

This is a typical stochastic control problem of interest in economic theory.

F. ANOTHER OPTIMAL STOCHASTIC ALLOCATION MODEL

Let the assets at time t be A_t and consist of two parts, those based in secure bonds (as in banks) B_t and the investment in equities (e.g., stocks) K_t. Then

$$A_t = B_t p_t + K_t q_t, \tag{15.39}$$

where p_t is the price of a unit bond and q_t the price of an equity. We assume as a first approximation that q_t fluctuates according to

$$\frac{dq_t}{q_t} = \pi \, dt + \sigma \, dB \tag{15.40a}$$

with a random component, while

$$\frac{dp_t}{p_t} = r \, dt \tag{15.40b}$$

involves no random component.

The problem is to choose B_t, K_t, and C_t (C_t is the part of the holdings put into consumption) to optimize an appropriate utility. We stipulate the dynamic stochastic equation

$$dA_t = B_t \, dp_t + K_t \, dq_t - C_t \, dt,$$

and substituting from (15.40),

$$dA_t = A_t\{[\alpha r + (1 - \alpha)\pi] \, dt + (1 - \alpha)\sigma \, dB\} - C_t \, dt,$$

where by definition $\alpha = B_t p_t / A_t$. In this set-up A_t (wealth) is given at every instance and the problem is to choose α and C_t to maximize the utility expectation

$$E\left[\int_0^\infty U(C_t) e^{-pt}\, dt\right].$$

G. A SIMPLE SIGNAL DETECTION MODEL

It is common practice to model various electrical dynamical systems by means of stochastic differential equations. The underlying signal process is assumed to evolve following

$$dX = F(t)X(t)\, dt + G(t)\, dB(t) \tag{15.41}$$

for the linear model and

$$dX = f(X, t)\, dt + \sigma(X, t)\, dB \tag{15.42}$$

for the nonlinear model. Usually (15.41) and (15.42) are vector processes. The observable processes are governed by the process

$$dZ = H(t)X\, dt + R(t)\, d\beta(t). \tag{15.43}$$

where B and β are independent Brownian motions. Observing $Z(t)$ at a finite number of time points the objective is to obtain various statistical estimates of the $X(t)$ process and thereby extrapolate, interpolate, or appropriately smoothen the observations. There is an elaborate development in the communications literature on these models, e.g., consult Wong [10].

16: A Preview of Stochastic Differential Equations and Stochastic Integrals

We devote this section to highlighting a number of important facts of stochastic differential equations *without proof.* This subject is very much in the modern spirit for both theoretical and applied investigations of stochastic phenomena.

The solutions of stochastic differential equations [driven by white noise $W(t) = dB/dt$] are constructed in terms of stochastic integrals of the form

$$\int_0^t f(\tau)\, dB(\tau), \tag{16.1}$$

where the $f(t)$ are suitable random functions determined from the process values $\{B(s), s \leq t\}$. In attaching meaning to (16.1) it is impossible to employ the standard calculus of integrals because almost every sample path of $B(t)$ is not of bounded variation. Stochastic integrals of the form (16.1) in which $f(\tau)$ is not random but deterministic were covered in Section 8 of Chapter 9.

Proceeding more exactly, the class of admissible functions $f(t)$ considered in (16.1) will be characterized in terms of the family of σ-fields $\{\mathscr{F}_t, t \geq 0\}$, where $\mathscr{F}_t$ consists of all events determined by the realizations of $B(s)$ over the time interval $0 \leq s \leq t$. (An elaborate discussion of such collections of σ-fields is given in Chapter 6, Section 8.) Accordingly, two realizations are identical in $\mathscr{F}_t$ if their paths coincide until time t, although they can differ afterward. Mathematically, $\mathscr{F}_t$ is the smallest σ-field generated by the family of random variables $\{B(s), 0 \leq s \leq t\}$. (Recall the concept of a σ-field generated by a random variable, Chapter 6, Section 7.)

Let ω signify a particular realization of the process. We write $f(t) = f(t, \omega)$ to denote a real-valued random function, ω serving to emphasize, where desirable, that $f(t)$ is random. We say that $f(t)$ is *progressively measurable*† if for every $t \geq 0$ and real numbers $a < b$ the set $\{(\tau, \omega) : 0 \leq \tau \leq t \text{ and } a < f(\tau, \omega) < b\}$ is in the product σ-field $B[0, t] \otimes \mathscr{F}_t$, where $B[0, t]$ denotes the Borel subsets of $[0, t]$. Intuitively, the value of $f(t)$ is determined by the Brownian path up to time t and not by future values.

In order to define (16.1) we shall further require

$$\int_0^T E[\{f(t)\}^2] \, dt < \infty. \tag{16.2}$$

The collection of all progressively measurable f obeying (16.2) is designated by $\mathscr{H}(0, T)$.

We shall define the integral (16.1) in such a manner that it possesses the following properties:

(i)   If f_1 and f_2 belongs to $\mathscr{H}(0, T)$, then

$$\int_0^T [\alpha_1 f_1(t) + \alpha_2 f_2(t)] \, dB(t) = \alpha_1 \int_0^T f_1(t) \, dB(t) + \alpha_2 \int_0^T f_2(t) \, dB(t). \tag{16.3}$$

Of course, this is to be understood as a statement of equality between random variables.

(ii)   If $[\alpha, \beta] \subset [0, T]$ and

$$I_{[\alpha, \beta)}(t) = \begin{cases} 1, & \alpha \leq t < \beta, \\ 0 & \text{otherwise,} \end{cases} \tag{16.4}$$

then

$$\int_0^T I_{[\alpha, \beta)}(t) \, dB(t) = B(\beta) - B(\alpha).$$

(iii)   If $f(t)$ belongs to $\mathscr{H}(0, T)$, then

$$E\left[\int_0^T f(t) \, dB(t)\right] = 0 \tag{16.5}$$

and

$$E\left[\left\{\int_0^T f(t) \, dB(t)\right\}^2\right] = \int_0^T E[\{f(t)\}^2] \, dt. \tag{16.6}$$

---

† The assumption that $f(t)$ is *nonanticipative* or *adapted to* $\mathscr{F}_t$ is often seen. In fact, the stronger assumption that $f(t, \omega)$ is progressively measurable is needed.

With (i)–(iii) in mind, we define (16.1) first in the case that $f(t)$ is a *finite step function* in $\mathcal{H}(0, T)$, that is, there exists a subdivision of $[0, T]$, $0 = t_0 < t_1 < \cdots < t_n = T$, such that

$$f(t) = f(t_i) \quad \text{on} \quad t_i \le t < t_{i+1}, \qquad i = 0, 1, 2, \ldots, n-1. \tag{16.7}$$

The subclass of all finite step functions of $\mathcal{H}(0, T)$ is denoted $\mathcal{H}_S(0, T)$. For f in $\mathcal{H}_S(0, T)$ (as in (16.7)), we define

$$\mathcal{I}(f) = \int_0^T f(t)\, dB(t) = \sum_{i=0}^{n-1} f(t_i)[B(t_{i+1}) - B(t_i)]. \tag{16.8}$$

The indefinite random integral $\mathcal{I}(s; f)$ [also written as $\mathcal{I}_f(s) = \mathcal{I}_f(s, \omega)$], $0 \le s \le T$, is defined for f in $\mathcal{H}_S(0, T)$ and is delineated as in (16.8) with

$$f(t) \qquad \text{replaced by} \qquad f_s(t) = \begin{cases} f(t), & t < s, \\ 0 & t \ge s. \end{cases} \tag{16.8a}$$

The construction of $\mathcal{I}(f)$ and the indefinite integral is endowed with the properties listed in the next lemma.

Lemma 16.1. *For f and g in $\mathcal{H}_S[0, T]$, we have*

(a) $E[\mathcal{I}(f)] = 0$, *and*
(b) $E[\mathcal{I}(f)\mathcal{I}(g)] = \int_0^T E[f(\tau)g(\tau)]\, d\tau.$ (16.9)

The basic theorem is now stated.

Theorem 16.1. *Consider the process generated by the indefinite integral of an element f of $\mathcal{H}_S(0, T)$ according to*

$$Y(t) = \mathcal{I}_f(t) = \int_0^t f(\tau)\, dB(\tau), \qquad 0 \le t \le T. \tag{16.10}$$

The process $\{Y(t), 0 \le t \le T\}$ is a continuous time martingale with respect to $\mathcal{F}_t$ having continuous sample paths.

The stochastic integral $\mathcal{I}(f)$ can be extended to all functions $f(t) = f(t, \omega)$ defined for $0 \le t \le T$, progressively measurable with respect to the family of σ-fields $\mathcal{F}_t$, and obeying the integrability condition (16.2). We have denoted the collection of all such functions by $\mathcal{H}(0, T)$. An elaborate approximation procedure founded on the class of stochastic step functions $\mathcal{H}_S(0, T)$ using more advanced real-variable analysis allows us to extend Theorem 16.1 as follows:

Theorem 16.2.   *Let f be in $\mathcal{H}[0, T]$. Then there exists a square integrable martingale process $\{X(t), t \geq 0\}$, for which $E[X(t)^2] < \infty$ for all $t > 0$, and such that*

$$X(t) = \mathcal{I}_f(t) = \mathcal{I}_f(t, \omega) = \int_0^t f(\tau, \omega) \, dB(\tau)$$

satisfies the properties (16.9). *Further, $X(t)$ has almost all its sample paths continuous.*

A useful comparison of the Ito integral with ordinary Riemann integration is valid for certain classes of smooth random functions.

Theorem 16.3.   *Suppose $f(t)$ is in $\mathcal{H}(0, T)$ and is continuous as a function of t with probability 1. Consider a sequence of partitions*

$$0 = t_0^{(n)} < t_1^{(n)} < t_2^{(n)} < \cdots < t_n^{(n)} = T$$

with maximum mesh size $\delta_n = \max_{1 \leq i \leq n} (t_i^{(n)} - t_{i-1}^{(n)})$. Then

$$\mathcal{I}(f) = \int_0^T f(\tau) \, dB(\tau) = \lim_{\delta_n \to 0, \, n \to \infty} \sum_{i=1}^n f(t_{i-1}^{(n)})[B(t_i^{(n)}) - B(t_{i-1}^{(n)})].$$

The collection of progressively measurable functions $\varphi(\tau, \omega)$, integrable in that

$$\int_0^T |\varphi(\tau, \omega)| \, d\tau < \infty \qquad \text{for almost every realization } \omega, \tag{16.11}$$

is denoted by $\mathcal{L}(0, t)$.

Let ψ be in $\mathcal{H}(0, T)$ and φ be in $\mathcal{L}(0, t)$. Consider the process

$$X(t) = \int_0^t \varphi(\tau) \, d\tau + \int_0^t \psi(\tau) \, dB(\tau), \qquad 0 \leq t \leq T, \tag{16.12}$$

which exhibits almost surely continuous paths. It is called a *semimartingale* because *the second integral is a bona fide martingale* (Theorem 16.2) and *the first integral is almost always a function of bounded variation.*

It is convenient for later purposes to represent the process $X(t)$ in the differential notation

$$dX(t) = \varphi(t) \, dt + \psi(t) \, dB(t). \tag{16.13}$$

THE ITO TRANSFORMATION FORMULA

The statement of the Ito transformation formula, some background material, and its formal operation are highlighted in Section 14. To ease the exposition (without omitting any essential ideas) we assume at first that $F(x, t) = F(x)$ is a function of a single real variable and is twice continuously differentiable.

Consider the semimartingale $X(t)$ defined in (16.12). Subject to mild restrictions on $F(x)$ the simplest version of the Ito transformation formula has the expression

$$dF(X(t)) = [F'(X(t))\varphi(t) + \tfrac{1}{2}F''(X(t))\psi^2(t)] \, dt + F'(X(t))\psi(t) \, dB(t). \quad (16.14)$$

The equivalent integrated form is

$$F(X(t)) - F(X(0)) = \int_0^t [F'(X(\tau)) \varphi(\tau) + \tfrac{1}{2}F''(X(\tau))\psi^2(\tau)] \, d\tau$$

$$+ \int_0^t F'(X(\tau))\psi(\tau) \, dB(\tau). \quad (16.15)$$

The precise statement is the following:

Theorem 16.4. Let $\varphi(t, \omega)$ be in $\mathscr{L}(0, T)$ and $\psi(t, \omega)$ be in $\mathscr{H}(0, T)$. Assume $F''(x)$ is bounded, and $F'(x)$ and $F(x)$ continuous. Then the Ito transformation formula (16.15) holds.

EXTENSIONS

Subject to the stipulations of Theorem 16.4 plus the condition that $F_t(x, t) = (\partial F/\partial t)(x, t)$ is continuous and $F_t(X(t), t) \in \mathscr{L}(0, T)$, where $X(t)$ is prescribed in (16.13), we have

$$F(X(t), t) - F(X(0), 0) = \int_0^t [F_t(X(\tau), \tau) + F_x(X(\tau), \tau)\varphi(\tau)$$

$$+ \tfrac{1}{2}F_{xx}(X(\tau), \tau)\psi^2(\tau)] \, dt$$

$$+ \int_0^t F_x(X(\tau), \tau)\psi(\tau) \, dB(\tau) \quad (16.16)$$

or in differential notation

$$dF(X(t), t) = [F_t(X(t), t) + F_x(X(t), t)\varphi(t) + \tfrac{1}{2}F_{xx}(X(t), t)\psi^2(t)] \, dt$$

$$+ F_x(X(t), t)\psi(t) \, dB(t). \quad (16.17)$$

There is a multivariable version of (16.16) which we now state for completeness. Consider the collection of processes

$$dX_i(t) = \varphi_i(t, \omega) \, dt + \psi_i(t, \omega) \, dB(t), \qquad i = 1, 2, \ldots, n,$$

and impose on $F(t, x_1, x_2, \ldots, x_n)$ the natural smoothness and integrability conditions consonant with those of Theorem 16.4. Then

$$dF(t, X_1(t), \ldots, X_n(t))$$

$$= \left[\frac{\partial F}{\partial t} + \sum_{i=1}^n \left(\frac{\partial F}{\partial x_i} \right) \varphi_i(t, \omega) + \tfrac{1}{2} \sum_{i, j=1}^n \left(\frac{\partial^2 F}{\partial x_i \, \partial x_j} \right) \psi_i(t, \omega)\psi_j(t, \omega) \right] dt$$

$$+ \left[\sum_{i=1}^n \frac{\partial F}{\partial x_i} \psi_i(t, \omega) \right] dB(t), \quad (16.18)$$

the argument of the partials being

$$\frac{\partial F}{\partial x_i} = \frac{\partial F}{\partial x_i}(t, X_1(t), \ldots, X_n(t)) \quad \text{and} \quad \frac{\partial^2 F}{\partial x_i \partial x_j} = \frac{\partial^2 F}{\partial x_i \partial x_j}(t, X_1(t), \ldots, X_n(t)).$$

EXISTENCE AND UNIQUENESS OF SOLUTIONS FOR STOCHASTIC DIFFERENTIAL EQUATION

We elaborate a number of conditions guaranteeing existence and uniqueness for solutions of the stochastic differential equation (16.13) where $\varphi(t, \omega) = \varphi(X(t, \omega), t)$ and $\psi(t, \omega) = \psi(X(t, \omega), t)$. With this refinement, (16.13) then becomes

$$dX(t) = \varphi(X(t), t) \, dt + \psi(X(t), t) \, dB(t), \tag{16.19}$$

equivalent to the integral equation

$$X(t) = X(0) + \int_0^t \varphi(X(\tau), \tau) \, d\tau + \int_0^t \psi(X(\tau), \tau) \, dB(\tau). \tag{16.20}$$

On this matter we should anticipate the usual restrictions imposed on $\varphi(x, t)$ and $\psi(x, t)$ arising in the ordinary differential equation framework and, perhaps, more stringent conditions for the stochastic setting. The customary stipulations are incorporated in paragraphs (a) and (b) that follow.

(a) Growth Condition

There exists a constant K independent of $0 \le t \le T$ and $-\infty < x < \infty$ such that

$$\varphi^2(x, t) + \psi^2(t, x) \le K(1 + x^2), \qquad -\infty < x < \infty. \tag{16.21}$$

Thus, the rate of growth of φ and ψ is at most linear in x as $|x| \to \infty$. This restriction is crucial for otherwise the solutions of (16.19) [that is, the process realizations of $X(t)$] can reach infinity in finite time with positive probability. Such occurrences are reminiscent of sample path behavior in birth processes, where with quadratic growth in the birth parameters the process $X(t)$ can attain an infinite boundary in finite time with positive probability (consult Chapter 4).

Again, in the context of ordinary differential equations some kind of growth restriction is essential in order to be assured that the solution can be continued for the total time horizon $0 \le t \le T$ without becoming infinite at an intermediate time point.

Clearly, (16.21) applies to the circumstance

$$\varphi(x, t) = xg(t), \tag{16.22}$$

with $g(t)$ bounded for $0 \le t \le T$.

(b) Lipschitz Conditions

There exists a constant L independent of t, $0 \le t \le T$, and of x, $-\infty < x$, $y < \infty$, such that

$$|\varphi(x, t) - \varphi(y, t)| + |\psi(x, t) - \psi(y, t)| \le L|x - y|. \tag{16.23}$$

The necessity of a Lipschitz condition in guaranteeing uniqueness for solutions of ordinary differential systems is classical.

We now state the first existence theorem.

Theorem 16.5. *Let φ and ψ satisfy conditions (16.21) and (16.23). Let $X(0)$ be prescribed obeying $E[\{X(0)\}^2] < \infty$. Then there exists a unique solution of (16.20) as a continuous process.*

In order to deal with (16.20) the method of successive approximation traditional in constructing a solution for a standard differential equation system is appropriately adapted. Extensive refinements are available (e.g., see Gikhman and Skorohod III [6, 7]).

SOME APPLICATIONS OF THE ITO TRANSFORMATION FORMULA

An Exponential Example

Let $X(t)$ be a solution to

$$dX(t) = \varphi(t)\, dt + \psi(t)\, dB(t),\tag{16.24}$$

where φ is in $\mathcal{L}(0, T)$ and ψ is in $\mathcal{H}(0, T)$, and assume throughout this section that $X(0) = x_0$ is constant. Consider $F(x) = e^x$. Observe that $F(x) = F'(x) = F''(x)$. If

$$\varphi(t, \omega) \quad\text{and}\quad e^{X(t, \omega)}\varphi(t, \omega) \text{ are in } \mathcal{L}(0, T),\tag{16.25a}$$

while

$$\psi(t, \omega) \quad\text{and}\quad e^{X(t, \omega)}\psi(t, \omega) \text{ are of class } \mathcal{H}(0, T),\tag{16.25b}$$

then formula (16.15) gives for $Y(t) = e^{X(t)}$ the stochastic differential

$$dY(t) = Y(t)\{[\varphi(t) + \tfrac{1}{2}\psi^2(t)]\, dt + \psi(t)\, dB(t)\}.\tag{16.26}$$

Except for elementary variants on Brownian motion the conditions (16.25) are often formidable to verify or, on occasion, may not apply. To overcome these problems a truncation procedure is commonly implemented.

Consider now the specialized process $Z(t)$ expressed in differential form by

$$dZ(t) = -\tfrac{1}{2}\psi^2(t)\, dt + \psi(t)\, dB(t)$$

and in integral form by

$$Z(t) - Z(0) = -\frac{1}{2}\int_0^t \psi^2(\tau)\, d\tau + \int_0^t \psi(\tau)\, dB(\tau).$$

Then, on account of the Ito transformation formula and (16.26), $R(t) = \exp[Z(t)]$ satisfies the stochastic differential equation

$$dR(t) = R(t)\psi(\tau)\, dB(t),\tag{16.27}$$

which entails

$$R(t) - R(0) = \int_0^t R(\tau)\psi(\tau) \, dB(\tau), \tag{16.28}$$

and the right-hand side should represent a continuous square integrable martingale with respect to $\{\mathscr{F}_t\}$ by Theorem 16.2.

It is useful to extend (16.28), introducing a real parameter λ. Accordingly, with ψ in $\mathscr{H}(0, T)$, then

$$R(t; \lambda) = \exp\left[-\tfrac{1}{2}\lambda^2 \int_0^t \psi^2(\tau) \, d\tau + \lambda \int_0^t \psi(\tau) \, dB(\tau) \right]$$

constitutes a square integrable martingale process. The formal justifications are quite formidable since a hierarchy of integrability properties need to be checked.

The special choice $\psi(\tau, \omega) \equiv 1$ (a constant) produces the family of martingale processes

$$R(t; \lambda) = e^{-\lambda^2 t/2 + \lambda B(t)}$$

already encountered in Eq. (5.1) in Chapter 7.

Parenthetically, the solution of

$$dR(t) = R(t) \, dB(t)$$

in the Ito sense is the martingale process

$$R(t) = R(0)e^{B(t) - t/2}$$

and *not* the solution obtained by integrating $dR(t)/R(t) = dB(t)$ along each sample path.

THE DIFFUSION COEFFICIENTS FOR A STOCHASTIC DIFFERENTIAL EQUATION

A diffusion process is characterized through its infinitesimal conditional mean displacement and infinitesimal variance coefficient determined by

$$\lim_{h \downarrow 0} \frac{1}{h} E[X(t + h) - X(t) \,|\, X(t) = x] = \mu(x, t),$$

$$\lim_{h \downarrow 0} \frac{1}{h} E[\{X(t + h) - X(t)\}^2 \,|\, X(t) = x] = \sigma^2(x, t), \tag{16.29}$$

and for any $\varepsilon > 0$,

$$\lim_{h \downarrow 0} \frac{1}{h} \Pr\{|X(t + h) - X(t)| > \varepsilon \,|\, X(t) = x\} = 0, \tag{16.30}$$

the convergence is uniform with respect to x and t confined to a finite segment.

A more accurate formulation replaces (16.29). It involves truncated moments endowed with the same limiting relations and circumvents the need to postulate the existence of moments. In most practical applications of diffusions the conditions (16.29) plus (16.30) directly apply.

As pointed out in Section 1, property (16.30) is implied by the moment inequality

$$E[|X(t + h) - X(t)|^{2+\delta}|X(t) = \xi] \leq M(\xi)h^{1+\gamma} \tag{16.31}$$

for δ and γ positive where $M(\xi)$ is bounded over any compact region of ξ.

Consider the diffusion process $\{X(t), t \geq 0\}$ that solves the stochastic differential equations (16.19) concomitant with the smoothness restrictions of (16.21) and (16.23). Its infinitesimal coefficients are described in the next theorem.

Theorem 16.6. *Let* $\{X(t), t \geq 0\}$ *be the diffusion that arises as a solution of the stochastic differential equation* (16.19), *where the coefficients* $\varphi(x, t)$ *and* $\psi(x, t)$ *satisfy the growth and smoothness conditions* (16.21) *and* (16.23), *respectively. Then*

$$\lim_{h \downarrow 0} \frac{1}{h} E[(X(t + h) - X(t)|X(t) = x] = \varphi(x, t), \tag{16.32}$$

$$\lim_{h \downarrow 0} \frac{1}{h} E[\{X(t + h) - X(t)\}^2|X(t) = x] = \psi^2(x, t), \tag{16.33}$$

and the limits exist uniformly for t *and* x *ranging over any bounded region. Moreover,*

$$E[\{X(t + h) - X(t)\}^4|X(t) = x] \leq h^2 C(t, x), \tag{16.34}$$

where $C(t, x)$ *is uniformly bounded for* t *and* x *traversing a bounded region.*

A FAMILY OF MARTINGALES

Let $X(t)$ solve uniquely the stochastic differential equation

$$dX(t) = \varphi(X(t), t) dt + \psi(X(t), t) dB(t) \quad \text{with} \quad X(0) = x_0, \tag{16.35}$$

where φ and ψ fulfill the growth and Lipschitz conditions (16.21) and (16.23), respectively.

We can write (16.35) equivalently in the form

$$X(t) - X(0) = \int_0^t \varphi(X(\tau), \tau) d\tau + \int_0^t \psi(X(\tau), \tau) dB(\tau). \tag{16.36}$$

We now form the process

$$Z_\lambda(t) = \lambda X(t) - \lambda \int_0^t \varphi(X(\tau), \tau) d\tau - \tfrac{1}{2}\lambda^2 \int_0^t \psi^2(X(\tau), \tau) d\tau \tag{16.37}$$

with λ a fixed real parameter, $t > 0$, such that

$$dZ_\lambda(t) = \lambda\, dX(t) - [\lambda\varphi(X(t), t) + \tfrac{1}{2}\lambda^2\psi^2(X(t), t)]\, dt$$
$$= -\tfrac{1}{2}\lambda^2\psi^2(X(t), t)\, dt + \lambda\psi(X(t), t)\, dB(t). \qquad (16.38)$$

It is possible to apply the Ito formula (16.16) for $F(x) = e^x$ [cf. the discussion of (16.26)–(16.28)] using recondite truncation arguments to obtain the following important theorem.

Theorem 16.7. *Let $\varphi(x, t)$ and $\psi(x, t)$ satisfy the conditions of Theorem 16.5 and suppose $X(t)$ is the unique diffusion process solving (16.35) with coefficients $\varphi(x, t)$ and $\psi^2(x, t)$. Then for each real λ, the process*

$$U_\lambda(t) = \exp\left[\lambda X(t) - \lambda \int_0^t \varphi(X(\tau), \tau)\, d\tau - \tfrac{1}{2}\lambda^2 \int_0^t \psi^2(X(\tau), \tau)\, d\tau\right] \qquad (16.39)$$

constitutes a continuous path square integrable martingale.

There is a converse to Theorem 16.7 known as the *martingale property characterization of a diffusion process*. It is delicate and technical and can be stated as follows. Under suitable regularity conditions on $\varphi(x, t)$ and $\psi(x, t)$, if $U_\lambda(t)$ defined in (16.39) constitute a continuous path martingale for each real λ then $\{X(t), t \geq 0\}$ constitutes a diffusion process with mean infinitesimal displacement $\varphi(x, t)$ and infinitesimal diffusion coefficient $\psi^2(x, t)$.

With respect to Sections 14–16 we encourage the reader to tackle Problems 36–38.

Elementary Problems

1. Calculate the covariance $K(s, t) = E[Y(s)Y(t) | Y(0) = 0]$ for an Ornstein–Uhlenbeck process having infinitesimal parameters

$$\mu(y) = -\beta y \quad \text{and} \quad \sigma^2(y) = 1.$$

Solution: $K(s, t) = e^{-\beta(t-s)}(1 - e^{-2\beta s})/2\beta$, for $s < t$.

2. Consider a diffusion process on the interval $[0, 1]$ in natural scale $[s(\xi) \equiv 1]$ and with a speed density $m(\xi)$ which is continuous and positive. Determine the diffusion coefficients $\mu(x)$ and $\sigma^2(x)$.

3. Let $\{B(t)\}$ be a standard Brownian motion and consider the transformation obtained from the function $f(x, t) = x + t$. Then $Y(t) = f(X(t), t)$ is a diffusion. Identify it.

4. Consider a regular diffusion $X(t)$ on $(0, 1)$ with variance coefficient $\sigma^2(x) = x^2(1 - x)^2$. Show that the diffusion

$$Y(t) = \log \frac{X(t)}{1 - X(t)}$$

on $(-\infty, +\infty)$ has a constant infinitesimal variance.

5. If $X(t)$ is a diffusion process on $(0, \infty)$ with infinitesimal parameters $\mu(x) = bx + c$, $c > 0$, and $\sigma^2(x) = 4x$, what are the infinitesimal parameters of the diffusion process $Y(t) = \sqrt{X(t)}$?

Solution: $\mu_Y(y) = (by^2 + c - 1)/2y$, $\sigma_Y^2(y) = 1$.

6. Let $X(t)$ be a diffusion process on $(0, \infty)$ with diffusion coefficients $\mu(x) = cx$ and $\sigma^2(x) = x^\alpha$, with $c > 0$ and $\alpha \neq 2$. Consider the diffusion $Y(t) = [X(t)]^\beta$. What choice of β will give a constant infinitesimal variance for $Y(t)$?

Solution: $\beta = -(\tfrac{1}{2}\alpha - 1)$.

7. Consider a diffusion process $X(t)$ on $[0, \infty)$ for which $\mu(x) = \mu x$ and $\sigma^2(x) = \sigma^2$ for fixed positive parameters μ and σ^2. The left boundary 0 is prescribed to be an absorbing state. In the financial analysis literature, this process is sometimes called a *compounded Brownian motion*, where $X(t)$ can be interpreted as assets which increase through interest earnings at a mean rate μ but perturbed by Brownian motion fluctuations. (i) Verify that 0 is a regular boundary, consonant with the prescription of 0 as an absorbing state. (ii) Find the probability $u(x)$ of ruin, that is, the probability that the process reaches 0 from an initial level $x > 0$.

8. Consider the infinitesimal parameters

$$\mu(x) = \begin{cases} -\alpha x & \text{for} \quad x \geq 0, \\ -\beta x & \text{for} \quad x < 0, \end{cases}$$

$\sigma^2(x) \equiv 1$ on $I = (-\infty, +\infty)$ where $\alpha, \beta > 0$. Compute the scale and speed functions $s(\xi)$, $m(\xi)$ and the stationary density $\varphi(\xi)$.

9. A quite general diffusion population growth model is characterized by the infinitesimal coefficients

$$\mu(y) = \alpha y, \quad \sigma^2(y) = \tau y + \omega y^2, \quad I = (0, \infty).$$

Classify the boundary 0 under various assumptions on $\tau \geq 0$ and $\omega \geq 0$ with $\tau + \omega > 0$ and $-\infty < \alpha < \infty$.

10. Consider a diffusion process on $[0, \infty)$ with infinitesimal drift $\mu(y) = \alpha y$ and variance $\sigma^2(y) = \beta y$ with $\alpha, \beta > 0$. What is the probability of absorption at 0?

11. Consider a diffusion on $[0, 1]$ with infinitesimal coefficients

$$\mu(x) = (v - r)(1 - 2x)x(1 - x), \quad \sigma^2(x) = [1 + 2(v - r)x(1 - x)](1 - x)x,$$

$$0 < |r| < v.$$

Let $E^*(x)$ be the expected time to absorption at the boundaries 0 or 1 starting from $X(0) = x$.
Establish the formula

$$E^*(x) = -2 \int_0^x s(\xi) \ln\left(\frac{\xi}{1 - \xi}\right) d\xi \quad \text{for} \quad 0 \leq x \leq \tfrac{1}{2},$$

with $s(x) = 1/[1 + 2(v - r)x(1 - x)]$.

12. Consider a standard Brownian motion $\{B(t)\}$ and let τ be the random time that the linear barrier $at + b$ is first reached by $B(t)$. Show that $E[e^{-\theta \tau}] = \exp\{-b[\sqrt{a^2 + 2\theta} + a]\}$.

Hint: Use the martingale

$$Z_\lambda(t) = \exp\{\lambda B(t) - \tfrac{1}{2}\lambda^2 t\}, \qquad -\infty < \lambda < \infty.$$

13. Let $\{X(t)\}$ be an Ornstein–Uhlenbeck process with infinitesimal parameters $\mu(x) = -\alpha x$ and $\sigma^2(x) = 1$, $\alpha > 0$. Show that the infinitesimal parameters of the diffusion process defined by $Y(t) = e^{X(t)}$ are $\mu(x) = \tfrac{1}{2}x - \alpha x \log x$ and $\sigma^2(x) = x^2$.

14. Let $\{X(t)\}$ be reflected Brownian motion on $[0, \infty)$. Show that the infinitesimal parameters for $Y(t) = X(t)/[1 + X(t)]$ are $\mu(y) = (y - 1)^3$ and $\sigma^2(y) = (y - 1)^4$.

15. Let $\{U(t)\}$ be an Ornstein–Uhlenbeck process with infinitesimal parameters $\mu(x) = -\gamma x$ and $\sigma^2(x) \equiv 1$. Show that $X(t) = |U(t)|$ defines a diffusion process and determine the infinitesimal parameters. What should the boundary condition at the origin be?

16. Let $\{X(t)\}$ be a time inhomogeneous diffusion process with drift coefficient $\mu(x, t)$ and diffusion coefficient $\sigma^2(x, t)$. Let $u(x, t)$ be the probability of hitting b before a starting from $X(t) = x$. Formally,

$$u(x, t) = \Pr\{X(T) = b \mid X(t) = x\}, \qquad a \le x \le b,$$

where $T = \inf\{s \ge t, X(s) = a \text{ or } X(s) = b\}$.
  (i) In the spirit of Section 3, establish that $u(x, t)$ satisfies

$$-\frac{\partial u}{\partial t} = \mu(x, t)\frac{\partial u}{\partial x} + \frac{1}{2}\sigma^2(x, t)\frac{\partial^2 u}{\partial x^2}, \qquad a < x < b,$$

and $u(a, t) = 0$, $u(b, t) = 1$.
  (ii) What is the corresponding differential equation for the mean time $v(x, t) = E[T \mid X(t) = x]$?

17. Let $X(t)$ be a regular diffusion process on $[a, b]$ with both boundaries exit. Derive a differential equation for

$$\varphi_n(x) = E\left[\int_0^T t^n g(X(t))\, dt \mid X(0) = x\right], \qquad n \ge 1,$$

where T is the first passage time to the boundaries.

Solution: $\mu(x)\varphi_n'(x) + \tfrac{1}{2}\sigma^2(x)\varphi_n''(x) = -n\varphi_{n-1}(x)$, $a < x < b$, with $\varphi_n(a) = \varphi_n(b) = 0$, and

$$\varphi_0(x) = w(x) = E\left[\int_0^T g(X(t))\, dt \mid X(0) = x\right].$$

18. Let $\{X(t)\}$ be a diffusion process with infinitesimal parameters $\mu(x)$ and $\sigma^2(x)$. Define $Y(t) = F(X(t), t)$ where $F(x, t)$ is smooth (possesses whatever derivatives are necessary) and $\partial F/\partial x = F_x(x, t) \ne 0$ for all x and t. Argue that $\{Y(t)\}$ is a (time inhomogeneous) diffusion and determine the infinitesimal coefficients.

Solution:

$$\mu_Y(y, t) = \frac{\partial F}{\partial t}(x, t) + \mu(x)\frac{\partial F}{\partial x}(x, t) + \tfrac{1}{2}\sigma^2(x)\frac{\partial^2 F}{\partial x^2}(x, t),$$

$$\sigma_Y^2(y, t) = \left[\frac{\partial F}{\partial x}(x, t)\right]^2 \sigma^2(x),$$

where x is determined uniquely (why?) such that $F(x, t) = y$.

19. Let $\{X(t)\}$ be a diffusion process on $(-\infty, +\infty)$ with infinitesimal parameters

$$\mu(x) = -\operatorname{sgn}(x) = \begin{cases} -1, & x > 0, \\ 0, & x = 0, \\ 1, & x < 0, \end{cases}$$

and $\sigma^2(x) = 1$ for all x.

(i) Verify the scale functions $s(\xi) = e^{2|\xi|}$, $-\infty < \xi < \infty$ and $S(x) = (\operatorname{sgn} x)(e^{2|x|} - 1)$.

(ii) Show that

$$\varphi(x) = e^{-2|x|}, \qquad -\infty < x < \infty,$$

is a stationary distribution for the process.

20. Consider the diffusion on $(0, 1)$ with infinitesimal parameters

$$\mu(x) = x^2(1 - x)(\alpha - \beta x^2) \qquad \text{and} \qquad \sigma^2(x) = \beta x^4(1 - x)^2.$$

If $0 < \alpha < \tfrac{1}{2}\beta$, show that there exists a unique stationary density $\psi(x)$ and compute it.

Hint: Use Eq. (5.34) and show $C_1 = 0$.

Solution:

$$\psi(x) = c\frac{e^{-2\alpha x/\beta}}{(1 - x)^{2\alpha/\beta}x^{4 - 2\alpha/\beta}}, \qquad 0 < x < 1,$$

where c is a normalizing constant.

21. Let $\{X(t), -\infty < t < \infty\}$ be a stationary Ornstein–Uhlenbeck process with infinitesimal parameters $\mu(x) = -x$ and $\sigma^2(x) = 1$. Characterize the diffusion process $W(t) = \sqrt{t}X(\tfrac{1}{2}\ln t)$.

Solution: $W(t)$ is standard Brownian motion.

22. What is the transition density for geometric Brownian motion (see Example D, Section 2.

Solution:

$$p(t, x, y) = \frac{1}{y\sigma\sqrt{t}}\varphi\left[\frac{\ln(y/x) - \mu t}{\sigma\sqrt{t}}\right], \qquad \text{where} \quad \varphi(z) = \frac{1}{\sqrt{2\pi}}\exp(-\tfrac{1}{2}z^2).$$

23. What is the probability that a standard Brownian motion in n dimensions starting at $\mathbf{x} = (x_1, \ldots, x_n)$, with $\|\mathbf{x}\| = r$, will achieve a magnitude r_2 before getting closer to the origin than r_1. Of course, $r_1 < r < r_2$.

Hint: The problem reduces to calculating the scale function for the radial process $\{R(t)\}$, which is a one-dimensional diffusion on $[0, \infty)$ with infinitesimal parameters $\mu(r) = (n - 1)/2r$ and $\sigma^2(r) = 1$.

24. Show that the boundaries in the approximating diffusion for the Wright–Fisher random sampling model with mutation correspond to reflecting barriers (Section 2.F(a)). (The time at the boundary is negligible relative to the speed measure.)

25. Consider the instantaneous return process $Z(t)$ (Example E, Section 8) induced by a regular diffusion $X(t)$ on (l, r) relative to an interval $[a, b] \subset (l, r)$. Suppose that when $X(t)$ first attains a or b it is returned to the interior of (a, b) according to a fixed probability density function $f(x)$, $a < x < b$. What is the limiting (stationary) distribution of $Z(t)$?

Solution:

$$\varphi(y) = \frac{\int_a^b f(x)G(x, y)\, dx}{\int_a^b \left[\int_a^b f(x)G(x, \eta)\, dx\right] d\eta},$$

where G is the Green function on $[a, b]$.

26. Let $\{X(t)\}$ be a regular diffusion on $[0, 1]$ with absorption at 0 and 1. The process $\{X^\#(t)\}$ is derived by restarting the $X(t)$ process at a if 0 is hit, and restarting at b if 1 is hit. Find the stationary distribution $\alpha(x)$ of $\{X^\#(t)\}$.

Solution:

$$\alpha(x) = \frac{\pi_0(b)G(a, x) + \pi_1(a)G(b, x)}{\int_0^1 \left[\pi_0(b)G(a, \xi) + \pi_1(a)G(b, \xi)\right] d\xi},$$

where $G(y, x)$ is the Green function corresponding to $\{X(t)\}$ on $[0, 1]$ and $\pi_i(x) = \Pr\{X(t)$ hits $i \,|\, X(0) = x\}$.

27. Consider the gene frequency selection model (Section 2.F(b)). For the state x in $(0, 1)$ the function $l(x) = s(1 - x)$ is called the *load* of the fraction of deleterious gene, where s is the corresponding unit cost of carrying each such individual in the population. Compute the *cumulative expected load until fixation*

$$L(x) = E\left[\int_0^T s(1 - X(t))\, dt \,|\, X(0) = x\right],$$

where T is the fixation time (hitting time to 0 or 1).

28. Consider a diffusion process on $(0, \infty)$ with infinitesimal parameters $\mu(x) = cx^\alpha$, $\sigma^2(x) = dx^\beta$, $\alpha > 0$, $\beta > 0$, $-\infty < c < \infty$, and $d > 0$. Classify the boundaries 0 and ∞ in terms of α, β, c, and d.

Solution: The classification involves the breaking down into cases and the calculation of integrals. Some results are that the 0 boundary is

Regular	if $\beta < 2$ and	$\frac{1}{2}d(\beta - 1) < c < \frac{1}{2}d,$	
Exit	if $\beta < 2$ and	$c < \frac{1}{2}d(\beta - 1),$	$\beta - \alpha = 1.$
Entrance	if $\beta < 2$ and	$c \geq \frac{1}{2}d,$	

Problems

1. A quality control model leads to a diffusion process $\{X(t)\}$ on $[0, \infty)$ having the infinitesimal parameters $\mu(x) = 1 + x$ and $\sigma^2(x) = x^2$. (i) Show that 0 is an entrance boundary. (ii) Evaluate $v(0) = E[T_\lambda | X(0) = 0]$ where $T_\lambda = \inf\{t \geq 0, X(t) = \lambda\}$.

Answer: $v(0) = 2 \int_0^\lambda [\int_0^\eta m(\xi) \, d\xi] s(\eta) \, d\eta$ where $m(\xi) = e^{-2/\xi}$ and $s(\eta) = \eta^{-2} e^{2/\eta}$. This simplifies to $v(0) = 2 e^{2/\lambda} E_1^*(2/\lambda)$ where $E_1^*(x)$ is the exponential integral

$$E_1^*(x) = \int_x^\infty u^{-1} e^{-u} \, du.$$

2. A certain population is modeled as a diffusion process $\{X(t)\}$ on $(0, \infty)$ with drift $\mu(x) = cx$, c fixed, and diffusion $\sigma^2(x) = d^2 x$. Observe that according to (1.14),

$$U_\lambda(t) = \exp\left[\lambda X(t) - \lambda c \int_0^t X(s) \, ds - \tfrac{1}{2}\lambda^2 \, d^2 \int_0^t X(s) \, ds \right]$$

is a martingale for every $\lambda > 0$. Suppose c and d are such that absorption at 0 (extinction) occurs with probability 1. Let T be the time of extinction and $Z = \int_0^T X(s) \, ds$ a measure of the aggregate population size. Determine the distribution of Z using the martingale $U_\lambda(t)$.

3. Consider a time inhomogeneous diffusion process $\{X(t)\}$ with diffusion coefficients $\mu(x, t) = \alpha t$ and $\sigma^2(x, t) = \beta t$, $\alpha \geq 0$, $\beta > 0$. Let $T = T_a$ be the first passage time to the level a and evaluate $E[e^{-\lambda T^2} | X(0) = 0]$, $\lambda > 0$.

Hint: Use the martingale

$$Y_\theta(t) = \exp\left[\theta X(t) - \theta \int_0^t \mu(X(s), s) \, ds - \tfrac{1}{2}\theta^2 \int_0^t \sigma^2(X(s), s) \, ds \right], \qquad \theta \text{ real.}$$

4. Consider an n-dimensional standard Brownian motion and let $\bar{D}$ be the closure of a bounded open set D in E^n (Euclidean n-space). Let T be the first exit time from D. Argue by the informal methods of Section 3 that

$$u(\mathbf{x}) = E[f(\mathbf{X}(T)) | \mathbf{X}(0) = \mathbf{x}], \qquad \mathbf{x} \text{ in } \bar{D}$$

satisfies

$$\tfrac{1}{2}\Delta u = 0 \qquad \text{for} \quad \mathbf{x} \text{ in } D \qquad \left(\Delta = \frac{\partial^2}{\partial x_1^2} + \cdots + \frac{\partial^2}{\partial x_n^2} \right)$$

and

$$u(\mathbf{x}) = f(\mathbf{x}) \qquad \text{for} \quad \mathbf{x} \text{ in } \partial D, \quad \text{the boundary of } D.$$

5. Suppose that a diffusion $\{X(t)\}$ with killing time ζ has infinitesimal coefficients $\mu(x) = \theta$, $\sigma^2(x) = x$, and $k(x) = \theta x$ on the state space $(0, \infty)$. Show that $v(x) = E[\zeta | X(0) = x]$ solves the differential equation

$$\tfrac{1}{2}xv''(x) + \theta v'(x) - \theta x v(x) = -1, \qquad 0 < x < \infty,$$

with the boundary condition $v(x) \downarrow 0$ as $x \uparrow \infty$.

6. Consider a discrete generation population growth process following $N_{n+1} = Z_n N_n$, $n = 0, 1, \ldots$, where N_n is the population size in generation n and $\{Z_k\}$ are independent identically distributed positive random variables with $E[Z_0] < \infty$ and $E[|\ln Z_0|] < \infty$. Set $\mu = E[\ln Z_0]$ and $\sigma^2 = \text{var}(\ln Z_0)$. Determine in what sense N_n can be approximated by the process $N_t = N_0 \exp[\mu t + \sigma B(t)]$, where $\{B(t)\}$ is standard Brownian motion.

7. For each integer N consider the discrete recursion

$$X_{n+1}^{(N)} - X_n^{(N)} = \left(\frac{\alpha}{N} + \frac{\sigma}{\sqrt{N}} \eta_n^{(N)} \right) X_n^{(N)}, \qquad X_0^{(N)} = 1,$$

where $\{\eta_n^{(N)}\}$ are independent random variables with mean 0 and variance 1. Let $X^{(N)}(t) = X_{[Nt]}^{(N)}$, where $[Nt]$ is the largest integer not exceeding Nt. Prove that $X^{(N)}(t) \to X(t)$ where $\log X(t)$ is a Brownian motion with drift $\alpha - \frac{1}{2}\sigma^2$ and variance parameter σ^2.

Hint: Iteration produces

$$X_{[Nt]}^{(N)} = \prod_{k=1}^{[Nt]} \left(1 + \frac{\alpha}{N} + \frac{\sigma}{\sqrt{N}} \eta_k^{(N)} \right).$$

Taking logarithms

$$\ln X_{[Nt]}^{(N)} = \sum_{k=1}^{[Nt]} \ln \left(1 + \frac{\alpha}{N} + \frac{\sigma}{\sqrt{N}} \eta_k^{(N)} \right)$$

$$= \sum_{k=1}^{[Nt]} \left[\frac{\alpha}{N} + \frac{\sigma}{\sqrt{N}} \eta_k^{(N)} - \frac{1}{2} \frac{\sigma^2}{N} (\eta_k^{(N)})^2 + o\left(\frac{1}{N}\right) \right].$$

Now invoke the law of large numbers and the central limit theorem.

8. Let $\{X(t)\}$ be a process whose sample paths contain both deterministic and jump components. During the deterministic part, $X(t)$ grows exponentially in the manner $dX(t)/dt = \alpha X(t)$. Random jumps also occur such that in the time period $(t, t + h)$ then $X(t)$ changes to $\beta X(t)$ with probability $\frac{1}{2}\lambda h + o(h)$, $X(t)$ changes to $(2 - \beta)X(t)$ with probability $\frac{1}{2}\lambda h + o(h)$, and no jump occurs with probability $1 - \lambda h + o(h)$. Consider the process $X(t) = X(t; \lambda, \alpha, \beta)$, displaying the dependence on parameters. Let $\beta \to 1$ and $\lambda \to \infty$ such that $\lambda(\beta - 1)^2 \to a > 0$. Show that $X(t)$ converges to a geometric Brownian motion with infinitesimal parameters $\mu(x) = \alpha x$ and $\sigma^2(x) = ax^2$.

Hint: Calculate $E[(\Delta X)^n | X(t) = x]$ for $n = 1, 2$, and 4 where $\Delta X = X(t + h) - X(t)$.

9. Consider a sequence of branching processes $\{Z_N(t), t \geq 0\}$ with initial size $Z_N(0) = N$. The offspring per individual has probability generating function $f_N(s)$ (see Chapter 8) and moments

$$E[Z_N(1) | Z_N(0) = 1] = f_N'(1) = 1 + \frac{\alpha}{N} + o\left(\frac{1}{N}\right),$$

$$\text{var}[Z_N(1) | Z_N(0) = 1] = \dot{\beta}, \qquad \text{independent of } N.$$

Consider the normalized process $Y_N(t) = Z_N([Nt])/N$. Show that

$$\lim_{N \to \infty} NE\left[Y_N\left(t + \frac{1}{N}\right) - Y_N(t) \mid Y_N(t) = y \right] = \alpha y,$$

$$\lim_{N \to \infty} NE\left[\left\{ Y_N\left(t + \frac{1}{N}\right) - Y_N(t) \right\}^2 \mid Y_N(t) = y \right] = \beta y,$$

and the fourth-moment displacement converges to zero. The calculations suggest that $Y_N(t)$ converges to a limiting diffusion on $[0, \infty)$ with infinitesimal parameters $\mu(y) = \alpha y$ and $\sigma^2(y) = \beta y$. What assumptions on $f_N(s)$ insure that

$$NE\left[\left\{ Y_N\left(t + \frac{1}{N}\right) - Y_N(t) \right\}^4 \mid Y_N(t) = y \right] = o\left(\frac{1}{N}\right)$$

uniformly for y in bounded intervals?

10. Consider a sequence of branching processes $\{Z_N(t), t \geq 0\}$ with initial size $Z_N(0) = N$. The offspring per individual has probability generating function $f_N(s)$ and moments

$$E[Z_N(1) \mid Z_N(0) = 1] = f'_N(1) = 1 + o\left(\frac{1}{N}\right);$$

$$\text{var}[Z_N(1) \mid Z_N(0) = 1] = 1, \qquad \text{independent of } N,$$

and the fourth central moment is $o(1/N)$. Suppose that a random number of immigrants enter the population each generation at an average rate of c/N with variance $o(1/N)$. By adapting the procedure of Problem 9 show formally that $Y_N(t) = N^{-1} Z_N([Nt])$ can be approximated by a diffusion on $[0, \infty)$ with $\sigma^2(x) = x$ and $\mu(x) = c$.

11. Let $\{Y_i(t)\}_{i=1,2}$ be independent diffusion processes on $I = (0, \infty)$ having infinitesimal coefficients $\mu_i(y) = c_i > 0$ and $\sigma_i^2(y) = y$. Show that $X(t) = Y_1(t) + Y_2(t)$ defines a diffusion process on $(0, \infty)$ for which $\mu_X(x) = c_1 + c_2$ and $\sigma_X^2(x) = x$.

Hint: Establish the corresponding property for the branching processes of Problem 10 and pass to the limit.

12. Consider a family of diffusion processes $X_a(t)$ on $(0, \infty)$ with coefficients

$$\mu(x) = ax + \alpha, \qquad a > 0, \qquad \alpha > 0;$$
$$\sigma^2(x) = \sigma^2 x, \qquad \sigma^2 > 0.$$

   (i) Classify the boundaries 0 and ∞.
   (ii) Let $X_\alpha(t)$ and $X_\beta(t)$ be independent processes with $X_\alpha(0) = x_\alpha$ and $X_\beta(0) = x_\beta$. Establish that for every t

$$\Pr\{X_\alpha(t) + X_\beta(t) \leq x\} = \Pr\{X_{\alpha+\beta}(t) \leq x \mid X_{\alpha+\beta}(0) = x_\alpha + x_\beta\}.$$

Hint: Refer to Problems 9–11.

13. Consider the diffusion process $\{X(t)\}$ on $[0, \infty)$ having infinitesimal coefficients $\sigma^2(x) = 2x$, $\mu(x) = 0$. (The endpoint 0 is an exit boundary.) Establish the Laplace transform

$$E[e^{-\lambda X(t)} \mid X(0) = x] = \exp\left(\frac{-\lambda x}{1 + \lambda t}\right)$$

by showing that the right-hand side $\varphi(\lambda; x, t)$ satisfies the differential equation

$$\frac{\partial \varphi}{\partial t} = x \frac{\partial^2 \varphi}{\partial x^2}, \qquad x > 0, \quad t > 0$$

and the initial condition $\varphi(x, 0) = e^{-\lambda x}$.

14. Let $\{B(t)\}$ be a standard Brownian motion and suppose $A(x)$ is a positive function for which $\int_0^\infty [A(B(\tau))]^{-1} \, d\tau = \infty$ with probability 1. For each t define the random variable $C(t)$ by

$$\int_0^{C(t)} \frac{1}{A(B(\tau))} \, d\tau = t$$

and form the process $Z(t) = B(C(t))$. Show that $\{Z(t)\}$ is a diffusion with infinitesimal parameters $\mu_Z(z) = 0$ and $\sigma_Z^2(z) = A(z)$.

Hint: Observe that

$$\int_{C(t)}^{C(t+h)} \frac{1}{A(B(\tau))} \, d\tau = h$$

so that

$$C(t + h) - C(t) = A(B(C(t)))h + o(h).$$

Then evaluate

$$E[\{Z(t + h) - Z(t)\}^n \,|\, Z(t) = z] = E[\{B(C(t + h)) - B(C(t))\}^n \,|\, B(C(t)) = z]$$

$$= 0 \qquad \text{for} \quad n = 1$$

$$= A(z)h + o(h) \qquad \text{for} \quad n = 2$$

$$= o(h) \qquad \text{for} \quad n = 4.$$

15. A regular diffusion process $\{X(t)\}$ on $[0, \infty)$ has continuous infinitesimal parameters $\mu(x)$ and $\sigma^2(x) > 0$ for $x \geq 0$, and 0 is a regular boundary. Suppose that reflecting barrier motion is prescribed at the boundary 0 and consider the problem of evaluating

$$w(x) = E\left[\int_0^T g(X(s)) \, ds \,|\, X(0) = x\right] \qquad \text{for} \quad 0 \leq x \leq a$$

for a given function $g(x)$ where $T = \inf\{t \geq 0, X(t) = a\}$ is the hitting time to a. Suppose that $|\mu(0)| < \infty$ and $0 < \sigma^2(0) < \infty$. Show that $w(x)$ is the solution to the differential equation

$$\mu(x)w'(x) + \tfrac{1}{2}\sigma^2(x)w''(x) = -g(x) \qquad \text{for} \quad 0 < x < a$$

with boundary conditions $w(a) = 0$, $w'(0) = 0$.

Hint: Let $s(x) = \exp\{-\int_0^x [2\mu(\xi)/\sigma^2(\xi)] \, d\xi\}$ and $m(x) = 1/[\sigma^2(x)s(x)]$ for $x \geq 0$ be the scale and speed densities for the process, and let $\{Y(t)\}$ be the diffusion process on $(-\infty, +\infty)$ having scale density $s_Y(y) = s(|y|)$ and speed density $m_Y(y) = m(|y|)$. Then $\{Y(t)\}$ has infinitesimal parameters $\mu_Y(y) = (\text{sgn } y)\mu(|y|)$ and $\sigma_Y^2(y) = \sigma^2(|y|)$.

Use the fact that the reflecting barrier phenomena is equivalent to setting $X(t) = |Y(t)|$. Let $U = \inf\{t \geq 0, |Y(t)| = a\}$ and

$$w_Y(y) = E\left[\int_0^U g(|Y(s)|)\,ds\,\middle|\,Y(0) = y\right].$$

Then $\frac{1}{2}(d/dM)[(dw_Y/dS)](y) = -g(|y|)$ for $-a < y < a$. Use the symmetry $w_Y(y) = w(|y|)$ and the smoothness of $w_Y(y)$, $s_Y(y)$, and $m_Y(y)$ at zero to deduce $w'_Y(0) = w'(0) = 0$.

16. Let $\{X(t)\}$ be a Brownian motion on $I = [0, \infty)$ with drift μ and variance parameter $\sigma^2 = 1$ where 0 is a reflecting boundary. Let T_a be the hitting time to level $a > 0$ and set $v(x) = E[T_a | X(0) = x]$ for $0 \leq x \leq a$. Obtain $v(x)$ by solving the differential equation

$$\tfrac{1}{2}v''(x) + \mu v'(x) = -1, \qquad 0 \leq x \leq a,$$

subject to $v'(0) = 0$, $v(a) = 0$.

Answer: When $\mu \neq 0$, then

$$v(x) = \frac{1}{\mu}\left[(a - x) - \frac{1}{2\mu}(e^{-2\mu x} - e^{-2\mu a})\right], \qquad 0 \leq x \leq a;$$

when $\mu = 0$, then

$$v(x) = (a^2 - x^2).$$

17. Let $\{X(t)\}$ be a diffusion process on $[0, \infty)$ with infinitesimal parameters $\mu(x) = -\alpha < 0$ and $\sigma^2(x) = 1$ for $x > 0$. Suppose 0 is a reflecting barrier. Show that $\varphi(\xi) = 2\alpha e^{-2\alpha\xi}$, $\xi > 0$, is a stationary density.

18. Let $\{X(t)\}$ be a Brownian motion with drift μ and variance parameter σ^2, but modified as follows: (i) 0 is a reflecting barrier; (ii) whenever the process reaches $a > 0$, it is instantaneously restarted at 0. Determine the stationary density $\varphi(\xi)$.

Answer: When $\mu \neq 0$, then $\varphi(\xi) = K^{-1}\{1 - \exp[-2\mu(a - \xi)/\sigma^2]\}$, $0 \leq \xi \leq a$, where $K = a - (\sigma^2/2\mu)[1 - \exp(-2\mu a/\sigma^2)]$. When $\mu = 0$, then

$$\varphi(\xi) = \frac{2}{a}\left(1 - \frac{\xi}{a}\right), \qquad 0 \leq \xi \leq a.$$

19. Define $X(t) = |B(t)|^3$ where $\{B(t)\}$ is standard Brownian motion on $[0, \infty)$.

    (i)   Show that $X(t)$ is a diffusion process for which 0 is a reflecting boundary.

    (ii) Show that $v(x) = E[T_\lambda | X(0) = x]$, where $T_\lambda = \inf\{t \geq 0, X(t) = \lambda\}$, is explicitly $v(x) = \lambda^{2/3} - x^{2/3}$ for $0 \leq x \leq \lambda$.

    (iii) Verify that $v'(0) = -\infty$ but $dv/dS|_{x=0} = 0$, where $S(x)$ is the scale function of the process.

20. Consider a regular diffusion process on the interval $[0, 1]$ where 0 is an absorbing state attainable in finite expected time and 1 is a reflecting barrier. Consider the differential operator

$$L\varphi(x) = \tfrac{1}{2}\sigma^2(x)\frac{d^2\varphi}{dx^2}(x) + \mu(x)\frac{d\varphi}{dx}(x), \qquad 0 < x < 1,$$

corresponding to the diffusion coefficients $\mu(x)$ and $\sigma^2(x)$. Show that the associated Green function (analog of (3.15)) is calculated by the formula

$$
G(x, \xi) = \begin{cases} \dfrac{u_0(x)u_1(\xi)}{W(\xi)\sigma^2(\xi)} & \text{for } 0 < x < \xi < 1, \\[2ex] \dfrac{u_1(x)u_0(\xi)}{W(\xi)\sigma^2(\xi)} & \text{for } 0 < \xi < x < 1, \end{cases}
$$

where $u_0(x)$, up to a scalar multiple, is a positive solution of $Lu_0(x) = 0, 0 < x < 1$ and $u_0(0) = 0$, and $u_1(x)$ solves $Lu_1(x) = 0, 0 < x < 1$ and $u_1'(1) = 0$, and $W(\xi) = u_0'(\xi)u_1(\xi) - u_0(\xi)u_1'(\xi)$.

21. Let $\{X(t)\}$ and $\{Y(t)\}$ be independent regular diffusion processes on the same interval I and having the same transition function $P(t, x, E)$ for x in I, $E \subset I$, and $t \geq 0$. Suppose that $X(0) = x$, $Y(0) = y$ with $x < y$, and let A and B be sets in I with $A < B$ in the sense that $a < b$ for all a in A and b in B.

    (i)   For a fixed time t, use a reflection argument to show that

$$
\Pr\{X(t) \in A, \ Y(t) \in B \text{ and } X(s) = Y(s) \text{ for some } s \leq t \,|\, X(0) = x, \ Y(0) = y\}
$$
$$
= \Pr\{X(t) \in B \text{ and } Y(t) \in A \,|\, X(0) = x, \ Y(0) = y\}.
$$

    (ii) Show that the determinant

$$
D = \det \begin{Vmatrix} P(t, x, A) & P(t, x, B) \\ P(t, y, A) & P(t, y, B) \end{Vmatrix}
$$

evaluates $\Pr\{X(t) \in A, \ Y(t) \in B \text{ and } X(s) \neq Y(s) \text{ for all } s \leq t\}$.

    (iii) Conclude that $D > 0$ whenever $P(t, x, E)$ is the transition function of a regular diffusion process and $x < y$, $A < B$.

    (iv) Let $p(t, x, y)$ be the density function corresponding to $P(t, x, E)$. Assume $p(t, x, y)$ is continuous in y for each x. Show that if $x_1 < x_2$ and $y_1 < y_2$, then

$$
\det \begin{Vmatrix} p(t, x_1, y_1) & p(t, x_1, y_2) \\ p(t, x_2, y_1) & p(t, x_2, y_2) \end{Vmatrix} > 0.
$$

22. Let $X^*(t)$ be the conditioned diffusion process discussed at the start of Section 9. Show that the boundary point 1 remains an exit boundary for $X^*(t)$.

23. A Wright–Fisher diffusion $X(t)$ with no mutation or selection (cf. Section 4.A(a)) has infinitesimal parameters $\mu(x) = 0$ and $\sigma^2(x) = x(1 - x)$ for $0 < x < 1$. Consider the related diffusion process $X^*(t)$, which is $X(t)$ conditioned on absorption at 1. (Consult the beginning of Section 9.) Let T^* be the absorption time in the conditioned process. Compute $\varphi(x) = E[(T^*)^2 \,|\, X^*(0) = x]$.

Answer:

$$
\varphi(x) = -\frac{4}{x}\left\{(1 - x)\int_0^x \log(1 - \xi)\frac{d\xi}{\xi} + \int_x^1 (1 - \xi)\log(1 - \xi)\frac{d\xi}{\xi}\right\}.
$$

24. Consider a regular diffusion process $X(t)$ on $[0, 1]$ where both boundaries are exit. Let $X_i^*(t)$ be the process conditioned on absorption at i for $i = 0, 1$. Prove that the expected occupation time of an interval $J = (\alpha, \beta)$ by $X_1^*(t)$ starting from $X_1^*(0) < \alpha$ is the same as that of $X_0^*(t)$ starting from $X_0^*(0) > \beta$.

Hint: Let $G_0^*(x, y)$ and $G_1^*(x, y)$ be the Green functions corresponding to the two conditioned processes. Show that $G_0^*(x, y) = G_1^*(y, x)$ by direct calculation from the infinitesimal parameters of the conditioned processes.

25. Let $\{X(t), 0 \le t \le 1\}$ be a standard Brownian motion but conditioned to be normally distributed with mean zero and variance σ^2 at time $t = 1$. Show that $\{X(t)\}$ is a time inhomogeneous diffusion with infinitesimal parameters

$$\mu(x, t) = \frac{x(\sigma^2 - 1)}{1 + t(\sigma^2 - 1)}, \qquad \sigma^2(x) = 1.$$

26. Let $\{X(t), t \ge 0\}$ be a standard Brownian motion and define $Y(t) = [X(t)]^3$.

    (a)   Argue that $\{Y(t)\}$ is a regular diffusion process on $(-\infty, +\infty)$.

Hint: $g(x) = x^3$ is strictly monotonic and continuous.

    (b)   Derive the infinitesimal parameters $\mu_Y(y)$ and $\sigma_Y^2(y)$. Is there a contradiction between the facts (i) $\mu_Y(0) = \sigma_Y^2(0) = 0$, and (ii) the $\{Y(t)\}$ process is regular on $(-\infty, +\infty)$?

Answer: $\mu_Y(y) = 3y^{1/3}$; $\sigma_Y^2(y) = 9y^{4/3}$.

    (c)   Consider the differential operator $L = \mu_Y(y)\, d/dy + \frac{1}{2}\sigma_Y^2(y)\, d^2/dy^2$ operating on twice continuously differentiable functions f. Show that $Lf(0) = 0$ for all such f. (Compare with Theorem 12.1.)

    (d)   Evaluate the scale and speed functions.

Answer: $S_Y(y) = y^{1/3}$, $M_Y(y) = y^{1/3}$.

    (e)   Let $T = T_{-1,1} = \min\{t \ge 0 : |Y(t)| \ge 1\} = \min\{t \ge 0 : |X(t)| \ge 1\}$. Determine $v(y) = E[T \mid Y(0) = y]$.

Answer: $v(y) = 1 - y^{2/3}$, $-1 \le y \le 1$.

    (f)   Then v is not twice continuously differentiable and hence, not in the domain of the differential operator L. Show that v is in the domain of the differential operator $A = \frac{1}{2}(d/dM)(d/dS)$ and that $Av(y) = -1$ for $-1 < y < 1$.

27. Let $\varphi(\xi)$, $-\infty < \xi < \infty$, be a bounded twice continuously differentiable function with $\varphi''(\xi) \to 0$ as $|\xi| \to \infty$. Let $\{B(t)\}$ be standard Brownian motion. Show that the process

$$Z(t) = \varphi(B(t)) - \frac{1}{2}\int_0^t \varphi''(B(s))\, ds$$

is a martingale.

Hint: Use Dynkin's formula (11.37).

28. Consider any solution $\varphi(x)$ to $\frac{1}{2}\varphi''(x) + q(x)\varphi(x) = 0$, $-\infty < x < \infty$, where $q(x)$ is bounded and continuous. Show that

$$M(t) = \varphi(B(t)) \exp\left[\int_0^t q(B(s))\, ds\right]$$

is a martingale with respect to the standard Brownian motion $\{B(t)\}$.

Hint: Using the Ito stochastic calculus formula $d\varphi(B(t)) = \varphi'(B(t))\, dB(t) + \frac{1}{2}\varphi''(B(t))\, dt$, show that $dM(t) = \varphi'(B(t)) \exp[\int_0^t q(B(s))\, ds]\, dB(t)$, whence $M(t)$ is an Ito integral.

29. Let $\{R(t), t \geq 0\}$ denote the distance of three-dimensional Brownian motion from the origin (i.e., the Bessel process for $d = 3$). Show that

$$Z_\alpha(t) = \frac{\sinh(\alpha R(t))}{\alpha R(t)} \exp(-\tfrac{1}{2}\alpha^2 t)$$

is a martingale for each fixed $\alpha > 0$.

Hint: The function $\varphi(r, t) = [\alpha r]^{-1} \sinh[\alpha r \exp(-\tfrac{1}{2}\alpha^2 t)$ satisfies the time reversed backward differential equation

$$-\frac{\partial \varphi}{\partial t} = \frac{1}{2}\frac{\partial^2 \varphi}{\partial r^2} + \frac{1}{r}\frac{\partial \varphi}{\partial r}.$$

30. The following diffusion on $I = (0, 1)$ arises in modeling gene frequency fluctuations:

$$\mu(x) = x(1 - x)[s_1 x - s_2(1 - x) - v_1 x^3 + v_2(1 - x)^3 + rx(1 - x)(2x - 1)],$$

$$\sigma^2(x) = x^2(1 - x)^2[v_1 x^2 + v_2(1 - x)^2 - 2rx(1 - x)],$$

where $v_1 > 0$, $v_2 > 0$, and $r^2 < v_1 v_2$. Verify the following:
   (i) If $s_1/v_1 > \tfrac{1}{2}$ and $s_2/v_2 < \tfrac{1}{2}$, then 0 is an entrance boundary while 1 is an exit boundary;
   (ii) If $s_1/v_1 < \tfrac{1}{2}$ and $s_2/v_2 > \tfrac{1}{2}$, then 0 is an exit boundary while 1 is entrance;
   (iii) If $s_1/v_1 > \tfrac{1}{2}$ and $s_2/v_2 > \tfrac{1}{2}$, then both boundaries are entrance;
   (iv) If $s_1/v_1 < \tfrac{1}{2}$ and $s_2/v_2 < \tfrac{1}{2}$, then both boundaries are exit.

31. Let $\{B(t)\}$ be a standard Brownian motion and $f(x)$, $-\infty < x < \infty$, an integrable function with $\int_{-\infty}^{+\infty} f(x) \, dx \neq 0$. For each $\lambda > 0$ consider the process

$$Y_\lambda(t) = \frac{1}{\sqrt{\lambda}} \int_0^{\lambda t} f(B(\tau)) \, d\tau.$$

Show that the first four moments of $Y_\lambda(t)$ converge as $\lambda \to \infty$ for each fixed t. [It can be shown that $Y_\lambda(t)$ converges in distribution as $\lambda \to \infty$ to $c(f)l_{\{0\}}(t)$, where $l_{\{0\}}(t)$ is the local time process at zero of a standard Brownian motion and $c(f)$ is a constant.]

32. Suppose the market price $S(t)$ of a stock fluctuates according to a diffusion process having infinitesimal parameters $\mu(s) = \mu s$ and $\sigma^2(s) = \sigma^2 s^2$ for $0 < s < \infty$. Suppose that the value of an option on the stock is a function $W = F(S(t), T - t)$ of $S(t)$ and the remaining time $T - t$ (T is fixed) to exercise the option. In Section 15.D it is established that F satisfies the differential equation $-F_t + rsF_s + \tfrac{1}{2}\sigma^2 s^2 F_{ss} = rF$, subject to $F(s, 0) = (s - a)^+ = \max\{0, s - a\}$ where a is the fixed cost to exercise the option and r is the secure interest rate in the economy.
   (i) Validate the representation

$$F(x, \tau) = E[e^{-r\tau}(X(\tau) - a)^+ \mid X(0) = x]$$

where $X(t)$ is geometric Brownian motion with parameters $\mu_X(x) = rx$ and $\sigma_X^2(x) = \sigma^2 x^2$.
   (ii) Calculate the explicit formula

$$F(x, \tau) = x\Phi\left(\frac{\ln(x/a) + (r + \tfrac{1}{2}\sigma^2)\tau}{\sigma\sqrt{\tau}}\right) - ae^{-r\tau}\Phi\left(\frac{\ln(x/a) + (r - \tfrac{1}{2}\sigma^2)\tau}{\sigma\sqrt{\tau}}\right),$$

where $\Phi(z) = (2\pi)^{-1/2} \int_{-\infty}^z \exp(-\tfrac{1}{2}\xi^2) \, d\xi$.

Hint: (i) Show that $F(x, \tau)$ as given solves the requisite differential equation and boundary condition.

(ii) $\ln X(\tau)$ is normally distributed with mean $\ln x + (r - \frac{1}{2}\sigma^2)\tau$ and variance $\sigma^2\tau$.

33. Consider a diffusion $\{X(t), Y(t)\}$ in two dimensions on the triangle $0 \le x, y$ and $x + y \le 1$, and having the infinitesimal parameters

$$\mu_X(x, y) = \alpha - (\alpha + \gamma)x, \qquad \mu_Y(x, y) = \beta - (\beta + \gamma)y;$$

$$\sigma^2_{X,X}(x, y) = x(1 - x), \quad \sigma^2_{Y,Y}(x, y) = y(1 - y) \qquad \sigma^2_{X,Y}(x, y) = -2xy.$$

(This is the analog of the Wright–Fisher diffusion model (Section 2.F(a)) under reversible mutation pressures involving three allele types and with no selection.) Establish that the Dirichlet density

$$f(x, y) = \frac{\Gamma(\alpha + \beta + \gamma)}{\Gamma(\alpha)\Gamma(\beta)\Gamma(\gamma)} x^{\alpha - 1} y^{\beta - 1}(1 - x - y)^{\gamma - 1}$$

is a stationary density by showing it satisfies the forward Kolmogorov equation

$$\frac{1}{2} \frac{\partial^2}{\partial x^2} [x(1 - x)f(x, y)] - \frac{\partial^2}{\partial x\, \partial y} [(xy)f(x, y)] + \frac{1}{2} \frac{\partial^2}{\partial y^2} [y(1 - y)f(x, y)]$$

$$- \frac{\partial}{\partial x} \{[\alpha - (\alpha + \beta + \gamma)x]f(x, y)\}$$

$$- \frac{\partial}{\partial y} \{[\beta - (\alpha + \beta + \gamma)y]f(x, y)\} = 0.$$

34. Let $Y(t)$ be a zero mean process whose covariance function $R(s, t) = E[Y(s)Y(t)]$ has a continuous derivative $\rho(s, t) = \partial^2 R(s, t)/\partial s\, \partial t$. For $\alpha > 0$ define the approximate derivative $\dot{Y}_\alpha(t) = [Y(t + \alpha) - Y(t)]/\alpha$.

(i) Show $E[\dot{Y}_\alpha(t)\dot{Y}_\beta(t)] \to \rho(t, t)$ as $\alpha \to 0$, $\beta \to 0$.

Hint: Use (14.9).

(ii) Show that $\|\dot{Y}_\alpha(t) - \dot{Y}_\beta(t)\|^2 \to 0$ as $\alpha \to 0$, $\beta \to 0$ where $\|Z\|^2 = E[Z^2]$ is mean square distance. Hence $\{\dot{Y}_\alpha(t)\}$ satisfies the Cauchy criterion for convergence in mean square and there exists a random variable $\dot{Y}(t)$ for which $\dot{Y}_\alpha(t) \to_{m.s.} \dot{Y}(t)$ as $\alpha \to 0$.

35. For any regular diffusion process $\{X(t)\}$ on (l, r) having continuous infinitesimal coefficients show that there exists an $\alpha > 0$ for which

$$E[\exp\{\alpha(T_a \wedge T_c)\}\,|\,X(0) = b] < \infty, \qquad l < a < b < c < r.$$

Hint: Refer to Eq. (11.50).

36. Consider a population size $N(t)$ whose fluctuations are governed by the differential equation $dN(t) = N(t)\, dW(t)$, where $\{W(t)\}$ is Brownian motion with drift μ and variance σ^2. Show that the solution in the *Stratonovich* sense has transition density

$$p(t, x, y) = \frac{1}{y\sqrt{2\pi\sigma^2 t}} \exp\left[-\frac{(\ln(y/x) - \mu t)^2}{2\sigma^2 t} \right],$$

and moments

$$E[N(t)|N(0) = n_0] = n_0 \exp[(\mu + \tfrac{1}{2}\sigma^2)t]$$
$$\text{var}[N(t)|N(0) = n_0] = n_0^2 \exp[(2\mu + \sigma^2)t][e^{\sigma^2 t} - 1].$$

37. Solve (in the Stratonovich sense) the following equation describing random growth under a restraining force:

$$dN(t) = N(t)[1 - N(t)/K]\,d\tilde{B}(t), \qquad 0 < N(0) < K,$$

where $\{\tilde{B}(t)\}$ is a Brownian motion with drift μ and variance σ^2.

Hint: Show that

$$Y(t) = \int_0^{N(t)} \frac{d\xi}{\xi(1 - \xi/K)}$$

is a Brownian motion with drift μ and variance σ^2.

38. Given the *Stratonovich* stochastic differential equation

$$dX = -\beta g(X(t)) \left\{ \int_0^{X(t)} [g(\xi)]^{-1}\,d\xi \right\} dt + g(X(t))\,dB,$$

where $g(\xi) > 0$ for all real ξ, $\beta > 0$ and $\{B(t)\}$ is standard Brownian motion, we define

$$Y(t) = \int_0^{X(t)} [g(\xi)]^{-1}\,d\xi.$$

(i) Show that $dY = -\beta Y\,dt + dB$, i.e., Y is an Ornstein–Uhlenbeck process.
(ii) Find the transition density function for $\{X(t)\}$.

39. Let $\{X(t)\}$ be a diffusion process with infinitesimal operator A and suppose that the function $f(x) \equiv 1$ is in $\mathcal{D}(A)$, the domain of A, and that $Af(x) = 0$. Show that if u is in $\mathcal{D}(A)$, and $u(x_0) = \max u(x) > u(x)$ for all $x \neq x_0$, then $Au(x_0) \leq 0$.

Hint: Use the Dynkin form (11.46) of the infinitesimal operator.

40. For a fixed $\alpha \leq 0$, let $\mathcal{D}$ be the set of all twice continuously differentiable functions f defined on $[0, \infty)$, and where the limits of $f(x), f'(x)$, and $f''(x)$ as $x \to \infty$ all exist, with $f''(\infty) = 0$, and $\alpha f'(0) + f''(0) = 0$. Define the operator A on $\mathcal{D}$ by $Af(x) = \tfrac{1}{2}f''(x)$. Show that A defined on $\mathcal{D}$ generates a positive semigroup on $C[0, \infty)$.

Solution: Check that for each g in $C[0, \infty)$ the equation $-\tfrac{1}{2}f'' + \lambda f = g$ admits a unique solution f in $\mathcal{D}$ as prescribed, and that if $g \geq 0$, then $f \geq 0$.

41. Consider $\mu(x)$ and $\sigma^2(x)$ continuously differentiable on $[0, 1]$ with $\sigma^2(x) > 0$, $0 \leq x \leq 1$. Let $\mathcal{D}$ be the space of all twice continuously differentiable functions on $[0, 1]$ with $f'(0) = f'(1) = 0$. Define the operator A on $\mathcal{D}$ by $Af(x) = \tfrac{1}{2}\sigma^2(x)f''(x) + \mu(x)f'(x)$. Show that A determines the infinitesimal operator of a diffusion process.

Hint: Check the conditions of the Hille-Yosida theorem.

42.   Let $\mathscr{D}$ be the set of all twice continuously differentiable functions f defined on the circle of circumference 1, (i.e., f is twice continuously differentiable and is periodic of period 1). Define the operator A on $\mathscr{D}$ by $Af(x) = \tfrac{1}{2}f''(x)$.

(i)   Check the hypotheses of the Hille–Yosida theorem affirming that A qualifies as an infinitesimal operator.

(ii)   Establish that the transition density of the induced process possesses the spectral representation

$$p(t, x, y) = 1 + \frac{2}{\pi} \sum_{n=1}^{\infty} \exp\left(-\frac{n^2\pi^2 t}{2}\right) \cos n\pi x \cos n\pi y.$$

(iii)   What is the stationary distribution of this process, commonly called Brownian motion on the circle?

43.   Consider a semigroup U_t for a Markov process. Let A be the infinitesimal generator with domain $\mathscr{D}$. If v is in $\mathscr{D}$, show that

$$\sup_x |\lambda R_\lambda v(x) - v(x)| \to 0 \qquad \text{as} \quad \lambda \to \infty.$$

44.   Consider Brownian motion restricted to the closed interval $I = [0, 1]$ with 0 and 1 acting as reflecting barriers. Show that $Af(x) = \tfrac{1}{2}f''(x)$ with $f'(0) = f'(1) = 0$ describes the infinitesimal operator of this diffusion.

45.   For reflecting Brownian motion show that the infinitesimal generator A is characterized by the domain $\mathscr{D}$ of all functions $f(x)$ on $[0, \infty)$ with $f(x), f'(x)$, and $f''(x)$ continuous on the interval $[0, \infty]$ with $f''(\infty) = 0$ and $f'(0) = 0$.

Hint: Using the explicit form of the corresponding semigroup U_t, show that $[U_h f(0) - f(0)]/h$ behaves like

$$f'(0)\sqrt{\frac{2}{\pi h}} + \frac{1}{2}f''(0) \qquad \text{as} \quad h \to 0.$$

46.   Consider Brownian motion on $[0, \infty)$ such that when 0 is first attained, the process is killed. Show that $f(0) = 0$ for every function f in the domain of the infinitesimal operator.

47.   Consider absorbing Brownian motion on $[0, \infty)$ where 0 is a trap state. Show that the domain $\mathscr{D}$ of the infinitesimal operator A has f, f', and f'' continuous on $[0, \infty]$ and $f''(0) = 0$.

48.   Let $\mathscr{D}$ be the space of functions $f(x)$ defined on $[0, \infty)$ for which $f(x), f'(x)$, and $f''(x)$ are continuous and bounded, $\lim_{x \to \infty} f''(x) = 0$ and $f'(0) = 0$. For f in $\mathscr{D}$, define the operator A by $Af(x) = \tfrac{1}{2}f''(x)$. Show that A satisfies the conditions of the Hille–Yosida theorem on the space $C[0, \infty]$.

Hint: Check that (i) $\mathscr{D}$ is dense in $C[0,\infty]$ and (ii) for each g in $C[0, \infty]$ and $\lambda > 0$ there exists a unique f in $\mathscr{D}$ for which

$$\lambda f(x) - \tfrac{1}{2}f''(x) = g(x).$$

49. Show that the spectral representation of the transition density for absorbing Brownian motion is proportional to the Fourier sine transform

$$\int_0^\infty e^{-\lambda^2 t} \sin x\lambda \sin y\lambda \, d\lambda.$$

50. Determine the spectral representation for standard Brownian motion in N dimensions.

Answer: The inverse Fourier transform of the multinormal characteristic function.

51. Determine the spectral representation for Brownian motion with drift.

52. Define $Z(t) = \int_t^{t+1} B(s) \, ds - B(t)$ where $\{B(t)\}$ is standard Brownian motion. Show that $Z(t)$ has stationary increments.

53. Let $\{X(t), t \geq 0\}$ be a Bessel process of order α on $I = [0, \infty)$. (Consult Example 6 on page 236.) Then $Au = \frac{1}{2}u'' + [\frac{1}{2}(\alpha - 1)/x]u'$. Assume that $\alpha \geq 2$, or equivalently, that 0 is an entrance boundary. (See Example 3 of Section 6.) Then

$$\mathcal{D}(A) = \{u: u \in C^2[0, \infty), u(x) \to 0 \text{ and } Au(x) \to 0 \text{ as } x \to \infty; x^{\alpha-1}u'(x) \to 0 \text{ as } x \to 0\}.$$

Let σ_K be the first passage time out of $[0, K]$.
   (i) Show that $E_x[\sigma_K] = (K^2 - x^2)/\alpha$ for $0 \leq x \leq K$.
   (ii) Show that $E_x[\sigma_K^2] = 4K^2(K^2 - x^2)/\alpha^2(\alpha + 2) + (K^2 - x^2)^2/\alpha(\alpha + 2)$ for $0 \leq x \leq K$.

Hint: (i) Apply the Dynkin formula as in Section 11. As in Eq. (11.41), take

$$u(x) = \begin{cases} K^2 - x^2 & \text{for } 0 \leq x \leq K, \\ \text{very smooth} & \text{for } x > K. \end{cases}$$

Show that u is in $\mathcal{D}(A)$ and follow the method of (11.40) to (11.43).
   (ii) Take

$$v(x) = \begin{cases} (K^2 - x^2)^2 & \text{for } 0 < x < K, \\ \text{very smooth} & \text{for } x > K, \end{cases}$$

and follow the approach of (12.24) to (12.25).

54. Consider a regular diffusion on (l, r). Give reasons that if both ends are entrance boundaries, then the process converges to a unique limit stationary distribution. Provide examples showing that, when both boundaries are natural, a limiting distribution may or may not exist.

55. Consider a diffusion process $\{X(s), s \geq 0\}$ on the state space $I = [0, \infty)$ with $\mu(x) = 0$ and $\sigma^2(x) = x$. Define the Kac functional $E_x[\exp\{-\lambda \int_0^t X(s) \, ds\}] = v(t, x)$. Show by direct verification that $w(t, x) = \exp(-x\sqrt{2\lambda} \tanh t\sqrt{\lambda/2})$ is a solution to

$$\frac{\partial v}{\partial t} = \frac{x}{2} \frac{\partial^2 v}{\partial x^2} - \lambda x v \quad \text{and} \quad v(t, 0) = 1.$$

Conclude that $w(t, x) = v(t, x)$.

56. Let $B(t)$ be standard Brownian motion. Show that the Kac functional

$$v(t, x, \lambda) = E_x\left[\exp\left\{-\lambda \int_0^t B^2(\tau)\, d\tau\right\}\right], \qquad \lambda > 0$$

is

$$v(t, x, \lambda) = \frac{\exp\{-\sqrt{\lambda}(x^2/2)\tanh 2t\sqrt{\lambda}\}}{\sqrt{\cosh 2t\sqrt{\lambda}}}$$

57. Establish that a probability density $\varphi(x)$ for a diffusion $X(t)$ of semigroup U_t with infinitesimal generator A is a stationary density if and only if

$$\int Au(x)\varphi(x)\, dx = 0 \qquad \text{for all} \quad u \in \mathcal{D}(A).$$

58. Consider a sequence of $M/M/1$ queueing systems indexed by $n = 1, 2, \ldots$. Thus, for the nth system customers arrive according to a Poisson process of parameter λ_n and served in order of their arrival. The service times are exponentially distributed having mean μ_n. Define $\rho_n = \lambda_n/\mu_n$. Let $Q_n(t)$ be the number of customers waiting for and receiving service at time epoch t in the nth queueing system. Form the rescaled process

$$Y_n(t) = \frac{Q_n(nt)}{\sqrt{n}} \qquad \text{for} \quad 0 \leq t \leq T, \qquad T \text{ fixed}.$$

Let $X(t)$, $t \geq 0$ be a Brownian motion with drift process of mean coefficient ct and variance coefficient σ^2, that is, $X(t) = \sigma B(t) + ct$ with $B(t)$ standard Brownian motion. Let $R(t) = X(t) - \inf\{X(s); 0 \leq s \leq t\}$ which is $X(t)$ reflected at the origin. Let $R_T(t)$ denote the restriction of $R(t)$ to $[0, T]$. Suppose as $n \to \infty$, that $\lambda_n \to \lambda$, $\mu_n \to \mu$, $\lambda < \mu$, $[\lambda_n - \mu_n]\sqrt{n} \to c < 0$. Show that the process $Y_n(t)$ for each t, $0 \leq t \leq T$, converges in the sense of distributions to $R_T(t)$ based on $X(t)$ with parameters c and $\sigma^2 = \lambda + \mu$.

59. Let $X(t)$ be Brownian motion with drift parameter $c > 0$ and variance coefficient, 1. We will refer to $X(t)$ as an income process with $X(0) = 0$. Define the corresponding assets process (for an initial endowment x) by

$$Y(t) = e^{\beta t}x + \int_0^t e^{\beta(t-s)}\, dX(s), \qquad t \geq 0$$

where the money earns at an interest rate β. By integration by parts, observe that

$$Y(t) = e^{\beta t}x + X(t) + \beta \int_0^t e^{\beta(t-s)}X(s)\, ds.$$

Show that, in terms of standard Brownian motion $B(t)$, we have

$$Y(t) = \frac{1}{\sqrt{2\beta}} B(e^{2\beta t} - 1) + \frac{c}{\beta}(e^{\beta t} - 1).$$

Find the infinitesimal mean and variance of $Y(t)$ (cf. (5.23) and (14.19)).
   Let $\{T = \inf t \geq 0: Y(t) < 0\}$ which is called the ruin time. Find

$$r(y) = \Pr\{T < \infty \mid Y(0) = y\} \qquad \text{for} \quad y > 0.$$

60. Find the spectral representation analogous to (13.14) on $[0, \infty)$ for

$$\frac{\partial u}{\partial t} = \frac{1}{2} \frac{1}{r^{N-1}} \frac{\partial}{\partial r} \left(r^{N-1} \frac{\partial u}{\partial r} \right) - \gamma r \frac{\partial u}{\partial r}, \qquad r > 0, \quad t > 0$$

(γ a positive parameter). What is the nature of the boundaries at 0 and ∞?

NOTES

Ito and McKean's advanced treatise [1] is virtually the definitive source on the theory of one-dimensional diffusion processes.

Some of the most interesting ideas in diffusion processes find their genesis in Lévy's inspiring book [2].

A wide ranging presentation of Markov processes is the two volume work by Dynkin [3], where diffusion processes occupy a prominent role.

A compact monograph highlighting the essential concepts and techniques of stochastic integrals and the stochastic calculus in the Ito sense is McKean [4].

Important classes of stochastic processes converging to diffusion are discussed in the basic book of Billingsley [5].

The multivolume works of Gikman and Skorokhod [6, 7] present extensive treatments of the theory of stochastic differential equations covering one and higher dimensions.

Diffusion process formulations and methodology play an important role in modeling dynamical systems for physical and engineering phenomena. In this vein, Jazwinski [8] and Wong [10] highlight a wide spectrum of applications.

Friedman [13] elaborates the theory of stochastic differential equations, mainly in the context of stochastic control problems, and in their relation to the solutions of parabolic partial differential equations. The motivation of stochastic control and statistical decision processes also underlies the developments in Lipster and Shiryaev [15]. A brief account is contained in the monograph of Kushner [9].

Mandl [11] handles many facets of one-dimensional diffusions in a direct analytical framework.

A far ranging advanced mathematical exposition by Ikeda and Watanabe is in press which integrates the subjects of stochastic integration with respect to generalized martingales instead of Brownian motion, the theory of stochastic differential equations with special attention to diffusion processes on manifolds, featuring theorems on comparisons and approximations of different processes.

Arnold [14] has provided a succinct presentation of stochastic differential equations and applications. For another approach, see McShane [12].

Four chapters of Ewens [16] exhibit the pervasive and important role of diffusion stochastic processes in treating a number of important population genetics models.

REFERENCES

1. K. Ito and H. P. McKean, Jr., "Diffusion Processes and Their Sample Paths." Springer-Verlag, Berlin and New York, 1965.
2. P. Lévy, "Processes Stochastiques et Mouvement Brownian." 2nd ed. Gauthier-Villars, Paris, 1965.

3. E. B. Dynkin, "Markov Process, I and II." Springer-Verlag, Berlin and New York, 1965.
4. H. P. McKean, Jr., "Stochastic Integrals." Academic Press, New York, 1969.
5. P. Billingsley, "Convergence of Probability Measures." Wiley, New York, 1968.
6. I. Gikman and A. V. Skorokhod, "Introduction to the Theory of Random Processes." Saunders, Philadelphia, 1969.
7. I. Gikman and A. V. Skorokhod, "The Theory of Stochastic Processes." Springer-Verlag, Berlin and New York, 1974–71979.
8. A. H. Jazwinski, "Stochastic Processes and Filtering Theory." Academic Press, New York, 1970.
9. H. J. Kushner, "Introduction to Stochastic Control." Holt, New York, 1971.
10. E. Wong, "Stochastic Processes in Information and Dynamical Systems." McGraw-Hill, New York, 1971.
11. P. Mandl, "Analytical Treatment of One-Dimensional Markov Processes." Springer, Berlin and New York, 1968.
12. E. J. McShane, "Stochastic Calculus and Stochastic Models." Academic Press, New York, 1974.
13. A. Friedman, "Stochastic Differential Equations and Applications. I and II." Academic Press, New York, 1975–1976.
14. L. Arnold, "Stochastic Differential Equations: Theory and Applications." Wiley, New York, 1974.
15. R. Lipster and A. N. Shiryaev, "Statistics of Random Processes. I and II." Springer-Verlag, Berlin and New York, 1977–78.
16. W. J. Ewens, "Mathematical Population Genetics." Springer-Verlag, Berlin and New York, 1979.

Chapter 16

COMPOUNDING STOCHASTIC PROCESSES

In this chapter we treat a series of isolated stochastic models, partly motivated by applications to astronomy, biology, engineering, and physics. These processes are formed by compounding various classical processes, including Poisson, branching, and growth processes of the diffusion type. In each case, a secondary process feeds into a primary process whose state variable is the object of study. In Section 1 we shall characterize multidimensional Poisson processes and in the following section give an application to astronomy. The concept of multi-dimensional Poisson processes will play an important part in the formulation of cascade or compound stochastic processes. Some of these will be studied in the later sections of this chapter (e.g., see Section 2).

In Section 3 we examine a stochastic model involving growth and immigration. In Section 4 we formulate a stochastic process of growth involving two types, a normal type and a mutant type. The population of wild (i.e., normal) types grows deterministically whereas the mutant type grows in accordance with the laws of a Markov branching process. Moreover, each normal type at its death (lifetime is exponentially distributed) changes into a mutant type and then reproduces like any other mutant individual.

Sections 5 and 6 treat stochastic models of growth which take account of the factor of geographical distribution and spread of the population as well as their natural growth behavior.

The stochastic processes investigated are typical of a large class of general cascade processes. The purpose of this chapter is to introduce the student to the richness of applications and subtleties of analysis of problems involving combinations of stochastic processes.

Sections 7 and 8 are devoted to a review of some deterministic models of population growth, taking account of the age structure of the population. The important compound Poisson process is introduced with applications in Section 9.

1: Multidimensional Homogeneous Poisson Processes

Poisson processes were introduced in Chapter 4 where the index parameter is identified as a real positive number $t \geq 0$, usually referred to as time. We now formulate a version of the Poisson process where the parameter value is determined by the measure of a set in the plane or in space or even in more general spaces. The objective of this section is to define some versions of multidimensional Poisson processes and to describe some examples of these processes and their applications.

In Chapter 4 the Poisson process $\{X(t), t \geq 0\}$, was characterized axiomatically. It was proved that $X(t)$ has the probability distribution

$$\Pr\{X(t) = k\} = e^{-\lambda t} \frac{(\lambda t)^k}{k!} \quad \text{for} \quad t \geq 0, \quad k = 0, 1, 2, \ldots,$$

where λ is a positive constant interpreted as the average rate at which events are happening per unit time. In this section we shall introduce a set of postulates that characterize a generalized homogeneous spatial Poisson process $X(S)$, where the parameter S denotes a bounded region of the plane or space and $X(S)$ has the probability distribution

$$\Pr\{X(S) = k\} = e^{-\lambda A(S)} \frac{[\lambda A(S)]^k}{k!} \quad \text{for} \quad k = 0, 1, 2, \ldots. \quad (1.1)$$

Here, λ is a positive constant called the intensity parameter of the process and $A(S)$ represents the area or volume of S, depending on whether S is a region in the plane or space. The constant λ is called the parameter of the multidimensional Poisson process. The required postulates are the following:

(i) For $X(S)$ only nonnegative integer values are assumed and $0 < \Pr\{X(S) = 0\} < 1$ if $A(S) > 0$.

(ii) The probability distribution of $X(S)$ depends on S only through the value of $A(S)$ with the further property that if $A(S) \to 0$ then $\Pr\{X(S) \geq 1\} \to 0$.

(iii) If $S_1, S_2, \ldots, S_n, (n \geq 1)$ are disjoint regions, then $X(S_1), \ldots, X(S_n)$ are mutually independent random variables and

$$X(S_1 \cup \cdots \cup S_n) = X(S_1) + \cdots + X(S_n).$$

(iv)

$$\lim_{A(S) \to 0} \frac{\Pr\{X(S) \geq 1\}}{\Pr\{X(S) = 1\}} = 1.$$

Prior to offering further descriptive discussion of these axioms it is worthwhile to highlight some suggestive examples.

(a)   In three dimensions $X(S)$ could represent the number of stars located in the region S.

(b)   On a two-dimensional plate $X(S)$ may represent the number of bacteria of a certain species contained in the region S.

The motivation and interpretation of the above axiom system is quite evident. Postulate (ii) asserts that $X(S)$ does not depend on the shape of S but only on its area or volume. This appears to be reasonable with reference to examples (a) and (b). According to Postulate (iii), to use the terminology of Example (a), the number of stars contained in disjoint regions are independent random variables and the value of $X(S)$ for the combined region is the sum of $X(\cdot)$ for the component regions. The independence assumption seems to be a reasonable approximation to the situation of the actual distribution of stars. Postulate (iv) is rather intuitive and self-explanatory.

Our main objective in this section is to prove the following theorem:

Theorem 1.1.   *If a random process $X(S)$ defined with respect to regions S of Euclidean n space satisfies Postulates* (i)–(iv), *then $X(S)$ has the distribution given in* (1.1).

Proof.   Consider an arbitrary finite region S of positive area (volume) $A(S) > 0$. Divide S into disjoint regions $S_1, S_2, \ldots, S_n$ of equal area (volume), i.e.,

$$S_1 \cup S_2 \cup \cdots \cup S_n = S,$$

$$S_i \cap S_j = \varnothing, \qquad i \neq j \quad (\varnothing = \text{empty set}),$$

$$A(S_i) = \frac{1}{n} A(S) \qquad \text{for all} \quad i = 1, 2, \ldots, n.$$

Then by Postulate (iii)

$$\begin{aligned}
\Pr\{X(S) = 0\} &= \Pr\{X(S_1 \cup \cdots \cup S_n) = 0\} \\
&= \Pr\{X(S_1) + \cdots + X(S_n) = 0\}.
\end{aligned}$$

But Postulate (i) tells us that $\sum_{i=1}^{n} X(S_i) = 0$ can occur if and only if $X(S_i) = 0$ for all $i = 1, 2, \ldots, n$. Then, using the independence of $X(S_i)$, $i = 1, 2, \ldots, n$, as postulated in (iii), we have

$$\Pr\{X(S) = 0\} = \Pr\{X(S_i) = 0, i = 1, 2, \ldots, n\} = \prod_{i=1}^{n} \Pr\{X(S_i) = 0\}.$$

According to Postulate (ii) $\Pr\{X(S_i) = 0\}$ depends only on $A(S_i) = (1/n)A(S)$; hence

$$\Pr\{X(S_1) = 0\} = \Pr\{X(S_2) = 0\} = \cdots = \Pr\{X(S_n) = 0\}.$$

Thus, we have

$$\Pr\{X(S) = 0\} = [\Pr\{X(S_n) = 0\}]^n. \tag{1.2}$$

Further

$$\Pr\{X(S_n) = 0\} = 1 - \Pr\{X(S_n) \geq 1\}.$$

Taking logarithms on both sides of (1.2), which is permissible because of Postulate (i), yields

$$-\log \Pr\{X(S) = 0\} = -n \log[1 - \Pr\{X(S_n) \geq 1\}]$$
$$= n[\Pr\{X(S_n) \geq 1\} + \tfrac{1}{2}(\Pr\{X(S_n) \geq 1\})^2 + \cdots], \tag{1.3}$$

where we have used the expansion

$$-\log(1 - x) = x + \tfrac{1}{2}x^2 + \tfrac{1}{3}x^3 + \cdots,$$

valid when $0 \leq x < 1$. It is clear that $0 \leq \Pr\{X(S_n) \geq 1\} < 1$, since otherwise we would have $\Pr\{X(S_n) = 0\} = 0$, which implies $\Pr\{X(S) = 0\} = 0$. This, however, is impossible by Postulate (i), as we assume $A(S) > 0$. Now, formula (1.3) may be rewritten in the form

$$-\log \Pr\{X(S) = 0\} = n \Pr\{X(S_n) = 1\}$$

$$\times \left[\frac{\Pr\{X(S_n) \geq 1\}}{\Pr\{X(S_n) = 1\}} (1 + O(\Pr\{X(S_n) \geq 1\})) \right]. \tag{1.4}$$

The interpretation of $O(\Pr\{X(S_n) \geq 1\})$ is as usual: The quantity

$$\frac{O(\Pr\{X(S_n) \geq 1\})}{\Pr\{X(S_n) \geq 1\}}$$

is bounded as $n \to \infty$. Note by Postulate (ii) that

$$\Pr\{X(S_n) \geq 1\} \to 0 \qquad \text{since} \qquad A(S_n) = \frac{1}{n} A(S) \to 0, \quad n \to \infty.$$

Further, it follows by Postulate (iv) that

$$\lim_{n \to \infty} \frac{\Pr\{X(S_n) \geq 1\}}{\Pr\{X(S_n) = 1\}} = \lim_{A(S_n) \to 0} \frac{\Pr\{X(S_n) \geq 1\}}{\Pr\{X(S_n) = 1\}} = 1.$$

Hence, letting $n \to \infty$ on both sides of (1.4) yields

$$-\log \Pr\{X(S) = 0\} = \lim_{n \to \infty} n \Pr\{X(S_n) = 1\}. \tag{1.5}$$

Because of Postulate (i) the left side must necessarily be positive and finite.

Consider the generating function of $X(S)$ and $X(S_n)$:

$$g(s) = E[s^{X(S)}] = \sum_{k=0}^{\infty} \Pr\{X(S) = k\}s^k$$

and

$$g_n(s) = E[s^{X(S_n)}] = \sum_{k=0}^{\infty} \Pr\{X(S_n) = k\}s^k.$$

Then, by Postulates (ii) and (iii)

$$g(s) = E[s^{X(S)}] = E[s^{X(S_1)+\cdots+X(S_n)}] = \prod_{i=1}^{n} E[s^{X(S_i)}] = (E[s^{X(S_n)}])^n;$$

that is,

$$g(s) = [g_n(s)]^n. \tag{1.6}$$

We may write $g_n(s)$ in the form

$$g_n(s) = \Pr\{X(S_n) = 0\} + \Pr\{X(S_n) = 1\}s + \Pr\{X(S_n) > 1\}\theta(s),$$

where $|\theta(s)| \leq 1$. But

$$\Pr\{X(S_n) = 0\} = 1 - \Pr\{X(S_n) = 1\} - \Pr\{X(S_n) > 1\}.$$

Hence, substituting for $\Pr\{X(S_n) = 0\}$ we get

$$g_n(s) = 1 + (s - 1)\Pr\{X(S_n) = 1\} + (\theta(s) - 1)\Pr\{X(S_n) > 1\}. \tag{1.7}$$

We will now use Postulate (iv) which asserts that

$$\frac{\Pr\{X(S_n) > 1\}}{\Pr\{X(S_n) = 1\}} \to 0, \qquad n \to \infty, \tag{1.8}$$

and we also need the fact that $\Pr\{X(S_n) = 1\} \to 0$ as $n \to \infty$. Indeed, we established above (see (1.6)) that

$$g(s) = (E[s^{X(S_n)}])^n$$

or, what is the same,

$$[g(s)]^{1/n} = \sum_{k=0}^{\infty} \Pr\{X(S_n) = k\}s^k, \qquad 0 \leq s \leq 1.$$

By hypothesis (i) we know that $g(0) = \Pr\{X(S) = 0\} > 0$. Therefore,

$$\Pr\{X(S_n) = 0\} = \frac{d}{ds}[g(s)]^{1/n}\Big|_{s=0} = \frac{1}{n}\frac{g'(s)}{g(s)}[g(s)]^{1/n}\Big|_{s=0} = \frac{1}{n}\frac{g'(0)}{g(0)}[g(0)]^{1/n},$$

which clearly tends to zero as $n \to \infty$.

Now, referring to (1.6)–(1.8) and expanding the logarithm in the form

$$\log(1 + z) = z - \frac{z^2}{2} + \frac{z^3}{3} - \cdots = z + o(z), \qquad |z| \to 0,$$

we obtain the formula

$$\log g(s) = n \log g_n(s)$$
$$= n[(s - 1) \Pr\{X(S_n) = 1\} + (\theta(s) - 1) \Pr\{X(S_n) > 1\}$$
$$+ o(\Pr\{X(S_n) = 1\})]$$

Taking limits on both sides as $n \to \infty$, again using Postulate (iv) (specifically relation (1.8)), yields

$$\log g(s) = (s - 1) \lim_{n \to \infty} n \Pr\{X(S_n) = 1\}. \tag{1.9}$$

We infer from (1.5) and (1.9) that

$$\log g(s) = -(s - 1) \log \Pr\{X(S) = 0\}$$

or

$$g(s) = \exp[(s - 1)(-\log \Pr\{X(S) = 0\})]. \tag{1.10}$$

This expression is the probability generating function of a Poisson law whose expectation is

$$E[X(S)] = -\log \Pr\{X(S) = 0\}.$$

But we also know that the expectation is a nonnegative additive function that depends only on $A(S)$, and this implies

$$-\log \Pr\{X(S) = 0\} = \lambda A(S). \tag{1.11}$$

A formal proof of this last statement goes as follows. Let f denote the function satisfying

$$E[X(S)] = f(A(S)).$$

This relation derives from Postulate (ii). We shall now prove that in fact it is linear. Let S_1 and S_2 be two bounded disjoint sets. Then

$$E[X(S_1 \cup S_2)] = f(A(S_1) + A(S_2)),$$

owing to the additivity of $A(S)$. On the other hand, by Postulate (iii), we have

$$E[X(S_1 \cup S_2)] = E[X(S_1)] + E[X(S_2)] = f(A(S_1)) + f(A(S_2)).$$

Since $A(S)$ can vary from 0 to ∞, this implies that

$$f(x + y) = f(x) + f(y) \qquad \text{for all} \quad x, y \geq 0. \tag{1.12}$$

Moreover, by its very meaning $f(x) \geq 0$ and clearly $f(0) = 0$. The only solution of (1.12) with these properties is the linear function $f(x) = \lambda x$ for some constant λ (cf. page 125 of Chapter 4). This proves (1.11).

In view of the remarks following formula (1.5), we conclude that λ is a positive real parameter. Substituting in (1.10) we have

$$g(s) = e^{\lambda A(S)(s - 1)}$$

or equivalently

$$g(s) = e^{-\lambda A(S)} \sum_{k=0}^{\infty} \frac{[\lambda A(S)]^k}{k!} s^k.$$

This confirms (1.1), that the probability distribution of $X(S)$ is Poisson, and the proof of Theorem 1.1 is complete. ∎

We next elaborate some further distribution properties of the stochastic process characterized by Postulates (i)–(iv). It is convenient to describe the event $\{X(S) = k\}$ by "there are exactly k points in the region S."

We now show that if the process $X(S)$ satisfies Postulates (i)–(iv), i.e., is a Poisson process in the plane or space, then under the condition that there is exactly one point in a region S of positive area, i.e., $X(S) = 1$, $A(S) > 0$, this point is uniformly distributed in S. Indeed, let $S = S_1 \cup S_2$, where S_1 and S_2 are disjoint regions. Then by virtue of Postulate (iii) we have

$$\begin{aligned} \Pr\{X(S_1) = 1 \mid X(S) = 1\} &= \frac{\Pr\{X(S_1) = 1, X(S) = 1\}}{\Pr\{X(S) = 1\}} \\ &= \frac{\Pr\{X(S_1) = 1, X(S_2) = 0\}}{\Pr\{X(S) = 1\}} \\ &= \frac{\Pr\{X(S_1) = 1\} \Pr\{X(S_2) = 0\}}{\Pr\{X(S) = 1\}} . \\ &= \frac{\exp[-\lambda A(S_1)]\lambda A(S_1) \exp[-\lambda A(S_2)]}{\exp[-\lambda A(S)]\lambda A(S)} . \end{aligned}$$

Since S_1 and S_2 are disjoint and their union is S, $A(S) = A(S_1) + A(S_2)$ and the exponential factors cancel. Thus

$$\Pr\{X(S_1) = 1 \mid X(S) = 1\} = \frac{A(S_1)}{A(S)},$$

and this is equivalent to saying that the point in S is uniformly distributed.

This result can be extended as follows:

Theorem 1.2. *If $X(S)$ fulfills Postulate (i)–(iv) then under the condition $X(S) = k$, for $A(S) > 0$, these k points are independent and uniformly distributed in S.*

Remark. The assertion that k points in S are independent and uniformly distributed shall mean that, given any n disjoint regions $S_1, S_2, \ldots, S_n$, with $\bigcup_{i=1}^{n} S_i = S$, and any integers $k_1, k_2, \ldots, k_n$, $\sum k_i = k$, then

$$\Pr\{k_1 \text{ points lie in } S_1; k_2 \text{ points lie in } S_2; \ldots, k_n \text{ points lie in } S_n \mid X(S) = k\}$$

$$= \frac{k!}{k_1! k_2! \cdots k_n!} \left[\frac{A(S_1)}{A(S)}\right]^{k_1} \left[\frac{A(S_2)}{A(S)}\right]^{k_2} \cdots \left[\frac{A(S_n)}{A(S)}\right]^{k_n}.$$

Proof. Let

$$S = S_1 \cup S_2 \cup \cdots \cup S_n,$$

where $S_1, S_2, \ldots, S_n$ are disjoint regions. Then for any nonnegative integers $k_1, k_2, \ldots, k_n$ with $k_1 + k_2 + \cdots + k_n = k$

$$\Pr\{X(S_1) = k_1, X(S_2) = k_2, \ldots, X(S_n) = k_n | X(S) = k\}$$

$$= \frac{\Pr\{X(S_1) = k_1, X(S_2) = k_2, \ldots, X(S_n) = k_n\}}{\Pr\{X(S) = k\}}$$

$$= \frac{\Pr\{X(S_1) = k_1\} \Pr\{X(S_2) = k_2\} \cdots \Pr\{X(S_n) = k_n\}}{\Pr\{X(S) = k\}}$$

$$= \frac{e^{-\lambda A(S_1)} \dfrac{[\lambda A(S_1)]^{k_1}}{k_1!} e^{-\lambda A(S_2)} \dfrac{[\lambda A(S_2)]^{k_2}}{k_2!} \cdots e^{-\lambda A(S_n)} \dfrac{[\lambda A(S_n)]^{k_n}}{k_n!}}{e^{-\lambda A(S)} \dfrac{[\lambda A(S)]^k}{k!}}$$

$$= \frac{k!}{k_1! k_2! \cdots k_n!} \left[\frac{A(S_1)}{A(S)}\right]^{k_1} \left[\frac{A(S_2)}{A(S)}\right]^{k_2} \cdots \left[\frac{A(S_n)}{A(S)}\right]^{k_n},$$

because $A(S_1) + A(S_2) + \cdots + A(S_n) = A(S)$. ∎

2: An Application of Multidimensional Poisson Processes to Astronomy

Consider stars distributed in space in accordance with a three-dimensional Poisson process $X(S)$ as described in Section 1. Let **x** and **y** designate general three-dimensional vectors. Assume that the light intensity exerted at **x** by a star located at **y** is $f(\mathbf{x}, \mathbf{y}, \alpha)$. Here α is a real random parameter depending on the intensity of the star at **y**. We assume that the parameters α associated with different stars are independent, identically distributed, random variables possessing a common density function $k(\cdot)$. We also assume that the combined intensity exerted at the point **x** due to light signals of different stars accumulates additively. Let $Y(\mathbf{x}, S)$ denote the total light intensity at the point **x** due to signals emanating from all the stars located in region S, i.e.,

$$Y(\mathbf{x}, S) = \sum_{\mathbf{y}_r \in S} f(\mathbf{x}, \mathbf{y}_r, \alpha_r).$$

Note that the summation contains a random number of terms which is actually finite with probability 1. We want to find the distribution of $Y(\mathbf{x}, S)$. We do this by determining the Laplace transform $g(z; \mathbf{x}, S)$ of $Y(\mathbf{x}, S)$, i.e.,

$$g(z; \mathbf{x}, S) = E[e^{-zY(\mathbf{x}, S)}] = \int_0^\infty e^{-z\xi} h(\xi, \mathbf{x}, S) \, d\xi, \qquad z > 0,$$

where $h(\cdot; \mathbf{x}, S)$ denotes the density function of $Y(\mathbf{x}, S)$.

Of course, in principle, from knowledge of the Laplace transform we can compute the moments of $Y(\mathbf{x}, S)$ routinely, and generally there is an inversion formula which calculates h in terms of g. The actual inversion expression takes on a rather complicated form in the case at hand and so we do not bother with it here.

Conditioning on the values of $X(S)$, we have

$$g(z; \mathbf{x}, S) = E[e^{-zY(\mathbf{x}, S)}] = \sum_{k=0}^{\infty} E[e^{-zY(\mathbf{x}, S)} | X(S) = k] \, \Pr\{X(S) = k\}.$$

But we know according to Theorem 1.2 that under the condition $X(S) = k$. i.e., k points in S, these k points are independently uniformly distributed in S. Hence

$$E[e^{-zY(\mathbf{x}, S)} | X(S) = k] = (E[e^{-zY(\mathbf{x}, S)} | X(S) = 1])^k.$$

To compute $E[e^{-zY(\mathbf{x}, S)} | X(S) = 1]$, note that if $X(S) = 1$, $Y(\mathbf{x}, S) = f(\mathbf{x}, \mathbf{y}, \alpha)$, where $\mathbf{y}$ is the position of a single star in S and α is the corresponding random parameter which reflects its intensity. Now, since the position of this star is uniformly distributed in S, we have

$$E[e^{-zY(\mathbf{x}, S)} | X(S) = 1] = \frac{1}{A(S)} \int_S \left[\int_{-\infty}^{\infty} e^{-zf(\mathbf{x}, \mathbf{y}, \alpha)} k(\alpha) \, d\alpha \right] d\mathbf{y},$$

where the integral with respect to $\mathbf{y}$ denotes a triple integral over the three-dimensional region S. Combining the relations above in the obvious manner, we obtain the Laplace transform of $Y(\mathbf{x}, S)$,

$$g(z; \mathbf{x}, S) = \sum_{k=0}^{\infty} \left\{ \frac{1}{A(S)} \int_S \left[\int_{-\infty}^{\infty} e^{-zf(\mathbf{x}, \mathbf{y}, \alpha)} k(\alpha) \, d\alpha \right] d\mathbf{y} \right\}^k$$

$$\times \exp[-\lambda A(S)] \frac{[\lambda A(S)]^k}{k!}$$

$$= e^{-\lambda A(S)} \exp\left\{ \lambda \int_S \left[\int_{-\infty}^{\infty} e^{-zf(\mathbf{x}, \mathbf{y}, \alpha)} k(\alpha) \, d\alpha \right] d\mathbf{y} \right\}$$

$$= \exp\left\{ \lambda \int_S \left[\int_{-\infty}^{\infty} e^{-zf(\mathbf{x}, \mathbf{y}, \alpha)} k(\alpha) \, d\alpha - 1 \right] d\mathbf{y} \right\},$$

because $\int_S d\mathbf{y} = A(S)$. We have determined $g(z; \mathbf{x}, S)$ in terms of $f(\mathbf{x}; \mathbf{y}, \alpha)$, $k(\alpha)$, and S, which can be regarded as known or calculable on the basis of other data.

3: Immigration and Population Growth

The model described in this section is that of a population composed of a single type evolving from an initial population whose growth behavior follows the laws of a continuous time Markov branching process. Moreover, in addition to the inherent growth of the population there is an influx of immigrants of the same type which contribute further descendants whose growth behavior is governed by the laws of the same branching process. The arrival pattern of immigrants into the colony is generally also a stochastic process. We formulate the process, for concreteness, as a model of growth of bacterial populations.

Consider a bacterial colony of n_0 individuals. Assume that each bacterium, independent of the others, produces descendants which in turn produce further offspring, etc. The population growth evolving from a single bacterium describes a continuous time Markov branching process. Let $F(s, t)$ be the probability generating function of the population size at time t derived from a single bacterium. Clearly, the population size at time t due to the existing colony of size n_0 at time 0 is a random variable which has the generating function $[F(s, t)]^{n_0}$. Assume, furthermore, that immigration of new bacteria occurs at times $t_j, j = 1, 2, \ldots, N$. Each immigrant will evolve a population following the same laws of reproduction as the original n_0 bacteria, independent of them and each other. The population size at time t derived from an immigrant arriving at time t_j has the probability generating function $F(s, t - t_j)$. The total population size at time t has the probability generating function

$$[F(s, t)]^{n_0} \prod_{j=1}^{N} F(s, t - t_j),$$

since each bacterium creates independent families. Assume, however, that immigration occurs not at fixed times t_j, but that the t_j constitute events of a Poisson process with parameter r. We want to calculate the probability generating function of the total population size in terms of $F(s, t)$ and r.

The immigration times t_j are random variables and their number $N(t)$ during the time interval $[0, t]$ is also a random variable whose distribution function is Poisson with parameter rt. Let $Y_j(t, t_j)$ denote the population size at time t derived from a single immigrant into the colony at time t_j for $j = 1, 2, \ldots, N(t)$. Then

$$Y(t) = \sum_{j=1}^{N(t)} Y_j(t, t_j)$$

is the population size at time t evolved from immigrants arriving during the time epoch $[0, t]$.

The probability generating function of $Y(t)$ can be computed by conditioning on $N(t)$ in the usual way. This leads to the expression

$$E[s^{Y(t)}] = \sum_{k=0}^{\infty} E[s^{Y(t)} | N(t) = k] \Pr\{N(t) = k\}. \tag{3.1}$$

By virtue of the developments of Section 2, Chapter 4, we know that the joint distribution in $[0, t]$ of the arrival times $t_j, j = 1, 2, \ldots N(t)$, given the number of arrivals $N(t) = k$, is the same as the distribution of the order statistics of k independent, uniformly distributed, random variables over $[0, t]$. Thus

$$E[s^{Y(t)} | N(t) = k] = \frac{k!}{t^k} \int_0^t dt_1 \int_{t_1}^t dt_2 \cdots \int_{t_{k-1}}^t dt_k \, E[s^{\sum_{j=1}^k Y_j(t, t_j)}]$$

$$= \frac{1}{t^k} \int_0^t dt_1 \int_0^t dt_2 \cdots \int_0^t dt_k \, E[s^{\sum_{j=1}^k Y_j(t, t_j)}],$$

because the integrand is a symmetric function of $t_1, t_2, \ldots, t_k$. Further, since different immigrants create independent histories, we get

$$E[s^{\sum_{j=1}^k Y_j(t, t_j)}] = \prod_{j=1}^k E[s^{Y_j(t, t_j)}] = \prod_{j=1}^k F(s, t - t_j).$$

Therefore

$$E[s^{Y(t)} | N(t) = k] = \frac{1}{t^k} \int_0^t dt_1 \int_0^t dt_2 \cdots \int_0^t dt_k \prod_{j=1}^k F(s, t - t_j)$$

$$= \frac{1}{t^k} \prod_{j=1}^k \int_0^t F(s, t - t_j) \, dt_j$$

$$= \left[\frac{1}{t} \int_0^t F(s, t - \tau) \, d\tau \right]^k.$$

Inserting this formula into (3.1) and taking account of the fact that $N(t)$ is a Poisson process we obtain

$$E[s^{Y(t)}] = \sum_{k=0}^\infty \left[\frac{1}{t} \int_0^t F(s, t - \tau) \, d\tau \right]^k e^{-rt} \frac{(rt)^k}{k!}$$

$$= \exp\left\{ r \int_0^t [F(s, t - \tau) - 1] \, d\tau \right\}. \tag{3.2}$$

Hence, the probability generating function of the total population size at time t is

$$G(s, t) = [F(s, t)]^{n_0} \exp\left\{ r \int_0^t [F(s, t - \tau) - 1] \, d\tau \right\}. \tag{3.3}$$

Example. As an example we assume that each individual bacterium follows a growth law described by the Yule process $X(t)$ of parameter $\beta > 0$ (see Section 2 of Chapter 4). Then the size of the population at time t derived from a single bacterium at time $t = 0$ is governed by the probability law

$$P_k(t) = \Pr\{X(t) = k | X(0) = 1\},$$

where

$$P_k(t) = e^{-\beta t}(1 - e^{-\beta t})^{k-1}, \qquad k = 1, 2, \ldots.$$

Equivalently the generating function of the Yule process is given by

$$F(s, t) = e^{-\beta t} \sum_{k=1}^{\infty} (1 - e^{-\beta t})^{k-1} s^k = \frac{se^{-\beta t}}{1 - (1 - e^{-\beta t})s}.$$

Since we assumed that each immigrant follows the same law of growth as the original population, the generating function of the population size emanating from immigrant bacteria, in accordance with (3.2), has the form

$$
\begin{aligned}
E[s^{Y(t)}] &= \exp\left\{ r \int_0^t \frac{se^{-\beta(t-\tau)} \, d\tau}{1 - s + se^{-\beta(t-\tau)}} - rt \right\} \\
&= \exp\left[\frac{-r}{\beta} \log(1 - s + se^{-\beta t}) - rt \right] \\
&= e^{-rt}(1 - s + se^{-\beta t})^{-r/\beta} = e^{-rt}[1 - (1 - e^{-\beta t})s]^{-r/\beta}.
\end{aligned}
$$

If we take account of the original population of bacteria, then by (3.3) the generating function of population size at time t evolving from the initial population in addition to the immigrant population is

$$G(s, t) = \exp[-(r + \beta n_0)t][1 - (1 - e^{-\beta t})s]^{-(n_0 + r/\beta)}.$$

The expected population size at time t is $(\partial/\partial s)G(s, t)|_{s=1}$, which reduces to

$$\left(n_0 + \frac{r}{\beta}\right)(e^{\beta t} - 1).$$

Higher-order moments can also be evaluated by further differentiation of the generating function.

4: Stochastic Models of Mutation and Growth

Often in microbiological populations, initially homogeneous, one or more individuals change into a mutant form which then continues reproducing in that form. The mutation may correspond, for instance, to immunity from virus attack which the descendants inherit or generally to some property distinguishing the mutant form from the original, "wild-type" colony. We now examine a model describing stochastic fluctuations of mutant growth. We assume that the mother or parent colony starts with N individuals at time $t = 0$ and grows deterministically in such a manner that its size at time t is Ne^t. Further we assume that each "wild-type" has a probability $ph + o(h)$ of changing into a mutant form during the time span $[t, t + h]$. Since the parent population at time t is of size Ne^t and taking account of the fact that individuals behave independently, the probability of the formation of some mutant during $[t, t + h]$ is

$$\rho Ne^t h + o(h).$$

Moreover, we postulate that the probability of two or more mutations occurring during the time span $[t, t + h]$ is $o(h)$. It is clear from the formulation above that the number of mutant types as a function of time describes a nonhomogeneous Poisson process with intensity function $r(t) = \rho N e^t$ (see Elementary Problem 12 of Chapter 4).

The developments of Elementary Problems 11 and 12 of Chapter 4, further inform us that the probability generating function of the number of events occurring during the time epoch $[0, t]$ for the nonhomogeneous Poisson process of parameter $r(t)$ is

$$\varphi(t, s) = e^{-m(t)(1 - s)},$$

where

$$m(t) = \int_0^t r(\tau) \, d\tau = \rho N(e^t - 1).$$

Thus, in our particular case

$$\varphi(t, s) = \exp[\rho N(e^t - 1)(s - 1)]. \tag{4.1}$$

Now suppose that each mutant evolves its own growth process, and let $F(s, t)$ denote the probability generating function of the number of descendants of a single mutant at time t after the creation of the mutant. We shall assume in this model that the mutant population undergoes only birth and no death, i.e., $F(0, t) \equiv 0$.

Let $H(s, t; N)$ denote the probability generating function of the number of mutants at time t given that the parent colony consisted of N individuals at time $t = 0$ and there were no mutants present in the population at that time. We want to determine $H(s, t; N)$ in terms of $F(s, t)$ and the parameters ρ and N, which for the problem at hand are regarded as known. To this end, we introduce the probabilities

$$P_k(t) = \Pr\left\{\begin{matrix}\text{there are exactly } k \text{ descendants at time } t \text{ from a} \\ \text{single mutation at time } t = 0\end{matrix}\right\},$$

$$h_k(t; N) = \Pr\left\{\begin{matrix}\text{there are exactly } k \text{ mutants by time } t \text{ provided the} \\ \text{parent population at time } t = 0 \text{ was of size } N \\ \text{and included no mutants}\end{matrix}\right\}.$$

Then

$$F(s, t) = \sum_{k=1}^{\infty} P_k(t)s^k, \tag{4.2}$$

since $P_0(t) = 0$ by assumption, and

$$H(s, t; N) = \sum_{k=0}^{\infty} h_k(t; N)s^k. \tag{4.3}$$

Clearly, from (4.1)

$$h_0(t; N) = \Pr\{\text{first mutation occurs after time } t\} = \varphi(t, 0)$$
$$= \exp[-\rho N(e^t - 1)] = 1 - K(t; N), \qquad (4.4)$$

where $K(t; N)$ is the distribution function of the time of birth of the first mutant for an initial population of N parents. Its density function is

$$-\frac{dh_0(t, N)}{dt} = \rho N e^t \exp[-\rho N(e^t - 1)].$$

The event that there exist exactly $k + 1$ mutants ($k = 0, 1, 2, \ldots$) at time t occurs if the first mutation happens at time τ ($0 \le \tau \le t$), and the mutant and the parent population, now of size $N e^\tau$, together produce k further mutants during the remaining time span $t - \tau$. The probability of no mutation before time τ is $\exp[-\rho N(e^\tau - 1)]$. The occurrence of a mutation during the time interval $(\tau, \tau + d\tau)$ has probability $\rho N e^\tau \, d\tau + o(d\tau)$. Finally, the probability that the mutant form and the parent colony, starting with size $N e^\tau$, will produce together exactly k mutants in time $t - \tau$ is

$$\sum_{l=0}^{k} P_{k-l}(t - \tau) h_l(t - \tau; N e^\tau)$$

(recall $P_0(t) \equiv 0$). But the time τ can be anywhere between 0 and t. Then, by the law of total probability conditioning at the time τ of the occurrence of the first mutant and integrating with respect to the possible values of τ yields

$$h_k(t; N) = \int_0^t \exp[-\rho N(e^\tau - 1)] \rho N e^\tau \left[\sum_{l=0}^{k} P_{k-l}(t - \tau) h_l(t - \tau; N e^\tau) \right] d\tau$$

$$\text{for} \quad k = 1, 2, \ldots.$$

With this quantity in hand and (4.4) we may pass to the corresponding probability generating functions. This leads to the formula

$$H(s, t; N) = \exp[-\rho N(e^t - 1)] + \sum_{k=1}^{\infty} s^k \int_0^t \exp[-\rho N(e^\tau - 1)]$$

$$\times \rho N e^\tau \left[\sum_{l=0}^{k} P_{k-l}(t - \tau) h_l(t - \tau; N e^\tau) \right] d\tau$$

$$= \exp[-\rho N(e^t - 1)] + \int_0^t \exp[-\rho N(e^\tau - 1)]$$

$$\times \rho N e^\tau \left[\sum_{l=0}^{\infty} h_l(t - \tau; N e^\tau) s^l \sum_{k=l}^{\infty} P_{k-l}(t - \tau) s^{k-l} \right] d\tau,$$

where we have used the hypothesis $P_0(t) = 0$. Referring to (4.2) and (4.3) we can write this relation in the simpler form

$$H(s, t; N) = \exp[-\rho N(e^t - 1)] + \rho N \int_0^t e^y \exp[-\rho N(e^y - 1)]$$

$$\times F(s, t - y)H(s, t - y; Ne^y) \, dy. \tag{4.5}$$

This is an integral equation for $H(s, t; N)$ of a rather complicated form. We may solve it, however, by employing the following device. Let $\xi(t; N)$ denote the number of mutants at time t, when the parent population consists of N individuals at time $t = 0$. Since mutations occur according to the laws of a nonhomogeneous Poisson process and individuals act independently, we easily infer that $\xi(t; N)$ satisfies the functional equation

$$\xi(t; N_1) + \xi(t; N_2) = \xi(t; N_1 + N_2). \tag{4.6}$$

Because of the independence of $\xi(t; N_1)$ and $\xi(t; N_2)$ and by the definition

$$H(s, t; N) = E[s^{\xi(t; N)}],$$

we conclude that

$$H(s, t; N_1)H(s, t; N_2) = H(s, t; N_1 + N_2)$$

for all nonnegative integers $N_1, N_2, \ldots$. This plainly implies that

$$H(s, t; N) = [H(s, t; 1)]^N,$$

i.e.,

$$H(s, t; N) = e^{NL(s, t)}, \tag{4.7}$$

where $L(s, t) = \log H(s, t; 1)$. We still must determine the function $L(s, t)$. To this end, we substitute the formula (4.7) into (4.5) and divide by N. This gives

$$\frac{\exp[NL(s, t)] - \exp[-\rho N(e^t - 1)]}{N} = \rho \int_0^t e^y \exp[-\rho N(e^y - 1)]$$

$$\times F(s, t - y)\exp[Ne^y L(s, t - y)] \, dy. \tag{4.8}$$

Now (4.8) presumably is only correct for N a nonnegative integer. However, we will operate with (4.8) as if it were valid for all real $N > 0$. (This can be justified by appropriately varying ρ. We do not enter into details on this point, which is rather delicate.)

Now, let $N \to 0$ in (4.8). Then the right-hand side will approach

$$\rho \int_0^t e^y F(s, t - y) \, dy.$$

For the left-hand side we have

$$\lim_{N \to 0} \frac{\exp[NL(s, t)] - \exp[-\rho N(e^t - 1)]}{N}$$

$$= \lim_{u \to 0} \left\{ \frac{\exp[uL(s, t)] - 1}{u} + \frac{1 - \exp[-\rho u(e^t - 1)]}{u} \right\}$$

$$= \frac{d}{du} \exp[uL(s, t)] \bigg|_{u=0} - \frac{d}{du} \exp[-\rho u(e^t - 1)] \bigg|_{u=0}$$

$$= L(s, t) + \rho(e^t - 1).$$

Hence, formally we have

$$L(s, t) = -\rho(e^t - 1) + \rho \int_0^t e^\tau F(s, t - \tau) \, d\tau \tag{4.9}$$

and now $H(s, t; N)$ is also determined through (4.7).

To compute the expected number of mutants at time t, let

$$v(t) = \frac{\partial F(s, t)}{\partial s} \bigg|_{s=1}, \tag{4.10}$$

which is the expected number of descendants of a single mutant at time t after its creation. Then from (4.7) and (4.10) we have

$$E[\xi(t; N)] = \frac{\partial H(s, t; N)}{\partial s} \bigg|_{s=1} = e^{NL(1, t)} N \frac{\partial L(s, t)}{\partial s} \bigg|_{s=1}.$$

But

$$L(1, t) = -\rho(e^t - 1) + \rho \int_0^t e^\tau \, d\tau = 0,$$

since

$$F(1, t - \tau) = \sum_{k=0}^{\infty} p_k(t - \tau) = 1,$$

and from (4.9) and (4.10)

$$\frac{\partial L(s, t)}{\partial s} \bigg|_{s=1} = \rho \int_0^t e^\tau v(t - \tau) \, d\tau.$$

Hence

$$E[\xi(t; N)] = \rho N \int_0^t e^\tau v(t - \tau) \, d\tau. \tag{4.11}$$

If for not very large t we may approximate

$$v(t) \sim n_0 e^t \qquad (n_0 = \text{const.}),$$

then from (4.11)

$$E[\xi(t; N)] \sim \rho N n_0 \, t e^t.$$

5: *One-Dimensional Geometric Population Growth*

Another example of geometric population growth is the following. Nuclear particles are situated on an infinitely long line. When they split their "offspring" are scattered according to some probability law. More exactly, we assume that an offspring of a particle situated at x will fall in the interval $(x + y, x + y + dy)$ with probability density $f(y)$, i.e.,

$$f(y) \geq 0, \quad -\infty < y < \infty, \quad \int_{-\infty}^{\infty} f(y) \, dy = 1.$$

Note that $f(y)$ depends only on y, the distance between the parent and child and not on the actual location of the parent. Further, for ease of exposition, assume at first that each particle splits into exactly two new particles. Starting with one particle at $x = 0$ we call the offspring of this particle the first generation; the offspring of the first generation form the second generation, etc. We define the random variable

$Z_n(x; 0)$ = number of particles in the nth generation that are in $(-\infty, x]$,

starting with one particle at $x = 0$ at the zeroth generation. Put

$$p_k^{(n)}(x) = \Pr\{Z_n(x; 0) = k\}.$$

Suppose we shift the position of the initial particle to u. Let $Z_n(x; u)$ denote the number of descendants in the nth generation located in $[-\infty, x]$ evolving from one particle initially at u. Because of the spatial homogeneity of the distribution law for the dispersion of offspring it is intuitively clear that

$$\Pr\{Z_n(x; u) = k\} = \Pr\{Z_n(x - u; 0) = k\} = p_k^{(n)}(x - u). \tag{5.1}$$

The reader should supply the formal proof.

We introduce the generating function

$$g_n(s; x) = \sum_{k=0}^{\infty} p_k^{(n)}(x) s^k \tag{5.2}$$

and the mean

$$E_n[x] = E[Z_n(x; 0)] = \sum_{k=1}^{\infty} k p_k^{(n)}(x) = g_n'(1; x), \tag{5.3}$$

where the prime indicates the derivative with respect to s.

The event that there will be exactly k particles in the $(n + 1)$th generation in $(-\infty, x]$ will occur if the two offspring of the original particle at $x = 0$ are located in $(u, u + du)$ and $(v, v + dv)$ $(-\infty < u, v < \infty)$, respectively, and each will have such a number of descendants n generations later that the total will be exactly k particles in $(-\infty, x]$. The probability that the two particles of the first generation are in $(u, u + du)$ and $(v, v + dv)$, respectively, is

$$f(u)f(v) \, du \, dv$$

The probability that these two particles give rise to a total of k descendants in $(-\infty, x]$ n generations later is

$$\sum_{l=0}^{k} p_l^{(n)}(x - u)p_{k-l}^{(n)}(x - v)$$

[see (5.1)]. Further, we observe that u and v may be anywhere on the real line, independently of each other. Hence

$$p_k^{(n+1)}(x) = \int_{-\infty}^{\infty} \int_{-\infty}^{\infty} du \, dv \, f(u)f(v)$$

$$\times \sum_{l=0}^{k} p_l^{(n)}(x - u)p_{k-l}^{(n)}(x - v) \qquad \text{for} \quad k = 0, 1, 2, \dots.$$

Passing to the probability generating function we have

$$g_{n+1}(s; x) = \sum_{k=0}^{\infty} s^k \int_{-\infty}^{\infty} \int_{-\infty}^{\infty} f(u)f(v) \sum_{l=0}^{k} p_l^{(n)}(x - u)p_{k-l}^{(n)}(x - v) \, du \, dv$$

$$= \int_{-\infty}^{\infty} \int_{-\infty}^{\infty} f(u)f(v) \sum_{l=0}^{\infty} p_l^{(n)}(x - u)s^l \sum_{k=l}^{\infty} p_{k-l}^{(n)}(x - v)s^{k-l} \, du \, dv$$

$$= \int_{-\infty}^{\infty} \int_{-\infty}^{\infty} g_n(s; x - u)g_n(s; x - v)f(u)f(v) \, du \, dv$$

$$= \int_{-\infty}^{\infty} g_n(s; x - u)f(u) \, du \int_{-\infty}^{\infty} g_n(s; x - v)f(v) \, dv;$$

i.e.,

$$g_{n+1}(s; x) = \left[\int_{-\infty}^{\infty} g_n(s; x - u)f(u) \, du \right]^2. \qquad (5.4)$$

We now generalize the assumption that each particle splits into two new particles and assume instead that each split produces r new particles, where r is a fixed positive integer. In place of (5.4) we obtain the formula

$$g_{n+1}(s; x) = \left[\int_{-\infty}^{\infty} g_n(s; x - u)f(u) \, du \right]^r.$$

If we further generalize the model so that each particle may split into r new particles with probability a_r, then the same method leads to the formula

$$g_{n+1}(s; x) = A\left(\int_{-\infty}^{\infty} g_n(s; x - u)f(u)\, du\right),$$

where

$$A(z) = \sum_{r=0}^{\infty} a_r z^r$$

is the generating function of the number of new particles produced in each split.

The expectation $E_{n+1}[x]$ of the number of particles in the $(n + 1)$th generation can be computed in the usual way. It becomes

$$E_{n+1}[x] = g'_{n+1}(1; x)$$

$$= A'\left(\int_{-\infty}^{\infty} g_n(1; x - u)f(u)\, du\right) \int_{-\infty}^{\infty} g'_n(1; x - u)f(u)\, du. \quad (5.5)$$

Examination of (5.1) indicates that

$$g_n(1; x - u) = \sum_{k=0}^{\infty} p_k^{(n)}(x) = 1;$$

and

$$A'(1) = \sum_{r=0}^{\infty} ra_r = m$$

is the expected number of new particles produced in each split. The formula (5.5) simplifies to the expression

$$E_{n+1}[x] = m \int_{-\infty}^{\infty} E_n[x - u]f(u)\, du. \quad (5.6)$$

We may easily solve this recursive relation for $E_n[x]$. Simply regard the original particle at $x = 0$ as the zeroth generation. Then obviously

$$E_0[x] = \begin{cases} 0 & \text{if} \quad x < 0, \\ 1 & \text{if} \quad x \geq 0. \end{cases}$$

From (5.6)

$$E_1[x] = m \int_{-\infty}^{x} f(u)\, du = mF(x),$$

where $F(x)$ is the cumulative distribution function corresponding to $f(x)$. Further

$$E_2[x] = m^2 \int_{-\infty}^{\infty} F(x - u)f(u)\, du = m^2 F^{(2)}(x),$$

where $F^{(2)}(x)$ is the convolution of $F(x)$ with itself. By induction we plainly infer that

$$E_n[x] = m^n F^{(n)}(x), \tag{5.7}$$

where $F^{(n)}(x)$ is the n-fold convolution of $F(x)$ with itself.

If the density function $f(y)$ possesses a variance σ^2 and mean μ then the central limit theorem tells us that for each fixed ξ

$$F^{(n)}(n\mu + \xi\sigma\sqrt{n}) \to \Phi(\xi), \qquad n \to \infty,$$

where $\Phi(\xi)$ is the standard normal distribution function, i.e.,

$$\Phi(\xi) = \frac{1}{\sqrt{2\pi}} \int_{-\infty}^{\xi} \exp(-\eta^2/2) \, d\eta.$$

This obviously leads to an asymptotic formula for $E_n(x)$, namely,

$$\frac{E_n[n\mu + \xi\sigma\sqrt{n}]}{m^n} \to \Phi(\xi) \qquad \text{as} \quad n \to \infty,$$

which is of some independent interest.

6: Stochastic Population Growth Model in Space and Time

Suppose certain plants are distributed in space in accordance with a two-dimensional Poisson process with intensity parameter λ. (We consider a model for the distribution of plants in two-dimensional space but an entirely parallel development would apply in a three-dimensional formulation.) We assume that each parent plant whose location is described by the two-dimensional vector $\mathbf{r}_0$ gives birth, independently of other plants, to a random number of progeny with a probability generating function $H(s)$; that is,

$$H(s) = \sum_{k=0}^{\infty} h_k s^k$$

and h_k denotes the probability that a parent plant will produce k offspring. Assume further that the progeny of one parent plant located at $\mathbf{r}_0$ are distributed independently in space around $\mathbf{r}_0$ in accordance with the two-dimensional density function $f(\mathbf{r} - \mathbf{r}_0)$ which depends only on the vector $\mathbf{r} - \mathbf{r}_0$; e.g., the two-dimensional normal density function

$$f(\mathbf{r} - \mathbf{r}_0) = \left(\frac{1}{\sqrt{2\pi\sigma^2}}\right)^2 \exp\left\{-\frac{1}{2\sigma^2}[(x - x_0)^2 + (y - y_0)^2]\right\}.$$

Thus the probability that a given progeny of a parent at $\mathbf{r}_0$ will be found in the region R is

$$p = \int_R f(\mathbf{r} - \mathbf{r}_0) \, d\mathbf{r}. \tag{6.1}$$

If a parent plant at $\mathbf{r}_0$ has exactly n progeny, then the number of offspring in region R due to this single parent will follow the binomial distribution with parameters p and n, where p is given by (6.1). However, the number of progeny of a single parent is a random variable with probability generating function $H(s)$. The usual method of compounding a generating function by using the law of total probability shows that the probability generating function of the number of progeny in R due to a single parent plant at $\mathbf{r}_0$ is $H[1 + p(s - 1)]$, where p is given by (6.1). Our objective is to calculate the probability generating function of the number of progeny in region R produced by all the parent plants located in a region S. For this purpose, we introduce the following notation. Let

$X(S)$ = number of parent plants in region S,

$Y(\mathbf{r}_0, R)$ = number of progeny in R due to a single parent plant at $\mathbf{r}_0$,

$Y(S, R)$ = number of progeny in R produced by all parents in S.

Then

$$Y(S, R) = \sum_{\mathbf{r}_0 \in S} Y(\mathbf{r}_0, R).$$

If S is a bounded region the sum is finite with probability one since the number of parents in S follows a Poisson distribution with parameter $\lambda A(S)$ [$A(S)$ denotes the area of S]. Further, $X(S)$ describes a two-dimensional Poisson process with intensity parameter λ. The generating function of $Y(S, R)$ will be calculated by conditioning on the values of $X(S)$. Thus

$$g(s) = E[s^{Y(S, R)}] = \sum_{k=0}^{\infty} E[s^{Y(S, R)} | X(S) = k] \Pr\{X(S) = k\}. \quad (6.2)$$

Since the parent plants produce independently, we have

$$E[s^{Y(S, R)} | X(S) = k] = \{E[s^{Y(S, R)} | X(S) = 1]\}^k. \quad (6.3)$$

Moreover, from the theory of spatial Poisson processes we know that under the condition $X(S) = 1$,

$$Y(S, R) = Y(\mathbf{r}_0, R),$$

where $\mathbf{r}_0$ is uniformly distributed in S. Then

$$E[s^{Y(S, R)} | X(S) = 1] = E[s^{Y(\mathbf{r}_0, R)} | \mathbf{r}_0 \text{ uniformly distributed in } S]$$

$$= \frac{1}{A(S)} \int_S H[1 + p(s - 1)] \, d\mathbf{r}_0, \quad (6.4)$$

where p is given by (6.1). The final equality is valid because we have shown $H[1 + p(s - 1)]$ to be the probability generating function of the number of progeny in R due to a single parent located at a point $\mathbf{r}_0$. Here, however, $\mathbf{r}_0$ is uniformly distributed in S and so (6.4) results. Now combining (6.2)–(6.4) yields

$$g(s) = \sum_{k=0}^{\infty} \left\{ \frac{1}{A(S)} \int_S H[1 + p(s - 1)] \, d\mathbf{r}_0 \right\}^k e^{-\lambda A(S)} \frac{[\lambda A(S)]^k}{k!},$$

as $X(S)$ is Poisson distributed with intensity parameter λ. This simplifies to the formula

$$g(s) = \exp\left\{ \lambda \int_S (H[1 + p(s - 1)] - 1) \, d\mathbf{r}_0 \right\}, \tag{6.5}$$

where

$$p = \int_R f(\mathbf{r} - \mathbf{r}_0) \, d\mathbf{r}_0.$$

In the final expression it is meaningful to let S be the whole two-dimensional space.

Formula (6.5) also holds if the plants are distributed in space with an underlying three-dimensional spatial Poisson distribution and $\mathbf{r}$, $\mathbf{r}_0$ and R, S denote three-dimensional vectors and regions, respectively. In either the two- or the three-dimensional case we may approximate p by

$$f(\mathbf{r} - \mathbf{r}_0)A(R),$$

provided the region R is very small. Then (6.5) reduces to

$$g(s) \approx \exp\left\{ \lambda \int_S (H[1 + f(\mathbf{r} - \mathbf{r}_0)A(R)(s - 1)] - 1) \, d\mathbf{r}_0 \right\}.$$

If S is the whole space (two or three dimensional) then we may write

$$g(s) \approx \exp\left\{ \lambda \int (H[1 + f(\mathbf{u})A(R)(s - 1)] - 1) \, d\mathbf{u} \right\}, \tag{6.6}$$

where the integration (double or triple) is over the whole space.

As an example we take the normal distribution in the plane for the function f, i.e.,

$$f(\mathbf{u}) = \frac{1}{2\pi\sigma^2} \exp\left[-\frac{1}{2\sigma^2} (x^2 + y^2) \right], \qquad \text{where} \quad \mathbf{u} = (x, y),$$

and let the probability distribution of the number of progeny of a single parent be

$$h_k = \mu^k(1 - \mu), \qquad k = 0, 1, 2, \ldots,$$

where μ is a constant, $0 < \mu < 1$. Then $H(s) = (1 - \mu) \sum_{k=0}^{\infty} \mu^k s^k = (1 - \mu)/(1 - \mu s)$. Substituting into (6.6) and simplifying, we obtain

$$g(s) \approx \exp\left\{ \lambda \int_{-\infty}^{\infty} \int_{-\infty}^{\infty} \frac{\mu A(R)(s - 1) \exp[-(2\sigma^2)^{-1}(x^2 + y^2)] \, dx \, dy}{2\pi\sigma^2(1 - \mu) - \mu A(R)(s - 1) \exp[-(2\sigma^2)^{-1}(x^2 + y^2)]} \right\},$$

which is an appropriate approximation for $A(R)$ small. After we switch to polar coordinates r and θ, the expression for $g(s)$ becomes

$$g(s) \approx \exp\left\{ 2\pi\lambda \int_0^\infty \frac{\mu A(R)(s-1)r \exp(-(2\sigma^2)^{-1}r^2)\, dr}{2\pi\sigma^2(1-\mu) - \mu A(R)(s-1)\exp[-(2\sigma^2)^{-1}r^2]} \right\}$$

$$= \exp\left\{ 2\pi\lambda \int_0^1 \frac{\mu A(R)(s-1)\sigma^2\, dz}{2\pi\sigma^2(1-\mu) - \mu A(R)(s-1)z} \right\} \quad \text{(where } z = \exp(-r^2/2\sigma^2)$$

$$= \exp\left\{ -2\pi\lambda\sigma^2 \log\left[1 - \frac{\mu A(R)(s-1)}{2\pi\sigma^2(1-\mu)} \right] \right\}$$

$$= [1 - \beta(s-1)]^{-\kappa},$$

where

$$\beta = \frac{\mu A(R)}{2\pi\sigma^2(1-\mu)} \qquad \text{and} \qquad \kappa = 2\pi\lambda\sigma^2.$$

This is the probability generating function of a negative binomial distribution.

7: Deterministic Population Growth with Age Distribution

In this section we shall discuss some simple deterministic models of population growth which take account of the age structure of the population. The stochastic version of these growth processes is quite complicated and beyond the scope of the text.

A. A SIMPLE GROWTH MODEL

We begin by considering a single species, and we let

$N(t) = $ population size at time t;

$v(t)\, dt = $ number of offspring produced by each individual in the "short" interval $(t, t + dt)$. More precisely, the number of offspring produced by an individual in the interval $(t, t + h)$ is $v(t)h + o(h)$.

Then

$$N(t + h) = N(t) + N(t)v(t)h + o(h)$$

$$\frac{N(t + h) - N(t)}{h} = v(t)N(t) + \frac{o(h)}{h}$$

Taking the limit of both sides as $h \to 0$, we obtain

$$\frac{dN(t)}{dt} = v(t)N(t). \tag{7.1}$$

Its solution is

$$N(t) = N(0) \exp\left(\int_0^t v(\tau)\, d\tau \right), \tag{7.2}$$

where $N(0)$ denotes the initial population size. If the integral $\int_0^t v(\tau)\, d\tau$ diverges as $t \to \infty$, then the population grows to infinity. If $v(\tau)$ is constant then $N(t) = N(0)e^{vt}$, and the population grows to infinity exponentially at the rate v.

B. A MODEL IN WHICH POPULATION SIZE DETERS GROWTH

In the above model an increase in size of the population does not deter growth. We now take population size into account by letting $v(t)$ depend upon $N(t)$. Specifically, suppose

$$v(t) = \begin{cases} \beta\left(1 - \dfrac{N(t)}{\alpha}\right) & \text{for} \quad N(t) \le \alpha, \\ 0 & \text{otherwise,} \end{cases}$$

where α and β are positive numbers. Note that the population size cannot grow beyond α. In this case (7.1) becomes

$$\frac{dN(t)}{dt} = N(t)\beta\left[1 - \frac{N(t)}{\alpha}\right] = \beta N(t) - \frac{\beta}{\alpha}N^2(t). \tag{7.3}$$

We can separate variables and the solution of (7.3) becomes

$$N(t) = \frac{\alpha N(0)e^{\beta t}}{\alpha + N(0)(e^{\beta t} - 1)} = \frac{\alpha N(0)}{\alpha e^{-\beta t} + N(0) - N(0)e^{-\beta t}}. \tag{7.4}$$

Inspection of the second expression of (7.4) reveals that $N(t) \to \alpha$ as $t \to \infty$.

C. EFFECT OF AGE STRUCTURE

We shall now consider the effect of age structure on a growing population. We need the following notation:

$\rho(u, t)$ = the frequency function of individuals of age u in the population at time t, i.e., $\rho(u, t)$ has the property that $\int_{u_1}^{u_2} \rho(u, t)\, du$ = proportion of individuals in the population at time t who are in the age range (u_1, u_2). The actual number of individuals in this age bracket is, of course, $N(t)$ times this proportion. (7.5)

$b(t)$ = the rate at which new individuals are being created in the population at time t. More explicitly, $\int_{t_1}^{t_2} b(t)\, dt$ = number of new individuals created in the time interval (t_1, t_2). (7.6)

$\lambda(u) \, dt$ = expected number of progeny of a single individual
of age u in the next dt units of time.   (7.7)

$l(u)$ = probability that an individual will survive, from
birth, at least u units of time.   (7.8)

$c(u)$ = the infinitesimal death rate, i.e., the probability
that an individual of age u will die in the next h
units of time is $c(u)h + o(h)$.   (7.9)

The relation between $l(\cdot)$ and $c(\cdot)$ may be derived as follows: For given u, $h \geq 0$, an individual will survive, from birth, at least $u + h$ units of time if and only if he survives, from birth, at least u units of time, and then does not die in the following h units of time. Thus

$$l(u + h) = l(u)[1 - c(u)h] + o(h)$$

and

$$\frac{l(u + h) - l(u)}{h} = - \frac{l(u)c(u)h + o(h)}{h}.$$

Taking limits as $h \to 0$, we have

$$\frac{dl(u)}{du} = -l(u)c(u).$$

Solving, we obtain

$$l(u) = l(0) \exp\left[- \int_0^u c(\xi) \, d\xi \right] = \exp\left[- \int_0^u c(\xi) \, d\xi \right], \qquad (7.10)$$

since $l(0) = 1$.

In considering the effect of age structure on a growing population, our interest will center on $b(t)$, i.e., we may regard $\lambda(u)$, $l(u)$, and $c(u)$ as known, and the problem is to determine $b(t)$. The rate at which new individuals are being created in the population at time t has two components. One component, $b_0(t)$, say, is the rate of creation due to those individuals in the population at time t who already existed at time zero. The density of individuals of age u in the population at time zero is $\rho(u, 0)$. The probability that an individual of age u at time zero will survive to time t [at which time he will be of age $(t + u)$] is $l(t + u)/l(u)$. Hence the proportion of those of age u that survive to time t is $[l(t + u)/l(u)]\rho(u, 0)$. The rate of births for individuals of age $t + u$ is $\lambda(t + u)$. Now, adding over all ages, we obtain

$$b_0(t) = N(0) \int_0^\infty \lambda(t + u) \frac{l(t + u)}{l(u)} \rho(u, 0) \, du. \qquad (7.11)$$

The other component of $b(t)$ is the rate of creation of new individuals at time t due to those individuals in the population who were born after time zero. The rate at which new individuals are being created in the population at time τ is $b(\tau)$. For $0 < \tau \leq t$, the probability that an individual born at time τ will survive to time t, at which time he will be of age $t - \tau$, is $l(t - \tau)$. The rate of births for individuals of age $t - \tau$ is $\lambda(t - \tau)$. It follows that

$$b(t) = b_0(t) + \int_0^t \lambda(t - \tau)l(t - \tau)b(\tau)\, d\tau, \tag{7.12}$$

where $b_0(t)$ is given by (7.11).

The relation (7.12) is a renewal equation (see Chapter 5). Its solution can be obtained by iteration.

Example. Assume that both the birth rate and the infinitesimal death rate are constants, independent of age, i.e., $\lambda(u) = \lambda$, $c(u) = c$. Then, by (7.10), the probability of an individual living to age u is

$$l(u) = e^{-cu}. \tag{7.13}$$

Suppose that the population started with the creation of a single individual at time zero. Then,

$$N(0) \int_0^\infty \rho(u, 0)\, du = 1. \tag{7.14}$$

Actually, the age density $\rho(u, 0)$ should be replaced by a degenerate distribution concentrated at $u = 0$. From (7.11), (7.13), and (7.14), we conclude that

$$b_0(t) = \lambda e^{-ct}. \tag{7.15}$$

Hence by (7.13) and (7.15), (7.12) becomes

$$b(t) = \lambda e^{-ct} + \lambda \int_0^t e^{-c(t-\tau)}b(\tau)\, d\tau. \tag{7.16}$$

We wish to solve (7.16) for the function $b(\cdot)$. Multiply both sides of the equation by e^{ct} to obtain

$$e^{ct}b(t) = \lambda + \lambda \int_0^t e^{c\tau}b(\tau)\, d\tau$$

and let

$$f(\tau) = e^{c\tau}b(\tau). \tag{7.17}$$

Then the equation, to be solved now for $f(\cdot)$, becomes

$$f(t) = \lambda + \lambda \int_0^t f(\tau)\, d\tau.$$

Clearly $f(0) = \lambda$, and upon differentiating both sides with respect to t we obtain $f'(t) = \lambda f(t)$. Thus $f(t) = \lambda e^{\lambda t}$, and, inserting this expression in (7.17), we have

$$b(t) = \lambda e^{(\lambda - c)t}. \tag{7.18}$$

Having determined $b(t)$ in this example, we now utilize this result to determine the age structure as given by $N(t)\rho(u, t)$. Since we have assumed, in deriving (7.18), that the population started with the creation of a single individual at time zero, we need only consider the case $u \le t$. An individual is of age u at time t if and only if it was created at time $t - u$. The rate at which new individuals are being created in the population at time $t - u$ is $b(t - u) = \lambda e^{(\lambda - c)(t - u)}$. The probability that an individual will live to at least age u is, by (7.13), e^{-cu}. It follows that

$$N(t)\rho(u, t) = e^{-cu}b(t - u), \qquad u \le t. \tag{7.19}$$

Substituting (7.18) into (7.19), we obtain

$$N(t)\rho(u, t) = e^{-cu}\lambda e^{(\lambda - c)(t - u)} = \lambda e^{-\lambda u}e^{(\lambda - c)t}, \qquad u \le t. \tag{7.20}$$

We return now to the general formulation, keeping the assumption that the population started with the birth of a single individual. Then the derivation of (7.19) holds in general and for $u \le t$ we have

$$N(t)\rho(u, t) = b(t - u)l(u). \tag{7.21}$$

Here, $b(t)$ is determined as the solution of the renewal equation (7.12) under the special circumstance $b_0(t) = 0$, since there were no individuals living at time $t = 0$. Thus,

$$b(t) = \int_0^t \lambda(t - \tau)l(t - \tau)b(\tau)\, d\tau = \int_0^t b(t - u)\lambda(u)l(u)\, du.$$

Let $\varphi(u) = \lambda(u)l(u)$. Then, the equation becomes

$$b(t) = \int_0^t b(t - u)\varphi(u)\, du. \tag{7.22}$$

As a tentative solution of (7.22), we try

$$b(t) = e^{\gamma t} \tag{7.23}$$

where γ is a constant to be determined so that (7.22) holds for t large. Inserting (7.23) into (7.22) leads to the condition

$$e^{\gamma t} = \int_0^t e^{\gamma(t - u)}\varphi(u)\, du = e^{\gamma t}\int_0^t e^{-\gamma u}\varphi(u)\, du$$

or, equivalently,

$$\int_0^t e^{-\gamma u}\varphi(u)\, du = 1. \tag{7.24}$$

We are interested in the asymptotic age structure of the population as $t \to \infty$. Letting $t \to \infty$ in (7.24) leads to the condition

$$R(\gamma) = \int_0^\infty e^{-\gamma u}\varphi(u)\, du = 1. \tag{7.25}$$

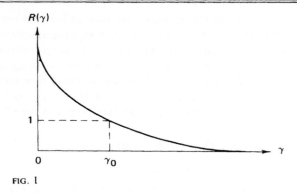

FIG. 1

From its definition we find that $R(\gamma)$ is a strictly decreasing function of γ; hence (7.25) has at most one positive root. Let

$$R = \int_0^\infty \varphi(u) \, du = \int_0^\infty \lambda(u)l(u) \, du.$$

R is referred to as the reproductive value of an individual; it is his expected number of offspring during his lifetime. R is sometimes called the Malthusian rate.

If $R > 1$, then $R(\gamma)$ has the form shown in Fig. 1 and a solution $\gamma_0 > 0$ of (7.25) exists.

In this case $b(t)$ is asymptotically proportional to $\exp(\gamma_0 t)$, and the population grows exponentially. If $R < 1$, then $R(\gamma)$ has the form shown in Fig. 2 and a solution $\gamma_0 < 0$ of (7.25) exists. In this case $b(t)$ is asymptotically proportional to $\exp(\gamma_0 t)$, so that the population dies out exponentially fast.

If $R = 1$, then the problem must be studied stochastically.

The results presented above are rather heuristic. The assumptions and analysis necessary to rigorize the arguments are beyond the scope of this book. The main conclusion is that under suitable conditions we expect the growth

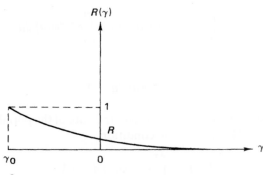

FIG 2

rate of the population to behave asymptotically like an exponential. Further evidence of this phenomenon will be forthcoming when we discuss a discrete model below.

We continue in the same heuristic manner. For the case $R > 1$, we shall determine the asymptotic age structure as given by the density function $\rho(u, t)$. By (7.21)

$$\rho(u, t) = \frac{b(t - u)l(u)}{N(t)}, \qquad u \leq t.$$

Now $b(t - u)$ is asymptotically proportional to $\exp[\gamma_0(t - u)]$; hence, $N(t)$ is asymptotically proportional to $\exp(\gamma_0 t)$. Therefore $\rho(u, t)$ is asymptotically proportional to

$$\frac{\exp[\gamma_0(t - u)]l(u)}{\exp(\gamma_0 t)} = \exp(-\gamma_0 u)l(u).$$

The factor of proportionality is determined from the fact that $\rho(u, t)$ is the density function of a probability distribution. Therefore, the age structure of the population for large t is given by the asymptotic density function

$$\rho(u, t) \sim \frac{\exp(-\gamma_0 u)l(u)}{\int_0^\infty \exp(-\gamma_0 x)l(x) \, dx}.$$

8: A Discrete Aging Model

We now give a precise treatment of the problem of the previous section for a discrete time model. Let $t = 0, 1, 2, \ldots$, and let

$n_x^{(t)}$ = number of individuals of age x at time t,
P_x = proportion of individuals of age x surviving to age $x + 1$,
F_x = number born, in the next unit of time, to each parent of age x.

Here F_x is assumed independent of t and P_x and F_x are positive for $x = 0, 1, 2, \ldots, m$. Assume that no-one lives beyond age m; i.e., set $P_m = 0$. Then the transition relationship for age structure between $t = 0$ and $t = 1$ is given by

$$b_0^{(1)} = \sum_{x=0}^m n_x^{(0)} F_x,$$

$$n_1^{(1)} = P_0 n_0^{(0)},$$

$$n_2^{(1)} = P_1 n_1^{(0)}$$

$$\vdots$$

$$n_m^{(1)} = P_{m-1} n_{m-1}^{(0)}.$$

We write this transition relationship in matrix form as

$$\mathbf{n}^{(1)} = \mathbf{M}\mathbf{n}^{(0)}, \qquad \begin{aligned} \mathbf{n}^{(0)} &= (n_0^{(0)}, n_1^{(0)}, \ldots, n_m^{(0)}), \\ \mathbf{n}^{(1)} &= (n_0^{(1)}, n_1^{(1)}, \ldots, n_m^{(1)}), \end{aligned} \tag{8.1}$$

where the matrix $\mathbf{M}$ is

$$\mathbf{M} = \begin{Vmatrix} F_0 & F_1 & F_2 & \cdots & \cdots & F_m \\ P_0 & 0 & \cdots & \cdots & \cdots & 0 \\ 0 & P_1 & 0 & \cdots & \cdots & 0 \\ \vdots & & & & & \vdots \\ 0 & 0 & 0 & \cdots & P_{m-1} & 0 \end{Vmatrix}. \tag{8.2}$$

$\mathbf{M}$ is a matrix with nonnegative elements. Properties of such matrices are given in Section 2 of the Appendix in *A First Course*. Since P_x and F_x do not depend upon time, the same transition relationship holds between any two consecutive times. Thus we may iterate formula (8.1) and obtain

$$\mathbf{n}^{(t)} = \mathbf{M}^t \mathbf{n}, \quad \mathbf{n}^{(t)} = (n_0^{(t)}, n_1^{(t)}, \ldots, n_m^{(t)}), \qquad t = 1, 2, 3, \ldots, \tag{8.3}$$

where we wrote

$$n_x = n_x^{(0)} \qquad \text{for} \quad x = 1, 2, \ldots, m.$$

For sufficiently large t, all of the elements of $\mathbf{M}^t$ are strictly positive. Also, there exists an eigenvalue $\lambda_0 > 0$ which is strictly greater in absolute value than any other eigenvalue (Theorem 2.2 of the Appendix). For any vector $\mathbf{n}$, $\mathbf{M}^t \mathbf{n}$ is asymptotically equal to $\lambda_0^t \mathbf{z}$, where $\mathbf{z}$ is a certain multiple of the unique right-hand eigenvector corresponding to the eigenvalue λ_0. Asymptotically, as $t \to \infty$, (8.3) becomes

$$\mathbf{n}^{(t)} = \mathbf{M}^t \mathbf{n} \sim \lambda_0^t \mathbf{z} = [\exp(t \log \lambda_0)]\mathbf{z}$$

and $\log \lambda_0$ corresponds to the critical value γ_0 introduced in the heuristic continuous time analysis. The population grows exponentially if $\lambda_0 > 1$, and dies at an exponential rate if $\lambda_0 < 1$.

9: Compound Poisson Processes

Let $\{Y_i\}_{i=1}^\infty$ be a sequence of independent identically distributed random variables having the common distribution function F and characteristic function φ. Let $\{N(t), t \geq 0\}$ be a Poisson process with parameter $\lambda > 0$, and independent of the sequence $\{Y_i\}$. The stochastic process

$$X(t) = \sum_{k=1}^{N(t)} Y_k, \qquad t \geq 0$$

is called a *compound Poisson process*.

Compound Poisson processes are used to model a large variety of physical situations. Typically, "events" occur in accordance with a Poisson process and each event has some randomly determined "value" associated with it. Here are some examples:

(a)   Let $N(t)$ be the number of claims against an insurance company up to time t and suppose Y_i is the amount of the ith claim. Then $X(t)$ is the accumulation of money claims up to time t.

(b)   Let $N(t)$ be the number of transactions in a publically traded security, say a share of stock, up to time t, and suppose Y_i is the number of shares traded in the ith transaction. Then $X(t)$ is the total number of shares traded up to time t.

(c)   Let $N(t)$ be the number of shocks that have occurred in a system up to time t and suppose Y_i is the damage or wear caused by the ith shock. Postulating that damage is additive, $X(t)$ is the cumulative damage to the system up to time t.

STATIONARY INDEPENDENT INCREMENTS

A compound Poisson process has stationary independent increments. We confirm this property in two steps, first showing that nonoverlapping increments are independent. Let $0 \le t_0 < t_1 < \cdots < t_n$ be given. Since $Y_1, Y_2, \ldots$ are mutually independent and independent of the Poisson process, which itself has independent increments, the random variables

$$\{Y_1, \ldots, Y_{N(t_0)}, N(t_0)\},$$

$$\{Y_{N(t_0)+1}, \ldots, Y_{N(t_1)}, N(t_1) - N(t_0)\},$$

$$\vdots$$

$$\{Y_{N(t_{n-1})+1}, \ldots, Y_{N(t_n)}, N(t_n) - N(t_{n-1})\}$$

are independent. Hence the increments

$$X(t_0) - X(0) = Y_1 + \cdots + Y_{N(t_0)},$$

$$X(t_1) - X(t_0) = Y_{N(t_0)+1} + \cdots + Y_{N(t_1)},$$

$$\vdots$$

$$X(t_n) - X(t_{n-1}) = Y_{N(t_{n-1})+1} + \cdots + Y_{N(t_n)}$$

in the compound Poisson process are independent.

To show that the distribution of the increments is stationary, let $t > 0$ and $h > 0$ be given. Then $N(t + h) - N(t)$ has the same distribution as $N(h)$ and $Y_1, \ldots, Y_{N(h)}$ have the same distribution as $Y_{N(t)+1}, \ldots, Y_{N(t+h)}$. Thus

$$X(t + h) - X(t) = Y_{N(t)+1} + \cdots + Y_{N(t+h)}$$

has the same distribution as

$$X(h) = Y_1 + \cdots + Y_{N(h)}.$$

This completes the proof that a compound Poisson process has stationary independent increments.

Thus, a compound Poisson process is Markov (Elementary Problem 7, Chapter 1). The process is a Markov chain only if the possible values of the summands form a discrete set, say, $\{0, \pm 1, \pm 2, \ldots\}$. If the random variables $Y_1, Y_2, \ldots$ are not discrete, we have a general Markov process with an uncountable state space.

THE CHARACTERISTIC FUNCTION

Since $N(t)$ follows a Poisson distribution with parameter λt, its generating function is

$$g_{N(t)}(s) = e^{-\lambda t(1-s)}.$$

Note that $X(t)$ is a random sum of random variables. Referring to the results of Section 1E in Chapter 1, we have

$$\varphi_{X(t)}(u) = g_{N(t)}[\varphi(u)] \qquad \text{(where} \quad \varphi(u) = E[e^{iuY_1}])$$
$$= \exp\{-\lambda t[1 - \varphi(u)]\}, \qquad -\infty < u < +\infty. \tag{9.1}$$

Next we write down the joint characteristic function of the process at any finite set of time points $0 < t_1 < \cdots < t_n$. Reliance on the property that $\{X(t), t \geq 0\}$ has stationary independent increments is vital. The joint characteristic function becomes

$$\varphi_{X(t_1), \ldots, X(t_n)}(u_1, \ldots, u_n) = E[\exp\{iu_1 X(t_1) + \cdots + iu_n X(t_n)\}]$$

$$= E\left[\exp\left\{i \sum_{k=1}^{n} a_k[X(t_k) - X(t_{k-1})]\right\}\right]$$

$$\text{(where} \quad a_k = u_n + \cdots + u_k)$$

$$= \prod_{k=1}^{n} E[\exp\{ia_k[X(t_k) - X(t_{k-1})]\}]$$

$$= \prod_{k=1}^{n} E[\exp\{ia_k[X(t_k - t_{k-1})]\}]$$

$$= \prod_{k=1}^{n} \varphi_{X(t_k - t_{k-1})}(a_k). \tag{9.2}$$

Since the joint distributions of arbitrary finite sets $\{X(t_1), \ldots, X(t_n)\}$ characterize completely the stochastic process, we have established the following theorem.

Theorem 9.1. *Let $\{X(t), t \geq 0\}$ be a stochastic process having stationary independent increments and for which $X(0) = 0$. Then $\{X(t), t \geq 0\}$ is a compound Poisson process if and only if the characteristic function $\varphi_{X(t)}$ of $X(t)$ is of the form*

$$\varphi_{X(t)}(u) = \exp\{-\lambda t[1 - \varphi(u)]\}, \qquad -\infty < u < +\infty$$

where $\lambda > 0$ and φ is a characteristic function.

The mean and variance of $X(t)$ may be computed routinely by differentiating the characteristic function. The results are

$$E[X(t)] = \mu\lambda t,$$

and

$$\text{variance of } X(t) = (\sigma^2 + \mu^2)\lambda t$$

where μ and σ^2 are the mean and variance, respectively, of Y_1.

THE DISTRIBUTION FUNCTION OF $X(t)$

The distribution function for $X(t)$ can be represented explicitly after conditioning on the values of $N(t)$. Thus we obtain

$$\Pr\{X(t) \le x\} = \Pr\left\{\sum_{i=1}^{N(t)} Y_i \le x\right\}.$$

$$= \sum_{n=0}^{\infty} \Pr\left\{\sum_{i=1}^{N(t)} Y_i \le x \,|\, N(t) = n\right\} \frac{(\lambda t)^n e^{-\lambda t}}{n!}$$

$$= \sum_{n=0}^{\infty} \frac{(\lambda t)^n e^{-\lambda t}}{n!} F^{(n)}(x) \qquad \text{(since } N(t) \text{ is independent of } \{Y_i\}),$$

$$(9.3)$$

where

$$F^{(n)}(x) = \Pr\{Y_1 + \cdots + Y_n \le x\}$$

with

$$F^{(0)}(x) = \begin{cases} 1 & \text{for } x \ge 0, \\ 0 & \text{for } x < 0. \end{cases}$$

Example 1. Let $N(t)$ be the number of shocks to a system up to time t and let Y_i be the damage or wear sustained by the ith shock. We assume that damage is positive, that is, $\Pr\{Y_i \ge 0\} = 1$, and that the damage accumulates additively so that $X(t)$ is the total damage up to time t. Suppose the system continues to operate as long as the total damage is strictly less than some critical value a and fails in the contrary circumstance. Let T be the time of failure. Then

$$\{T > t\} \qquad \text{if and only if} \qquad \{X(t) < a\}. \tag{9.4}$$

In view of (9.3) and (9.4), we have

$$\Pr\{T > t\} = \sum_{n=0}^{\infty} \frac{(\lambda t)^n e^{-\lambda t}}{n!} F^{(n)}(a). \tag{9.5}$$

All summands are nonnegative, so we may certainly interchange integration and summation to get

$$E[T] = \int_0^\infty \Pr\{T > t\}\, dt$$

$$= \sum_{n=0}^\infty \left(\int_0^\infty \frac{(\lambda t)^n e^{-\lambda t}}{n!}\, dt \right) F^{(n)}(a) = \lambda^{-1} \sum_{n=0}^\infty F^{(n)}(a),$$

an expression for the mean time to failure.

Consider the special case where $Y_1, Y_2, \ldots$ each follow the exponential distribution,

$$\Pr\{Y_i \le a\} = 1 - e^{-\mu a}, \qquad a > 0.$$

Then $Y_1 + \cdots + Y_n$ follows a gamma distribution, with

$$F^{(n)}(a) = 1 - \sum_{k=0}^{n-1} \frac{(\mu a)^k e^{-\mu a}}{k!} = \sum_{k=n}^\infty \frac{(\mu a)^k e^{-\mu a}}{k!}, \qquad n \ge 0$$

and

$$\sum_{n=0}^\infty F^{(n)}(a) = \sum_{n=0}^\infty \sum_{k=n}^\infty \frac{(\mu a)^k e^{-\mu a}}{k!}$$

$$= \sum_{k=0}^\infty \sum_{n=0}^k \frac{(\mu a)^k e^{-\mu a}}{k!}$$

$$= \sum_{k=0}^\infty (k + 1) \frac{(\mu a)^k e^{-\mu a}}{k!}$$

$$= 1 + \mu a.$$

In the special case at hand, we have

$$E[T] = \lambda^{-1} \sum_{n=0}^\infty F^{(n)}(a)$$

$$= (1 + \mu a)/\lambda.$$

SUM OF INDEPENDENT COMPOUND POISSON PROCESSES

Compound Poisson processes have many desirable properties. Here we display one of these: that the sum of two independent compound Poisson processes is itself a compound Poisson process. For $k = 1, 2$, let $\{X_k(t), t \ge 0\}$ be compound Poisson processes, mutually independent, and with Poisson rate parameter λ_k. Let φ_k be the characteristic function of the "value" sequence associated with the kth process. Form $X(t) = X_1(t) + X_2(t)$. We claim that $\{X(t), t \ge 0\}$ is a compound Poisson process with Poisson rate $\lambda = \lambda_1 + \lambda_2$ and associated characteristic function

$$\varphi(u) = \frac{\lambda_1}{\lambda_1 + \lambda_2}\,\varphi_1(u) + \frac{\lambda_2}{\lambda_1 + \lambda_2}\,\varphi_2(u).$$

This characteristic function corresponds to that of a random variable Y which with probability $\lambda_1/(\lambda_1 + \lambda_2)$ assumes the value $Y(1)$ and with probability $\lambda_2/(\lambda_1 + \lambda_2)$ assumes the value $Y(2)$, where $Y(1)$ and $Y(2)$ are random variables with characteristic functions φ_1 and φ_2, respectively.

Clearly $\{X(t),\, t \geq 0\}$ has stationary independent increments, since both $\{X_1(t),\, t \geq 0\}$ and $\{X_2(t),\, t \geq 0\}$ do. In view of Theorem 9.1 we need only compute the characteristic function of $X(t)$. Thus,

$$\begin{aligned}
\varphi_{X(t)}(u) &= \varphi_{X_1(t)}(u)\varphi_{X_2(t)}(u) \\
&= \exp\{-\lambda t[1 - p_1\varphi_1(u) - p_2\varphi_2(u)]\},
\end{aligned}$$

where $\lambda = \lambda_1 + \lambda_2$ and $p_i = \lambda_i/\lambda$. Thus $\varphi_{X(t)}(u)$ has the desired form and $\{X(t),\, t \geq 0\}$ is a compound Poisson process.

Example 2. Taxis arrive at a stand according to a Poisson process with rate λ_1. Customers arrive at the stand according to an independent Poisson process with rate λ_2. If a taxi arrives and individuals are waiting, the first person in line is served; if no individuals are waiting, the taxi waits. If a person arrives and there are taxis waiting, the person requisitions the first taxi; if no taxis are available, the person waits.

Let $X(t)$ be the number of taxis waiting at time t if taxis are waiting for people, and be minus the number of people waiting at time t if people are waiting for taxis. Then $\{X(t), t \geq 0\}$ is a stochastic process with state space $\{0, \pm 1, \pm 2, \ldots\}$ which may be written, assuming $X(0) = 0$, in the form $X(t) = X_1(t) + X_2(t)$, where $\{X_1(t), t \geq 0\}$ is a Poisson process with rate λ_1 and $\{X_2(t), t \geq 0\}$ is the negative of a Poisson process having rate λ_2. The result on the sum of compound Poisson processes tells us that $X(t)$ is a compound Poisson process having rate $\lambda = \lambda_1 + \lambda_2$ and whose "values" $Y_1, Y_2, \ldots$ have the common discrete distribution

$$p(k) = \begin{cases} \lambda_1/(\lambda_1 + \lambda_2) & \text{if } k = +1, \\ \lambda_2/(\lambda_1 + \lambda_2) & \text{if } k = -1, \\ 0 & \text{otherwise.} \end{cases}$$

If $S_n = Y_1 + \cdots + Y_n$, then $(n + S_n)/2$ is distributed binomially with parameters n and $p = \lambda_1/(\lambda_1 + \lambda_2)$ so that for $(n + k)/2 = 0, 1, \ldots, n$ the jump of the distribution at k is

$$\begin{aligned}
F^{(n)}(k) - F^{(n)}(k-) &= \Pr\{S_n = k\} \\
&= \binom{n}{\frac{n+k}{2}} p^{(n+k)/2}(1 - p)^{(n-k)/2}.
\end{aligned}$$

We apply (9.3) to compute the probability distribution of $X(t)$. Since $F^{(n)}(k) - F^{(n)}(k-)$ is zero unless $(n - k)/2 = 0, 1, \ldots, n$, we change variables, replacing n by $v = (n - k)/2$, to get

$$\Pr\{X(t) = k\} = \sum_{n=0}^{\infty} \frac{(\lambda t)^n e^{-\lambda t}}{n!} [F^{(n)}(k) - F^{(n)}(k-)]$$

$$= \sum_{v=0}^{\infty} \frac{(\lambda t)^{2v+k} e^{-\lambda t} p^{v+k} (1 - p)^v}{v!(v + k)!}.$$

In terms of the modified Bessel function,

$$I_r(x) = \sum_{v=0}^{\infty} \frac{1}{v! \Gamma(v + r + 1)} \left(\frac{x}{2}\right)^{2v+r}$$

this becomes

$$\Pr\{X(t) = k\} = \left(\frac{p}{1 - p}\right)^{k/2} e^{-\lambda t} I_k(2\lambda t \sqrt{p(1 - p)}).$$

LÉVY PROCESSES

A stochastic process having stationary independent increments and continuous in probability in the sense that for every positive ε,

$$\lim_{s \to t} \Pr\{|X(t) - X(s)| > \varepsilon\} = 0, \qquad t \geq 0,$$

is called a Lévy process. A compound Poisson process is a Lévy process; so is Brownian motion and the so-called "uniform translation," the process $U(t) = \alpha t$ for $t \geq 0$, where α is a fixed constant. While beyond the scope of this book, it is a fact that the general Lévy process can be represented as a sum of a Brownian motion, a uniform translation, and a limit (actually, an integral) of a one-parameter family of compound Poisson processes, where all the contributing basic processes are mutually independent.

Let $\{X(t), t > 0\}$ be a compound Poisson process. Consulting Eq. (9.1) we see that

$$\varphi_{X(t)}(u) = [\varphi_{X(1)}(u)]^t,$$

or

$$\varphi_{X(1)}(u) = [\varphi_{X(t)}(u)]^{1/t}. \tag{9.6}$$

A random variable X is called *infinitely divisible* if for every positive integer n, there are n independent and identically distributed random variables $X_1^{(n)}, \ldots, X_n^{(n)}$ such that their sum $X_1^{(n)} + \cdots + X_n^{(n)}$ has the same distribution as X. In terms of characteristic functions, the requirement becomes that for every positive integer n, there exists a characteristic function φ_n for which

$$E[e^{iuX}] = [\varphi_n(u)]^n.$$

Where $\{X(t), t \geq 0\}$ is a compound Poisson process, inspection of (9.6) reveals that $X(1)$ is infinitely divisible. Both Brownian motion $\{W(t), t \geq 0\}$ and uniform translation $\{U(t), t \geq 0\}$ possess the property expressed in (9.6), so that $B(1)$ and $U(1)$ are infinitely divisible as well. It is readily checked that the sum of a finite set of infinitely divisible random variables is itself infinitely divisible and the limit, in distribution, of a sequence of infinitely divisible random variables is infinitely divisible. The latter fact follows from Lévy's convergence criterion, mentioned in Chapter 1.

Since every Lévy process $\{L(t), t \geq 0\}$ may be written as a sum of a Brownian motion, a uniform translation and a limit of compound Poisson processes, it follows that $L(1)$ for such a process is always an infinitely divisible random variable.

A converse is also valid. For every infinitely divisible distribution, there exists a Lévy process $\{L(t), t \geq 0\}$ for which $L(1)$ has the specified distribution.

Since Eq. (9.6), or in this case

$$\varphi_{L(t)}(u) = [\varphi_{L(1)}(u)]^t,$$

is satisfied for every Lévy process, it follows that for an arbitrary such process

$$E[L(t)] = tE[L(1)]$$

and

$$\text{variance of } L(t) = t \times \text{variance of } L(1),$$

whenever these moments exist. This was argued from first principles in Chapter 1.

DECOMPOSITION OF POISSON PROCESSES

Let $X(t) = \sum_{k=1}^{N(t)} Y_k, t \geq 0$, be a compound Poisson process, where $\{N(t), t \geq 0\}$ is a Poisson process with parameter λ, and $Y_1, Y_2, \ldots$ are independent identically distributed random variables, independent of $\{N(t), t \geq 0\}$. Construct two new processes as follows: Let

$$U_k = \begin{cases} Y_k & \text{if } Y_k > 0, \\ 0 & \text{if } Y_k \leq 0, \end{cases} \qquad V_k = \begin{cases} 0 & \text{if } Y_k > 0, \\ Y_k & \text{if } Y_k \leq 0. \end{cases}$$

Then set

$$X_1(t) = \sum_{k=0}^{N(t)} U_k \quad \text{and} \quad X_2(t) = \sum_{k=0}^{N(t)} V_k, \qquad t \geq 0. \tag{9.7}$$

Then $\{U_i\}$ is a sequence of independent and identically distributed random variables, independent of the Poisson process $\{N(t), t > 0\}$, and thus $\{X_1(t), t > 0\}$ is a compound Poisson process. Similarly $\{X_2(t), t \geq 0\}$ is a compound Poisson process. What is also important, and more interesting, is that the two processes are independent of one another. By this it is meant that for every finite set of ordered time points $t_{11}, t_{12}, \ldots, t_{1n}$ and $t_{21}, t_{22}, \ldots, t_{2m}$, the vectors

$$(X_1(t_{11}), X_1(t_{12}), \ldots, X_1(t_{1n})) \quad \text{and} \quad (X_2(t_{21}), X_2(t_{22}), \ldots, X_2(t_{2m}))$$
(9.8)

are independent. We leave it to the reader to verify that only the case $m = n$ and $t_{11} = t_{21} = t_1 < t_{12} = t_{22} = t_2 < \cdots < t_{1n} = t_{2m} = t_n$ need be considered (combine both sets of time points). The random vectors in (9.8) are independent if and only if their joint characteristic function factors into the marginal characteristic functions. (See Problem 11, Chapter 1.) That is, we need to show for all real numbers $u_{11}, \ldots, u_{1n}, u_{21}, \ldots, u_{2n}$, that

$$E\left[\exp\left\{i \sum_{k=1}^{n} [u_{1k} X_1(t_k) + u_{2k} X_2(t_k)]\right\}\right] = E\left[\exp\left\{i \sum_{k=1}^{n} u_{1k} X_1(t_k)\right\}\right]$$
$$\times E\left[\exp\left\{i \sum_{k=1}^{n} u_{2k} X_2(t_k)\right\}\right]. \quad (9.9)$$

Since the vector-valued process $\{(X_1(t), X_2(t)); t \geq 0\}$ also has stationary independent increments, we may execute the computations in the manner preceding Theorem 9.1 and deduce thereby

$$E\left[\exp\left\{i \sum_{k=1}^{n} u_{1k} X_1(t_k) + i \sum_{k=1}^{n} u_{2k} X_2(t_k)\right\}\right]$$
$$= E\left[\exp\left\{i \sum_{k=1}^{n} a_{1k}[X_1(t_k) - X_1(t_{k-1})] + i \sum_{k=1}^{n} a_{2k}[X_2(t_k) - X_2(t_{k-1})]\right\}\right]$$

where $a_{jk} = u_{jn} + \cdots + u_{jk}$.

$$= \prod_{k=1}^{n} E[\exp\{ia_{1k}[X_1(t_k) - X(t_{k-1})] + ia_{2k}[X_2(t_k) - X_2(t_{k-1})]\}]$$
$$= \prod_{k=1}^{n} E[\exp\{ia_{1k}[X_1(t_k - t_{k-1})] + ia_{2k}[X_2(t_k - t_{k-1})]\}] \quad (9.10)$$

Similarly, since $\{X_j(t), t > 0\}$ each exhibit stationary independent increments, we obtain (see (9.2))

$$E\left[\exp\left\{i \sum_{k=1}^{n} u_{jk} X_j(t_k)\right\}\right] = \prod_{k=1}^{n} \varphi_{X_j(t_k - t_{k-1})}(a_{jk}), \quad (9.11)$$

By noting the correspondence between the individual terms in the products of (9.10) and (9.11) we see that to prove (9.9) it suffices to consider the case $n = 1$ and prove that

$$E[\exp\{i[u_1 X_1(t) + u_2 X_2(t)]\}] = E[\exp\{iu_1 X_1(t)\}]E[\exp\{iu_2 X_2(t)\}]. \quad (9.12)$$

We condition now on $N(t) = n$. Then $u_1 X_1(t) + u_2 X_2(t) = W_1 + \cdots + W_n$, where $W_1, \ldots, W_n$ are independent and identically distributed:

$$W_k = \begin{cases} u_1 Y_k & \text{if } Y_k > 0, \\ u_2 Y_k & \text{if } Y_k \leq 0. \end{cases}$$

Thus

$$E[\exp\{i[u_1 X_1(t) + u_2 X_2(t)]\} | N(t) = n] = E[\exp\{i(W_1 + \cdots + W_n)\}]$$
$$= \{E[\exp(iW_k)]\}^n$$
$$= [\varphi_W(1)]^n,$$

where $\varphi_W(1) = E[\exp(iW_k)]$. Removing the condition $N(t) = n$ gives

$$E[\exp\{i[u_1 X_1(t) + u_2 X_2(t)]\}] = \sum_{k=0}^{\infty} [\varphi_W(1)]^k \frac{e^{-\lambda t}(\lambda t)^k}{k!}$$
$$= \exp\{-\lambda t[1 - \varphi_W(1)]\}. \qquad (9.13)$$

Now

$$\varphi_W(1) = E[\exp(iW_k)]$$
$$= E[\exp\{i(u_1 Y_k^+ + u_2 Y_k^-)\}]$$
$$= \int_{-\infty}^{0} \exp(iu_2 y)\, dF_Y(y) + \int_{0^+}^{\infty} \exp(iu_1 y)\, dF_Y(y)$$
$$= \int_{-\infty}^{+\infty} \exp(iu_2 y^-)\, dF_Y(y) + \int_{-\infty}^{+\infty} \exp(iu_1 y^+)\, dF_Y(y) - 1$$

(the notation $y^+ = \max(y, 0)$, $y^- = \min(y, 0)$ is used)

$$= \varphi_2(u_2) + \varphi_1(u_1) - 1, \qquad (9.14)$$

where

$$\varphi_1(u) = E[\exp(iuU_k)] = E[\exp(iuY_k^+)]$$

and

$$\varphi_2(u) = E[\exp(iuV_k)] = E[\exp(iuY_k^-)].$$

Then, substituting (9.14) in (9.13) gives

$$E[\exp\{i[u_1 X_1(t) + u_2 X_2(t)]\}] = \exp\{-\lambda t[1 - \varphi_1(u_1) - \varphi_2(u_2) + 1]\}$$
$$= \exp\{-\lambda t[1 - \varphi_1(u_1)]\} \exp\{-\lambda t[1 - \varphi_2(u_2)]\}$$
$$= E[\exp\{iu_1 X_1(t)\}]E[\exp\{iu_2 X_2(t)\}],$$

which completes the proof of independence.

The restriction to the partition between $Y_k > 0$ versus $Y_k \leq 0$ served only for convenience in exposition. The same steps will verify the following theorem, which decomposes an arbitrary compound Poisson process into independent compound Poisson subprocesses.

Theorem 9.2. *Let* $X(t) = \sum_{k=1}^{N(t)} Y_k$, $t \geq 0$ *be a compound Poisson process, where* $\{N(t), t \geq 0\}$ *is a Poisson process with parameter* λ *and* $\{Y_k\}$ *is a sequence of independent and identically distributed random variables. Let* $A_1, \ldots, A_m$ *be a partition of the space of possible values for* Y_k, *that is* $A_i \cap A_j = \emptyset$ *if* $i \neq j$ *and* $\Pr\{Y_k \in A_1 \cup \cdots \cup A_m\} = 1$. *Let*

$$Y_k^i = \begin{cases} Y_k & \text{if} \quad Y_k \in A_i, \\ 0 & \text{if} \quad Y_k \notin A_i, \end{cases}$$

for $i = 1, \ldots, m$, *and define*

$$X_i(t) = \sum_{k=1}^{N(t)} Y_k^i, \qquad i = 1, \ldots, m.$$

Then each $\{X_i(t), t \geq 0\}$ *is a compound Poisson process and the processes*

$$\{X_1(t), t \geq 0\}, \quad \{X_2(t), t \geq 0\}, \quad \ldots, \quad \{X_m(t), t \geq 0\}$$

are mutually independent.

THE POISSON POINT PROCESS ASSOCIATED WITH A COMPOUND POISSON PROCESS

Consider events occurring on the positive time axis $(0, \infty)$ in accordance with a Poisson process having parameter λ. Let T_k be the time of the kth event. Then

$$T_1, \quad T_2 - T_1, \quad T_3 - T_2, \quad \ldots$$

are independent random variables, each exponentially distributed with parameter λ. Let Y_k be a value associated with the kth event and suppose $Y_1, Y_2, \ldots$ are independent and identically distributed, independent of the Poisson process, and follow the common distribution function F.

We plot the points of the coordinate pairs (T_1, Y_1), (T_2, Y_2), ... Fig. 3 illustrates the plotting.

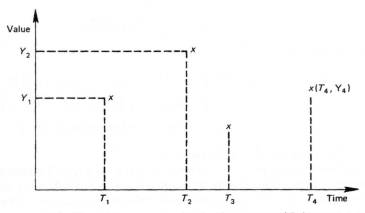

FIG 3   The point process associated with a compound Poisson process.

For each set A in the positive half plane, let $K(A)$ be the number of plotted points that fall in A. The process $\{K(A), A \subset (-\infty, +\infty) \times (0, \infty)\}$ is a *point process* (see Chapter 1) called the *point process associated with a compound Poisson process*.

Theorem 9.3. *Let $A, A_1, \ldots, A_m$ be subsets of the positive half plane. Then*

(i) *$K(A)$ has a Poisson distribution with mean*

$$\mu(A) = \iint_A \lambda \, dF(y) \, dt$$

and

(ii) *If $A_1, \ldots, A_m$ are disjoint, then $K(A_1), \ldots, K(A_m)$ are independent random variables.*

Proof. We will prove this only when $A, A_1, \ldots, A_m$ are rectangles. The general case may be obtained by approximating an arbitrary set with a grid of rectangles. First suppose

$$A = (s, t] \times (x, y].$$

Then

$$\mu(A) = \lambda(t - s)[F(y) - F(x)].$$

Let $p = F(y) - F(x)$, and let N be the number of time points T_i falling in $(s, t]$. Then N has the Poisson distribution with mean $\lambda(t - s)$ and, conditioned on $N = n$, $K(A)$ has a binomial distribution with parameters n and p. Thus for $0 \leq |u| < 1$,

$$E[u^{K(A)}] = \sum_{n=0}^{\infty} E[u^{K(A)} | N = n] \Pr\{N = n\}$$

$$= \sum_{n=0}^{\infty} (1 - p + pu)^n \frac{[\lambda(t - s)]^n e^{-\lambda(s-t)}}{n!}$$

$$= \exp[-\lambda(t - s)p(1 - u)].$$

This is the generating function of a Poisson random variable having mean $\mu(A) = \lambda(t - s)p$, which completes the proof of part (i).

For part (ii) let us consider the case in which the rectangles $A_1, \ldots, A_m$ completely partition the space $(0, t] \times (-\infty, +\infty)$. This is illustrated in Fig. 4. By appending rectangles, if necessary, the general case can be subsumed in this one. We condition on $N(t) = n$. Then, according to Theorem 2.3 of Chapter 4, the times $T_1, \ldots, T_m$ have the distribution of the ordered values of n independent

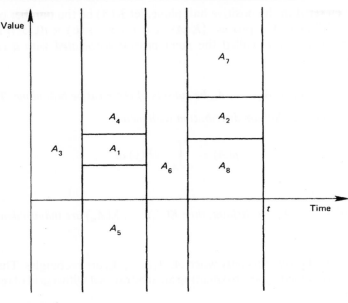

FIG. 4

observations, uniformly distributed on $[0, t]$. The probability that any single such observation would result in a point in $A_i = (s_i, t_i] \times (x_i, y_i]$ is

$$p_i = \frac{(t_i - s_i)[F(y_i) - F(x_i)]}{t} = \frac{1}{\lambda t}\mu(A_i).$$

Since the uniform observations are independent and the values are independent, the n points so scattered will result in a multinomial distribution for

$$K(A_1), \quad \ldots, \quad K(A_m)$$

with parameters $N = n$, $p_1, \ldots, p_m$. Thus the joint generating function for $(K(A_1), \ldots, K(A_m))$, under the condition $N = n$, is the multinomial generating function

$$E[u_1^{K(A_1)} \times \cdots \times u_m^{K(A_m)} | N = n] = (p_1 u_1 + \cdots + p_m u_m)^n,$$

and

$$E[u_1^{K(A_1)} \times \cdots \times u_m^{K(A_m)}] = \sum_{l=0}^{\infty} (p_1 u_1 + \cdots + p_m u_m)^l \frac{(\lambda t)^l e^{-\lambda t}}{l!}$$

$$= \exp[-\lambda t(1 - p_1 u_1 - \cdots - p_m u_m)]$$

$$= \prod_{l=1}^{m} \exp[-\lambda t p_l(1 - u_l)]$$

$$= \prod_{l=1}^{m} \exp[-\mu(A_l)(1 - u_l)].$$

Thus, the joint generating function of $K(A_1), \ldots, K(A_m)$ factors into the appropriate marginal generating functions and the proof of independence is complete. ∎

Example 3. It was noted that the logarithm of mean tensile strength of brittle fibers, such as boron filaments, in general varied linearly with the length of the filament, but that this relation did not hold for short filaments. It was suspected that the breakdown in the log–linear relation might be due to testing or measurement problems, rather than be an inherent property of short filaments. Evidence supporting this was the observation that short filaments would break in the test clamps, rather than between them as desired, more often than would long filaments. Some means of correcting observed mean strengths to account for filaments breaking in, rather than between, the clamps was desired. It was decided to compute the ratio between the actual mean strength and an ideal mean strength, obtained under the assumption of no stress in the clamps, as a correction factor.

Since the molecular bonding strength is several orders of magnitude higher than generally observed strengths, it was felt that failure typically was caused by

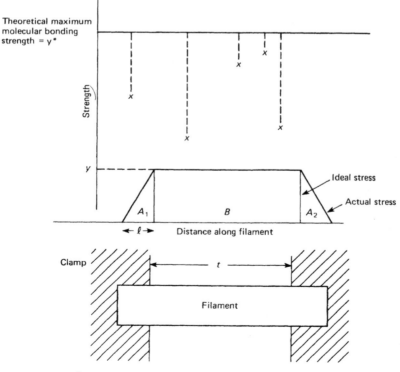

FIG. 5

flaws. There are a number of different types of flaws, both internal flaws such as voids, inclusions, and weak grain boundaries, and external or surface flaws, such as notches and cracks, which cause stress concentrations. Let us suppose that flaws occur independently in a Poisson manner along the length of the filament. We let Y_k be the reduction in strength caused by the kth flaw and suppose Y_k has the probability density function $f(y)$, $y > 0$. We have plotted this in Fig. 5. The flaws reduce the strength. Opposing the strength is the stress in the filament. Ideally, the stress should be constant along the filament between the clamp faces, and zero within the clamp. In practice the stress tapers off to zero over some positive length in the clamp. As a first approximation it is reasonable to assume that the stress decreases linearly. Let l be the length of the clamp and t the distance between the clamps, called the *gauge length*, as illustrated in Fig. 5.

The filament holds as long as the stress has not exceeded the strength as determined by the weakest flaw. That is, the filament will support a stress of y as long as no flaw points fall in the stress trapezoid of Figure 5. The number of points in this trapeziod has a Poisson distribution with mean $\mu(B) + 2\mu(A)$. In particular, no points fall there with probability

$$e^{-[\mu(B) + 2\mu(A)]}.$$

If we let S be the strength of the filament, then

$$\Pr\{S > y\} = e^{-2\mu(A) - \mu(B)}.$$

We compute

$$\mu(A) = \int_0^l \left[\int_0^{xt/l} f(y^* - s) \, ds \right] \lambda \, dx,$$

$$= \int_0^y l\left(1 - \frac{s}{y}\right) f(y^* - s) \, ds$$

and

$$\mu(B) = \lambda t[1 - F(y)],$$

where $F(y) = \int_y^{y^*} f(s) \, ds$. Finally, the mean strength of the filament is

$$E[S] = \int_0^\infty \Pr\{S > y\} \, dy$$

$$= \int_0^\infty \exp\left\{ -\lambda t[1 - F(y)] - 2\int_0^y l\left(1 - \frac{s}{y}\right) f(y^* - s) \, ds \right\} dy.$$

For an ideal filament we use the same expression but with $l = 0$.

Elementary Problems

1. It is assumed that the sizes of firms are quantized and that in each time increment h the unit of a firm has a probability $h + o(h)$ of increasing one unit, independently of the previous history of growth and its current size. Given that a firm is initially of size 1, what is the probability that during the elapsed time t the firm grows to total size n?

Hint: This is a Yule process; see Section 1 of Chapter 4.

Solution: $e^{-t}(1 - e^{-t})^{n-1}$.

2. N bacteria are spread independently with uniform distribution on a microscope slide of area A. An arbitrary region of area a is selected for observation. Determine the probability of k bacteria within the region of area a.

Solution:

$$p(k) = \binom{N}{k}\left(\frac{a}{A}\right)^k\left(1 - \frac{a}{A}\right)^{N-k}.$$

3. Show that as $N \to \infty$ and $a \to 0$ such that $(a/A)N \to c$ $(0 < c < \infty)$ then $p(k) \to e^{-c}c^k/k!$.

4. Let $\{X_i(t), t \geq 0\}_{i=1}^n$ be independent Poisson processes with the same parameter λ. Find the distribution of the time until at least one event has occurred in every process.

Solution: $P\{T \leq t\} = (1 - e^{-\lambda t})^n$.

5. An experiment has N possible outcomes each occurring with probability p_i, $\sum_{i=1}^N p_i = 1$. Let T be the number of trials necessary for all different outcomes to have occurred. Show that

$$E(T) = \int_0^\infty t\, d\left[\prod_{i=1}^n (1 - e^{-tp_i})\right].$$

Hint: Consider performing the experiment at the event times of a Poisson process with parameter $\lambda = 1$.

Problems

1. Consider a two-dimensional Poisson process of particles in the plane with intensity parameter v. Determine the distribution $F_D(x)$ of the distance between a particle and its nearest neighbor. Compute the mean distance.

Answer: $F_D(x) = 1 - \exp(-v\pi x^2)$; $E[D] = 1/(2\sqrt{v})$.

2. Solve the preceding problem for a three-dimensional Poisson process.

Answer: $F_D(x) = 1 - \exp(-v\frac{4}{3}\pi x^3)$; $E[D] = \Gamma(\frac{1}{3})/(36v\pi)^{1/3}$.

3. Suppose a device is exposed to one of k possible environments $E_1, E_2, \ldots, E_k$ which can occur with respective probabilities $c_1, c_2, \ldots, c_k$ ($\sum_{j=1}^k c_j = 1$). In each environment dangerous peaks occur according to a Poisson process with parameter $\lambda_j, j = 1, 2, \ldots, k$. Within the environment E_j the conditional probability that the device fails, given that a peak occurs, is p_j. Find the probability that the device fails within a given length of time t.

Answer: $\Pr\{T < t\} = 1 - \sum_{j=1}^{k} c_j \exp(-\lambda_j p_j t)$.

4.   A group of n engineers is engaged in a project. The time before an engineer makes an error has the probability distribution $F(t)$. If an error is made it can either be a type I error with probability p or of type II with probability $1 - p$. A type I error is so serious that if anybody commits it at any given time the whole project is sure to run afoul. A type II error, however, is so slight that the only way it can ruin the whole project is if all engineers independently commit this error. Compute the probability that the project is still on the right track at time t.

Hint: Compute the probability that exactly k engineers have made type II errors and the others have no error by time t. Show that this probability is

$$\binom{n}{k}[(1 - p)F(t)]^k[1 - F(t)]^{n-k}.$$

Answer: $[1 - pF(t)]^n - [(1 - p)F(t)]^n$.

5.   Consider a circuit consisting of m subsystems in parallel, each subsystem consisting of n like components in series. Assume that component lives are independently and identically distributed according to $F(t)$.

Show that the probability that the circuit survives until time t is given by $1 - \{1 - [1 - F(t)]^n\}^m$.

6.   Consider a sequence $i = 1, 2, \ldots, m$ of electrical components of a complex system S. Let $F_i(t)$ denote the distribution function of the time until failure of the ith component. Let $1 - p_i$ denote the conditional probability that if the ith component fails it will render the whole system inoperative.

   (i)   A system of components is said to be *semiparallel* whenever the system fails only if all components fail or a single component fails (say the ith) and with probability $1 - p_i$ the system becomes inoperative.

   (ii)   A system of components is said to be in series provided the system fails if a single component fails.

The reliability of S at time t is defined to be the probability that the whole system S is operative.

   (1)   Suppose $F_i(t) = F(t)$, $p_i = p$ ($i = 1, 2, \ldots, m$), and consider a semiparallel system. Prove that

$$\text{reliability at time } t = [1 - F(t) + pF(t)]^m - [pF(t)]^m.$$

   (2)   Let the conditions of (1) hold. Suppose $F(t) = 1 - e^{-\lambda t}$ ($\lambda > 0$). Prove

$$\text{expected time until failure} = \frac{1}{\lambda} \sum_{k=1}^{m} \frac{p^{m-k}}{k}.$$

Hint for (2): Let $u_m =$ expected time until failure for m components in a semiparallel system. Deduce the recursion formula

$$u_m = \frac{1}{m\lambda} + pu_{m-1}.$$

7. Consider a collection of circles in the plane whose centers are distributed according to a spatial Poisson process with parameter $\lambda|A|$, where $|A|$ denotes the area of the set A. (In particular, the number of centers $\xi(A)$ in the set A follows the distribution law $\Pr\{\xi(A) = k\}$ $= e^{-\lambda|A|}[(\lambda(|A|)^k/k!].$) The radius of each circle is assumed to be a random variable independent of the location of the center of the circle with density function $f(r)$ and finite second moment. Show that the family of random variables $C(r)$ the number of circles which cover the origin and have centers at a distance less than r from the origin, determines a variable time Poisson process where the time variable is now taken to be the distance r (cf. Elementary Problem 12, Chapter 4).

Hint: Prove that an event occurring between r and $r + dr$ (i.e., there is a circle of center in the ring of radius r to $r + dr$ which covers the origin) has probability $\lambda 2\pi r\, dr \int_r^\infty f(\rho)\, d\rho$ $+ o(dr)$ and events occurring over disjoint intervals constitute independent r.v.'s. Show that $C(r)$ is a variable time (inhomogeneous) Poisson process with parameter

$$\lambda(r) = 2\pi\lambda r \int_r^\infty f(\rho)\, d\rho.$$

8. Show that the number of circles which cover the origin is a Poisson random variable with parameter $\lambda \int_0^\infty \pi r^2 f(r)\, dr$.

9. Consider sphere in three-dimensional space with centers distributed according to a Poisson distribution with parameter $\lambda|A|$ where $|A|$ now represents the volume of the set A. If the radii of all spheres are distributed according to $F(r)$ with density $f(r)$ and finite third moment, show that the number of spheres which cover a point t is a Poisson random variable with parameter $\frac{4}{3}\lambda\pi \int_0^\infty r^3 f(r)\, dr$.

10. Suppose there is a reaction between the bacteria whenever two or more bacteria have their centers less than a distance r apart. Find the distribution function for the number of reactions in the area A, valid in the limit as $r \to 0, N \to \infty$ with $\pi r^2 N^2/A \to \lambda, 0 < \lambda < \infty$.

Answer: $p(l) = e^{-\lambda}(\lambda^l/l!)$.

11. Suppose new mutant types arise according to a Poisson process of parameter v. The population generated by each new mutant fluctuates in accordance with the laws of a birth and death process of infinitesimal parameters $\lambda_n = n\lambda$ and $\mu_n = n\mu$, where $\mu > \lambda$. Distinct mutants create independent lines of growth. (Recall that since $\mu > \lambda$ each line becomes extinct in finite time with probability 1; see Section 7 of Chapter 4.) Show that the number of mutant lines L_t existing at time t has a Poisson distribution.

Answer: Let $\Omega(\xi)$ be the distribution function of the time to extinction of a linear growth birth and death process $(\lambda_n = n\lambda, \mu_n = n\mu)$ starting with one individual. The parameter of the Poisson distribution is

$$v \int_0^t [1 - \Omega(\xi)]\, d\xi.$$

12 (*continuation of Problem 11*). Determine the limiting distribution of L_t as $t \to \infty$.

Answer: Poisson distribution of parameter

$$v \int_0^\infty [1 - \Omega(\xi)]\, d\xi = v \frac{\mu}{\lambda}\left[-\log\left(1 - \frac{\lambda}{\mu}\right)\right].$$

13. Suppose particles arrive in a pattern of a Poisson process with parameter λ. On arrival each particle enters one of $r\,(\geq 1)$ states with probabilities $p_1, p_2, \ldots, p_r$, respectively. After arrival a specific particle undergoes changes of its state according to the laws of a time homogeneous Markov process with transition probabilities $P_{ij}(t) = \Pr\{$state at time t is $j|$ state at time 0 was $i\}$, $i, j = 1, \ldots, r$. Let $\mathbf{Y}(t) = \{X_1(t), \ldots, X_r(t); t \geq 0\}$ be a vector-valued stochastic process where $X_i(t)$ is the number of particles in state i at time t. Prove that $\{\mathbf{Y}(t), t \geq 0\}$ is a time-homogeneous Markov process. (The states could be interpreted as different stages of an illness.)

14 (*continuation of Problem* 13).   Consider a single particle of type i at time 0. Let

$$\delta_{ij}(t) = 1 \qquad \text{if at time } t \text{ this particle is in state } j$$
$$= 0 \qquad \text{otherwise.}$$

Prove the generating function relation

$$E[z_1^{\delta_{i1}(t)} z_2^{\delta_{i2}(t)} \cdots z_r^{\delta_{ir}(t)}] = \sum_{j=1}^{r} z_j P_{ij}(t).$$

15 (*continuation of Problem* 14).   Establish that the probability generating function of $\{X_1(t), \ldots, X_r(t)\}$ is

$$\varphi(z_1, z_2, \ldots, z_r; t) = E[z_1^{X_1(t)} z_2^{X_2(t)} \cdots z_r^{X_r(t)}]$$

$$= \exp\left[\lambda \sum_{i=1}^{r} p_i \sum_{k=1}^{r} (z_k - 1) \int_0^t P_{ik}(\tau)\, d\tau\right].$$

Hint (cf. the analysis of Section 3):

1.   Condition on the number of particles arriving up to time t.

2.   Use the fact that in a Poisson process, given the number of events up to time t, the times at which the events occur are independent and identically distributed uniformly in $[0, t]$.

3.   Use the fact that individual particles act independently of each other.

16.   Consider the following two-type population growth model. The two types of particles are either of normal or mutant form. A normal type lives a random length of time with exponential distribution with mean λ^{-1} and then gives birth to two normal types with probability p or to one normal and one mutant type with probability $q = 1 - p$. Each normal offspring acts the same as the parent. A mutant type lives a random length of time following an exponential distribution with mean μ^{-1} and then gives birth to exactly two mutant types which behave the same way as their parent. Assume that all types act independently. We start with one normal type. Let $\{X(t), Y(t)\}$ denote the number of normal and mutant types in the system at time t. Find the probability generating functions $\psi_1(z, t) = E[z^{X(t)}]$ and $\psi_2(z, t) = E[z^{Y(t)}]$.

Hint: (a)   Show that $\{X(t)\}$ is a Yule process with parameter λp.

(b)   Derive the following integral equation for $\psi = \psi_2$ by conditioning on the time of the first split,

$$\psi(z, t) = e^{-\lambda t} + \int_0^t \lambda e^{-\lambda \tau}[p\psi^2(z, t - \tau) + q\psi(z, t - \tau)g(z, t - \tau)]\, d\tau,$$

where

$$g(z, t) = ze^{-\mu t}[1 - z(1 - e^{-\mu t})]^{-1}.$$

Make a change of variables, $t - \tau = u$, in the integral and then differentiate with respect to t to get

$$\frac{\partial \psi}{\partial t} = -\lambda \psi + \lambda \psi(p\psi + qg).$$

Let $\varphi = 1/\psi$. Then

$$\frac{\partial \varphi}{\partial t} = \lambda \varphi - \lambda p - \lambda qg\varphi.$$

Solve this to show that

$$\psi = \frac{F(z, t)}{1 - \lambda p \int_0^t F(u; t) \, du},$$

where $F(z, t) = e^{-\lambda t}[1 - z(1 - e^{-\mu t})]^{-\lambda q/\mu}$. In the special case $\lambda = \mu$, show that

$$\frac{1}{\psi} = 1 + \frac{1 - z}{z} e^{\mu t}[1 - (1 - z + ze^{-\mu t})^q].$$

17. Customers arrive at a counter according to a Poisson process having rate λ. Each customer pays a dollar at the time of his arrival. If a customer arrives at time τ, his dollar is discounted to a present value of $e^{-\beta \tau}$. Compute the mean total discounted payments of all customers arriving in the first T time units of operation. That is, if τ_i is the time of arrival of the ith customer, and $N(T)$ is the number of customers arriving in the first T units of time, then find $E[\sum_{i=1}^{N(T)} \exp(-\beta \tau_i)]$.

18. Let $\{N(A), A \subset R^2\}$ be a homogenous Poisson point process in the plane having parameter λ. That is, (i) if $A_1, A_2, \ldots, A_k$ are disjoint subsets of the plane, then $N(A_1)$, $N(A_2), \ldots, N(A_k)$ are independent random variables and (ii) if A is a subset of the plane, then $N(A)$ has a Poisson distribution with mean $\lambda |A| = \lambda \times$ area of A.

(a) Let A_1 and A_2 be arbitrary subsets of the plane. Compute $K(A_1, A_2) = $ covariance of $N(A_1), N(A_2)$. Note: If X and Y are random variables having finite second moments, the covariance of X, Y is given by $E[(X - E[X])(Y - E[Y])]$.

(b) Show that for arbitrary sets $A_1, A_2, \ldots, A_k$ and arbitrary real numbers $\alpha_1, \alpha_2, \ldots, \alpha_k$ that

$$\sum_{i, j=1}^{k} \alpha_i \alpha_j K(A_i, A_j) \geq 0.$$

19. The following model is proposed to describe the statistical distribution of the times of lightning discharges in a localized storm. Very briefly, the model proposes a succession of static electric charge buildups relieved by lightning discharges.

More precisely, we suppose that immediately after a discharge the static charge in the atmosphere is zero. The charge then builds up in a deterministic manner in which $r(t)$ is the charge in the atmosphere t time units after the previous lightning flash.

A triggering event is required to initiate a lightning flash. Such events occur according to a Poisson process with parameter λ. A given triggering event may or may not cause the flash. Whether it does or does not is a random event whose probability depends on the current charge level $r(t)$. Let us suppose that, given a charge level $r(t) = r$, a triggering event, should one occur, will cause a flash with probability $p(r)$. Then $0 \leq p(r) \leq 1$ and it is reasonable to assume that $p(r)$ is an increasing function of r. It is also reasonable to assume that $r(t)$ is an increasing function of t.

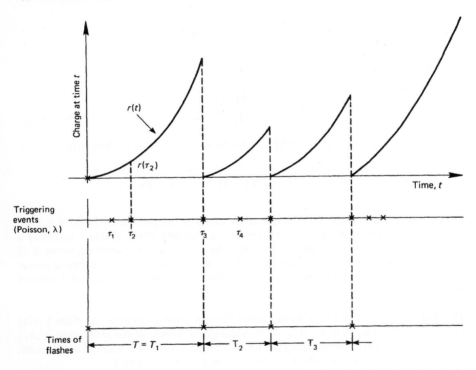

The triggering event at time τ_k causes a flash with probability $p[r(\tau_k)]$ and no flash with probability $1 - p[r(\tau_k)]$. A flash did not occur for τ_2 pictured here but did occur at τ_3.

(a) Assume that at time $t = 0$, $r(0) = 0$, and let T be the random time of the first flash. Find the distribution function

$$F(t) = \Pr\{T \leq t\}.$$

(b) Immediately after the first flash, say at time $T + 0$, again the charge $r(T + 0) = 0$, and the process repeats, generating a sequence, $T = T_1, T_2, T_3, \ldots$ of random times between flashes, independent and each with distribution function $F(t)$. In terms of $F(t)$, what is the mean number of flashes observed up to time t_0?

20. Let $\{X(t), t \geq 0\}$ be a stochastic process having stationary independent increments and finite second moments $E[X(t)^2] < \infty$. Suppose $X(0) = 0$. We showed in Chapter 1 that the mean value and variance function for such a process were linear, say,

$$E[X(t)] = \mu t$$

and

$$E[\{X(t) - \mu t\}^2] = \sigma^2 t.$$

Fix $h > 0$ and consider the process $Z(t) = X(t + h) - X(t)$. In terms of μ and σ^2, compute the mean value function, variance function and covariance function for the Z process. That is, find

$$m(t) = E[Z(t)]$$

$$V(t) = E[\{Z(t) - m(t)\}^2]$$

and

$$K(s, t) = E[\{Z(s) - m(s)\}\{Z(t) - m(t)\}].$$

21. Let $0 = \tau_0 < \tau_1 < \tau_2 \cdots$ be times of successive events of a Poisson process with parameter λ. Suppose each event (independently of others) is recorded with probability p and erased with probability $1 - p$. Show that the random variables $X_1(t)$ and $X_2(t)$ of recorded and erased events respectively constitute independent Poisson processes with parameters λp and $\lambda(1 - p)$.

Hint: For any $t \geq 0$, let $X_1(t)$ and $X_2(t)$ denote the number of recorded and erased events respectively in $(0, t)$. Then their joint probability generating function

$$
\begin{aligned}
f(z_1, z_2) &= E[z_1^{X_1(t)} z_2^{X_2(t)}] \\
&= E[E[z_1^{X_1(t)} z_2^{X_2(t)} | X_1(t) + X_2(t)]] \quad \text{(by conditioning on } X_1(t) + X_2(t)) \\
&= E[(pz_1 + qz_2)^{X_1(t) + X_2(t)}] \\
&\quad \text{(since given } X_1(t) + X_2(t), X_1(t) \text{ has a binomial distribution)} \\
&= e^{\lambda t(pz_1 + qz_2 - 1)} = e^{\lambda pt(z_1 - 1)} e^{\lambda qt(z_2 - 1)}.
\end{aligned}
$$

22. In the above problem, let each event be classified, independently of others, into k categories with probabilities p_i $i = 1, 2, \ldots, k$, $\sum_{i=1}^{k} p_i = 1$. Let $X_i(t)$ be the number of events that happen in $(0, t)$ and belong to the ith category $i = 1, 2, \ldots, k$. Show that $\{X_1(t), X_2(t), \ldots, X_k(t)\}$ for $i = 1, 2, \ldots, k$ are independent Poisson processes, with parameters respectively λp_i, $i = 1, 2, \ldots, k$.

23. Consider a Poisson process in $(0, \infty)$ with parameter λ. Suppose an event that occurs at time t is classified, independently of other events, into one of k categories with probability $p_i(t)$, for $i = 1, 2, \ldots, k$, $\sum_{i=1}^{k} p_i(t) = 1$. Assume each $p_i(t)$ is continuous in t. Let $X_i(t)$ denote the number of events during the time duration $(0, t)$ belonging to the ith category. Show that for each i, $\{X_i(t), t \geq 0\}$ is a variable time (i.e., inhomogeneous) Poisson process (cf. Problem 12 of Chapter 4).

Hint: Show that Pr{an event of category i occurs in $(t, t + h)$} = $\lambda p_i(t)h + o(h)$ and the event in braces is independent of the values of $X_i(\tau)$ for $\tau \le t$.

24. In Problem 23 show that the processes $\{X_i(t), t \ge 0\}$ $i = 1, 2, \ldots, k$ are independent.

Hint: Let (a_i, b_i), $i = 1, 2, \ldots, k$ be any k intervals in $[0 \, \infty)$. Define $Y_i = X_i(b_i) - X_i(a_i)$. Show that $Y_1, Y_2, \ldots, Y_k$ are independent r.v.'s. To do this, condition on the number $N(T)$ of events of all categories in $(0, T)$ where $T = \max_{1 \le i \le k} b_i$. Given $N(T)$, the instants at which the events occur in $(0, T)$ constitute n independent observations from a uniform distribution in $(0, T)$. Under the condition $N(T) = n$, $(Y_1, Y_2, \ldots, Y_k, n - \sum_{i=1}^{k} Y_i)$ has a multinomial distribution with associated probabilities

$$p_i = \frac{1}{T} \int_{a_i}^{b_i} p_i(t) \, dt, \qquad i = 1, 2, \ldots, k,$$

$$p_{k+1} = 1 - \sum_{i=1}^{k} p_i.$$

Therefore,

$$E[z_1^{Y_1} z_2^{Y_2} \cdots z_k^{Y_k} | N(T) = n] = (p_1 z_1 + p_2 z_2 + \cdots + p_k z_k + 1 - p_1 - \cdots - p_k)^n.$$

But $N(T)$ is distributed as a Poisson random variable with mean λT. Hence

$$E[z_1^{Y_1} \cdots z_k^{Y_k}] = \exp\left\{\mu T\left[\sum_{i=1}^{k} p_i(z_i - 1)\right]\right\}$$

$$= \prod_{i=1}^{k} \exp[\lambda_i(z_i - 1)],$$

where

$$\lambda_i = \lambda \int_{a_i}^{b_i} p_i(t) \, dt.$$

The same argument applies if each interval (a_i, b_i) is replaced by any finite union of intervals.

25. In Problem 23, let $\int_0^\infty p_i(t) \, dt < \infty$ for $i = 1, 2, \ldots, r$, where $r \le k - 1$. Prove that the joint distribution of $(X_1(t), \ldots, X_r(t))$ tends in the limit $(t \to \infty)$ to that of r independent Poisson variables with means respectively $\lambda \int_0^\infty p_i(t) \, dt < \infty$, $i = 1, 2, \ldots, r$.

26. Let $0 = \tau_0 < \tau_1 < \tau_2 \cdots$ be events of a Poisson process with parameter λ. Let at time 0 a particle commence executing a Brownian motion ($\sigma^2 = 1$) with initial position τ_i. We say the event of the Poisson process occurring at the instant τ_i is erased if the Brownian particle starting at τ_i has position to the left of $-\alpha$ ($\alpha > 0$) at the instant t_0 and is recorded otherwise. Show that the processes of recorded and erased points are independent, variable time (i.e., inhomogeneous) Poisson processes with parameters $\lambda\{1 - \Phi[(t + \alpha)/\sqrt{t_0}]\}$ and $\lambda\Phi[(t + \alpha)/\sqrt{t_0}]$, respectively, where Φ is the standard normal distribution function.

27. Let $0 = \tau_0 < \tau_1 < \cdots$ be the times of successive events of a Poisson process with parameter λ. Let $\varphi(x, y)$ be a real-valued function for $x \ge 0$. Consider $\tilde{\tau}_i = \varphi(\tau_i, \xi(\tau_i))$ where

$\{\xi(t), t \geq 0\}$ is a stochastic process independent of the $\{\tau_i\}$ process and such that for any n and any $t_1, t_2, \ldots, t_n$, the random variables $\xi(t_1), \ldots, \xi(t_n)$ are independent. Define $F(x, t) = \Pr\{\varphi(t, \xi(t)) \leq x\}$. For each $t > 0$ and r disjoint intervals $(a_i, b_i), i = 1, 2, \ldots, r$ on $(-\infty, +\infty)$, let $X_i(t)$ denote the number of τ_j in $(0, t)$ satisfying $a_i < \tau_i < b_i$. Show that $\{X_i(t), t \geq 0\}$ for $i = 1, 2, \ldots, r$ and r independent time varying (inhomogeneous) Poisson processes. Show that $E[X_i(t)] = \lambda \int_0^t [F(b_i, u) - F(a_i, u)] \, du$, assuming, of course, that for each x, $F(x, u)$ is an integrable function of u over any finite interval.

Hint: Use the results and methods of Problems 23 and 24.

28. In Problem 27 assume $\int_0^\infty F(x, u) \, du < \infty$ for each x. Show that the distribution function of $X_i(t)$ tends to a Poisson random variable as $t \to \infty$. Show also that $\{N(a), -\infty < a < \infty\}$ where $N(a_2) - N(a_1)$ is the number of $\tilde{\tau}_j$'s in (a_1, a_2) determines a process with independent but not necessarily stationary increments.

29. In Problem 23 assume there exists a function $h(x)$ such that for any $x_1 \leq x_2$

$$\int_0^\infty F(x_2, u) \, du - \int_0^\infty F(x_1, u) \, du = \int_{x_1}^{x_2} h(t) \, dt.$$

Show that the process $\{N(a), -\infty < a < \infty\}$ defined in Problem 28 is a Poisson process not necessarily time homogeneous with parameter $\lambda h(t)$.

30. In Problem 29 assume $h(t) \equiv \mu$. Show that $\{N(a), -\infty < a < \infty\}$ is a one-dimensional spatial Poisson process with parameter $\lambda \mu$.

31. Let the process $\{\xi(t), t \geq 0\}$ of Problem 27 be such that for any n and any $t_1, t_2, \ldots, t_n \geq 0$, $\xi(t_1), \ldots, \xi(t_n)$ are positive independent r.v.'s and possess a common distribution function $G(x)$. Specialize the results of Problems 27–30 for the following cases:
  (a) $\varphi(x, y) = xy$,
  (b) $\varphi(x, y) = x + y$.
  In (a) assume $1/w = \int_0^\infty dG(v)/v < \infty$.

32. Deduce from 31b above that for an infinite queue $(M|G|\infty)$, see Problem 10 of Chapter 18) with a homogeneous Poisson input and arbitrary service time distribution $G(x)$ that the output process $\{U(t), t \geq 0\}$ (i.e., the number of customers served by time t) is a time-dependent Poisson process provided at $t = 0$ there are no customers in the system. The parameter is $\lambda G(t)$.

33. Suppose cars enter a highway at instants $0 = \tau_0 < \tau_1 < \tau_2 < \cdots$ which form a Poisson process with parameter λ. The highway is semi-infinite and cars enter it from one end and move in the same positive direction. The car entering at time τ_k picks a velocity v_k and travels with constant velocity v_k. We assume the v_k are independent positive random variables with common distribution function $F(x)$. Show that in the stationary case, i.e., after an infinite length of time, the spatial distribution of cars along the highway constitute a homogeneous Poisson process, provided $1/w = \int_0^\infty dF(v)/v < \infty$. Find the parameter.

Hint: Let (a_1, b_1) and (a_2, b_2) be two disjoint intervals on $(0, \infty)$. Consider the time axis as extending from the present to the infinite past and measure the position of the cars now from the entrance to the highway. The instants at which the cars entered the highway form a Poisson process in this time axis. Any τ_i will be of category 1 if $a_1 \leq \tau_i v_i \leq b_1$, of category 2 if $a_2 \leq \tau_i v_i \leq b_2$, and of category 3 otherwise. Then in the notation of Problem 24

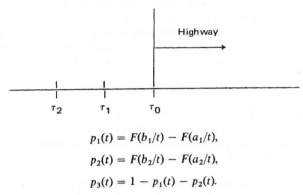

$$p_1(t) = F(b_1/t) - F(a_1/t),$$

$$p_2(t) = F(b_2/t) - F(a_2/t),$$

$$p_3(t) = 1 - p_1(t) - p_2(t).$$

Now use the results of Problems 23–25. Note that $\int_0^\infty F(x/t)\,dt = x/w$, where $w^{-1} = \int_0^\infty dF(v)/v$. A similar argument applies for any finite number (a_i, b_i) of nonoverlapping intervals.

NOTES

Material related to this chapter is contained in Bartlett [1] and Harris [2].

Further discussion of compounding stochastic processes with emphasis on applications is contained in Bartlett [3].

Elegant presentations of topics from the rapidly developing field of coupled and interacting stochastic processes are introduced in Preston [4] and Griffeath [5].

The pervasive occurrence of point processes in studies of traffic flow, in descriptions of the distributions of astronomical bodies, in systematics for biological species, in reference to geological formations and in reliability systems is amply illustrated in Lewis [6].

REFERENCES

1. M. S. Bartlett, "Stochastic Population Models in Ecology and Epidemiology." Wiley, New York, 1960.
2. T. E. Harris, "The Theory of Branching Processes." Springer-Verlag, Berlin, 1963.
3. M. S. Bartlett, "An Introduction to Stochastic Processes with Special Reference to Methods and Applications." Cambridge Univ. Press, London and New York, 1955.
4. C. Preston, "Random Fields." Springer-Verlag, New York, 1976.
5. D. Griffeath, "Additive and Cancellative Interacting Particle Systems," Springer Lecture Notes in Math. No. 724.
6. P. A. Lewis, "Stochastic Point Processes: Statistical Analysis, Theory, and Applications." Wiley-Interscience, New York, 1972.
7. K. Matthes, J. Kerstan and J. Mecke, "Infinitely Divisible Point Processes," Wiley, New York, 1978.

Chapter 17

FLUCTUATION THEORY OF PARTIAL SUMS OF INDEPENDENT IDENTICALLY DISTRIBUTED RANDOM VARIABLES

1: The Stochastic Process of Partial Sums

Consider independent, identically distributed real-valued random variables $X_1, X_2, \ldots$ (not necessarily positive) and define the partial sums

$$S_n = X_1 + X_2 + \cdots + X_n, \qquad n = 1, 2, \ldots,$$
$$S_0 = 0 \qquad \text{(by convention).} \tag{1.1}$$

These partial sums S_n may be graphed against n and the points (n, S_n) connected by straight lines as in Fig. 1.

The process of partial sums (1.1) plays a fundamental role in many diverse areas of applications of stochastic processes. Thus, the analysis of embedded recurrent events in Markov processes, renewal phenomena (see Chapter 5), queueing and dam systems, calculation of risk and ruin probabilities, etc. all revolve on discerning properties of certain random functionals connected to the process (1.1). We can succinctly characterize fluctuation theory as the study of random variables of the form $f(S_0, S_1, S_2, \ldots, S_n)$ defined on the partial sums (1.1). It is worth highlighting several random functionals of importance in the theory and its applications. Set

$$M_n = \max(0, S_1, S_2, \ldots, S_n), \qquad m_n = \min(0, S_1, S_2, \ldots, S_n). \tag{1.2}$$

The random variables M_n keep track of the maximum of the partial sums evolving over time. When $M_n = S_n$, then a maximum point of the realization

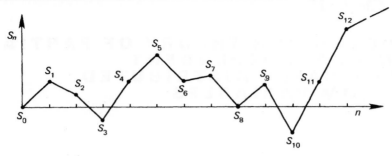

is manifested at the nth occurring partial sum, and in this case,

$$S_j \leq S_n, \qquad j = 0, 1, 2, \ldots, n - 1.$$

Define also†

$$P_n^+ = \text{number of } S_i > 0, \qquad i = 1, 2, \ldots, n,$$

$$P_n = \text{number of } S_i \geq 0, \qquad i = 1, 2, \ldots, n,$$

$$Q_n^- = \text{number of } S_i < 0, \qquad i = 1, 2, \ldots, n, \tag{1.3}$$

$$Q_n = \text{number of } S_i \leq 0, \qquad i = 1, 2, \ldots, n.$$

The number P_n^+/n evaluates the fraction of time (in the first n time units) that the process $\{S_k\}$ spends on the positive axis. This quantity is commonly referred to as the *occupation time random variable of the positive axis*. The sojourn time or occupation time of certain states are basic random variables which often underlie the structure of the process. This is of special relevance for the analyses of diffusion processes on the line (see Chapter 15). Sometimes we write $P_n(X)$, $X = (X_1, X_2, \ldots, X_n)$, instead of P_n to indicate that P_n is evaluated for the sequence of partial sums 0, X_1, $X_1 + X_2$, $X_1 + X_2 + X_3, \ldots, X_1 + X_2 + \cdots + X_n$. Similarly we will sometimes write $P_n^+(X), Q_n(X), M_n(X)$, etc.

   Of importance also are the positions of the first and last maximum (minimum) terms among the collection of partial sums $\{S_j\}_0^n$. We say that the partial sums $S_0, S_1, S_2, \ldots, S_n$ exhibit their last maximum at position k $(0 \leq k \leq n)$ if

$$S_k \geq S_j \qquad \text{for all} \quad j = 0, 1, \ldots, k - 1$$

and

$$S_k > S_j \qquad \text{for all} \quad j = k + 1, k + 2, \ldots, n.$$

Analogously the index of the first maximum is that value l satisfying

$$S_l > S_i, \qquad i = 0, 1, 2, \ldots, l - 1,$$

$$S_l \geq S_i, \qquad i = l + 1, \ldots, n.$$

† It is more traditional to use the notation N_n in place of P_n.

The first and last minima are defined in a parallel fashion or equivalently by observing that the position k is a first (last) minimum for the partial sums S_0, $S_1, \ldots, S_n$ if and only if k is a first (last) maximum for the corresponding partial sums of the sequence $\{-X_i\}_0^n$. It is now convenient to introduce some further notation. Define the random variables

$$L_{nn} = \text{position (i.e., the index) of the last maximum among } \{S_0, S_1, \ldots, S_n\},$$

$$L_{n0} = \text{position of the first maximum among } \{S_0, S_1, \ldots, S_n\},$$

$$K_{n0} = \text{position of the first minimum among } \{S_0, S_1, \ldots, S_n\},$$

$$K_{nn} = \text{position of the last minimum among } \{S_0, S_1, \ldots, S_n\}.$$

(1.4)

There are a number of key relationships among the random variables of (1.3) and (1.4) emanating from certain symmetry and combinatoric considerations. In the next section we develop a fundamental equivalence principle, and the main identities for the distributions of M_n, P_n^+, etc. are delineated in the succeeding sections. Several applications of the identities are set forth in Secs. 3 and 6.

One of the principal objectives of fluctuation theory is to ascertain the distribution of M_n in terms of the convolution distribution functions $F^{(j)}$, $j = 1, 2, \ldots, n$, of the random variables S_j, respectively. The task will be carried out in Theorems 5.2 and 5.3 of Sec. 5.

2: An Equivalence Principle

Distribution relationships for the random variables of (1.3) and (1.4) are developed first for a very special sample space designated as $\mathscr{B}$ and later the identities are validated in the general framework of an arbitrary sequence of i.i.d. (independent identically distributed) random variables. Specify $y = (y_1, y_2, \ldots, y_n)$ as a fixed ordered n-tuple of real numbers. Let $\Sigma = \{\sigma\}$ consist of the collection of all permutations $\sigma = (\sigma_1, \sigma_2, \ldots, \sigma_n)$ carrying $(1, 2, \ldots, n)$ into itself (the permutation group on n elements). Construct $\mathscr{B}$ to be comprised of the $n!$ points

$$\mathscr{B} = \{\sigma y = (y_{\sigma_1}, y_{\sigma_2}, \ldots, y_{\sigma_n}), \sigma \text{ traversing } \Sigma\}. \tag{2.1}$$

To illustrate, where $n = 3$ the set $\mathscr{B}$ consists of the triplets

$$\{(y_1, y_2, y_3), (y_1, y_3, y_2), (y_2, y_1, y_3), (y_2, y_3, y_1), (y_3, y_1, y_2), (y_3, y_2, y_1)\}.$$

Assign to each member of $\mathscr{B}$ the same probability $1/n!$, thereby converting $\mathscr{B}$ into a probability space. Each $\sigma y \in \mathscr{B}$ furnishes a possible value of the random vector $X = (X_1, X_2, \ldots, X_n)$, where $X_k(\sigma y) = y_{\sigma_k}$. Assuming $y_1, \ldots, y_n$ to be distinct values, X_k is uniformly distributed for any $k = 1, 2, \ldots, n$;

$$\Pr\{X_k(\sigma y) = y_i\} = \frac{1}{n}, \qquad i, k = 1, 2, \ldots, n, \qquad (2.2)$$

since all permutations are equally likely.

Patently, $X_1, \ldots, X_n$ are *not* independent random variables. It is well worth noting, however, that they are *exchangeable*; that is, $X_1, \ldots, X_n$ have the same joint distribution as any permutation $X_{k_1}, \ldots, X_{k_n}$.

The quantities $S_n(\sigma y), P_n(\sigma y), L_{nn}(\sigma y)$, etc. can be viewed as particular values of the random variables $S_n(X), P_n(X), L_{nn}(X)$, etc. For example, again assuming $y_1, \ldots, y_n$ to be distinct, $n! \Pr\{P_n(X) = k\}$ is just the number of permutations σ with the property that exactly k among the partial sums

$$S_1(\sigma y) = y_{\sigma_1}, \qquad S_2(\sigma y) = y_{\sigma_1} + y_{\sigma_2},$$

$$S_3(\sigma y) = y_{\sigma_1} + y_{\sigma_2} + y_{\sigma_3}, \qquad \ldots, \qquad S_n(\sigma y) = y_{\sigma_1} + \cdots + y_{\sigma_n},$$

(the partial sums induced by the point $\sigma y = (y_{\sigma_1}, y_{\sigma_2}, \ldots, y_{\sigma_n})$) are nonnegative. This may be written

$$n! \Pr\{P_n(X) = k\} = \sum_\sigma I_{[P_n = k]}(\sigma y), \qquad (2.3)$$

where the summation runs over all $n!$ permutations of $(1, 2, \ldots, n)$ and

$$I_{[P_n = k]}(\sigma y) = \begin{cases} 1 & \text{if } P_n(\sigma y) = k, \\ 0 & \text{otherwise}; \end{cases} \qquad (2.4)$$

i.e., $I_{[P_n = k]}(\cdot)$ is the indicator function of the set $\{P_n = k\}$.

Lemma 2.1 (*The Equivalence Principle for the Special Sample Space $\mathscr{B}$*).   Let $y = (y_1, y_2, \ldots, y_n)$ be fixed and define the random variables $X_1, X_2, \ldots, X_n$ on $\mathscr{B}$ by the prescriptions $X_k(\sigma y) = y_{\sigma_k}, k = 1, 2, \ldots, n$. Then

$$\Pr\{P_n = k\} = \Pr\{L_{nn} = k\}, \qquad (2.5)$$

$$\Pr\{P_n^+ = k\} = \Pr\{L_{n0} = k\}, \qquad (2.6)$$

$$\Pr\{Q_n^- = k\} = \Pr\{K_{n0} = k\}, \qquad (2.7)$$

$$\Pr\{Q_n = k\} = \Pr\{K_{nn} = k\}, \qquad (2.8)$$

$$\Pr\{L_{n0} = k\} = \Pr\{K_{nn} = n - k\}, \qquad (2.9)$$

$$\Pr\{L_{nn} = k\} = \Pr\{K_{n0} = n - k\}, \qquad (2.10)$$

where P_n means $P_n(X)$, P_n^+ abbreviates $P_n^+(X)$, etc.

Proof. We first prove (2.9) and (2.10). Let τ be the special permutation that reverses the order of $(1, 2, \ldots, n)$, i.e., that transforms it into $(n, n-1, \ldots, 1)$. If $x = (x_1, x_2, \ldots, x_n)$, then $\tau x = (x_n, x_{n-1}, \ldots, x_1)$. Now observe that

$$S_j(\tau x) = S_n(x) - S_{n-j}(x). \tag{2.11}$$

Further, by definition of $K_{nn}(\tau x)$, we have

$$S_{K_{nn}(\tau x)}(\tau x) < S_j(\tau x) \qquad \text{for all} \quad j = K_{nn}(\tau x) + 1, \ldots, n$$

and

$$S_{K_{nn}(\tau x)}(\tau x) \leq S_j(\tau x) \qquad \text{for all} \quad j = 0, 1, \ldots, K_{nn}(\tau x) - 1.$$

By means of (2.11) these inequalities may be recast as

$$S_{n-K_{nn}(\tau x)}(x) > S_j(x) \qquad \text{for all} \quad j = 0, 1, \ldots, n - K_{nn}(\tau x) - 1$$

and

$$S_{n-K_{nn}(\tau x)}(x) \geq S_j(x) \qquad \text{for all} \quad j = n - K_{nn}(\tau x) + 1, \ldots, n.$$

These relations determine that $n - K_{nn}(\tau x)$ is the index value delimiting $L_{n0}(x)$. Thus

$$L_{n0}(x) = n - K_{nn}(\tau x). \tag{2.12}$$

In a similar fashion, from the definitions of L_{nn} and K_{n0} and with the aid of (2.11), we deduce

$$L_{nn}(x) = n - K_{n0}(\tau x). \tag{2.13}$$

Since (2.12) and (2.13) persist for any n-tuple $x \in \mathscr{B}$, these identities apply in particular for $x = y$.

Now, as both events, $X = y$ and $X = \tau y$ have the same probability, namely, $1/n!$, (2.12) and (2.13) imply equations (2.9) and (2.10), respectively, of the theorem.

The proofs of (2.5)–(2.8) proceed by induction. For $n = 1$ these assertions are immediate. Assume that (2.5)–(2.8) prevail for any fixed $(n - 1)$-tuple of real numbers. We advance the induction by demonstrating these equations are maintained for any n-tuple $y = (y_1, \ldots, y_n)$ of size n. To this end, it is convenient to distinguish three separate cases.

Case 1: $S_n(y) = y_1 + \cdots + y_n < 0$. Then the random variables $P_n(X)$, $L_{n0}(X)$, $L_{nn}(X)$, and $P_n^+(X)$ cannot take the value n. Let us examine $\Pr\{P_n = k \mid X_n = y_j\}$. Under the conditioning the sample space mimics the original sample space but deals with a symmetric set of $(n - 1)$-tuples where each point carries probability $1/(n - 1)!$ The interpretation of $\Pr\{L_{nn} = k \mid X_n = y_j\}$

is similar. We can accordingly invoke the induction hypothesis for $0 \leq k \leq n - 1$, obtaining

$$\Pr\{P_n = k | X_n = y_j\} = \Pr\{L_{nn} = k | X_n = y_j\},$$

and similarly

$$\Pr\{P_n^+ = k | X_n = y_j\} = \Pr\{L_{n0} = k | X_n = y_j\}.$$

By the law of total probabilities, these equations immediately imply (2.5) and (2.6) for n-tuples. To establish (2.7) and (2.8), observe from the definitions that $\Pr\{Q_n^- = k\} = \Pr\{P_n = n - k\}$ and $\Pr\{Q_n = k\} = \Pr\{P_n^+ = n - k\}$ and then we use (2.5), (2.6), (2.9), and (2.10). This gives

$$\Pr\{Q_n^- = k\} = \Pr\{P_n = n - k\} = \Pr\{L_{nn} = n - k\} = \Pr\{K_{n0} = k\},$$
$$(2.14)$$

$$\Pr\{Q_n = k\} = \Pr\{P_n^+ = n - k\} = \Pr\{L_{n0} = n - k\} = \Pr\{K_{nn} = k\}.$$

Case 2: $S_n(y) = y_1 + \cdots + y_n > 0$. Now, $Q_n^-(X)$, $Q_n(X)$, K_{n0}, and K_{nn} cannot take the value n. Applying the induction assumption, we have

$$\Pr\{Q_n = k | X_n = y_j\} = \Pr\{K_{n0} = k | X_n = y_j\},$$
$$\Pr\{Q_n = k | X_n = y_j\} = \Pr\{K_{nn} = k | X_n = y_j\}, \qquad \text{for all } j.$$

Therefore (2.7) and (2.8) follow by the law of total probabilities. Now, with (2.7)–(2.10) already proved, we obtain

$$\Pr\{P_n = k\} = \Pr\{Q_n^- = n - k\} = \Pr\{K_{n0} = n - k\} = \Pr\{L_{nn} = k\},$$
$$(2.15)$$

$$\Pr\{P_n^+ = k\} = \Pr\{Q_n = n - k\} = \Pr\{K_{nn} = n - k\} = \Pr\{L_{n0} = k\}.$$

This proves (2.5) and (2.6).

Case 3: $S_n(y) = y_1 + \cdots + y_n = 0$. Here, $P_n^+(X)$, $Q_n^-(X)$, $L_{n0}(X)$, and $K_{n0}(X)$ cannot take the value n. Thus induction proves Eqs. (2.6) and (2.7) just as it did in Cases 1 and 2, respectively. Then (2.14) implies (2.8) and (2.15) entails (2.5). This completes the proof of Lemma 2.1. ∎

Let $g(x_1, x_2, \ldots, x_n)$ denote a function symmetric in the variables $(x_1, x_2, \ldots, x_n)$, which in formal language asserts the equation

$$g(x) = g(\sigma x)$$

for every permutation $\sigma \in \Sigma$.

Lemma 2.1 dealt with a fixed n-tuple and the sample space generated by the permutations thereof. The following theorem extends the pertinent identities to general n-vectors composed from independent identically distributed random variables (i.i.d. r.v.'s).

Theorem 2.1 (*The Equivalence Principle*). Let $X_1, \ldots, X_n$ be independent identically distributed random variables and let $g_n(x)$ be a symmetric function of $x = (x_1, \ldots, x_n)$. Then for any k, $0 \le k \le n$

$$E[g_n; P_n = k] = E[g_n; L_{nn} = k],$$

$$E[g_n; P_n^+ = k] = E[g_n; L_{n0} = k],$$

$$E[g_n; Q_n^- = k] = E[g_n; K_{n0} = k],$$

$$E[g_n; Q_n = k] = E[g_n; K_{nn} = k], \tag{2.16}$$

$$E[g_n; L_{n0} = k] = E[g_n; K_{nn} = n - k],$$

$$E[g_n; L_{nn} = k] = E[g_n; K_{n0} = n - k],$$

Note. The above expectations mean that they are taken not over the original sample space but only over the indicated subset. Specifically, if $\mathscr{E}$ is any event, then

$$E[g_n; \mathscr{E}] = \int_{\mathscr{E}} g_n(x) \, dF(x),$$

where $F(x)$ is the distribution of the random vector $X = (X_1, \ldots, X_n)$, and the expectation of $g_n(x)$ is extended only over those realizations of X which belong to $\mathscr{E}$.

Proof. Because the sequel makes primary use only of the first equation of (2.16), we present the proof of this identity in full detail. The other equations are proved completely analogously.

Consider

$$E[g_n; P_n = k] = \int_{\{P_n(x) = k\}} g_n(x) \, dF(x) = \int g_n(x) I_{\{P_n = k\}}(x) \, dF(x),$$

where the notation of (2.4) is employed. Make the change of variable $x \to \sigma x$ for some prescribed permutation σ. Since X is composed from independent identically distributed random variables, it is clear that $F(\sigma x) = F(x)$ and $g_n(\sigma x) = g_n(x)$ because by stipulation $g_n(x)$ is symmetric.

Therefore, we have

$$E[g_n; P_n = k] = \int g_n(x) I_{\{P_n = k\}}(\sigma x) \, dF(x)$$

for every permutation σ. Averaging with respect to $\sigma \in \Sigma$ (running over the set of all permutations) leads to the equation

$$E[g_n; P_n = k] = \frac{1}{n!} \sum_{\sigma} \int g_n(x) I_{\{P_n = k\}}(\sigma x) \, dF(x).$$

Next, interchanging summation and integration yields

$$E[g_n; P_n = k] = \int g_n(x) \frac{1}{n!} \sum_\sigma I_{[P_n = k]}(\sigma x) \, dF(x). \tag{2.17}$$

Now consult Eq. (2.3) and then appeal to formula (2.5) of Lemma 2.1 to validate the equations

$$\frac{1}{n!} \sum_\sigma I_{[P_n = k]}(\sigma x) = \Pr\{P_n = k\} = \Pr\{L_{nn} = k\} = \frac{1}{n!} \sum_\sigma I_{[L_{nn} = k]}(\sigma x).$$

The last equality ensues on the same basis as (2.3). Inserting the above formula into (2.17) and working backwards as leading to (2.17), we obtain

$$E[g_n; P_n = k] = \int g_n(x) \frac{1}{n!} \sum_\sigma I_{[L_{nn} = k]}(\sigma x) \, dF(x)$$

$$= \sum_\sigma \frac{1}{n!} \int g_n(x) I_{[L_{nn} = k]}(\sigma x) \, dF(x)$$

$$= \int g_n(x) I_{[L_{nn} = k]}(x) \, dF(x) = E[g_n; L_{nn} = k].$$

The first identity of (2.16) is established. The other equations are proved in the same way. ∎

Remark 2.1. The assumption that the components of $X = (X_1, \ldots, X_n)$ are i.i.d. was needed in the proof of Theorem 2.1 only to the extent that the joint distribution function $dF(X)$ is symmetric in the sense that $F(\sigma X) = F(X)$ holds for every $\sigma \in \Sigma$ (see (2.17)). A random vector X with this property is said to be *exchangeable*.

We recover a generalized form of the identities in Lemma 2.1 by prescribing $g_n(x) = 1$. For example, we have

$$\Pr\{P_n(X) = k\} = \Pr\{L_{nn}(X) = k\}, \qquad \Pr\{P_n^+(X) = k\} = \Pr\{L_{n0}(X) = k\},$$

etc., where the probabilities are computed based on a random vector X symmetrically distributed.

The most important specification of g_n in (2.16) to be featured in our later developments is the function

$$g_n(x) = e^{iu(x_1 + x_2 + \cdots + x_n)} \tag{2.18}$$

where u is a real parameter.

3: *Some Fundamental Identities of Fluctuation Theory and Direct Applications*

A remarkable series of identities will be proved in the remainder of this chapter. To state them we introduce the generating function

$$P(u;t) = \sum_{n=0}^{\infty} E[e^{iuS_n}; L_{nn} = n]t^n = \sum_{n=0}^{\infty} t^n \int_{\{L_{nn}=n\}} e^{iuS_n}\, dF_{(n)}(x), \qquad |t| < 1,$$

$$\text{(3.1)}$$

where $F_{(n)}$ is the joint distribution function of the random vector $X = (X_1, X_2, \ldots, X_n)$ and u is an arbitrary real parameter. It is customary to regard (3.1) as a double generating function since the characteristic function $E(e^{iuS_n}; L_{nn} = n)$ is also a sort of "generating" function of the random variable S_n appropriately restricted. The event $L_{nn} = n$ imposes a confinement of the realizations to obey the conditions $\{S_i \leq S_n, i = 0, 1, 2, \ldots, n\}$. We define parallel to (3.1)

$$Q(u;t) = \sum_{n=0}^{\infty} E[e^{iuS_n}; L_{nn} = 0]t^n \qquad \text{(here } L_{00} = 0 \text{ by definition)}. \quad \text{(3.2)}$$

The content of the next theorem embraces two typical identities of fluctuation theory.

Theorem 3.1. *For every real u and t, $|t| < 1$ we have*

$$P(u;t) = \exp\left\{ \sum_{k=1}^{\infty} \frac{t^k}{k} E[e^{iuS_k}; S_k \geq 0] \right\}, \tag{3.3}$$

$$Q(u;t) = \exp\left\{ \sum_{k=1}^{\infty} \frac{t^k}{k} E[e^{iuS_k}; S_k < 0] \right\}. \tag{3.4}$$

The right-hand sides involve simply the distributions $F^{(k)}(x)$ of S_k, the k-fold convolution of $F(x)$, suitably restricting the range of outcomes. Explicitly,

$$E[e^{iuS_k}; S_k \geq 0] = \int_{0-}^{\infty} e^{iu\xi}\, dF^{(k)}(\xi)$$

(notice that the integration includes the contribution at 0) and

$$E[e^{iuS_k}; S_k < 0] = \int_{-\infty}^{0-} e^{iu\xi}\, dF^{(k)}(\xi)$$

(the last integral does not involve the probability at zero). In principle, the series appearing in the exponent on the right of (3.3) and (3.4) are accessible to calculation.

The proof of Theorem 3.1 is elaborated in Section 5. For the moment, it is convenient to denote *the right-hand side of* (3.3) *by* $f_+(u, t)$ *and that of* (3.4) *by* $f_-(u; t)$.

The next identity, called Spitzer's identity, involving the maximal process $\{M_n\}$, see (1.2), is basic in the study of a wide assortment of applied probability models, including queueing theory, branching processes, and renewal events, among others.

***Theorem* 3.2.** *For $|t| < 1$, t and u real*

$$\sum_{n=0}^{\infty} E[e^{iuM_n}]t^n = \exp\left(\sum_{k=1}^{\infty} \frac{t^k}{k} E[e^{iuS_k}; S_k \geq 0]\right)\exp\left(\sum_{k=1}^{\infty} \frac{t^k}{k} \Pr\{S_k < 0\}\right)$$

$$= f_+(u, t)f_-(0, t). \tag{3.5}$$

The proof of Theorem 3.2 appears in Section 5.

It is useful to record the following elementary fact:

$$f_+(u, t)f_-(u, t) = \frac{1}{1 - t\varphi(u)} \tag{3.6}$$

where $\varphi(u) = E[e^{iuX_i}]$ is the common characteristic function of X_i. We verify (3.6) directly. Indeed,

$$f_+(u, t)f_-(u, t) = \exp\left(\sum_{k=1}^{\infty} \frac{t^k}{k} E[e^{iuS_k}; S_k \geq 0]\right)\exp\left(\sum_{k=1}^{\infty} \frac{t^k}{k} E[e^{iuS_k}; S_k < 0]\right)$$

$$= \exp\left\{\sum_{k=1}^{\infty} \frac{t^k}{k} (E[e^{iuS_k}; S_k \geq 0] + E[e^{iuS_k}; S_k < 0])\right\}$$

$$= \exp\left(\sum_{k=1}^{\infty} \frac{t^k}{k} E[e^{iuS_k}]\right)$$

$$= \exp\left\{\sum_{k=1}^{\infty} \frac{t^k}{k} \left(\prod_{l=1}^{k} E[e^{iuX_l}]\right)\right\} \qquad \text{(since } X_1, X_2, \dots \text{ are i.i.d.)}$$

$$= \exp\left\{\sum_{k=1}^{\infty} \frac{t^k}{k} [\varphi(u)]^k\right\}$$

$$= e^{-\log[1 - t\varphi(u)]} \qquad \left(\text{since } -\log(1 - x) = \sum_{k=1}^{\infty} \frac{x^k}{k}\right)$$

$$= \frac{1}{1 - t\varphi(u)}.$$

In the remainder of this section we shall prove one key lemma and content ourselves with extracting a series of direct applications of the identities hitherto displayed.

Lemma 3.1. *For each* $k = 0, 1, 2, \ldots, n$

$$\Pr\{L_{nn} = k\} = \Pr\{L_{kk} = k\} \Pr\{L_{n-k,n-k} = 0\}, \qquad (3.7)$$

$$\Pr\{P_n = k\} = \Pr\{P_k = k\} \Pr\{P_{n-k} = 0\}. \qquad (3.8)$$

Proof. We concentrate first on (3.7). (Figure 2 is helpful.) The first step comes from the definition

$$
\begin{aligned}
\Pr\{L_{nn} = k\} &= \Pr\{S_k \geq S_j, j = 0, 1, \ldots, k - 1; S_k > S_i, i = k + 1, \ldots, n\} \\
&= \Pr\{S_k \geq S_j, j = 0, 1, \ldots, k; S_i - S_k < 0, i = k + 1, \ldots, n\} \\
&= \Pr\{S_k \geq S_j, j = 0, 1, \ldots, k\} \\
&\quad \times \Pr\{X_{k+1} + \cdots + X_i < 0, i = k + 1, \ldots, n\} \\
&\qquad\qquad\qquad\qquad\qquad\qquad\qquad\qquad \text{(by independence)} \\
&= \Pr\{L_{kk} = k\} \Pr\{S_j < 0, j = 1, 2, \ldots, n - k\} \\
&\qquad\qquad\qquad \text{(since } X_i \text{ are identically distributed)} \\
&= \Pr\{L_{kk} = k\} \Pr\{L_{n-k,n-k} = 0\},
\end{aligned}
$$

and (3.7) is validated. The identity (3.8) emanates by virtue of the equivalence principle equation (2.16) for $g_n \equiv 1$. ∎

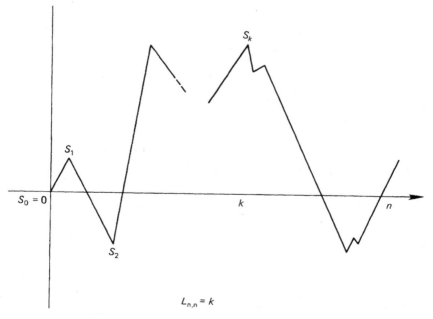

FIG. 2

Paraphrasing the above argument mutatis mutandis or applying the reversal permutation τ to $(X_1, \ldots, X_n)$ (see the proof of Lemma 2.1), we secure the following result:

Lemma 3.2

$$\Pr\{L_{n0} = k\} = \Pr\{L_{k0} = k\} \Pr\{L_{n-k,0} = 0\}$$

and

$$\Pr\{P_n^+ = k\} = \Pr\{P_k^+ = k\} \Pr\{P_{n-k}^+ = 0\}, \qquad k = 0, 1, 2, \ldots, n. \qquad (3.9)$$

To dramatize the utility and agility of the identities (3.3)–(3.5) a number of immediate applications will be highlighted.

A. COMPUTATION OF $\Pr\{P_n = k\}$ IN CERTAIN CASES

We concentrate on the special but important situation of

$$\Pr\{S_k \geq 0\} = \tfrac{1}{2}, \qquad k = 1, 2, 3, \ldots. \qquad (3.10)$$

For example, when $\{X_i\}$ are real symmetric random variables having continuous density function then certainly the random variables S_k inherit the same property and (3.10) prevails.

Setting $u = 0$ in the basic relation (3.3) for $P(u; t)$ yields

$$\sum_{n=0}^{\infty} \Pr\{L_{nn} = n\} t^n = \exp\left(\sum_{k=1}^{\infty} \frac{t^k}{k} \Pr\{S_k \geq 0\} \right). \qquad (3.11)$$

Since $\Pr\{S_k \geq 0\} = \tfrac{1}{2}$, by assumption, and taking account of the elementary formula $1 - t = \exp(-\sum_{k=1}^{\infty} t^k/k)$ the previous equation reduces to

$$\sum_{n=0}^{\infty} \Pr\{L_{nn} = n\} t^n = (1 - t)^{-1/2}. \qquad (3.12)$$

The familiar binomial expansion reads

$$(1 - t)^{-1/2} = \sum_{n=0}^{\infty} \binom{-\tfrac{1}{2}}{n} (-1)^n t^n. \qquad (3.13)$$

Equating coefficients in (3.12) and (3.13) produces the evaluation

$$\Pr\{L_{nn} = n\} = \binom{-\tfrac{1}{2}}{n} (-1)^n$$

$$= \frac{\left(-\tfrac{1}{2}\right)\left(-\tfrac{3}{2}\right) \cdots \left(\frac{-2n+1}{2}\right)}{1 \cdot 2 \cdot \ \cdots \ \cdot n} (-1)^n$$

$$= \frac{(2n)!}{2^{2n}(n!)^2} = \frac{1}{2^{2n}} \binom{2n}{n}. \qquad (3.14)$$

The equivalence principle (2.16) also gives

$$\Pr\{P_n = n\} = \Pr\{L_{nn} = n\} = \binom{2n}{n}\frac{1}{2^{2n}}. \tag{3.15}$$

Following the same course, by means of (3.4), setting $u = 0$ in $Q(u, t)$ with obvious evaluations leads to the result

$$\Pr\{L_{nn} = 0\} = \Pr\{P_n = 0\} = \binom{2n}{n}\frac{1}{2^{2n}}. \tag{3.16}$$

With the expressions (3.15) and (3.16) at hand, we can determine the complete distribution of P_n (subject to the stipulation of (3.10)) exploiting the identity of (3.8). Thus,

$$\begin{aligned}
\Pr\{P_n = k\} &= \Pr\{L_{nn} = k\} \\
&= \Pr\{L_{kk} = k\} \Pr\{L_{n-k,n-k} = 0\} \\
&= \frac{1}{2^{2n}}\binom{2k}{k}\binom{2(n-k)}{n-k}.
\end{aligned} \tag{3.17}$$

B. THE EXPECTATION OF THE MAXIMUM M_n

We write the identity (3.5) in the form

$$\sum_{n=0}^{\infty} E[e^{iuM_n}]t^n = \exp\left(\sum_{k=1}^{\infty} \frac{t^k}{k} E[e^{iuS_k^+}]\right),$$

where $S_k^+ = \max\{S_k, 0\}$, and differentiate both sides with respect to u, and afterwards set $u = 0$ to obtain

$$\sum_{n=0}^{\infty} E[M_n]t^n = \exp\left(\sum_{k=1}^{\infty} \frac{t^k}{k}\right)\sum_{k=1}^{\infty} \frac{t^k}{k} E[S_k^+]. \tag{3.18}$$

Now $\exp(\sum_{k=1}^{\infty}(t^k/k)) = (1-t)^{-1} = \sum_{k=0}^{\infty} t^k$, and the power series multiplication formula

$$\left(\sum_{i=0}^{\infty} a_i t^i\right)\left(\sum_{j=0}^{\infty} b_j t^j\right) = \sum_{n=0}^{\infty}\left(\sum_{k=0}^{n} a_k b_{n-k}\right)t^n$$

applied in (3.18) shows that

$$\begin{aligned}
\sum_{n=0}^{\infty} E[M_n]t^n &= \sum_{j=0}^{\infty} t^j \sum_{k=1}^{\infty} \frac{t^k}{k} E[S_k^+] \\
&= \sum_{n=0}^{\infty}\left(\sum_{k=1}^{n} \frac{1}{k} E[S_k^+]\right)t^n.
\end{aligned}$$

We equate the corresponding coefficients to conclude

$$E[M_n] = \sum_{k=1}^{n} \frac{1}{k} E[S_k; S_k \geq 0] = \sum_{k=1}^{n} \frac{1}{k} E[S_k^+].$$

As n increases, the random variable M_n increases to the random variable

$$M = \sup\{S_0, S_1, S_2, \ldots\} \qquad (\infty \text{ values are possible})$$

and in all cases the limit relation

$$E[M] = \sum_{k=1}^{\infty} \frac{1}{k} E[S_k; S_k \geq 0]$$

obtains. In particular, if $\sum_{k=1}^{\infty} (1/k)E[S_k; S_k \geq 0] < \infty$, then $E[M] < \infty$.

4: The Important Concept of Ladder Random Variables

We say n is a ladder point for the partial sums $S_j, j = 0, 1, \ldots, n$ if S_n is at least as great as any previous partial sum, i.e., if

$$S_n \geq S_j \qquad \text{for all} \quad j = 0, 1, \ldots, n.$$

Let N be the number of ladder points in the series $\{S_n\}$. If the summands have a positive mean μ, by the law of large numbers $S_n/n \to \mu > 0$ and $S_n \to \infty$ as $n \to \infty$; so in this case $N = \infty$. If the summands have a negative mean μ, $S_n \to -\infty$; so here N is finite. But $N \geq 1$ because 0 is a ladder point.

Define W_1 to be the waiting time to the first ladder point after time zero, with $W_1 = \infty$ should no such ladder point exist. Let W_2 be the waiting time from the first to the second ladder point, with $W_2 = \infty$ should the second not exist. In general, define W_r as the waiting time from the $(r - 1)$st to the rth ladder point for $r < N$ and $W_N = \infty$ when N is finite. In symbols, this reads:

$$W_1 = T_1 = \inf\{n; n > 0, S_n \geq S_j \quad \text{for all} \quad j = 0, 1, \ldots, n\},$$

$$T_2 = \inf\{n; n > W_1, S_n \geq S_j \quad \text{for all} \quad j = 0, 1, \ldots, n\}, \qquad (4.1)$$

$$W_2 = T_2 - T_1,$$

and in general,

$$T_r = \inf\{n; n > T_{r-1}, S_n \geq S_j \quad \text{for all} \quad j = 0, 1, \ldots, n\},$$

$$W_r = T_r - T_{r-1}, \qquad 1 \leq r < N.$$

Further, let Z_1 be the value of the partial sum S_j at the first ladder point, Z_2 the value of the partial sum S_j at the second ladder point less Z_1, and in general Z_k

the value of the partial sum S_j at the kth ladder point less $Z_1 + \cdots + Z_{k-1}$, for $k < N$. In symbols,

$$Z_1 = S_{T_1}, \qquad Z_2 = S_{T_2} - S_{T_1},$$

and in general

$$Z_r = S_{T_r} - S_{T_{r-1}}, \qquad r = 1, 2, \ldots . \tag{4.2}$$

We may add for convenience the definitions

$$T_0 = W_0 = 0, \qquad Z_0 = 0, \qquad \text{and} \qquad T_N = W_N = \infty \quad \text{when} \quad N < \infty.$$

We now state a lemma about the distribution of N and the random vectors (W_k, Z_k).

Lemma 4.1. *N has the geometric distribution in which*

$$\Pr\{N \geq k\} = \beta^{k-1}, \qquad k = 1, 2, \ldots,$$

where $\beta = \Pr\{W_1 < \infty\} = \Pr\{S_n \geq 0 \text{ for some } n = 1, 2, \ldots\}$. *(N* $= \infty$ *corresponds to* $\beta = 1$.) *Given* $N = n > 0$, *the random vectors* (W_r, Z_r), $1 \leq r < N$ *are independent and identically distributed.*

Note. The statement of the lemma is highly intuitive in view of the fact that $X_1, X_2, \ldots$ are random variables such that the process starts afresh with the value $S_{T_{r-1}}$ and this corresponds to zero with reference to the next waiting period W_r. We encourage the student to supply a formal proof.

The next theorem will be of considerable value once Theorem 3.1 is validated. Tentatively it is a vehicle toward our developments of fluctuation theory.

Theorem 4.1. *For u and t real and* $|t| < 1$,

$$\frac{1}{1 - E[e^{iuZ_1}t^{W_1}]} = P(u; t) \tag{4.3}$$

(for the definition of $P(u; t)$ *see* (3.1)).

Proof. Expanding the geometric series

$$\frac{1}{1 - E[e^{iuZ_1}t^{W_1}]} = \sum_{k=0}^{\infty} (E[e^{iuZ_1}t^{W_1}])^k \tag{4.4}$$

is permissible since

$$|E[e^{iuZ_1}t^{W_1}]| < E[|e^{iuZ_1}t^{W_1}|] < 1 \qquad \text{for} \quad |t| < 1.$$

Moreover, by Lemma 4.1

$$E[\exp\{iu(Z_1 + \cdots + Z_k)\}t^{W_1 + \cdots + W_k}] = E\left[\prod_{j=1}^{k} e^{iuZ_j}t^{W_j}\right]$$

$$= \prod_{j=1}^{k} E[e^{iuZ_j}t^{W_j}]$$

$$= (E[e^{iuZ_1}t^{W_1}])^k. \qquad (4.5)$$

Further, decomposing for fixed k the values of $W_1 + \cdots + W_k$ into their mutually exclusive possible values allows the left-hand side to be written in the form

$$E[\exp\{iu(Z_1 + \cdots + Z_k)\}t^{W_1 + \cdots + W_k}]$$

$$= \sum_{n=0}^{\infty} \int_{W_1 + \cdots + W_k = n} \exp\{iu(Z_1 + \cdots + Z_k)\}t^{W_1 + \cdots + W_k} \, dF$$

$$= \sum_{n=0}^{\infty} t^n E[\exp\{iuS_n\}; W_1 + \cdots + W_k = n] \qquad (4.6)$$

because

$$Z_1 + \cdots + Z_k = S_n \qquad \text{when} \quad W_1 + \cdots + W_k = n.$$

Substituting (4.6), with cognizance of (4.5), into (4.4) produces

$$\frac{1}{1 - E[e^{iuZ_1}t^{W_1}]} = \sum_{k=0}^{\infty} \sum_{n=0}^{\infty} t^n E[e^{iuS_n}; W_1 + \cdots + W_k = n]$$

$$= \sum_{n=0}^{\infty} t^n \sum_{k=0}^{\infty} E[e^{iuS_n}; W_1 + \cdots + W_k = n], \qquad (4.7)$$

where the interchange of summations is justified, since the double sum is absolutely convergent. Now observe that the event $L_{nn} = n$, asserting that the last maximum occurs at n, may be expanded as the union of the mutually exclusive events that the index n is a ladder point for some k, $k = 1, 2, 3, \ldots$.
    Therefore

$$E[\exp\{iuS_n\}; L_{nn} = n] = \sum_{k=0}^{\infty} E[e^{iuS_n}; W_1 + \cdots + W_k = n].$$

Then placing this representation into (4.7) confirms (4.3). The theorem is proved. ∎

As a sample application, let us quickly develop an expression for the $\beta = \Pr\{W_1 < \infty\}$ that appears in Lemma 4.1. Observe that $\beta = \Pr\{W_1 < \infty\} = \lim_{t \uparrow 1} E[t^{W_1}]$. Hence, using (4.3) with $u = 0$,

$$\frac{1}{1 - \beta} = \lim_{t \uparrow 1} P(0; t)$$

$$= \exp\left(\sum_{k=1}^{\infty} \frac{1}{k} \Pr\{S_k \geq 0\}\right) \qquad \text{(using (3.3))}$$

or

$$\beta = 1 - \exp\left(-\sum_{k=1}^{\infty} \frac{1}{k} \Pr\{S_k \geq 0\}\right).$$

A companion identity to that of (4.3) is incorporated in the next theorem.

Theorem 4.2. *For u and t real and $|t| < 1$, we have*

$$\frac{1 - E[e^{iuZ_1} t^{W_1}]}{1 - t\varphi(u)} = Q(u; t) \tag{4.8}$$

(see (3.2) for the definition of $Q(u; t)$).

Proof. Consider the expression

$$[1 - t\varphi(u)]Q(u; t) = [1 - t\varphi(u)] \sum_{n=0}^{\infty} E[\exp(iuS_n); L_{nn} = 0]t^n$$

$$= \sum_{n=0}^{\infty} E[\exp(iuS_n); L_{nn} = 0]t^n$$

$$- \sum_{n=0}^{\infty} \varphi(u)E[\exp(iuS_n); L_{nn} = 0]t^{n+1}. \tag{4.9}$$

With cognizance of the hypothesis that the X_n are independent and identically distributed and that therefore X_{n+1} is independent of any event determined by the random variables $\{X_1, \ldots, X_n\}$, we find that

$$\varphi(u)E[\exp(iuS_n); L_{nn} = 0] = E[\exp(iuX_{n+1})]E[\exp(iuS_n); L_{nn} = 0]$$

$$= E[\exp(iuS_{n+1}); L_{nn} = 0]. \tag{4.10}$$

Combining (4.9) and (4.10), a change of variable and the obvious cancellation of adjacent terms ultimately produces the equations

$$[1 - t\varphi(u)] \sum_{n=0}^{\infty} E[\exp(iuS_n); L_{nn} = 0]t^n$$

$$= \sum_{n=0}^{\infty} E[\exp(iuS_n); L_{nn} = 0]t^n - \sum_{n=1}^{\infty} E[\exp(iuS_n); L_{n-1,n-1} = 0]t^n$$

$$= 1 - \sum_{n=1}^{\infty} E[\exp(iuS_n); L_{n-1,n-1} = 0, L_{nn} \neq 0]t^n. \tag{4.11}$$

Now observe that the last maximum will be at position zero if and only if the waiting time to the first ladder point exceeds n. In symbols, $L_{nn} = 0$ if and only if $W_1 > n$. Thus $L_{n-1,n-1} = 0$ and $L_{nn} \neq 0$ simultaneously hold if and only if $W_1 = n$.

Using this fact in (4.11) we obtain

$$[1 - t\varphi(u)] \sum_{n=0}^{\infty} E[\exp(iuS_n); L_{nn} = 0]t^n = 1 - \sum_{n=1}^{\infty} E[\exp(iuS_n); W_1 = n]t^n$$

$$= 1 - E[\exp(iuZ_1)t^{W_1}]$$

because $S_n = Z_1$ when $W_1 = n$. This establishes (4.8). ∎

5: Proof of the Main Fluctuation Theory Identities

The task of this section is to furnish the proofs of Theorems 3.1 and 3.3 and of additional interesting identities, formulas, and general byproducts. We commence with the derivation of another double generating function identity of importance.

Theorem 5.1. For $|x| \leq 1, |t| < 1$, x and t real, we have

$$\sum_{n=0}^{\infty} E[\exp(iuS_n)x^{L_{nn}}]t^n = P(u; tx)Q(u; t). \tag{5.1}$$

Proof. Observe that the partial sums of the finite set of variables $X_1, X_2, \ldots, X_n$ have their last maximum at the position k ($0 \leq k \leq n$) if and only if the partial sums of $X_1, X_2, \ldots, X_k$ have their last maximum at k and the partial sums based on $X_{k+1}, X_{k+2}, \ldots, X_n$ exhibit their last maximum at zero. In symbols,

$$L_{nn}(X_1, \ldots, X_n) = k \tag{5.2}$$

if and only if

$$L_{kk}(X_1, \ldots, X_k) = k \qquad \text{and} \qquad L_{n-k,n-k}(X_{k+1}, \ldots, X_n) = 0. \tag{5.3}$$

Let $I_{\{L_{kk}=k\}}$ denote the indicator function of the first event of (5.3) and $I_{\{\tilde{L}_{n-k,n-k}=0\}}$ bear the corresponding connotations for the last event in (5.3). These two events of (5.2) and (5.3) are independent, since their characterizations involve independent sets of random variables. It follows that

$$E[\exp(iuS_n); L_{nn} = k] = E[\exp(iuS_k)$$

$$\times \exp\{iu(S_n - S_k)\}; L_{kk} = k, \tilde{L}_{n-k,n-k} = 0] \tag{5.4}$$

[where $\tilde{L}_{n-k,n-k} = L_{n-k,n-k}(X_{k+1}, \ldots, X_n)$]

$$= \int (e^{iuS_k}I_{\{L_{kk}=k\}})(e^{iu(S_n - S_k)}I_{\{L_{n-k,n-k}(X_{k+1},\ldots,X_n)=0\}}) \, dF_{(n)}(x)$$

The expression (5.4) equals

$$E[\exp(iuS_k); L_{kk} = k]E[\exp\{iu(S_n - S_k)\}; \tilde{L}_{n-k, n-k} = 0]$$

because $S_n - S_k$ and $\{\tilde{L}_{n-k, n-k} = 0\}$ are determined by random variables independent of $X_1, X_2, \ldots, X_k$. Next multiply both sides of the resulting equation from (5.4) by x^k and sum on k from 0 to n. We obtain

$$E[\exp(iuS_n)x^{L_{nn}}] = \sum_{k=0}^{n} x^k E[\exp(iuS_k); L_{kk} = k]$$

$$\times E[\exp\{iu(S_n - S_k)\}; \tilde{L}_{n-k, n-k} = 0]$$

$$= \sum_{k=0}^{n} x^k a_k b_{n-k} \qquad (5.5)$$

with the evident definitions for a_k and b_{n-k}. Note that b_{n-k} depends unambiguously solely on $n - k$ since $X_{k+1}, \ldots, X_n$ have the same distribution as $X_1, \ldots, X_{n-k}$.

Recall the elementary power series equality

$$\sum_{n=0}^{\infty} t^n \sum_{k=0}^{n} a_k b_{n-k} = \left(\sum_{k=0}^{\infty} a_k t^k \right) \left(\sum_{k=0}^{\infty} b_k t^k \right).$$

Comparing this to (5.5), we infer that

$$\sum_{n=0}^{\infty} E[\exp(iuS_n)x^{L_{nn}}]t^n = \sum_{k=0}^{\infty} E[\exp(iuS_k); L_{kk} = k]x^k t^k$$

$$\times \sum_{n=k}^{\infty} E[\exp(iu\tilde{S}_{n-k}); \tilde{L}_{n-k, n-k} = 0]t^{n-k}$$

$$= P(u; tx)Q(u; t) \qquad \text{(defined in (3.1) and (3.2))},$$

since $\tilde{S}_{n-k} = X_{k+1} + \cdots + X_n$ has the same distribution as S_{n-k}. ∎

Note. Theorems 4.1, 4.2, and 5.1 jointly express the fact that

$$\sum_{n=0}^{\infty} E[\exp(iuS_n)x^{L_{nn}}]t^n = \frac{P(u; tx)}{[1 - t\varphi(u)]P(u; t)}. \qquad (5.6)$$

The import of Theorem 3.1 asserts

Theorem 5.2 (*Basic Identity*). *For* $|t| < 1$

$$P(u; t) = \exp\left\{ \sum_{k=1}^{\infty} \frac{t^k}{k} E[\exp(iuS_k); S_k \geq 0] \right\} \stackrel{\text{def}}{=} f_+(u; t) \qquad (5.7)$$

and

$$Q(u; t) = \exp\left\{ \sum_{k=1}^{\infty} \frac{t^k}{k} E[\exp(iuS_k); S_k < 0] \right\} \stackrel{\text{def}}{=} f_-(u; t). \qquad (5.8)$$

Proof. First we show that (5.7) and (5.8) are equivalent. Setting $x = 1$ in (5.1) gives

$$P(u; t)Q(u; t) = \sum_{n=0}^{\infty} E[e^{iuS_n}]t^n$$

$$= \sum_{n=0}^{\infty} E\left[\prod_{j=1}^{n} e^{iuX_j}\right]t^n = \sum_{n=0}^{\infty} [\varphi(u)]^n t^n$$

(owing to the independence of the r.v.'s $\{X_i\}$)

$$= \frac{1}{1 - t\varphi(u)} = f_+(u; t)f_-(u; t), \tag{5.9}$$

where the last equation was proved earlier in (3.6). It follows from (5.9) that either of (5.7) and (5.8) implies the other.

We now prove (5.7). Differentiate equation (5.6) with respect to x and set $x = 1$ to yield

$$\psi(t) = \sum_{n=0}^{\infty} E[e^{iuS_n}L_{nn}]t^n = \frac{tP'(u; t)}{[1 - t\varphi(u)]P(u; t)}, \tag{5.10}$$

where $P'(u; t)$ indicates $[dP(u; t)]/dt$.

With u fixed we regard (5.10) as a first-order differential equation

$$\frac{P'(u; t)}{P(u; t)} = \frac{1 - t\varphi(u)}{t} \psi(t) \overset{\text{def}}{=} A(t)$$

(i.e., $A(t)$ is the symbol for the middle expression). Integration gives

$$\log\left[\frac{P(u; t)}{P(u; 0)}\right] = \int_0^t A(\tau)\, d\tau. \tag{5.11}$$

Note that

$$A(t) = \frac{1}{t} [1 - t\varphi(u)] \sum_{n=0}^{\infty} E[e^{iuS_n}L_{nn}]t^n$$

$$= \frac{1}{t} [1 - t\varphi(u)] \sum_{n=1}^{\infty} E[e^{iuS_n}L_{nn}]t^n \quad (\text{since } L_{00} = 0)$$

$$= \sum_{n=1}^{\infty} E[e^{iuS_n}L_{nn}]t^{n-1} - \sum_{n=0}^{\infty} E[e^{iuS_{n+1}}L_{nn}]t^n$$

$$= \sum_{n=1}^{\infty} E[e^{iuS_n}(L_{nn} - L_{n-1,n-1})]t^{n-1},$$

and termwise integration gives

$$\int_0^t A(\tau)\, d\tau = \sum_{n=1}^{\infty} E[e^{iuS_n}(L_{nn} - L_{n-1,n-1})]\frac{t^n}{n}. \tag{5.12}$$

From the definition we see that $P(u; 0) = 1$ and therefore (5.11) and (5.12) produces the representation

$$P(u; t) = \exp\left\{ \sum_{n=1}^{\infty} E[e^{iuS_n}(L_{nn} - L_{n-1,n-1})] \frac{t^n}{n} \right\}. \tag{5.13}$$

Now, we make use of the fundamental equivalence, Theorem 2.1, with $g_n(x) = \exp(i \sum_{j=1}^{n} u_j x_j)$:

$$E[\exp(iuS_n)L_{nn}] = \sum_{k=1}^{\infty} E[\exp(iuS_n)L_{nn}; L_{nn} = k]$$

(mutually exclusive possibilities)

$$= \sum_{k=1}^{\infty} kE[\exp(iuS_n); L_{nn} = k]$$

$$= \sum_{k=1}^{\infty} kE[\exp(iuS_n); P_n = k]$$

(by the equivalence principle (2.16))

$$= \sum_{k=1}^{\infty} E[\exp(iuS_n)P_n; P_n = k] = E[\exp(iuS_n)P_n]. \tag{5.14}$$

A similar identity holds with L_{nn} and P_n replaced by $L_{n-1,n-1}$ and P_{n-1}, respectively. Comparing (5.13) and (5.14), we conclude that

$$P(u; t) = \exp\left\{ \sum_{n=1}^{\infty} E[\exp(iuS_n)(P_n - P_{n-1})] \frac{t^n}{n} \right\}. \tag{5.15}$$

Observe, however, that the number of nonnegative partial sums of X_1, $X_2, \ldots, X_n$ is the same as that of $X_1, X_2, \ldots, X_{n-1}$ unless S_n itself is nonnegative. More precisely,

$$P_n = \begin{cases} P_{n-1} & \text{if } S_n < 0, \\ P_{n-1} + 1 & \text{if } S_n \geq 0. \end{cases}$$

Thus

$$E[\exp(iuS_n)(P_n - P_{n-1})] = E[\exp(iuS_n)(P_n - P_{n-1}); S_n < 0]$$
$$+ E[\exp(iuS_n)(P_n - P_{n-1}); S_n \geq 0]$$
$$= E[\exp(iuS_n); S_n \geq 0].$$

Substitution of this into (5.15) gives (5.7), which completes the proof of Theorem 5.2. ∎

We are now prepared to give the proof of Theorem 3.2.

Proof of Theorem 3.2.   Consider

$$E[\exp(iuM_n)] = \sum_{k=1}^{n} E[\exp(iuM_n); L_{nn} = k] = \sum_{k=1}^{n} E[\exp(iuS_k); L_{nn} = k]$$

because $M_n = S_k$ if $L_{nn} = k$. By the relations (5.2) and (5.3) and continuing with the notation used in the proof of Theorem 5.1, we obtain

$$E[\exp(iuM_n)] = \sum_{k=1}^{n} E[\exp(iuS_k); L_{kk} = k, \tilde{L}_{n-k, n-k} = 0]$$

$$= \sum_{k=1}^{n} E[\exp(iuS_k); L_{kk} = k] E[1; \tilde{L}_{n-k, n-k} = 0].$$

Thus we have

$$c_n = E[\exp(iuM_n)] = \sum_{k=0}^{n} E[\exp(iuS_k); L_{kk} = k] E[1; \tilde{L}_{n-k, n-k} = 0]$$

$$= \sum_{k=0}^{n} a_k b_{n-k}, \tag{5.16}$$

the composition formula for coefficients of power series (cf. the analysis following (5.5)). Constructing the power series with coefficients c_n, it follows from (5.16)

$$\sum_{n=0}^{\infty} E[\exp(iuM_n)] t^n = \sum_{k=0}^{\infty} E[\exp(iuS_k); L_{kk} = k] t^k \sum_{n=k}^{\infty} E[1; L_{n-k, n-k} = 0] t^{n-k}$$

$$= P(u; t) Q(0; t).$$

Applying Theorem 5.2, then

$$\sum_{n=0}^{\infty} E[\exp(iuM_n)] t^n = \exp\left\{ \sum_{k=1}^{\infty} \frac{t^k}{k} E[\exp(iuS_k); S_k \geq 0] \right\}$$

$$\times \exp\left(\sum_{k=1}^{\infty} \frac{t^k}{k} \Pr\{S_k < 0\} \right),$$

as was to be proved.   ∎

We prove one more consequence of Theorem 5.2, namely,

Theorem 5.3.   *For u and t real with $|t| < 1$ and $|x| \leq 1$,*

$$\sum_{n=0}^{\infty} t^n \sum_{k=0}^{n} x^k \Pr\{L_{nn} = k\} = \exp\left\{ \sum_{k=1}^{\infty} \frac{(tx)^k}{k} \Pr\{S_k \geq 0\} \right\} \exp\left\{ \sum_{k=1}^{\infty} \frac{t^k}{k} \Pr\{S_k < 0\} \right\}.$$

Proof. Set $u = 0$ in (5.1) of Theorem 5.1. Then

$$P(0; tx)Q(0; t) = \sum_{n=0}^{\infty} E[x^{L_{nn}}]t^n. \tag{5.17}$$

But

$$E[x^{L_{nn}}] = \sum_{k=0}^{n} E[x^{L_{nn}}; L_{nn} = k] = \sum_{k=0}^{n} x^k E[1; L_{nn} = k] = \sum_{k=0}^{n} x^k \Pr\{L_{nn} = k\}.$$

Substitution of this into (5.17) yields

$$\sum_{n=0}^{\infty} t^n \sum_{k=0}^{n} x^k \Pr\{L_{nn} = k\} = P(0; tx)Q(0; t).$$

Now apply Theorem 5.2 with $u = 0$. Then the expectations are replaced by probabilities and Theorem 5.3 is proved. ∎

The identifications of Theorem 5.2 in conjunction with Theorems 4.1 and 4.2 produce the following important formula:

Theorem 5.4. *For u and t real and $|t| < 1$*

$$1 - E[e^{iuZ_1}t^{W_1}] = \exp\left(-\sum_{k=1}^{\infty} \frac{t^k}{k} E[e^{iuS_k}; S_k \geq 0]\right). \tag{5.18}$$

6: More Applications of Fluctuation Theory

A. LIMIT THEOREMS FOR THE NUMBER OF POSITIVE SUMS

Under the sole condition

$$\Pr\{S_k \geq 0\} = \Pr\{S_k > 0\} = \tfrac{1}{2} \tag{6.1}$$

(this requirement is fulfilled if $\{X_i\}$ are all symmetrical continuous random variables; cf. the discussion of (3.10)), we shall establish the famous *arcsine law*,

$$\lim_{n \to \infty} \Pr\left\{\frac{P_n}{n} \leq x\right\} = \int_0^x \frac{1}{\pi} \frac{d\xi}{\sqrt{\xi(1 - \xi)}}$$

$$= \frac{2}{\pi} \arcsin \sqrt{x}, \qquad 0 \leq x \leq 1.$$

This result is quite remarkable. The ratio P_n/n is the fraction of times (or periods) up to n for which $S_j \geq 0$, and one might expect that this fraction would most likely occur near $\tfrac{1}{2}$. Quite the contrary, as shown in Fig. 3. The density $f(x) = 1/(\pi\sqrt{x(1 - x)})$ is bimodal, which is interpreted as follows: for any particular realization of $S_0, S_1, S_2, \ldots$ it is far more likely to find the limiting fraction P_n/n close to 0 or 1 rather than $\tfrac{1}{2}$.

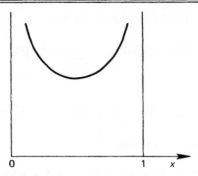

FIG. 3   The density $f(x) = 1/(\pi\sqrt{x(1-x)})$.

Recall the formulas (3.17)

$$\Pr\{P_n = k\} = \frac{1}{2^{2n}}\binom{2k}{k}\binom{2(n-k)}{n-k}. \tag{6.2}$$

To extract the limit distribution as n increases, we rely on Stirling's asymptotic approximation

$$m! \approx \left(\frac{m}{e}\right)^m\sqrt{2\pi m}$$

to obtain

$$\binom{2m}{m} = \frac{(2m)!}{(m!)^2} \sim \frac{1}{\sqrt{2\pi}}2^{2m}\sqrt{\frac{2}{m}}.$$

Thus, for large n, provided k and $n \to \infty$ in a manner implying $k/n \to x$, $0 < x < 1$, we secure the asymptotic relation

$$\Pr\{P_n = k\} \cong \frac{1}{\pi}\frac{1}{n}\frac{1}{\sqrt{\dfrac{k}{n} - \left(\dfrac{k}{n}\right)^2}} \cong \frac{1}{\pi}\frac{1}{n}\frac{1}{\sqrt{x(1-x)}}.$$

Compute next

$$\Pr\left\{\frac{P_n}{n} < x\right\} \cong \frac{1}{\pi}\sum_{k=0}^{[nx]}\frac{1}{\sqrt{\dfrac{k}{n} - \left(\dfrac{k}{n}\right)^2}}\frac{1}{n}.$$

This sum is the approximation to the integral (the student should formalize this comment)

$$\frac{1}{\pi}\int_0^x\frac{1}{\sqrt{\xi(1-\xi)}}\,d\xi = \frac{2}{\pi}\arcsin\sqrt{x}.$$

Thus

$$\lim_{n \to \infty} \Pr\left\{\frac{P_n}{n} < x\right\} = \frac{2}{\pi} \arcsin \sqrt{x}, \qquad 0 \le x \le 1. \tag{6.3}$$

It is striking to contrast the limit distribution (6.3) with the conditional limit law,

$$\lim_{n \to \infty} \Pr\left\{\frac{P_n}{n} \le x \mid S_n = 0\right\} = x, \qquad 0 < x < 1. \tag{6.4}$$

Consult Chapter 13, page 123, where the latter limit law is discussed.

B. SOME PATH FLUCTUATION RESULTS

Our aim is to prove the following list of affirmations:

$$\sum_{k=1}^{\infty} \frac{1}{k} \Pr\{S_k \ge 0\} = \infty \qquad \text{if and only if} \qquad \Pr\{\lim_{n \to \infty} M_n = \infty\} = 1$$

$$\text{if and only if} \qquad \Pr\{S_n \ge 0 \text{ i.o.}\} = 1$$

(i.o. stands for infinitely often), while

$$\sum_{k=1}^{\infty} \frac{1}{k} \Pr\{S_k \ge 0\} < \infty \qquad \text{if and only if} \qquad \Pr\left\{\sup_{n \ge 0} M_n < \infty\right\} = 1$$

$$\text{if and only if} \qquad \Pr\{S_n \ge 0 \text{ i.o.}\} = 0.$$

Recall that $\{S_n \ge 0 \text{ i.o.}\}$ is the event of $S_n \ge 0$ occurring for infinitely many n. Its converse, $\{S_n \ge 0 \text{ f.o.}\}$ (f.o. abbreviates finitely often) is the event of $S_n \ge 0$ happening for at most a finite number of values n.

We begin by defining

$$q_n = \Pr\{L_{nn} = 0\} = \Pr\{S_j < 0 \quad \text{for} \quad j = 1, \ldots, n\}.$$

Then manifestly $q_n \ge q_{n+1} \ge 0$, so that $\{q_n\}$ converges and

$$\lim_{n \to \infty} q_n = \Pr\{S_j < 0 \text{ for all } j \ge 1\}.$$

Now

$$(1 - t) \sum_{k=0}^{\infty} q_k t^k = \sum_{k=0}^{\infty} q_k t^k - \sum_{k=0}^{\infty} q_k t^{k+1} = q_0 - \sum_{k=1}^{\infty} (q_{k-1} - q_k) t^k. \tag{6.5}$$

Since the coefficients of the series in (6.5) are nonnegative, we may apply the elementary Tauberian theorem, part (b) of Lemma 5.1 in Chapter 2, to conclude

$$\lim_{t \to 1} (1 - t) \sum_{k=0}^{\infty} q_k t^k = \lim_{n \to \infty} \left[q_0 - \sum_{k=1}^{n} (q_{k-1} - q_k) \right]$$

$$= \lim_{n \to \infty} q_n$$

$$= \Pr\{S_j < 0 \text{ for all } j \ge 1\}.$$

At this point we display the basic identity (5.8):

$$Q(0; t) = \sum_{n=0}^{\infty} \Pr\{L_{nn} = 0\} t^n = \sum_{n=0}^{\infty} q_n t^n = \exp\left(\sum_{k=1}^{\infty} \frac{1}{k} \Pr\{S_k < 0\} t^k\right).$$

Multiply this expression by

$$1 - t = \exp\left(- \sum_{k=1}^{\infty} \frac{t^k}{k}\right)$$

to get

$$(1 - t) \sum_{n=0}^{\infty} q_n t^n = \exp\left(- \sum_{k=1}^{\infty} \frac{t^k}{k}\right) \exp\left(\sum_{k=1}^{\infty} \frac{t^k}{k} \Pr\{S_k < 0\}\right)$$

$$= \exp\left(- \sum_{k=1}^{\infty} \frac{t^k}{k} \Pr\{S_k \geq 0\}\right).$$

The left-hand side has a limit as $t \to 1$, and so must the right. Taking account of the nonnegative nature of the coefficients, it follows that (again consult Lemma 5.1 of Chapter 2)

$$\lim_{t \to 1} \sum_{k=1}^{\infty} \frac{t^k}{k} \Pr\{S_k \geq 0\} = \sum_{k=1}^{\infty} \frac{1}{k} \Pr\{S_k \geq 0\} \leq \infty.$$

Thus

$$\Pr\{S_j < 0 \text{ for all } j \geq 1\} = \lim_{t \to 1}(1 - t) \sum_{n=0}^{\infty} q_n t^n$$

$$= \lim_{t \to 1} \exp\left(- \sum_{k=1}^{\infty} \frac{t^k}{k} \Pr\{S_k \geq 0\}\right)$$

$$= \exp\left(- \sum_{k=1}^{\infty} \frac{1}{k} \Pr\{S_k \geq 0\}\right).$$

Hence

$$\sum_{k=1}^{\infty} \frac{1}{k} \Pr\{S_k \geq 0\} = \infty \text{ implies } \Pr\{S_j < 0 \text{ for all } j \geq 1\} = 0 \qquad (6.6)$$

and

$$\sum_{k=1}^{\infty} \frac{1}{k} \Pr\{S_k \geq 0\} < \infty \text{ implies } \Pr\{S_j < 0 \text{ for all } j \geq 1\} > 0. \qquad (6.7)$$

Next we aim to prove the assertion

$$\Pr\{S_n \geq 0 \text{ i.o.}\} = \begin{cases} 0 & \text{if } \sum_{k=1}^{\infty} \frac{1}{k} \Pr\{S_k \geq 0\} < \infty, \\[4mm] 1 & \text{if } \sum_{k=1}^{\infty} \frac{1}{k} \Pr\{S_k \geq 0\} = \infty. \end{cases}$$

To this end, suppose $\sum_{k=1}^{\infty} (1/k) \Pr\{S_k \geq 0\} = \infty$, so that by (6.6)

$$\Pr\{S_j < 0 \text{ for all } j \geq 1\} = 0. \tag{6.8}$$

Then

$$1 - \Pr\{S_n \geq 0 \text{ i.o.}\} = \Pr\{S_n \geq 0 \text{ f.o.}\}$$
$$\leq \Pr\{S_j < 0 \text{ for all } j \geq 1\}$$
$$+ \sum_{k=1}^{\infty} \Pr\{S_k \geq 0 \text{ and } S_{k+l} < 0 \text{ for all } l \geq 1\}.$$

The kth term of the series evaluates the probability of the event signifying that S_k is the last nonnegative partial sum among $\{S_i\}_{i=1}^{\infty}$. Continuing, this is

$$\leq 0 + \sum_{k=1}^{\infty} \Pr\{S_k \geq 0 \text{ and } S_{k+l} - S_k < 0 \text{ for all } l \geq 1\}$$

$$= \sum_{k=1}^{\infty} \Pr\{S_k \geq 0\} \Pr\{S_l < 0 \text{ for all } l \geq 1\}$$

(by virtue of the independence and identical distribution hypothesis)

$$= 0 \quad \text{(applying (6.8) to each term).}$$

Hence $\Pr\{S_n \geq 0 \text{ i.o.}\} = 1$.

Conversely, suppose $\sum_{k=1}^{\infty} (1/k) \Pr\{S_k \geq 0\} < \infty$. Then $\Pr\{S_n < 0 \text{ for all } n = 1, 2, \ldots\} > 0$ and $\Pr\{S_n \geq 0 \text{ i.o.}\} < 1$. But the Hewitt–Savage 0–1 law.[†] implies $\Pr\{S_n \geq 0 \text{ i.o.}\}$ has probability 0 or 1 exclusively. In the present circumstances it follows that it must be 0.

Next we show that for any $x > 0$

$$\Pr\{S_n \geq 0 \text{ i.o.}\} = \Pr\{S_n \geq x \text{ i.o.}\}.$$

Again appeal to the Hewitt–Savage 0–1 law, which affirms that both quantities have value either 0 or 1. Manifestly, if $\Pr\{S_n \geq 0 \text{ i.o.}\} = 0$ then $\Pr\{S_n \geq x \text{ i.o.}\} \leq \Pr\{S_n \geq 0 \text{ i.o.}\} = 0$. Next suppose $\Pr\{S_n \geq 0 \text{ i.o.}\} = 1$. By virtue of the 0–1 law we need merely show $\Pr\{S_n \geq x \text{ i.o.}\} > 0$.

---

† A function $g(x_1, x_2, \ldots)$ is called *exchangeable* if $g(x_1, \ldots x_N, x_{N+1}, \ldots) = g(x_{\sigma(1)}, \ldots, x_{\sigma(N)}, x_{N+1}, \ldots)$ for every N and finite permutation $(\sigma(1), \ldots, \sigma(N))$ of $(1, \ldots, N)$. An event A is *exchangeable* if its indicator function $I_A(x_1, x_2, \ldots)$ is exchangeable. The Hewitt–Savage 0–1 law states that the probability of every exchangeable event is 0 or 1 when $X_1, X_2, \ldots$ are independent and identically distributed random variables. For any x, the event $\{S_n \geq x \text{ i.o.}\}$ is exchangeable. Its occurrence does not vary with any finite permutation of the summands. Therefore its probability is 0 or 1 when the summands are independent and identically distributed.

Let N be the first n, if any, for which $S_n \geq x$. Excluding the trivial case $S_n \equiv 0$ for all n, $\Pr\{N < \infty\} > 0$. Let $F(y) = \Pr\{N < \infty \text{ and } S_N \leq y\}$. Then

$$\Pr\{S_n \geq x \text{ i.o.}\} = \int_x^\infty \Pr\{S_{N+k} > x \text{ i.o.} | N < \infty \text{ and } S_N = y\} \, dF(y)$$

$$\geq \int_x^\infty \Pr\{S_{N+k} - S_N > 0 \text{ i.o.} | N < \infty \text{ and } S_N = y\} \, dF(y)$$

$$= \int_x^\infty 1 \, dF(y) = \Pr\{N < \infty\} > 0.$$

Hence $\Pr\{S_n \geq x \text{ i.o.}\} = 1$.

Summing up, we have established the following important fluctuation properties:

$$\sum_{k=1}^\infty \frac{1}{k} \Pr\{S_k \geq 0\} = \infty \qquad \text{implies} \qquad \Pr\{S_n \geq 0 \text{ i.o.}\} = 1$$

$$\text{implies} \qquad \Pr\{S_n \geq x \text{ i.o.}\} = 1 \text{ for all } x > 0$$

$$\text{implies} \qquad \Pr\{M \geq x\} = 1 \quad \text{for all} \quad x > 0$$
$$\text{where} \quad M = \max(0, S_1, S_2, \ldots)$$

$$\text{implies} \qquad \Pr\{M = \infty\} = 1, \tag{6.9}$$

while

$$\sum_{k=1}^\infty \frac{1}{k} \Pr\{S_k \geq 0\} < \infty \qquad \text{implies} \qquad \Pr\{S_n \geq 0 \text{ i.o.}\} = 0$$

$$\text{implies} \qquad \Pr\{S_n \geq 0 \text{ f.o.}\} = 1$$

$$\text{implies} \qquad \Pr\{S_n \geq x \text{ f.o.}\} = 1 \quad \text{for all} \quad x > 0$$

$$\text{implies} \qquad \Pr\{M < \infty\} = 1. \tag{6.10}$$

It should be underscored that all the implications in (6.9) are mutually exclusive with those of (6.10). To illustrate, suppose $\Pr\{M < \infty\} > 0$. Then certainly an x sufficiently large exists with the property

$$\Pr\{S_n > x \text{ i.o.}\} < 1.$$

Therefore, $\Pr\{S_n > x \text{ i.o.}\} = 0$ and it follows from (6.9) that $\Pr\{M < \infty\} = 1$.

Assuming $E[X_i] = \mu$ exists, it is possible to correlate the criteria of (6.9) and (6.10) with the sign of the mean contributed by each observation X_i. The result is as follows: If $\mu > 0$ then

$$\sum_{k=1}^\infty \frac{\Pr\{S_k \geq 0\}}{k} = \infty \tag{6.11}$$

and (6.9) holds. If $\mu < 0$ then

$$\sum_{k=1}^{\infty} \frac{\Pr\{S_k \geq 0\}}{k} < \infty \tag{6.12}$$

and (6.10) applies.

Proof of (6.11). The strong law of large numbers asserts that

$$\Pr\{S_n > n(\mu - \varepsilon) \text{ i.o.}\} = 1$$

for any positive ε and when $\mu - \varepsilon > 0$; in particular,

$$\Pr\{S_n > 0 \text{ i.o.}\} = 1,$$

establishing (6.9).

Proof of (6.12). Application of the law of large numbers informs us that

$$\Pr\left\{\lim_{n \to \infty} \frac{S_n}{n} < \frac{\mu}{2} < 0\right\} = 1,$$

i.e., almost every path of the $\{S_n\}$ process exhibits the characteristic that $S_n < 0$ for all n beyond a certain time point depending on the realization. Equivalently, we have

$$\Pr\{S_n \geq 0 \text{ f.o.}\} = 1. \tag{6.13}$$

Comparing this result with (6.10) secures the conclusion of (6.12).

A much deeper fact supplementing (6.11) and (6.12) pertains to the circumstance when $\mu = 0$ and where X_i are nondegenerate random variables. Then necessarily both

$$\sum_{k=1}^{\infty} \frac{\Pr\{S_k \geq 0\}}{k} = \infty \quad \text{and} \quad \sum_{k=1}^{\infty} \frac{\Pr\{S_k \leq 0\}}{k} = \infty \tag{6.14}$$

are maintained. In particular, this means that in the case of X_i having mean zero, then

$$\Pr\{M = \infty\} = \Pr\{m = -\infty\} = 1 \tag{6.15}$$

with $M = \max(0, S_1, S_2, \ldots)$, $m = \min(0, S_1, S_2, \ldots)$, and the fluctuations of the $\{S_n\}$ process stretch continually over time from arbitrarily large positive to arbitrarily large negative values and also the other way around.

For the case of symmetric random variables, i.e., where

$$\Pr\{S_k \geq 0\} = \Pr\{S_k \leq 0\} = \tfrac{1}{2},$$

then obviously (6.14) holds and extreme fluctuations are present such that (6.15) is correct.

C. CALCULATION OF MEANS OF LADDER RANDOM VARIABLES

Let T be the time (number of periods) until the occurrence of the first *positive* partial sum in the collection $\{S_k\}$. Formally

$$T = \min\{n; S_n > 0\}. \tag{6.16}$$

Define $T = \infty$ whenever $S_i \leq 0$ for all i.

Set

$$U = S_T = \text{the first positive } S_n. \tag{6.17}$$

Manifestly, U is a positive random variable and T is positive and integer valued.

Caution. The ladder random variables defined in Sec. 4 involved the first *nonnegative* partial sum. Certainly if the underlying random variables X_i are continuous, which we stipulate henceforth, then

$$\{T, U\} \qquad \text{and} \qquad \{W_1, Z_1\}$$

share the same distribution. Intuitively where S_n first becomes nonnegative and X_i are continuous, then S_n will undoubtedly be positive.

Referring to Theorem 5.4 the joint generating function of $\{T, U\}$ can be evaluated using the identity

$$E[x^T e^{-\lambda U}] = \sum_{n=0}^{\infty} x^n E[e^{-\lambda S_n}; L_{n0} = n]$$

$$= 1 - \exp\left(-\sum_{k=1}^{\infty} \frac{x^k}{k} E[e^{-\lambda S_k}; S_k \geq 0]\right), \tag{6.18}$$

valid for every $|x| < 1$ and $\lambda > 0$. All the series and relevant integrals converge absolutely. We proceed to calculate $E[T]$ and $E[U]$. First insert $\lambda = 0$ in (6.18) and rearrange the right-hand side, yielding

$$E[x^T] = 1 - \exp\left(-\sum_{k=1}^{\infty} \frac{x^k}{k} \Pr\{S_k \geq 0\}\right)$$

$$= 1 - \exp\left(-\sum_{k=1}^{\infty} \frac{x^k}{k} \Pr\{S_k < 0\}\right)\exp\left(-\sum_{k=1}^{\infty} \frac{x^k}{k}\right)$$

$$= 1 - (1 - x)\exp\left(-\sum_{k=1}^{\infty} \frac{x^k}{k} \Pr\{S_k < 0\}\right). \tag{6.19}$$

We restrict attention henceforth to the case where $\sum_{k=1}^{\infty} \Pr\{S_k < 0\}/k < \infty$, which clearly implies the condition $\sum_{k=1}^{\infty} \Pr\{S_k \geq 0\}/k = \infty$. In this situation on account of (6.9) we know that $\Pr\{S_n > 0 \text{ i.o.}\} = 1$ and therefore T is well defined and finite valued. Differentiate (6.19) with respect to x, setting afterwards $x = 1$. This easily furnishes the result

$$E[T] = \exp\left(\sum_{k=1}^{\infty} \frac{\Pr\{S_k < 0\}}{k}\right). \qquad (6.20)$$

Returning to (6.18), we now put $x = 1$ and execute the following evident elementary manipulations:

$$E[e^{-\lambda U}] = 1 - \exp\left\{-\sum_{k=1}^{\infty} \frac{1}{k} E[e^{-\lambda S_k}; S_k \geq 0]\right\}$$

$$= 1 - \exp\left\{+\sum_{k=1}^{\infty} \frac{1}{k} E[e^{-\lambda S_k}; S_k < 0]\right\} \exp\left\{\sum_{k=1}^{\infty} \frac{1}{k} E[e^{-\lambda S_k}]\right\}$$

$$= 1 - \exp\left\{+\sum_{k=1}^{\infty} \frac{1}{k} E[e^{-\lambda S_k}; S_k < 0]\right\} \exp\left\{\sum_{k=1}^{\infty} \frac{1}{k} [\varphi(\lambda)]^k\right\} \qquad (6.21)$$

where $\varphi(\lambda) = E[e^{-\lambda X_1}]$. Here we tacitly postulated the existence of $\varphi(\lambda)$ for $\lambda > 0$.

$$= 1 - [1 - \varphi(\lambda)] \exp\left(\sum_{k=1}^{\infty} \frac{1}{k} E[e^{-\lambda S_k}; S_k < 0]\right).$$

Next differentiate the final expression of (6.21) and subsequently substitute $\lambda = 0$ (noting $\varphi'(0) = -\mu = E[X_1]$) to obtain

$$-E[U] = \varphi'(0) \exp\left(\sum_{k=1}^{\infty} \frac{1}{k} \Pr\{S_k < 0\}\right).$$

Thus we have achieved the formula

$$E[U] = \mu \exp\left(\sum_{k=1}^{\infty} \frac{1}{k} \Pr\{S_k < 0\}\right) = \mu E[T], \qquad (6.22)$$

where the last equation results by virtue of (6.20).

The outcome is a version of the Wald identity (cf. Chapter 6).

It is evident that (6.22) is impossible when $\mu = 0$. Indeed recall after consulting (6.9) in this case that necessarily $\sum_{k=1}^{\infty} (1/k) \Pr\{S_k < 0\} = \infty$ and the right-hand side is not well defined. A much deeper analysis is required which we do not undertake. It can be shown that for $\mu = 0$

$$E[U] = c\sigma^2$$

where $\sigma^2 = \operatorname{Var} X_1$ and c is a suitable positive constant. It follows that where $\mu = 0$, then $E[U]$ is finite if and only if $\sigma^2 < \infty$.

Suppose $E[X_i] = \mu > 0$. Continuing the constructions of (6.16), we define $T_0 = 0$ and

$$T_r = \min\{n; n > T_{r-1}, S_n > S_{T_{r-1}}\}, \qquad r = 1, 2, \ldots. \tag{6.23}$$

The indices T_r determine the successive times of new heights in the process $\{S_n\}$. Owing to the assumption $E[X_i] = \mu > 0$, the random variables

$$W_r = T_r - T_{r-1}, \qquad r = 1, 2, \ldots,$$

are positive integer valued i.i.d. random variables and generate a renewal sequence (see Chapter 5). The sequence

$$U_r = S_{T_r} - S_{T_{r-1}}, \qquad r = 1, 2, \ldots,$$

also induces a renewal process of positive i.i.d. random variables. Consider

$$\sum_{k=1}^{\infty} \Pr\{x < S_{T_k} < x + h\}.$$

This is obviously the expected number of ladder heights falling in the interval $(x, x + h)$. Applying the basic renewal limit, Theorem 5.2 of Chapter 5, we obtain

$$\lim_{x \to \infty} \sum_{k=1}^{\infty} \Pr\{x < S_{T_k} < x + h\} = \frac{h}{E[S_{T_{r+1}} - S_{T_r}]} = \frac{h}{E[U_1]}. \tag{6.24}$$

Consider next

$$\lim_{x \to \infty} \sum_{n=1}^{\infty} \Pr\{x < M_n < x + h, S_n > M_{n-1}\} = \frac{h}{E[U_1]}, \tag{6.25}$$

since the sum in (6.25) is clearly recognized as the same sum as that in (6.24). We conclude this topic by proving the following limit

$$\lim_{x \to \infty} \sum_{n=1}^{\infty} \Pr\{x < M_n \le x + h\} = \frac{h}{\mu}. \tag{6.26}$$

Decompose the event $\{x < M_n \le x + h\}$ in terms of the index time at which the maximum M_n is first attained. The equality of events

$$\{x < M_n \le x + h\} = \bigcup_{r=0}^{n} \{x < M_n = M_r < x + h, S_r > M_{r-1}\}$$

$$\bigcap \{S_n - S_k \le 0, k = r, r + 1, \ldots, n\} \tag{6.27}$$

is a familiar one already used in the proof of Theorem 5.3. Since $\{X_i\}$ are i.i.d., we infer from (6.27).

$$\Pr\{x < M_n \le x + h\} = \sum_{r=0}^{n} \Pr\{x < M_r \le x + h, S_r > M_{r-1}\} \Pr\{L_{n-r,0} = 0\}$$

$$= \sum_{r=0}^{n} a_r b_{n-r},$$

exhibiting the usual convolution structure. It follows that

$$\sum_{n=1}^{\infty} \Pr\{x < M_n \le x + h\} = \sum_{n=0}^{\infty} a_n \sum_{n=0}^{\infty} b_n$$

$$= \sum_{r=1}^{\infty} \Pr\{x < M_r \le x + h, S_r > M_{r-1}\}$$

$$\times \sum_{n=0}^{\infty} \Pr\{L_{n0} = 0\}. \tag{6.28}$$

Referring to the generating function (5.8), we find that

$$\sum_{n=0}^{\infty} \Pr\{L_{n0} = 0\} = \exp\left(\sum_{k=1}^{\infty} \frac{\Pr\{S_k < 0\}}{k}\right). \tag{6.29}$$

The limit of the first sum on the right of (6.28) is $h/E[U_1]$ according to (6.25). But, citing (6.22), we have

$$E[U_1] = \mu E[T_1] = \mu \exp\left(\sum_{k=1}^{\infty} \frac{\Pr\{S_k < 0\}}{k}\right). \tag{6.30}$$

Combining the results of (6.25), (6.29), and (6.30) in (6.28) yields (6.26). That (6.26) holds is rather remarkable since a corresponding renewal result asserts that (see page 192 of Chapter 5)

$$\lim_{x \to \infty} \sum_{n=1}^{\infty} \Pr\{x < S_n < x + h\} = \frac{h}{\mu}. \tag{6.31}$$

(This limit relation is a version of the extended renewal limit theorem and is considerably deeper.)

Comparing (6.26) and (6.31) reveals that the expected number of visits to a fixed far-off interval equals the expected number of ladder heights falling in this same interval.

E. AN INTEGRAL EQUATION FOR THE DISTRIBUTION OF $M = \max(0, S_1, S_2, \ldots)$

We define

$$G_n(x) = \Pr\{M_n \le x\}$$

with $M_n = \max(0, S_1, S_2, \ldots, S_n)$. Observe that $M_n \le x$ and $X_1 = \xi$ are equivalent to the relations

$$X_1 = \xi \le x$$

and

$$\tilde{M}_{n-1} = \max(0, X_2, X_2 + X_3, \ldots, X_2 + \cdots + X_n) \le x - \xi.$$

Conditioning on the values of X_1 with cognizance of the i.i.d. character of $\{X_i\}$, we obtain

$$G_n(x) = \int_{-\infty}^x G_{n-1}(x - \xi)\, dF(\xi).$$

By stipulating that F is derived from the density f, then

$$G_n(x) = \int_{-\infty}^x G_{n-1}(x - \xi) f(\xi)\, d\xi = \int_0^\infty G_{n-1}(\eta) f(x - \eta)\, d\eta,$$

where the last equation results by an obvious change of variable. Since

$$G(x) = \lim_{n \to \infty} G_n(x) = \Pr\{M \le x\},$$

we find that G satisfies the integral equation

$$G(x) = \int_0^\infty G(\eta) f(x - \eta)\, d\eta. \tag{6.32}$$

It should be emphasized that we are interested in solutions of (6.32) apart from $G(x) \equiv 0$. Actually, a nontrivial solution exists if and only if $\Pr\{M < \infty\} = 1$, and the necessary and sufficient condition that G be nontrivial is

$$\sum_{k=1}^\infty \frac{\Pr\{S_k \ge 0\}}{k} < \infty \qquad \text{(prove this)}.$$

The integral equation (6.32) features prominently in the theory of queueing and is commonly referred to as of Wiener–Hopf type.

Problems

1. Let $X_1, X_2, \ldots$ be a sequence of independent, identically distributed integer-valued random variables. Define the partial sums $S_0 = 0$ and $S_k = X_1 + X_2 + \cdots + X_k$ for $k \ge 1$. The subscript $n \ge 1$ is called a ladder index if $S_n > S_j$ for $j = 0, 1, \ldots, n - 1$. Call the event that n is a ladder index $\mathscr{E}$. Define $Y_0 = 0$ and Y_N as the time (i.e., the index) of the last occurrence of $\mathscr{E}$, where the present trial is the Nth. Let W denote the time of first occurrence of $\mathscr{E}$, with $W = \infty$ should $\mathscr{E}$ not occur. Suppose $\mathscr{E}$ occurs at trial n. Prove that the number of trials until the next occurrence of $\mathscr{E}$ is independent of n and distributed as W.

2. Under the hypothesis of Problem 1 prove the identity

$$\sum_{n=0}^{\infty} t^n E[x^{Y_n}] = \frac{1 - F(t)}{1 - t} \frac{1}{1 - F(xt)},$$

where $F(t) = \sum_{n=1}^{\infty} \Pr\{W = n\}t^n$. Assume $\Pr\{W < \infty\} = 1$.

Hint: Use the relation

$$\Pr\{Y_n = k\} = \Pr\{Y_k = k\} \Pr\{Y_{n-k} = 0\}.$$

3. Under the hypothesis of Problem 2 prove the exponential representation

$$U(t) = \exp\left\{\sum_{k=1}^{\infty} \frac{t^k}{k} (E[Y_k] - E[Y_{k-1}])\right\},$$

where $U(t) = 1/[1 - F(t)]$.

Hint: With the aid of Problem 2 derive the differential equation

$$\frac{U'(t)}{U(t)} = \sum_{n=1}^{\infty} E[Y_n - Y_{n-1}]t^{n-1}$$

and solve it.

4. Let $X, X_1, X_2, \ldots$ be independent identically distributed random variables with the distribution

$$\Pr\{X = 1\} = p, \qquad \Pr\{X = -1\} = q,$$

where $p + q = 1$. Let $S_0 = 0$, and for $n = 1, 2, \ldots$ let $S_n = X_1 + \cdots + X_n$.
   (i) For $p = q = \frac{1}{2}$ compute $\Pr(\max_{0 \le j \le n} S_j = k)$. Consider separately the two cases of n even and n odd.
   (ii) Suppose $p < q$, and set $M = \max_{j \ge 0} S_j$. Compute the distribution of M.

5. Establish the identity

$$\sum_{n=0}^{\infty} x^n \Pr\{S_n = M_n = 0\} = \exp\left(\sum_{k=1}^{\infty} \frac{x^k}{k} \Pr\{S_k = 0\}\right).$$

Hint: Examine the appropriate generating function identity for

$$\sum_{n=0}^{\infty} x^n E[e^{-\lambda S_n}; L_{nn} = 0] \qquad \text{for} \quad \lambda > 0$$

and let $\lambda \to \infty$.

6. Prove the convergence of the series

$$\sum_{k=1}^{\infty} \frac{1}{k} \Pr\{S_k = 0\}.$$

Hint: Define $u_k = \Pr\{S_k = 0, M_k = 0\}$ and show that the sequence is monotone. Then use the result of the previous problem.

7. Determine a generating function identity for the first nonpositive partial sum (consult first formula (5.18)).

8. Determine an identity for

$$\sum_{n=0}^{\infty} x^n E[e^{itM_n} e^{iu(M_n - S_n)}],$$

where $|x| < 1$ and t and u are real.

Hint: Condition on the index time of the first maximum in the sequence $\{0, S_1, S_2, \ldots, S_n\}$.

9. Assume $\Pr\{S_n \le 0\} = \frac{1}{2} = \Pr\{S_n \ge 0\}$. Determine the limit as $n \to \infty$ of $\Pr\{(n - L_{nn})/n \le x\}$.

Answer: $(2/\pi) \arcsin \sqrt{x}$.

10. Define T as the time of the first ladder point. Show that the following conditions are equivalent:

(i) $\Pr\{T < \infty\} = 1$,
(ii) $E[T] < \infty$,
(iii) $\Pr\{\lim_{n \to \infty} M_n = \infty\} = 1$,
(iv) $\Pr\{\overline{\lim}_{n \to \infty} S_n = \infty\} = 1$.

11. Prove that if

$$\lim_{n \to \infty} \frac{1}{n} \sum_{k=0}^{n} \Pr\{S_k \ge 0\} = \alpha, \qquad 0 < \alpha < 1,$$

and with P_n as defined in (1.3), then

$$\lim_{n \to \infty} \frac{E[P_n]}{n} = \alpha.$$

12. Establish the identity

$$\sum_{n=0}^{\infty} E[e^{-\lambda M_n}] = \frac{E[T]}{1 - E[e^{-\lambda Z}]},$$

where Z is the value of the first positive partial sum and T is the index of this partial sum.

13. Let X_i be i.i.d. r.v.'s with distribution law

$$\Pr\{X_i = 1\} = \Pr\{X_i = -1\} = \frac{1}{2},$$

and let

$$S_n = \sum_{i=1}^{n} X_i, \qquad S_0 = 0.$$

Define

$$U_{2n} = \text{number of } \{k \,|\, S_{k-1} \ge 0, S_k \ge 0, k = 1, 2, \ldots, n\},$$

$$T_{2n} = \max\{k \,|\, S_k = 0, \qquad k = 1, 2, \ldots, n\}.$$

Prove

$$\Pr\{U_{2n} = 2k\} = \Pr\{T_{2n} = 2k\} = \binom{2k}{k} \binom{2n - 2k}{n - k} \frac{1}{2^{2n}}.$$

14. Let $\{X_i; i = 1, 2, \ldots\}$ be independent, identically distributed random variables such that

$$\Pr\{X_i = +1\} = \Pr\{X_i = -1\} = \tfrac{1}{2}.$$

Let $S_n = \sum_{i=1}^{n} X_i$ for $1 \le n < \infty$ and $P(m, n) = \Pr\{S_{2j} = 0\}$ for some j satisfying $m \le j < m + n\}$.

Prove that $P(m, n) + P(n, m) = 1$ for $m \ge 1, n \ge 1$.

Hint: Note that $\Pr\{S_{2n} = 0\} = \Pr\{S_1 \ne 0, S_2 \ne 0, \ldots, S_{2n} \ne 0\}$. Now assume the result holds for $m = k$ and arbitrary $n \ge 1$. Then justify the following inequalities:

$$
\begin{aligned}
1 - \Pr\{k + 1, n\} &= \Pr\{S_{2j} \ne 0 \text{ for } k \le j < k + n + 1\} \\
&\quad + \Pr\{S_{2k} = 0 \text{ and } S_{2j} \ne 0 \text{ for } k + 1 \le j < k + n + 1\} \\
&= \Pr\{S_{2j} \ne 0 \text{ for } k \le j < k + n + 1\} \\
&\quad + \Pr\{S_{2k} = 0\} \Pr\{S_{2j} \ne 0 \text{ for } 1 \le j < n + 1\} \\
&= \Pr\{S_{2j} = 0 \text{ for some } j \text{ with } n + 1 \le j < k + n + 1\} \\
&\quad + \Pr\{S_{2j} \ne 0 \text{ for } 1 \le j < k + 1\} \Pr\{S_{2n} = 0\} \\
&= \Pr\{S_{2j} = 0 \text{ for some } j \text{ with } n + 1 \le j < k + n + 1\} \\
&\quad + \Pr\{S_{2n} = 0 \text{ and } S_{2j} \ne 0 \text{ for } n + 1 \le j < k + n + 1\} \\
&= \Pr(n, k + 1).
\end{aligned}
$$

15. Find conditions so that

$$\Pr\{P_n \to \infty\} = 1$$

holds in terms of the convergence or divergence of the series

$$\sum_{k=1}^{\infty} \frac{\Pr\{S_k \ge 0\}}{k}.$$

16. Let $X_1, \ldots, X_n$ be independent, symmetric random variables such that

$$\Pr\left\{\sum_{i=1}^{n} \varepsilon_i X_i = 0\right\} = 0$$

for all nontrivial choices of the ε_i, $\varepsilon_i = 0, +1$, or -1. For each nonvoid subset T of $\{1, \ldots, n\}$ let $S_T = \sum_{i \in T} X_i$ and let N be the number of nonvoid subsets T of $\{1, \ldots, n\}$ with the property that $S_T > 0$. Show that for each $m, 0 \le m \le 2^{n-1}$, $\Pr\{N = m\} = 1/2^n$.

Hint: Prove and apply the following result. Let $x_1, \ldots, x_n$ be real numbers such that $\sum_{i=1}^{n} \varepsilon_i x_i \ne 0$ for each nontrivial choice of the ε_i, $\varepsilon_i = 0, +1, -1$. Then for each $m, 0 \le m \le 2^n - 1$, there is a unique sequence $\{\eta_i\}$ of signs, $\eta_i = \pm 1$, such that exactly m of the $2^n - 1$ sums $\sum_{i \in T} \eta_i x_i$, as T ranges through the nonvoid subsets of $\{1, \ldots, n\}$, are positive.

To prove this result, notice that if

$$A(\eta_1, \ldots, \eta_n) = \left\{x : x = \sum_{i \in T} \eta_i x_i \quad \text{for some} \quad T \subseteq (1, \ldots, n)\right\},$$

where $\sum_{i \in \varnothing} \eta_i x_i$ is interpreted to be zero, then

$$A(\eta_1, \ldots, \eta_{j-1}, -\eta_j, \eta_{j+1}, \ldots, \eta_n) = \{x : x + \eta_j x_j \in A(\eta_1, \ldots, \eta_n)\}.$$

17. Let $X, X_1, X_2, \ldots$ be independent and identically distributed random variables with means $E[X_1] = -\mu < 0$ and variance $\mathrm{Var}(X_1) = \sigma^2 < \infty$. With $S_0 = 0$ and $S_n = X_1 + \cdots + X_n$, $n \geq 1$, let $M = \max_{k \geq 0} S_k$. Prove that M and $(X + M)^+$ share the same distribution, where $x^+ = \max\{x, 0\}$.

18 (*continuation*). Using the same notation as in Problem 17, establish the inequality

$$E[M] \leq \tfrac{1}{2}\sigma^2/\mu.$$

Assume $E[M^2] < \infty$.

Hint: M has the same distribution as $(X + M)^+$, whence

$$E[M] = E[(M + X)^+] \quad \text{and} \quad E[M^2] = E[\{(M + X)^+\}^2].$$

With the notation $x^- = (-x)^+$,

$$M + X = (M + X)^+ - (M + X)^-, \quad (M + X)^2 = \{(M + X)^+\}^2 + \{(M + X)^-\}^2,$$

and so

$$E[(M + X)^-] = -E[X] \quad \text{and} \quad E[\{(M + X)^-\}^2] = 2E[M]E[X] + E[X^2].$$

But

$$E[\{(M + X)^-\}^2] \geq \{E[(M + X)^-]\}^2,$$

and so

$$2E[M]E[X] + E[X^2] - (E[X])^2 \geq 0,$$

from which the result follows.

NOTES

The fundamental Spitzer formula and related functional identities provide powerful tools for determining distributions of a wide spectrum of natural functionals on the process of sums of i.i.d. random variables. Spitzer [1] and Feller [2] set forth this important topic with its numerous ramifications and applications to queuing and storage phenomena, probabilistic potential theory, and more elaborate compound renewal structures.

Takács [3] emphasizes the intrinsic combinatorial nature of many of the identities.

REFERENCES

1. F. Spitzer, "Principles of Random Walk." Van Nostrand-Reinhold, New York, 1964.
2. W. Feller, "An Introduction to Probability Theory and Its Applications," Vol. II. Wiley, New York, 1971.
3. L. Takács, "Combinatorial Methods in the Theory of Stochastic Processes." Wiley, New York, 1967.

Chapter 18

QUEUEING PROCESSES

1: General Description

In Chapter 3 we discussed two simple examples of discrete time queueing processes. In this chapter we develop more completely the underlying concepts and methods of several continuous time versions of queueing processes. The general queueing model embodies the following physical structure. Customers† arrive at random times to some facility and request service of some kind. Queueing processes are classified according to

(1) *Input distribution* (*input process*), the distribution of the pattern of arrivals of customers in time, or more specifically, the distribution of time between arrivals;

(2) *Service distribution*, the distribution of time to serve a customer; and

(3) *Queue discipline*, the number of servers and the organization of line and service. In most models the discipline is "first come, first served," with service for a customer beginning as soon as he reaches the head of the line. All the models we shall consider in this chapter are of this type. (For other types of queue disciplines we refer the reader to Problems 2–4 of this chapter.)

We always assume that the durations between successive arrivals of customers, called the interarrival times, are independent, identically distributed,

---

† Customer is a generic term which may refer, for example, to bona fide customers demanding service at a counter, ships entering a port, flow of messages into an office, broken machines awaiting repair, etc.

positive random variables. Such input processes, referred to as renewal processes, are extensively discussed in Chapter 5.

The term "random arrival" is sometimes used to indicate that the input distribution is Poisson, more precisely, the events corresponding to the arrival of customers form a Poisson process. This implies that the interarrival distribution is exponential.

It is also assumed that the lengths of service for individual customers are independent, identically distributed, random variables, and independent of the input process.

2: The Simplest Queueing Processes (M/M/1)†

The simplest and most extensively studied queueing processes are those where the input process is Poisson and the service distribution is exponential. We have already described these processes and shown that the queue size forms a birth and death process (cf. Example 2, Section 6, Chapter 4).

Let us examine the one-server case again. The density function of the inter-arrival time is

$$dF(t) = \lambda e^{-\lambda t} \, dt, \qquad \lambda > 0,$$

and the service density is

$$dG(t) = \mu e^{-\mu t} \, dt, \qquad \mu > 0.$$

Because of the "forgetfulness" property (Theorem 2.2, Chapter 4) of the exponential distribution we conclude easily that the process $X(t) = $ length of queue at time t is a time homogeneous Markov process and is in fact a birth and death process. Let $P_{ij}(t)$ be the transition probability function. Then $P_{i, i+1}(h)$ is the probability of a single new arrival occurring in the next h units of time with no services completed. Thus for small $h > 0$

$$P_{i, i+1}(h) = \lambda h + o(h), \qquad i \geq 0.$$

Similarly, we find

$$P_{i, i-1}(h) = \mu h + o(h), \qquad i \geq 1,$$

and

$$P_{i, i}(h) = 1 - (\lambda + \mu)h + o(h) \qquad (i \geq 1),$$
$$P_{0, 0}(h) = 1 - \lambda h + o(h).$$

† A standardized shorthand is used in much of the literature for identifying simple queueing processes. We shall include for reference the shorthand names of the processes we study. In the symbol $A/B/c$, c is the number of servers, while A and B indicate the arrival and service distribution, respectively. The symbols used in the first two places are $G = GI$, general independent; M, exponential interarrival or service times; E_k (Erlangian, interarrival or service times distributed as a gamma distribution of order k (so that $E_1 = M$); and D (deterministic), a schedule of arrivals or fixed service lengths.

The infinitesimal generator is

$$\mathbf{A} = \left\|\begin{array}{ccccc} -\lambda & \lambda & 0 & 0 & \cdots \\ \mu & -(\lambda + \mu) & \lambda & 0 & \cdots \\ 0 & \mu & -(\lambda + \mu) & \lambda & \cdots \\ \vdots & \vdots & \vdots & \vdots & \end{array}\right\|.$$

We indicated in Section 6 of Chapter 4 that

$$\lim_{t \to \infty} P_{ij}(t) = p_j = \begin{cases} (\lambda/\mu)^j(1 - \lambda/\mu), & \lambda < \mu, \\ 0, & \lambda \geq \mu. \end{cases}$$

This gives us the answer to many problems involving stationarity. If the process has been going on a long time and $\lambda < \mu$, the probability of being served immediately upon arrival is

$$p_0 = \left(1 - \frac{\lambda}{\mu}\right).$$

We can also calculate the distribution of waiting time in the stationary case when $\lambda < \mu$. If an arriving customer finds n people in front of him, his total waiting time, T, including his own service time, is the sum of the service times of himself and those ahead, all distributed exponentially with parameter μ, and since the service times are independent of the queue size, T has a gamma distribution of order $n + 1$ with scale parameter μ

$$\Pr\{T \leq t | n \text{ ahead}\} = \int_0^t \frac{\mu^{n+1}\tau^n e^{-\mu\tau}}{\Gamma(n+1)} d\tau. \tag{2.1}$$

By the law of total probabilities, we have

$$\Pr\{T \leq t\} = \sum_{n=0}^{\infty} \Pr\{T \leq t | n \text{ ahead}\}\left(\frac{\lambda}{\mu}\right)^n\left(1 - \frac{\lambda}{\mu}\right),$$

since $(\lambda/\mu)^n(1 - \lambda/\mu)$ is the probability that in the stationary case a customer on arrival will find n ahead in line, Now, substituting from (2.1), we obtain

$$\begin{aligned} \Pr\{T \leq t\} &= \sum_{n=0}^{\infty} \int_0^t \frac{\mu^{n+1}\tau^n e^{-\mu\tau}}{\Gamma(n+1)}\left(\frac{\lambda}{\mu}\right)^n\left(1 - \frac{\lambda}{\mu}\right) d\tau \\ &= \int_0^t \mu e^{-\mu\tau}\left(1 - \frac{\lambda}{\mu}\right)\left(\sum_{n=0}^{\infty} \frac{\tau^n \lambda^n}{\Gamma(n+1)}\right) d\tau \\ &= \int_0^t \left(1 - \frac{\lambda}{\mu}\right)\mu \exp\left[-\tau\mu\left(1 - \frac{\lambda}{\mu}\right)\right] d\tau \\ &= 1 - \exp\left[-t\mu\left(1 - \frac{\lambda}{\mu}\right)\right], \end{aligned}$$

which is also an exponential distribution.

If we wish to answer nonstationary questions, it is essential to determine $P_{ij}(t)$ for all t. This is a much harder problem but it has been solved. The details of this solution are beyond the scope of this book and we refer the interested student to any of the advanced books on queueing theory listed at the close of the chapter.

With the same exponential input and service distribution, when two servers are waiting on the line containing more than two customers, the expected time $1/\mu_n$ until some completion of service is half of that when only one server is engaged. Thus $\mu_n = 2\mu, n \geq 2$. But if $n = 1$, one server is inactive, so $\mu_1 = \mu$, and the infinitesimal generator of this birth and death process has the form

$$\mathbf{A} = \left\| \begin{matrix} -\lambda & \lambda & 0 & 0 & \cdots \\ \mu & -(\lambda + \mu) & \lambda & 0 & \cdots \\ 0 & 2\mu & -(\lambda + 2\mu) & \lambda & \cdots \\ 0 & 0 & 2\mu & -(\lambda + 2\mu) & \cdots \\ \vdots & \vdots & \vdots & \vdots & \end{matrix} \right\|.$$

3: Some General One-Server Queueing Models

We shall discuss aspects of three methods for analyzing particular cases of the general queueing process GI/GI/1. The first method, known as the integral equation method, reduces the problem of finding the limiting distribution of the waiting time of the nth customer ($n \to \infty$) to the problem of solving an integral equation of so-called Wiener–Hopf type.

If the input process is Poisson, a second method of attack is to examine the length of the line at just those moments when a person completes service. This *embedded process* can be seen to be a Markov chain (see Section 4 of this chapter). If the service distribution is exponential and the input distribution is general then the *embedded Markov chain* is obtained by inspecting the queue size at the instant of each new arrival. The resulting process forms a Markov chain of special structure.

The third method investigates the properties of the random variable $W(t)$, which is the time a customer would have to wait if he arrived at time t, regardless of whether or not there was an actual arrival at time t. This quantity is called the *virtual waiting time* of a fictitious customer arriving at time t.

We shall begin by considering the integral equation method and then go on to the more restricted models to which the method of the embedded Markov chain is applicable. Some aspects of the third method will be discussed in Section 8.

A. INTEGRAL EQUATION METHOD†

Let us define the quantities

W_r = the waiting time, excluding service, of the rth customer,

S_r = the service time of the rth customer, and

T_r = time between arrival of rth and $(r + 1)$th customers (i.e., the rth interarrival time),

with W_0, S_0, T_0 all taken to be zero. We shall assume the first person arrives at the instant $t = 0$ and finds no one ahead of him.

Clearly $W_r + S_r$ is the length of time that the rth customer spends in the system, i.e., waiting time plus service time. Therefore, if $T_r > W_r + S_r$, then the $(r + 1)$th customer will find on arrival no line ahead of him, and so his waiting time W_{r+1} is zero. In the case that $T_r \leq W_r + S_r$, then the length of his waiting time is clearly $W_r + S_r - T_r$. Therefore,

$$W_{r+1} = \begin{cases} W_r + S_r - T_r & \text{if } W_r + S_r - T_r \geq 0, \\ 0 & \text{if } W_r + S_r - T_r < 0. \end{cases}$$

Let us write

$$U_r = S_r - T_r.$$

Then clearly $\{U_r\}_{r=1}^{\infty}$ is a sequence of identically and independently distributed r.v.'s. Let $F_r(x)$ be the distribution function of W_r and $g(x) = G'(x)$ the density function of U_r, which by assumption is the same for all r.‡ Then for $x \geq 0$, since W_r and U_r are independent r.v.'s,

$$F_{r+1}(x) = \Pr\{W_{r+1} \leq x\} = \Pr\{\max(W_r + U_r, 0) \leq x\} = \Pr\{W_r + U_r \leq x\}$$

$$= \int_{-\infty}^{x} \Pr\{W_r + U_r \leq x \mid U_r = y\} g(y)\, dy$$

$$= \int_{y \leq x} F_r(x - y) g(y)\, dy \qquad (r \geq 1). \tag{3.1}$$

Now since the first person arrives at $t = 0$ and does not wait,

$$F_1(x) = \begin{cases} 1 & \text{for } x \geq 0, \\ 0 & \text{for } x < 0, \end{cases}$$

and since for $x < 0$ all $F_i(x) = 0$, then

$$F_1(x) - F_2(x) \geq 0, \qquad -\infty < x < \infty.$$

---

† The remaining material of this section can be skipped on first reading without loss of continuity.

‡ The restriction that $G(x)$ has a density function is imposed to ease the exposition and is not essential to the following arguments.

But

$$F_r(x) - F_{r+1}(x) = \int_{y \leq x} [F_{r-1}(x-y) - F_r(x-y)]g(y)\,dy,$$

and so by induction it follows that, for all r,

$$F_r(x) - F_{r+1}(x) \geq 0, \qquad -\infty < x < \infty.$$

This shows that the $F_r(x)$ are decreasing with respect to r for each x; and since $F_r(x) \geq 0$, it follows that $F_r(x)$ converges to, say, $F(x)$. Passing to the limit in (3.1) yields

$$F(x) = \int_{y \leq x} F(x-y)g(y)\,dy\dagger$$

or, setting $z = x - y$,

$$F(x) = \int_0^\infty F(z)g(x-z)\,dz.$$

We must now investigate the question of when the limit $F(x)$ is an honest distribution. It is clear that $F(x)$ is nondecreasing, but it could happen that $\lim_{x \to \infty} F(x) < 1$ rather than $\lim_{x \to \infty} F(x) = 1$. The former possibility can be interpreted to mean that the waiting time of the nth customer ($n \to \infty$) approaches ∞ with positive probability or, equivalently, that the length of the line becomes ∞ (i.e., arbitrarily large) with positive probability.

We shall first derive a new expression for $F(x)$.

Since

$$F_1(x) = \begin{cases} 1, & x \geq 0, \\ 0, & x < 0, \end{cases}$$

we have

$$F_2(x) = \int_{u \leq x} g(u)\,du = \Pr\{U_1 \leq x\}, \qquad x \geq 0.$$

Now

$$F_3(x) = \int_{u \leq x} F_2(x-u)g(u)\,du = \int_{u \leq x} \left[\int_{v \leq x-u} g(v)\,dv \right] g(u)\,du$$

$$= \int_{u_2 \leq x,\, u_1 + u_2 \leq x} g(u_2)g(u_1)\,du_2\,du_1$$

$$= \Pr\{U_2 \leq x,\, U_1 + U_2 \leq x\} = \Pr\{U_1 \leq x,\, U_1 + U_2 \leq x\},$$

---

† The justification of interchange of limit and integral requires knowledge of the Lebesgue integral, and should be accepted on faith if not already familiar to the reader.

where we have used the fact that U_1 and U_2 are independent, identically distributed r.v.'s. A straightforward induction now yields

$$F_{r+1}(x) = \Pr\{U_r \leq x, U_r + U_{r-1} \leq x, \ldots, U_r + \cdots + U_1 \leq x\}$$
$$= \Pr\{U_1 \leq x, U_1 + U_2 \leq x, \ldots, U_1 + \cdots + U_r \leq x\}$$

(since $U_1, U_2, \ldots, U_r$ are identically distributed). Thus if $\tilde{U}_r = \sum_{i=1}^{r} U_i$,

$$F_{n+1}(x) = \Pr\{\tilde{U}_r \leq x, r = 1, \ldots, n\}, \qquad x \geq 0.$$

Clearly $F_n(x)$ decreases monotonically with n (we also proved this analytically above), and so

$$F(x) = \Pr\{\tilde{U}_r \leq x \text{ for all } r\}, \qquad x \geq 0.$$

For $x < 0$, all $F_i(x) = 0$, and so trivially $F(x) = 0$ for this range.

Using the above result, we may determine when $F(x)$ is an honest distribution: Assuming $E[S] < \infty$ and $E[T] < \infty$, i.e., the r.v.'s S and T have finite expectations, we have the following.

Theorem 3.1. (i)   *If $E(U) \geq 0$, then $F(x) \equiv 0$.*   (ii)   *If $E(U) < 0$, then $F(x)$ is a distribution, i.e., $\lim_{x \to \infty} F(x) = 1$.*

Intuitively, this result is anticipated. It asserts that if the interarrival time on the average is shorter than the average length of service, the line is sure to grow without bound and $W_r \to \infty$ with probability 1. The proof falls into three parts.

Proof.   (1) $E[U] > 0$.
By the strong law of large numbers,

$$\lim_{n \to \infty} \frac{\tilde{U}_n}{n} = E[U] \qquad \text{with probability 1};$$

therefore for almost all realizations of the sequence $\tilde{U}_1, \tilde{U}_2, \tilde{U}_3, \ldots$, i.e., with probability 1,

$$\tilde{U}_n \geq \tfrac{1}{2} n E[U] \tag{3.2}$$

for sufficiently large n, where the choice of n depends on the realization. The event $\tilde{U}_r \leq x$, for all r, is part of the event complementary to (3.2), and so its probability $F(x)$ is zero.

(2) $E[U] < 0$.
Again, from the strong law of large numbers we know that, for any preassigned $\varepsilon > 0$ and arbitrary $\delta > 0$, there exists an integer $N_{\varepsilon, \delta}$ such that for $n \geq N_{\varepsilon, \delta}$ we have

$$\Pr\left\{ \left| \frac{\tilde{U}_n}{n} - E[U] \right| \leq \varepsilon \text{ for all } n \geq N_{\varepsilon, \delta} \right\} \geq 1 - \delta.$$

Now let ε be small enough so that $E[U] + \varepsilon < 0$. Then for any $\delta > 0$ there is an N_δ such that

$$1 - \delta \leq \Pr\{\tilde{U}_n \leq n(E[U] + \varepsilon) \text{ for all } n \geq N_\delta\}$$
$$\leq \Pr\{\tilde{U}_n \leq 0 \text{ for all } n \geq N_\delta\}$$

by virtue of the property that $E[U] + \varepsilon < 0$. Now since $G(x)$ is a bone fide distribution, for this same δ and N_δ we can choose x sufficiently large so that

$$\Pr\{B\} = \Pr\{\tilde{U}_r \leq x \text{ for all } 1 \leq r \leq N_\delta - 1\} \geq 1 - \delta$$

where the event B is defined in the obvious way.

Let A denote the event defined by the condition $\{\tilde{U}_n \leq 0 \text{ for all } n \geq N_\delta\}$. The event $\{\tilde{U}_r \leq x, \text{ for all } r\}$, includes the intersection of the two events A and B and so has probability $\geq \Pr\{A \cap B\} = \Pr\{A\} + \Pr\{B\} - \Pr\{A \cup B\} \geq 2 - 2\delta - 1 = 1 - 2\delta$. Therefore, for x sufficiently large, $F(x) \geq 1 - 2\delta$, and

$$\lim_{x \to \infty} F(x) \geq 1 - 2\delta.$$

But since δ is arbitrary this means that $\lim_{x \to \infty} F(x) = 1$.

(3)　$E[U] = 0$.

In this case the result follows from a rather deep theorem dealing with recurrence phenomena for sums of independent random variables whose presentation is beyond the level of this book.† A discrete version of the theorem is Theorem 1.1 of Chapter 12.

B. RECURRENCE OF THE EVENT THAT AN ARRIVING CUSTOMER HAS NO WAITING TIME

Looking at the discrete time process defined by the variables $W_n, n = 0, 1, 2, \ldots,$ with state space $[0, \infty]$, we might ask about the recurrence of the event that some customer finds an empty line. Formally, we say that A occurs at the nth step if $W_n = 0$. The statement that the event A occurs without further qualification shall mean that it occurs at some finite step. Notice that whenever event A occurs the process begins with the next value of W equal to 0.

Theorem 3.2.　(1)　*If $E[U] > 0$, the event A is transient (i.e., the probability of A is less than 1).*

(2)　*If $E[U] \leq 0$, the event A is recurrent (i.e., $\Pr\{A\} = 1$).*

(3)　*If $E[U] < 0$, the event A is positive recurrent (i.e., the expected time until occurrence of A is finite).*

Proof.　(1)　Using the same notation as before, we note that

$$W_{n+1} \geq \tilde{U}_n$$

† F. Spitzer, "Principles of Random Walk." Van Nostrand Reinhold, New York, 1964.

(recall that $\tilde{U}_n = U_1 + \cdots + U_n = S_1 + \cdots + S_n - T_1 - T_2 - \cdots - T_n =$ the difference between the total service times of the first n customers and the total interarrival times of the first $n + 1$ customers). Inequality holds provided the server has been busy throughout. Alternatively, if all $\tilde{U}_n > 0$, A does not occur. By the strong law of large numbers, for any ε and $\delta > 0$ there is an N such that

$$\Pr\left\{\left|\frac{\tilde{U}_n}{n} - E[U]\right| \le \varepsilon \text{ for all } n \ge N\right\} \ge 1 - \delta.$$

Thus, if $E[U] > 0$ and ε is chosen sufficiently small, in fact if $\varepsilon < E[U]$, there is an N so that

$$\Pr\{\tilde{U}_n > 0 \text{ for all } n \ge N\} > 0.$$

This says that A may occur only a finite number of times with positive probability, i.e., the event $\{W_r = 0 \text{ only finitely often}\}$ has positive probability. But an event is recurrent if and only if it occurs infinitely often with probability 1 (cf. Theorem 7.1, Chapter 2). Therefore, if $E[U] > 0$, A is transient.

(2)   If $E[U] < 0$, for appropriate ε and $\delta > 0$, there is an N such that

$$\Pr\left\{\left|\frac{\tilde{U}_n}{n} - E[U]\right| < \varepsilon \text{ for all } n \ge N\right\} \ge 1 - \delta$$

so

$$\Pr\{\tilde{U}_n \le 0 \text{ for all } n \ge N\} \ge 1 - \delta,$$

and since δ is arbitrary, a fortiori

$$\Pr\{\tilde{U}_n \le 0 \text{ for some } n\} = 1.$$

But if $\tilde{U}_n \le 0$, then some $W_i = 0$, in particular, if $\tilde{U}_k$ is the first $\tilde{U}_n$ which is ≤ 0, then $W_{k+1} = 0$, and so if $E[U] < 0$, A is recurrent.

If $E[U] = 0$, the result is rather deep, and we do not enter into the details of the proof; see, e.g., the references listed at the close of the chapter.

(3)   When $E[U] < 0$, we claim that A is positive recurrent. The proof will not be given. ■

4: Embedded Markov Chain Method Applied to the Queueing Model (M/GI/1)

We examine the special case of a one-server queue with Poisson input process with parameter λ. The service distribution is assumed to be any distribution $B(v)$ of a positive random variable V for which $E[V] < \infty$. For ease of exposition we will assume that $B(v)$ has a density function $b(v)$. We shall investigate this process by the method of the embedded Markov chain.

The associated embedded Markov chain is determined as follows.

Let $Z(t)$ denote the number of customers in the queue at time t ($t \geq 0$). Suppose we look at the $Z(t)$ process just when a customer finishes service. In this way we generate a sequence of integer values

$$Z(t_1), \quad Z(t_2), \quad Z(t_3), \quad \ldots, \tag{4.1}$$

where $t_1, t_2, t_3, \ldots$ are the successive times of completion of service. The sequence $\{Z(t_n)\}$ forms a discrete time process

$$X_0 = 0, \qquad X_n = Z(t_n), \quad n = 1, 2, \ldots \tag{4.2}$$

We shall show below that because of the Poisson nature of the input process the sequence (4.2) describes a Markov chain.

A more intuitive description of the Markov chain goes as follows:

The transitions of the Markov chain occur only at those times when a person completes service. Let the state of the chain, until the next person completes service, be the number of people the departing customer leaves behind him, including any person who begins service as he leaves.

It is easy to see that this process is Markov, for if X_n is the state of the system at time n, then

$$X_{n+1} = \begin{cases} X_n - 1 + N & \text{if } X_n \geq 1, \\ N & \text{if } X_n = 0, \end{cases} \quad n = 0, 1, 2, \ldots, \tag{4.3}$$

where N is the number of people who arrived during the service time V of the $(n + 1)$th customer. But the random variable V, by assumption, is independent of previous service times and the length of line. By the stationary character of the Poisson process, the number of arrivals N during a service period depends only on V and not on the length of the line or on the moment at which service began. These facts imply that $\{X_n\}$ is a Markov chain.

The probability distribution of N can be computed by conditioning on the values of V and invoking the law of total probabilities. Thus

$$\Pr\{N = n\} = \int_0^\infty \Pr\{N = n | V = v\} b(v) \, dv.$$

Now, the number of customers arriving in an interval v is a random variable following the Poisson law with parameter λv, and therefore we have

$$\Pr\{N = n | V = v\} = e^{-\lambda v} \frac{(\lambda v)^n}{n!},$$

so that if $j \geq i - 1 \geq 0$, then

$$P_{ij} = \Pr\{X_{n+1} = j | X_n = i\} = \Pr\{N = j - i + 1\}$$

$$= \int_0^\infty \Pr\{N = j - i + 1 | V = v\} b(v) \, dv = \int_0^\infty e^{-\lambda v} \frac{(\lambda v)^{j-i+1}}{(j - i + 1)!} b(v) \, dv$$

and $P_{ij} = 0$ for $j < i - 1$.

If a customer departs leaving no one behind him, the state remains zero until another has arrived and been served. Thus, as far as transitions to another state are concerned, the zeroth and first states are identical.

Thus if we set

$$k_r = \int_0^\infty \frac{e^{-\lambda v}(\lambda v)^r}{r!} b(v)\, dv, \qquad r = 0, 1, 2, \ldots$$

which is the probability of r people arriving during a service period, then

$$\mathbf{P} = \begin{Vmatrix} k_0 & k_1 & k_2 & \cdots \\ k_0 & k_1 & k_2 & \cdots \\ 0 & k_0 & k_1 & \cdots \\ 0 & 0 & k_0 & \cdots \\ \vdots & \vdots & \vdots & \end{Vmatrix}. \tag{4.4}$$

The study of the Markov chain (4.4) per se was carried out in some detail in Section 5 of Chapter 3. We proved there that $\mathbf{P}$ is positive recurrent, recurrent null, or transient according to whether

$$\rho = \sum_{r=0}^\infty r k_r = \int_0^\infty e^{-\lambda v} \sum_{r=0}^\infty \frac{r(\lambda v)^r}{r!} b(v)\, dv$$

$$= \lambda \int_0^\infty v b(v)\, dv = \lambda E[V] < 1, \;=1, \text{ or } >1.$$

The quantity

$$\rho = \frac{\text{expected length of service time per customer}}{\text{expected length of interarrival time}} = \frac{E[V]}{1/\lambda} = \lambda E[V]$$

is called the *traffic intensity*. In the case where $\rho < 1$ we shall now determine the limiting stationary distribution of the Markov chain.

A. THE STATIONARY DISTRIBUTION OF THE EMBEDDED MARKOV CHAIN

We wish to determine

$$\pi = (\pi_0, \pi_1, \ldots), \qquad \pi_i > 0, \qquad \sum_{i=0}^\infty \pi_i = 1,$$

such that

$$\sum_{i=0}^\infty \pi_i P_{ij} = \pi_j, \qquad j = 0, 1, 2, \ldots,$$

where $\|P_{ij}\| = \mathbf{P}$ is defined in (4.4). In terms of the k_i, these equations are

$$\pi_i = \pi_0 k_i + \sum_{r=1}^{i+1} \pi_r k_{i-r+1}, \qquad i = 0, 1, 2, \ldots$$

We solve for the generating function

$$\pi(s) = \sum_{i=0}^{\infty} \pi_i s^i$$

in terms of

$$K(s) = \sum_{i=0}^{\infty} k_i s^i.$$

Multiplying through by s^i, we get

$$s^i \pi_i = \pi_0 k_i s^i + \frac{1}{s} \sum_{r=0}^{i+1} \pi_r k_{i-r+1} s^{i+1} - \frac{\pi_0 k_{i+1} s^{i+1}}{s}, \qquad i = 0, 1, 2, \dots.$$

Summing over i, and recognizing $\sum_{r=0}^{i+1} \pi_r k_{i+1-r}$ as a convolution, we have

$$\sum_{i=0}^{\infty} \pi_i s^i = \pi(s) = \pi_0 K(s) + \frac{1}{s} [K(s)\pi(s) - \pi_0 k_0] - \frac{\pi_0}{s} [K(s) - k_0].$$

Solving for $\pi(s)$ gives

$$\pi(s) = \frac{\pi_0 K(s)(s - 1)}{s - K(s)}.$$

This formula determines the generating function of the stationary distribution to within a constant factor. Since it is clear from the definition of the k_i that

$$K(1) = \sum_{i=0}^{\infty} k_i = 1,$$

we notice that the expression for $\pi(s)$ contains in the denominator the factor

$$\frac{s - K(s)}{s - 1} = \frac{s - 1}{s - 1} - \frac{1 - K(s)}{1 - s},$$

which approaches $1 - K'(1)$ as $s \to 1$.

Let us evaluate $K'(1)$, the expected number of people to arrive during a service period.

$$K'(1) = \sum_{r=1}^{\infty} r k_r = \lambda E[V] = \frac{E[V]}{E[A]} = \rho$$

[ρ = (expected service time)/(expected interarrival time) is, of course, the traffic intensity], where $1/\lambda = E(A)$ is the expected duration of interarrival times. Since $\rho < 1$, a stationary distribution exists, and therefore $\pi(1) = 1$.

But our formula gives

$$\pi(1) = \frac{\pi_0}{1 - \rho},$$

and so

$$\pi_0 = 1 - \rho.$$

Thus the generating function of the stationary distribution is given by

$$\pi(s) = (1 - \rho) \frac{K(s)(s - 1)}{s - K(s)}.$$

The quantity $1 - \rho$ represents the stationary probability that the line is empty.

B. EXPECTED LENGTH OF LINE IN EQUILIBRIUM FOR THE (M/GI/1) QUEUE

We close this section by calculating explicitly the expected length of line and the expected waiting time for a customer just arriving under the condition that the queueing process has attained a stationary state.

Differentiation of the generating function $\pi(s)$ does not easily lead to an expression for $E[q]$ where q is the stationary r.v. of the embedded Markov chain. We can, however, find $E[q]$ by another method.

If q is the number of people in line after a customer departs, and q' is the number after the next departure, then

$$q' = q - 1 + \delta + N,$$

where N is the number of arrivals during this service period, and

$$\delta = \begin{cases} 1 & \text{if } q = 0, \\ 0 & \text{if } q > 0. \end{cases}$$

In the stationary state, q' has the same distribution as q. Thus $E[q'] = E[q]$, and

$$E[\delta] = 1 - E[N] = 1 - \rho. \tag{4.5}$$

From the same expression, since $\delta^2 = \delta$,

$$q'^2 = q^2 + \delta + (N - 1)^2 + 2q\delta + 2\delta(N - 1) + 2q(N - 1),$$

and since $q\delta = 0$ (by the definition of δ),

$$q'^2 = q^2 + \delta + N(N - 1) + (1 - N) + 2\delta(N - 1) + 2q(N - 1).$$

But N, the number of customers that arrive during a service period, is independent of q, and hence also of δ. Thus, taking expectations and referring to (4.5), we obtain

$$E[q'^2] = E[q^2] + 1 - \rho + E[N(N - 1)] + 1 - \rho + 2(1 - \rho)(\rho - 1)$$
$$+ 2E[q](\rho - 1).$$

Since $E[q'^2] = E[q^2]$ by the stationarity postulate, these terms drop out. Then solving for $E[q]$ gives

$$E[q] = \rho - \frac{E[N(N - 1)]}{2(\rho - 1)}.$$

Now

$$K(s) = \sum_{n=0}^{\infty} k_n s^n = \int_0^{\infty} e^{-\lambda v(1-s)} b(v) \, dv;$$

so

$$E[N(N-1)] = \sum_{n=0}^{\infty} n(n-1)k_n = K''(1)$$

and

$$K''(s) = \lambda^2 \int_0^{\infty} v^2 e^{-\lambda v(1-s)} b(v) \, dv.$$

Therefore

$$E[N(N-1)] = \lambda^2 \int v^2 b(v) \, dv = \lambda^2 \{\mathrm{var}(V) + [E[V]]^2\}$$

$$= \mathrm{var}(\lambda V) + \rho^2,$$

since

$$\rho = \lambda E[V].$$

Thus

$$E[q] = \rho + \frac{\mathrm{var}(\lambda V) + \rho^2}{2(1-\rho)}. \tag{4.6}$$

C. EXPECTED WAITING TIME

Still assuming equilibrium, we may find the expected waiting time for a single customer. Suppose a customer waits a length of time W until his service begins which lasts V units of time. Assume that when he departs he leaves q customers behind him. Then in time $W + V$, q customers arrived, in accordance with the laws of a Poisson process. For stationarity to hold, it follows that

expected waiting time + expected service time per individual
$$= E[W] + E[V]$$

should equal the

expected number of customers to arrive in this period multiplied by the average length of the interarrival time

$$= \frac{1}{\lambda} E[q].$$

But we are in an equilibrium situation, and so substituting from (4.6) gives

$$E[W] + E[V] = \frac{1}{\lambda}\left[\rho + \frac{\text{var}(\lambda V) + \rho^2}{2(1 - \rho)}\right].$$

Dividing by $\mu = E[V]$ and remembering that $\lambda\mu = \rho$, the last relation becomes

$$\frac{E[W]}{E[V]} = \frac{\text{var}(\lambda V) + \rho^2}{2\rho(1 - \rho)},$$

or after some simplification

$$E[W] = \rho\frac{\text{var}(V/\mu) + 1}{2(1 - \rho)}E[V]. \tag{4.7}$$

The formulas (4.6) and (4.7) express somewhat surprising facts. They say that for given average arrival and service times we can decrease the expected line length and expected waiting time by decreasing the variance of service time. The best possible case in this respect is clearly the constant service time.

D. DISTRIBUTION OF WAITING TIME

By the same approach as above we can find the Laplace transform of the waiting time distribution. Let $\{\pi_i\}$ be the equilibrium probabilities whose generating function $\pi(s)$ was described above. If a person waits a length of time W until service and is served in time V, the probability that he leaves q people behind him when he departs, which must be π_q, is the probability that q people arrived during the time $W + V$. Since arrivals are Poisson with parameter λ,

$$\pi_q = \int_0^\infty e^{-\lambda t}\frac{(\lambda t)^q}{q!}\,dC(t),$$

where $C(t)$ is the distribution function of $W + V$. Then

$$\pi(s) = \sum_{q=0}^\infty \pi_q s^q = \int_0^\infty e^{-\lambda t(1-s)}\,dC(t)$$
$$= \tilde{C}(\lambda(1 - s)),$$

where $\tilde{C}(s)$ is the Laplace transform of $C(t)$. But $C(t)$ is the distribution function of the sum of the independent random variables W and V with distribution functions $W(t)$ and $B(t)$, respectively. The Laplace transform of the sum of W and V is the product of their Laplace transforms. Therefore

$$\pi(s) = \tilde{W}(\lambda(1 - s))\tilde{B}(\lambda(1 - s)),$$

or

$$\tilde{W}(z) = \frac{\pi((\lambda - z)/\lambda)}{\tilde{B}(z)}.$$

5: *Exponential Service Times* (G/M/1)

Another model which can also be studied by the method of the embedded
Markov chain is one with a general input distribution $H(u)$ for interarrival
times, but exponential service; i.e., service times are distributed exponentially
with parameter μ.

In this case let the transitions of the embedded Markov chain be induced
at the times of arrival of new customers, and let the state until the next transition
be the number of people the new arrival found in front of him.

If q is the state of the system after one arrival and q' the state after the next
arrival, then

$$q' = \max\{q + 1 - N, 0\} \tag{5.1}$$

where N is the number served in the intervening period. Because of the exponential distribution's property of "forgetfulness" (see Theorem 2.2, Chapter 4,
the number N served during an epoch between arrivals depends only on the
length of that interval and q and not on the extent of service which the present
customer has already received. The interarrival times are, of course, independent
random variables. By virtue of the foregoing facts we conclude that the transition
law (5.1) generates a Markov chain. We next calculate its transition probability
matrix $\|P_{ij}\|$.

Since $N \geq 0$, if $j > i + 1$, then obviously $P_{ij} = 0$. If $i + 1 \geq j \geq 1$, then
$i + 1 - j$ individuals were served during an interarrival epoch. We denote
the probability of this event by a_{i+1-j}. Evidently if $i + 1 \geq j \geq 1, P_{ij} = a_{i+1-j}$.

It is worthwhile to determine a_k explicitly in terms of the interarrival distribution and the exponential service time distribution. To this end, we note that
for an interarrival time of length ξ the probability that exactly k services occur is

$$F^{(k)}(\xi) - F^{(k+1)}(\xi), \tag{5.2}$$

where $F(\xi) = 1 - e^{-\mu\xi}$ is the service time distribution and $F^{(k)}(\xi)$ denotes
the k-fold convolution of $F(\xi)$. In fact, let $\Xi_1, \Xi_2, \ldots, \Xi_r, \ldots$ denote the durations
of the first, second, etc. services. The Ξ_i are independent and identically distributed with distribution law $F(\xi)$. The probability that at least k services occur
in time ξ is the same as the probability that the time span until the finish of the
kth service does not exceed ξ, i.e.,

$$\Pr\{\Xi_1 + \Xi_2 + \cdots + \Xi_k \leq \xi\} = F^{(k)}(\xi).$$

Hence the probability of exactly k services in time ξ is equal to

$$\Pr\{\text{time required for at least } k \text{ services} \leq \xi\}$$
$$- \Pr\{\text{time required for at least } k + 1 \text{ services} \leq \xi\}$$

and (5.2) obtains. Now $F^{(k)}(\xi)$ is known explicitly:

$$F^{(k)}(\xi) = \int_0^\xi e^{-\mu t} \frac{t^{k-1} \mu^k}{\Gamma(k)} \, dt.$$

Integrating the corresponding formula $F^{(k+1)}(\xi)$ by parts, we reduce (5.2) to

$$F^{(k)}(\xi) - F^{(k+1)}(\xi) = e^{-\mu\xi} \frac{\xi^k \mu^k}{\Gamma(k+1)}.$$

By the law of total probabilities we have

$$a_k = \int_0^\infty e^{-\mu\xi} \frac{\xi^k \mu^k}{\Gamma(k+1)} \, dH(\xi),$$

where $H(\xi)$ is the distribution function of an interarrival period. The formula for a_k can be derived directly. However, the method described above has independent merit of value in other contexts.

Finally, P_{i0} is the probability that all i persons present were served, which is equal to the probability that at least i would have been served if more had been present. It is most conveniently expressed in the form $P_{i0} = 1 - \sum_{j=1}^\infty P_{ij}$, so that

$$\mathbf{P} = \begin{Vmatrix} r_0 & a_0 & 0 & 0 & 0 & \cdots \\ r_1 & a_1 & a_0 & 0 & 0 & \cdots \\ r_2 & a_2 & a_1 & a_0 & 0 & \cdots \\ \vdots & \vdots & \vdots & & & \end{Vmatrix}, \tag{5.3}$$

where $r_i = 1 - a_0 - a_1 - \cdots - a_i$.

We encountered the Markov chain with transition probability matrix (5.3) in Section 6, Chapter 3, and a rather complete analysis was made pertaining to its properties of positive recurrence, recurrence, and transientness. We proved, in particular, that if

$$\sum_{k=0}^\infty k a_k > 1$$

then the Markov chain process is positive recurrent and the stationary limiting distribution has the form

$$\pi_i = (1 - \xi_0)\xi_0^i, \qquad i = 0, 1, 2, \ldots,$$

where ξ_0 is the unique solution of $f(\xi_0) = \xi_0$ ($0 < \xi_0 < 1$) for

$$f(\xi) = \sum_{k=0}^\infty a_k \xi^k.$$

By the very meaning of a_k we have

$$f'(1) = \sum_{k=0}^\infty k a_k = \frac{\text{expected interarrival time}}{\text{expected service time}} = \frac{1}{\rho}.$$

Therefore the process is positive recurrent if and only if $\rho < 1$.

Waiting Time

Now when $f'(1) > 1$ and the distribution function of the length of line approaches a stationary distribution we now determine under the condition of stationarity the distribution of the waiting time W prior to service.

The probability of not waiting is clearly

$$\pi_0 = 1 - \xi_0.$$

If a customer arrives and finds $n \geq 1$ customers ahead of him, he must wait the sum of n independent identically distributed exponential service times before he can receive service. Such a sum is distributed according to a gamma distribution of order n with scale parameter μ: Thus,

$$\Pr\{W \leq t \,|\, n \text{ ahead}\} = \int_0^t \frac{\mu^n \tau^{n-1} e^{-\mu\tau}}{\Gamma(n)} \, d\tau, \qquad n \geq 1.$$

Therefore, since

$$\Pr\{n \text{ ahead}\} = \pi_n = (1 - \xi_0)\xi_0^n,$$

we have

$$W(t) = \Pr\{W \leq t\} = \sum_{n=1}^{\infty} \Pr\{W \leq t \,|\, n \text{ ahead}\} \Pr\{n \text{ ahead}\} + \pi_0$$

$$= (1 - \xi_0) \int_0^t \sum_{n=1}^{\infty} \frac{\mu^n \tau^{n-1} e^{-\mu\tau}}{\Gamma(n)} \xi_0^n \, d\tau + (1 - \xi_0)$$

$$= (1 - \xi_0) + \xi_0\{1 - \exp[-\mu t(1 - \xi_0)]\}.$$

This distribution is a mixture of an exponential distribution with parameter $\mu(1 - \xi_0)$ and a degenerate distribution whose only possible value is zero, the latter occurring with probability $1 - \xi_0$, which is the probability that a new arrival will not have to wait for service. The conditional distribution function of the length of wait, given that the server is busy, is then

$$\Omega(t) = 1 - \exp[-\mu t(1 - \xi_0)].$$

6: Gamma Arrival Distribution and Generalizations ($E_k/M/1$)

This is a special case of the previous model which can be attacked by a rather elegant trick, possessing quite wide applicability in other contexts. Consider a one-server queueing process with exponential service times with parameter μ

but with interarrival time distributed as a gamma distribution $H(u)$ of order k, the density being

$$
h(u) = \begin{cases} \dfrac{\lambda^k u^{k-1} e^{-\lambda u}}{\Gamma(k)}, & u > 0, \\ \\ 0, & u \leq 0. \end{cases}
$$

The distribution function $H(u)$ can be regarded as the distribution of the sum of k independent random variables all distributed exponentially (λ). (The notation signifies that the parameter of the exponential is λ.) Therefore, we may reduce the problem to a Markov process by considering each arrival as consisting of k stages $,0, 1, \ldots, k - 1$, in each of which the customer waits an exponentially distributed time (λ) before proceeding to the next stage. The physical arrival of a customer in the line corresponds to his reaching the kth stage. There is exactly one person in one of the stages $0, 1, \ldots, k - 1$ at all times, a new person entering stage zero just as his predecessor leaves stage $k - 1$.

The *state* of the system is defined to be the sum of the stages of the people in it. Thus if the system is in state $nk + l, l < k$, it means that n people are actually standing in line or being served and another is in the "lth stage of arriving." As a person completes service the state of the system decreases by k.

We have thus defined a continuous time Markov chain whose infinitesimal generator is

$$
\mathbf{A} = \begin{Vmatrix}
-\lambda & \lambda & 0 & 0 & 0 & 0 & 0 & 0 & 0 & 0 & 0 & \cdots \\
0 & -\lambda & \lambda & 0 & 0 & 0 & 0 & 0 & 0 & 0 & 0 & \cdots \\
0 & 0 & -\lambda & \lambda & 0 & 0 & 0 & 0 & 0 & 0 & 0 & \cdots \\
\vdots & \vdots & \vdots & & & & & & & & & \\
0 & 0 & 0 & \cdots & 0 & 0 & -\lambda & \lambda & 0 & 0 & 0 & \cdots \\
\mu & 0 & 0 & \cdots & 0 & 0 & 0 & -(\mu + \lambda) & \lambda & 0 & 0 & \cdots \\
0 & \mu & 0 & \cdots & 0 & 0 & 0 & 0 & -(\mu + \lambda) & \lambda & 0 & \cdots \\
0 & 0 & \mu & \cdots & 0 & 0 & 0 & 0 & 0 & -(\mu + \lambda) & \lambda & \cdots \\
\vdots & \vdots & \vdots & & \vdots & \vdots & & \vdots & & \vdots & \vdots &
\end{Vmatrix}
$$

The equilibrium properties of this continuous time Markov chain can be determined. We do not enter into this analysis because as mentioned above this case is a special example of G/M/1, which was discussed in Section 5. The advantage in the above formulation lies in the fact that the time-dependent behavior of the process can also be determined exploiting the Markov character of the process. Its discussion is too advanced for this text.

A. GAMMA SERVICE AND GENERAL INPUT†

We may combine the techniques of the past few sections to determine the stationary characteristics of a single-server queueing process with general

† The rest of Section 6 can be skipped on first reading without loss of continuity.

input distribution $H(v)$ and service distribution a gamma of order k and parameter μ.

Consider the service as consisting of k stages, $1, 2, \ldots, k$, in each of which the customer remains a length of time distributed exponentially (μ). Upon completion of the kth stage the customer is finished being served, and leaves.

We may construct an embedded Markov chain whose transitions are effected at each arrival of a customer. Let the state of the chain during an interarrival time be indicated by $kq - p + 1$, where q is the number of people in the system and p is the stage of service of the person being served at the moment of the last arrival. Since $k - p + 1$ is the number of stages (counting the present one) that the person being served has to pass through, the state of the system, which is $k(q - 1) + (k - p + 1) = kq - p + 1$, may be interpreted to be the number of exponential waiting times which all the customers ahead of the new arrival must undergo before the new arrival can begin service. If $q = 0$, the state of the system is defined as zero.

The one-step transition probabilities for the chain fall into several cases. For all i:

(i)   If $j > i + k$, then $P_{ij} = 0$,
(ii)   If $j \leq i + k, j \neq 0$, then $i + k - j$ exponential waiting times have passed in one interarrival time and then

$$P_{ij} = \eta_r = \int_0^\infty \frac{(\mu v)^r e^{-\mu v}}{r!} \, dH(v),$$

where $r = i + k - j$. The derivation of this expression is identical with that of a_k on page 505.

(iii)   Finally, $P_{i,0}$ is the probability that $i + k$ exponential waiting times have a sum denoted by S_{i+k} not exceeding an interarrival time,

$$P_{i,0} = \int_0^\infty \Pr\{S_{i+k} \leq v\} \, dH(v) = \int_0^\infty \int_0^v \frac{\xi^{i+k-1} \mu^{i+k} e^{-\xi \mu}}{(i + k - 1)!} \, d\xi \, dH(v).$$

B. STATIONARY PROBABILITIES

When the traffic intensity

$$\rho = \frac{E[\text{service time}]}{E[\text{interarrival time}]} = \frac{k}{\mu E[\text{interarrival time}]}$$

is less than 1, we expect that the probabilities of being in the various states will approach a limiting distribution. Such a stationary distribution is proportional to a nonnegative convergent series determined by the sequence $\mathbf{x} = (x_0, x_1, \ldots)$ satisfying

$$\mathbf{x} = \mathbf{x} \mathbf{P}. \tag{6.1}$$

By analogy with previous models, a trial solution of the form

$$x_i = \lambda^i$$

for some real number λ suggests itself. The component equation of (6.1) for $j \geq k$ simplifies to

$$\lambda^j = \sum_{i=0}^{\infty} \lambda^i P_{ij} = \sum_{i=j-k}^{\infty} \lambda^i \eta_{i+k-j} = \sum_{r=0}^{\infty} \lambda^{j-k+r} \eta_r = \lambda^{j-k} \sum_{r=0}^{\infty} \lambda^r \eta_r \qquad (6.2)$$

or

$$\lambda^k = F(\lambda),$$

where

$$F(\lambda) = \sum_{r=0}^{\infty} \lambda^r \eta_r = \int_0^{\infty} e^{-\mu v(1-\lambda)} \, dH(v).$$

It can be proved with the aid of Rouche's theorem of complex variables that $\lambda^k - F(\lambda)$ has k roots, counting multiplicities for $|\lambda| < 1$.

Rouche's theorem states in particular that, if $f(z)$ and $g(z)$ are analytic functions in a common domain D and $|f(z)| > |g(z)|$ for z on the boundary of D, then

$$f(z) \qquad \text{and} \qquad f(z) + g(z)$$

have the same number of zeros counting multiplicities. (For a proof of this theorem the reader should consult standard books in complex variables.) We shall now apply Rouche's theorem to the case of $D = \{z; |z| \leq 1 - \delta, \delta > 0\}$ and $f(z) = z^k$, $g(z) = -F(z)$. Indeed for $|z| = 1 - \delta$, we have $|z|^k = (1 - \delta)^k = 1 - k\delta + o(\delta)$. Now for $|z| = 1 - \delta$, $|F(z)| \leq F(1 - \delta)$ (since $F(z)$ is a power series with nonnegative coefficients). But

$$F(1 - \delta) = F(1) - \delta F'(1) + o(\delta)$$

as $\delta \to 0$. Moreover, a direct calculation shows that

$$F'(1) = \mu \int_0^{\infty} v \, dH(v) = \frac{k}{\rho} > k,$$

since we assumed $\rho < 1$. Hence it follows, provided δ is sufficiently small, that

$$|z^k| > |F(z)|, \qquad |z| = 1 - \delta.$$

By virtue of Rouche's theorem, we conclude that z^k and $z^k - F(z)$ have the same number k of zeros in $\{z; |z| \leq 1 - \delta\}$.

If the k roots are $\lambda_1, \lambda_2, \ldots, \lambda_k$, then $\{x_n = \lambda_r^n\}_{n=0}^{\infty}$ will satisfy Eqs. (6.2) for any r ($r = 1, 2, \ldots, k$). We might attempt to find a linear combination

$$\pi_n = \alpha_1 \lambda_1^n + \alpha_2 \lambda_2^n + \cdots + \alpha_k \lambda_k^n, \qquad \sum_{r=1}^{k} \alpha_r = 1,$$

such that the system of equations $\mathbf{x} = \mathbf{x}P$ is completely satisfied. Certainly the equations $x_j = \sum_{i=0}^{\infty} x_i P_{ij}$ $(j = 0, 1, \ldots)$ for $j \geq k$ are satisfied for all choices of α_r since each sequence $\{\lambda_r^n\}$ for $r = 1, \ldots, k$ provides a solution and linear combinations of solutions remain solutions. It remains to determine the constants $\alpha_1, \alpha_2, \ldots, \alpha_k$ so that the first k equations of $\mathbf{x} = \mathbf{x}P$ will also be satisfied. For the case where all the λ_j are distinct, some algebraic manipulations lead to the explicit solution (the normalization $\sum_{n=0}^{\infty} \pi_n = 1$ is employed)

$$\pi_n = \frac{\sum_{i=1}^{k} \alpha_i \lambda_i^n}{\sum_{i=1}^{k} \alpha_i/(1 - \lambda_i)}, \qquad n = 0, 1, \ldots$$

where

$$\alpha_i = \prod_{j \neq i, j=1}^{k} \left(\frac{\lambda_i}{\lambda_i - \lambda_j} \right).$$

The details of the algebra and the modifications necessary when roots coincide are tedious and will not be given.

C. WAITING TIMES

We pointed out above that the state of the system as defined is the number of exponential (μ) waiting times the new arrival has to wait. Therefore, if the state of the system is $n > 0$ the waiting time distribution of the person who just arrived is gamma of order n with parameter μ. If $n = 0$, he does not wait. Therefore

$$W(\xi) = \Pr\{W \leq \xi\} = \int_0^{\xi} \sum_{j=1}^{\infty} \frac{\mu^j w^{j-1} e^{-\mu w}}{(j-1)!} \pi_j \, dw + \pi_0.$$

For the case of distinct roots we may substitute the expression of π_j derived above and obtain

$$W(\xi) = \frac{1}{\sum_{i=1}^{k} \alpha_i/(1 - \lambda_i)} \left[\int_0^{\xi} \sum_{j=1}^{\infty} \frac{\mu^j w^{j-1} e^{-\mu w}}{(j-1)!} \sum_{i=1}^{k} \alpha_i \lambda_i^j \, dw + 1 \right]$$

$$= \frac{1}{\sum_{i=1}^{k} \alpha_i/(1 - \lambda_i)} \left\{ 1 + \int_0^{\xi} \sum_{i=1}^{k} \alpha_i \mu \lambda_i \exp[-\mu w(1 - \lambda_i)] \, dw \right\}$$

$$= \frac{1}{\sum_{i=1}^{k} \alpha_i/(1 - \lambda_i)} \left(1 + \sum_{i=1}^{k} \frac{\alpha_i \lambda_i}{(1 - \lambda_i)} \{1 - \exp[-\mu \xi(1 - \lambda_i)]\} \right)$$

$$= 1 - \frac{\sum_{i=1}^{k} [\alpha_i \lambda_i/(1 - \lambda_i)] \exp[-\mu \xi(1 - \lambda_i)]}{\sum_{i=1}^{k} \alpha_i/(1 - \lambda_i)}.$$

7: *Exponential Service with s Servers* (GI/M/s)

We indicate the generalization of the preceding techniques in dealing with the case of one server by considering an s-server queueing problem whose input distribution function is $H(v)$ and for which the distribution of service times is exponential with parameter μ. (We assume that the service time distributions for all the s servers are identical.)

As before, the process is not Markov, but we may investigate an embedded Markov chain. Let the transitions of the chain be effected at the instants of arrivals of new customers, and let q, the state of the system, be the number of customers, waiting and being served, that the last customer encounters on arrival.

We may calculate P_{ij} as follows:

(i) If $j > i + 1$, $P_{ij} = 0$ for all $i = 0, 1, 2, \ldots$

(ii) If $j \leq i + 1 \leq s$, everyone is being served, and there are exactly $i - j + 1$ departures during the interarrival period, where the probability of any given person departing by time t is $1 - e^{-\mu t}$. Thus

$$P_{ij} = \int_0^\infty \Pr\{i + 1 - j \text{ depart in time } t \,|\, i + 1 \text{ present originally}\} \, dH(t)$$

$$= \int_0^\infty \binom{i + 1}{j} (1 - e^{-\mu t})^{i + 1 - j} e^{-\mu t j} \, dH(t). \tag{7.1}$$

The integrand is the binomial distribution corresponding to $i + 1 - j$ successes (completions of service) in an interarrival epoch.

(iii) If $i + 1 \geq j \geq s$, and $i \geq s$, all servers are busy throughout the interarrival period. Therefore

$$P_{ij} = \Pr\{i + 1 - j \text{ depart}\} = \int_0^\infty \Pr\{i + 1 - j \text{ depart in time } t\} \, dH(t)$$

$$= \int_0^\infty \frac{e^{-\mu s t}(\mu s t)^{i + 1 - j}}{(i + 1 - j)!} \, dH(t).$$

(The derivation of the last identity is the same as that of (5.2). The distribution of the time of service completion now is exponential with parameter $s\mu$ since there are s busy servers in the present case.)

(iv) If $i + 1 \geq s > j$, there will be $m = i - s + 1$ customers waiting as well as s being served at the beginning of the interarrival period; but $n = s - j$ servers idle at the end. Let v denote the time until none are waiting, i.e., the time for m people to be served while all s servers are working. Each service time is distributed exponentially with parameter $s\mu$, so v is distributed as a gamma of order m with parameter $s\mu$. Suppose the duration of service for the m customers with all servers busy is v, while the remaining n customers are served in a period of length $u - v$, where u denotes the next interarrival time. Then

$$
\begin{aligned}
P_{ij} &= \int_0^\infty \Pr\{m + n \text{ people served in time } u\}\, dH(u) \\
&= \int_0^\infty \left[\int_0^\infty \Pr\{n \text{ people served in time } u - v\} \frac{e^{-s\mu v}(s\mu)^m v^{m-1}}{(m-1)!}\, dv \right] dH(u) \\
&= \frac{(s\mu)^m}{(m-1)!} \int_0^\infty \int_0^u v^{m-1} e^{-s\mu v} \binom{s}{n} e^{-\mu(n-v)(s-n)}(1 - e^{-\mu(u-v)})^n\, dv\, dH(u),
\end{aligned}
$$

where the last identity follows by the binomial distribution as in (7.1).

A. STATIONARY PROBABILITIES

We expect that if the traffic intensity is < 1, i.e.,

$$
\begin{aligned}
\rho &= \frac{E[\text{service time per customer when all servers are occupied}]}{E[\text{interarrival time}]} \\
&= \frac{1}{s\mu E[\text{interarrival time}]} < 1,
\end{aligned}
$$

then after a long time the probabilities of being in each state should stabilize. We look for a positive vector $\mathbf{x} = (x_0, x_1, x_2, \ldots)$ satisfying $\sum x_i < \infty$ and $\mathbf{x} = \mathbf{x}\mathbf{P}$. By comparison with the special case for one server, which was discussed earlier, we are led to consider a possible solution of the form

$$
\mathbf{x} = (\beta_0, \beta_1, \ldots, \beta_{s-2}, 1, \alpha, \alpha^2, \ldots).
$$

The jth component equation ($j > s$) of $\mathbf{x} = \mathbf{x}\mathbf{P}$ is

$$
x_j = \alpha^{j-s+1} = \sum_{i=0}^\infty x_i P_{ij} = \sum_{i=j-1}^\infty x_i P_{ij} = \sum_{i=j-1}^\infty \alpha^{i-s+1} P_{ij}
$$

$$
= \alpha^{j-s} \int_0^\infty e^{-\mu s u(1-\alpha)}\, dH(u).
$$

This equation is of the form $\alpha = F(\alpha)$, where

$$
F(\alpha) = \int_0^\infty e^{-\mu s u(1-\alpha)}\, dH(u)
$$

is a convex increasing function in $(0, 1)$, with $F(0) > 0$, and $F(1) = 1$. The convexity of F can be verified by differentiating twice. Therefore, there is a solution α in $(0, 1)$ if and only if $F'(1) > 1$. Since

$$
F'(1) = \mu s \int_0^\infty u\, dH(u),
$$

this is just the criterion $\rho < 1$.

Having found the solution α, we may find the remaining components, $\beta_{s-2}, \beta_{s-3}, \ldots, \beta_0$, from the recursion relations

$$\beta_j = \sum_{i=j-1}^{s-2} \beta_i P_{ij} + \sum_{i=s-1}^{\infty} \alpha^{i-s+1} P_{ij}, \qquad j = 0, 1, \ldots, s-2,$$

or

$$\beta_{j-1} = \frac{\beta_j - \sum_{i=j}^{s-2} \beta_i P_{ij} - \sum_{i=s-1}^{\infty} \alpha^{i-s+1} P_{ij}}{P_{j-1,j}}, \qquad j = 1, 2, \ldots, s-1$$

starting from $\beta_{s-1} = 1$. Normalizing, we have the final probabilities

$$\pi_j = \frac{x_j}{(1-\alpha)^{-1} + \sum_{i=0}^{s-2} \beta_i}.$$

B. WAITING TIMES UNDER CONDITIONS OF STATIONARITY

The probability that an arrival does not have to wait for service is the probability that at the arrival instant at least one server is free, which is

$$W(0) = \Pr\{q \le s-1\} = \sum_{i=0}^{s-1} \pi_i = \frac{1 + \sum_{i=0}^{s-2} \beta_i}{(1-\alpha)^{-1} + \sum_{i=0}^{s-2} \beta_i} = A \sum_{i=0}^{s-1} \beta_i,$$

where

$$A = \frac{1}{(1-\alpha)^{-1} + \sum_{i=0}^{s-2} \beta_i}, \qquad \beta_{s-1} = 1.$$

If the state of the system is $n \ge s$, the new arrival has to wait until $n - s + 1$ customers are served before he can be served. But since there are s servers working, the waiting time between completions of service is exponential (μs). Thus his waiting time is a gamma distribution of order $n - s + 1$ with scale parameter μs and

$$\Pr\{W \le \xi\} = W(\xi) = W(0) + A \int_0^\xi \sum_{n=s}^\infty \frac{(\mu s)^{n-s+1} w^{n-s} e^{-\mu s w}}{(n-s)!} \alpha^{n-s+1} \, dw$$

$$= A \left[\sum_{i=0}^{s-1} \beta_i + \int_0^\xi \alpha \mu s e^{-\mu s w(1-\alpha)} \, dw \right]$$

$$= 1 - \frac{\alpha}{1 + (1-\alpha) \sum_{i=0}^{s-2} \beta_i} e^{-\mu s \xi(1-\alpha)}.$$

8: *The Virtual Waiting Time and the Busy Period*

This section is devoted to a different approach to the problem of waiting times in the simple single-server queueing process with Poisson input and general service (M/GI/1). We shall illustrate the point of view by developing some

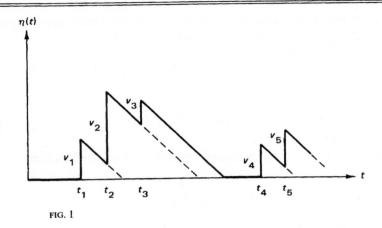

FIG. 1

results concerning the busy period of this process, utilizing for this purpose the discussion of Section 3, Chapter 13.

Since the queueing process as a whole is not Markov, our previous technique was to consider an embedded Markov chain and analyze the waiting times of customers in terms of the embedded chain. However, if we consider $\eta(t)$ [$\eta(t)$ is sometimes interpreted as the *virtual waiting time*], the time a customer would have waited had he arrived at the instant t, then $\eta(t)$ determines a continuous time Markov process. For if t_n and v_n are the instant of arrival and service time of the nth customer, then for $t_n < t < t_{n+1}$, we have

$$\eta(t) = [\eta(t_n+) - (t - t_n)]_+ \dagger$$

and

$$\eta(t_n+) = \eta(t_n-) + v_n,$$
$$[\eta(t+) = \lim_{\varepsilon \downarrow 0} \eta(t + \varepsilon), \text{ and } \eta(t-) = \lim_{\varepsilon \downarrow 0} \eta(t - \varepsilon)].$$

To fix the notation, the input distribution is assumed to be exponential with parameter λ and the service distribution is generally given by $H(v)$. Typically, $\eta(t)$ has the appearance of Fig. 1. It is clear that the future behavior of $\eta(t)$ does not depend on anything previous to its present value. Indeed, the values t_i being the successive events of a Poisson process, it follows that the time until the next arrival is independent of when the last arrival took place.

Another interpretation for $\eta(t)$ is that this is the time needed to complete service of all those customers who are present in the system at time t. The actual waiting time of the nth customer is $\eta(t_n) = \eta(t_n-)$.

† The symbol $[x]_+$ is defined to be

$$[x]_+ = \begin{cases} x & \text{if } x \geq 0, \\ 0 & \text{if } x < 0. \end{cases}$$

Now if

$$F(t, x) = \Pr\{\eta(t) \leq x\},$$

it is possible to derive a differential–integral equation satisfied by $F(t, x)$. This equation can be analyzed to study the properties of $F(t, x)$. We proved in Section 3 that the waiting time of the nth person,

$$F_n(x) = F(t_n, x),$$

converges to a limiting distribution as $n \to \infty$. By use of appropriate renewal theorems the convergence of $F(t, x)$ can also be proved and shown to possess the same limiting distribution as $F_n(x)$. The development of this proposition is beyond the scope of this book. (See references at the close of the chapter.)

The remainder of this chapter is devoted to a discussion of various random variables of interest connected with the process M/GI/1.

Notice that if $\eta(t) > 0$ then the server is busy at time t and if $\eta(t) = 0$ then the server is idle at time t. Let

$$P_0(t) = \Pr\{\eta(t) = 0\},$$

i.e., $P_0(t)$ is the probability that the server is idle at time t.

The *busy period* is defined as the time interval during which the server is continuously busy. If $\eta(0) > 0$, i.e., the server is busy at time $t = 0$, then there is an initial busy period which ends when $\eta(t)$ vanishes for the first time. Denote by $\hat{G}(x)$ the probability that the length of the initial busy period is $\leq x$. Following the initial busy period (if any) idle periods and busy periods alternate. The lengths of the busy periods following the initial busy period are identically distributed, mutually independent, random variables, since each subsequent busy period commences under identical conditions. Denote by $G(x)$ the probability that the length of a busy period other than the initial one is $\leq x$. The idle periods are also identically distributed, mutually independent, random variables whose distribution function is exponential with parameter λ.

The first principal task is the proof of Theorem 8.1 below. It depends on a result of Section 2, Chapter 13, which we record here for convenience as Lemma 8.1.

Lemma 8.1. *Let* $\chi_1, \chi_2, \ldots, \chi_n$ *be nonnegative, exchangeable, random variables with sum* $\chi_1 + \chi_2 + \cdots + \chi_n = y$. *Let* $\tau_1, \tau_2, \ldots, \tau_n$ *be the coordinates, arranged in increasing order, of n points distributed uniformly and independently of each other in the interval* $(0, t)$. *If* $\{\chi_k\}$ *and* $\{\tau_k\}$ *are independent sequences, then*

$$\Pr\{\chi_1 + \cdots + \chi_k \leq \tau_k \text{ for } k = 1, 2, \ldots, n\} = \begin{cases} 1 - y/t & \text{if } 0 \leq y \leq t, \\ 0 & \text{if } y > t. \end{cases}$$

$$(8.1)$$

The proof of Lemma 8.1 and other of its applications are given in Chapter 13, Section 2. We need one more lemma.

Lemma 8.2. *Let* $\chi_1, \chi_2, \ldots, \chi_n$ *be nonnegative, exchangeable, random variables with sum* $\chi_1 + \chi_2 + \cdots + \chi_n = t$ *and let* $\tau_1, \tau_2, \ldots, \tau_{n-1}$ *be the coordinates, arranged in increasing order, of* $n - 1$ *points distributed uniformly and independently of each other in the interval* $(0, t)$. *If* $\{\chi_k\}$ *and* $\{\tau_k\}$ *are independent sequences then*

$$\Pr\{\chi_1 + \cdots + \chi_k \le \tau_k \text{ for } k = 1, 2, \ldots, n - 1\} = \frac{1}{n}. \tag{8.2}$$

Proof. By Lemma 8.1 we have

$$\Pr\{\chi_1 + \cdots + \chi_k \le \tau_k \text{ for } k = 1, 2, \ldots, n - 1 | \chi_1 + \cdots + \chi_{n-1} = y\}$$

$$= \begin{cases} 1 - \dfrac{y}{t} & \text{if } 0 \le y \le t, \\ 0, & y \ge t. \end{cases}$$

Now by the law of total probabilities

$$\Pr\{\chi_1 + \cdots + \chi_k \le \tau_k, k = 1, 2, \ldots, n - 1\}$$

$$= \int_0^t \Pr\{\chi_1 + \cdots + \chi_k \le \tau_k, k = 1, 2, \ldots, n - 1 | \chi_1 + \cdots + \chi_{n-1} = y\}$$

$$\times d \Pr\{\chi_1 + \cdots + \chi_{n-1} \le y\}$$

$$= \int_0^t \left(1 - \frac{y}{t}\right) d \Pr\{\chi_1 + \cdots + \chi_{n-1} \le y\} = 1 - \frac{1}{t} E[\chi_1 + \cdots + \chi_{n-1}]$$

$$= 1 - \frac{1}{t} \left(\frac{n - 1}{n} t\right),$$

since $\chi_1, \ldots, \chi_n$ are exchangeable and their sum is t. This obviously reduces to

$$\Pr\{\chi_1 + \cdots + \chi_k \le \tau_k \text{ for } k = 1, 2, \ldots, n - 1\} = \frac{1}{n}. \qquad \blacksquare$$

Theorem 8.1. *If* $\eta(0) = c$ *(constant), then the probability that the initial busy period has duration* $\le x$ *is given by*

$$\hat{G}(x) = \sum_{n=0}^{\infty} \frac{c\lambda^n}{n!} \int_0^{x-c} e^{-\lambda(c+y)}(c + y)^{n-1} dH_n(y) \qquad \text{if } x \ge c,$$

$$\hat{G}(x) = 0 \qquad \text{if } x < c, \tag{8.3}$$

where $H_n(y)$ *denotes the n-fold convolution of* $H(y)$, *with the convention that* $H_0(y)$ *is the distribution with unit jump at* 0.

Proof. We make the calculation by conditioning in terms of the number of arrivals. The number of arrivals during the initial busy period may be $n = 0, 1, 2, \ldots$. If $n = 0$ then the initial busy period has length c and the probability that no customer arrives in the time interval $(0, c)$ is $e^{-\lambda c}$. This contributes the term for $n = 0$ in (8.3). If $n \geq 1$ then denote by $\tau_1, \tau_2, \ldots, \tau_n$ the arrival times and by $\chi_1, \chi_2, \ldots, \chi_n$ the service times of these customers. They must satisfy the conditions

$$\tau_j \leq \chi_1 + \cdots + \chi_{j-1} + c \qquad \text{for} \quad j = 1, 2, \ldots, n, \tag{8.4}$$

where the empty sum is equal to zero. In fact, relation (8.4) asserts that the cumulative service times for the $j - 1$ customers that arrive after time 0 plus the cumulative service times of those customers waiting at time 0 exceeds the time of arrival of the jth customer, $j = 1, 2, \ldots, n$. This condition plainly assures that the server stays busy at least until completing service for the nth arrival. Of course, $\Pr\{\chi_1 + \chi_2 + \cdots + \chi_n \leq x\}$ is $H_n(x)$.

If $\chi_1 + \cdots + \chi_n = y$, then the length of the initial busy period is $c + y$ and the probability that exactly n customers arrive during the time interval $(0, c + y)$ is $e^{-\lambda(c+y)}[\lambda(c + y)]^n/n!$. The arrival instants can be considered as the coordinates, arranged in increasing order, of n points distributed uniformly and independently of each other in the interval $(0, c + y)$ (see page 126, Chapter 4). Further $\chi_1, \chi_2, \ldots, \chi_n$ are nonnegative, exchangeable, random variables.

Now subtracting the inequalities (8.4) from $y + c$, we obtain the equivalent relations

$$y + c - \tau_j \geq y - \chi_1 - \chi_2 - \cdots - \chi_{j-1}, \qquad j = 1, 2, \ldots, n. \tag{8.5}$$

Let $\tau^*_{n+1-j} = y + c - \tau_j$, and since $\chi_1 + \cdots + \chi_n = y$ we may rewrite (8.5) in the form

$$\tau^*_{n+1-j} \geq \chi_n + \chi_{n-1} + \cdots + \chi_j, \qquad j = 1, 2, \ldots, n. \tag{8.6}$$

But the $\tau^*_{n+1-j}, j = 1, \ldots, n$ are again clearly distributed like the n order statistics of the uniform distribution on $(0, c + y)$. To convince ourselves of this we merely look at the values of τ_j from right to left, scanning the interval from $c + y$ to 0. Symmetry considerations yield the desired conclusions. Moreover, $\chi_n, \chi_{n-1}, \ldots, \chi_{n+1-j}$ are distributed jointly like $\chi_1, \chi_2, \ldots, \chi_j$ because they are exchangeable. Hence the event (8.6) has the same probability as the event

$$\tau_j \geq \chi_1 + \chi_2 + \cdots + \chi_j, \qquad j = 1, 2, \ldots, n. \tag{8.7}$$

Appealing to Lemma 8.1, we conclude that the probability of the event (8.7) is

$$1 - \frac{y}{c + y} = \frac{c}{c + y}.$$

Now putting these facts together with the aid of the law of total probabilities, we have

$$
\hat{G}(x) = \sum_{n=0}^{\infty} \int \Pr\left\{\text{initial busy period} \leq x \,\middle|\, \begin{array}{l} \eta(0) = c, \, n \text{ arrivals in this period} \\ \text{require total service time } y \end{array}\right\}
$$

$$
\times \Pr\left\{n \text{ arrivals in the busy period} \,\middle|\, \begin{array}{l} \eta(0) = c, \text{ total service} \\ \text{time of arrivals is } y \end{array}\right\} dH_n(y)
$$

$$
= \sum_{n=0}^{\infty} \int_0^{x-c} \frac{c}{c+y} e^{-\lambda(c+y)} \frac{(c+y)^n \lambda^n}{n!} \, dH_n(y)
$$

$$
= \sum_{n=0}^{\infty} c \frac{\lambda^n}{n!} \int_0^{x-c} e^{-\lambda(c+y)}(c+y)^{n-1} \, dH_n(y). \quad \blacksquare
$$

Our next theorem yields the distribution function of a busy period other than the initial busy period.

Theorem 8.2. *The probability that a busy period other than the initial period has length $\leq x$ is given by*

$$
G(x) = \sum_{n=1}^{\infty} \frac{\lambda^{n-1}}{n!} \int_0^x e^{-\lambda y} y^{n-1} \, dH_n(y), \qquad x \geq 0. \tag{8.8}
$$

Proof. If we suppose that a busy period consists of n, $n = 1, 2, \ldots$, services, then its length is $\chi_1 + \chi_2 + \cdots + \chi_n$ where $\chi_1, \chi_2, \ldots, \chi_n$ are identically distributed, mutually independent, random variables with the distribution function $\Pr\{\chi_i \leq x\} = H(x)$, $i = 1, 2, \ldots, n$. In this case exactly $n - 1$ customers arrive during this busy period. Measure time from the starting point of the busy period and denote by $\tau_1, \tau_2, \ldots, \tau_{n-1}$ the arrival times. They must satisfy the conditions

$$
\tau_j \leq \chi_1 + \cdots + \chi_j, \qquad j = 1, 2, \ldots, n - 1. \tag{8.9}
$$

If $\chi_1 + \cdots + \chi_n = y$, then the busy period has length y and the arrival instants can be considered as the coordinates, arranged in increasing order, of n points distributed uniformly and independently of each other in the interval $(0, y)$. Further, $\chi_1, \ldots, \chi_n$ are nonnegative, exchangeable random variables. If $\chi_1 + \cdots + \chi_n = y$, then (8.9) has the same probability as

$$
\chi_1 + \cdots + \chi_k \leq \tau_k, \qquad k = 1, 2, \ldots, n - 1, \tag{8.10}
$$

for in (8.9) we can replace χ_j by χ_{n+1-j} and τ_j by $y - \tau_{n-j}, j = 1, 2, \ldots, n - 1$ without altering the probability of the event. By Lemma 8.2 the probability of the event (8.10) or (8.9) is $1/n$. Since $\Pr\{\chi_1 + \cdots + \chi_n \leq y\} = H_n(y)$ and the probability that during the time interval $(0, y)$ exactly $n - 1$ customers arrive is $e^{-\lambda y}(\lambda y)^{n-1}/(n - 1)!$, we can apply the law of total probabilities to obtain

$$G(x) = \sum_{n=1}^{\infty} \frac{1}{n} \int_0^x e^{-\lambda y} \frac{(\lambda y)^{n-1}}{(n - 1)!} \, dH_n(y), \qquad (8.11)$$

which was to be proved. ■

Problems

1. Show for the $M/M/s$ system that the stationary queue size distribution $\{p_n, n = 0, 1, 2, \ldots\}$ is given by

$$p_0 = \left\{ \frac{(s\rho)^s}{s!(1 - \rho)} + \sum_{i=0}^{s-1} \frac{(s\rho)^i}{i!} \right\}^{-1},$$

$$p_n = \begin{cases} p_0 \dfrac{(s\rho)^n}{n!}, & 1 \le n \le s, \\[2mm] p_0 \rho^n \dfrac{s^s}{s!}, & s < n < \infty, \end{cases}$$

where $\rho = \lambda/s\mu < 1$. Let $Q = \max(n - s, 0)$ $(n = 0, 1, 2, \ldots)$ be the size of the queue not including those being served. Show that

(i) $\quad \gamma = \Pr\{Q = 0\} = \dfrac{\sum_{i=0}^{s} (s\rho)^i/i!}{\sum_{i=0}^{s} [(s\rho)^i/i!] + [(s\rho)^s \rho/s!(1 - \rho)]};$

(ii) $\quad E[Q] = (1 - \gamma)/(1 - \rho).$

2. Compare the $M/M/1$ system for a first-come first-served queue discipline with one of last-come first-served type (for example, articles for service are taken from the top of a stack). How do the queue size, waiting time, and busy period distribution differ, if at all?

Answer: Queue size and busy period do not differ but the waiting time distributions differ. Why?

3. Consider the $M/M/1$ system with queue discipline of last-come first-served type. Let $X(t)$ be the queue size at time t. Show that the process $\{X(t), t \ge 0\}$ is a birth and death process and determine its parameters.

Answer: $\lambda_n = \lambda, \mu_n = \mu.$

4. Consider an infinitely many server queue with an exponential service time distribution with parameter μ. Suppose customers arrive in batches with the interarrival time following an exponential distribution with parameter λ. The number of arrivals in each batch is assumed to follow the geometric distribution with parameter $\rho(0 < \rho < 1)$, i.e., $\Pr\{\text{number of arrivals in a batch is } k\} = \rho^{k-1}(1 - \rho)$ $(k = 1, 2, \ldots)$.

Formulate this process as a continuous time Markov chain and determine explicitly the infinitesimal matrix of the process.

Answer: $\mathbf{Q} = \|q_{ij}\|$ where

$$q_{i,\, i-1} = i\mu, \qquad\qquad\qquad i \geq 1,$$

$$q_{i,\, j} = \lambda \rho^{j-i-1}(1 - \rho), \qquad j > i,$$

$$q_{ij} = 0, \qquad\qquad\qquad\qquad j < i - 1,$$

and

$$q_{ii} = -\sum_{j \neq i} q_{ij}.$$

5. (continuation). Determine the probability generating function $\pi(s)$ of the equilibrium distribution of the process.

Answer:

$$\pi(s) = \left[1 + \frac{\rho}{1 - \rho}(1 - s) \right]^{-\lambda/\rho\mu}.$$

6. (queueing with balking). Customers, with independent and identically distributed service time distribution $H(x)$, arrive at a counter in the manner of a Poisson process with parameter λ. A customer who finds the server busy joins the queue with probability $p (0 < p < 1)$. Derive the transition probabilities of the Markov chain embedded at the points of departure of customers. Can you find the limiting distribution of queue size?

Answer:

$$\mathbf{P} = \begin{Vmatrix} p_0 & p_1 & p_2 & \cdots \\ p_0 & p_1 & p_2 & \cdots \\ 0 & p_0 & p_1 & \cdots \\ 0 & 0 & p_0 & \cdots \\ \vdots & \vdots & \vdots & \end{Vmatrix}, \qquad p_j = \int_0^\infty e^{-\lambda p x} \frac{(\lambda p x)^j}{j!} \, dH(x),$$

$$K(s) = \sum_{j=0}^\infty p_j s^j, \qquad \pi(s) = \frac{(1 - \rho)K(s)(s - 1)}{s - K(s)},$$

where

$$\rho = \alpha \lambda p < 1 \qquad \text{and} \qquad \alpha = \int_0^\infty x \, dH(x) < \infty.$$

7. Consider the $M/M/1$ queueing model with balking as in Problem 6. Now we assume that the interarrival distribution is exponential with parameter λ. The service time distribution is exponential with parameter μ. The balking parameter is p as in the example above. Formulate this model as a birth and death stochastic process.

Answer:

$$\lambda_0 = \lambda, \lambda_n = \lambda p \quad \text{for} \quad n \geq 1, \mu_n = \mu.$$

8. The following two birth and death processes (cf. Section 4, Chapter 4) can be viewed as models for queueing with balking.

(a) First consider a birth and death process with parameters

$$\lambda_n = \lambda q^n, \qquad 0 < q < 1, \quad \lambda > 0 \quad (n = 0, 1, 2, \ldots),$$

$$\mu_n = \mu, \qquad\qquad n > 0,$$

$$\mu_0 = 0.$$

(b) Let the parameters be

$$\lambda_n = \frac{\lambda}{n+1}, \qquad \mu_n = \mu \quad (n = 1, 2, \ldots), \qquad \mu_0 = 0.$$

Determine the stationary distribution in each case.

Answer: (a) $p_m = p_0(\lambda/\mu)^m q^{m(m-1)/2}$ for $m \geq 1$. (b) $p_m = p_0(\lambda/\mu)^m(1/m!)$ for $m \geq 0$, whence $p_0 = e^{-\lambda/\mu}$.

9. Consider the problem of pedestrians wishing to cross a one-way road at a given point. Suppose vehicles (of zero length), which have the right of way, pass the point in the manner of a Poisson process with parameter μ. All waiting pedestrians will cross the road whenever a gap of at least T seconds appears for the first time between vehicles in the road. What is (i) the distribution of the waiting time for a pedestrian who arrives at an arbitrary time and (ii) the distribution of the time from the end of one possible pedestrian crossing point to the beginning of the next? Give answers in terms of Laplace transforms. Find the mean wait of a pedestrian.

Answer: Both (i) and (ii) have the same Laplace transform

$$L(s) = [(\mu + s)e^{-\mu T}/(s + \mu e^{-(\mu+s)T})];$$

$$\text{mean wait} = \frac{e^{\mu T} - (1 + \mu T)}{\mu}.$$

10. (*M/G/∞* system). Suppose that at a counter there are infinitely many servers, so that there is no waiting time for customers, but we are interested in the number of busy servers. Customers arrive at times corresponding to a Poisson process with parameter λ. The service times for the customers are independent and identically distributed. Let $H(x)$ denote the service distribution. If there are initially no customers present find:

(i) $P_k(t) = \Pr\{\text{at time } t \text{ there are exactly } k \text{ customers being served}\}$;
(ii) $\lim_{t \to \infty} P_k(t) = P_k$, (Assume $\alpha = \int_0^\infty x \, dH(x) < \infty$.)

Hint: Use the law of total probability and the fact that, given n arrivals by time t, the instance of arrival are distributed like the order statistics of n independent observations from the uniform distribution on $(0, t]$.

Answer:

(i) $P_k(t) = \left(\exp\left\{ -\lambda \int_0^t \left[1 - H(x) \right] dx \right\} \right) \left(\lambda \int_0^t [1 - H(x)] \, dx \right)^k \Big/ k! \,;$

(ii) $P_k = e^{-\lambda \alpha} \dfrac{(\lambda \alpha)^k}{k!}, \qquad \alpha = \displaystyle\int_0^\infty x \, dH(x).$

11. In an $(M/G/\infty)$ queueing system customers arrive in a Poisson process with parameter λ and have identically and independently distributed service times with distribution function $H(x)$. Initially there are no customers being served. Show that the probability of n departures by time t is given by

$$\frac{1}{n!}\left[\lambda \int_0^t H(u) \, du \right]^n \exp\left[-\lambda \int_0^t H(u) \, du \right].$$

12. Consider the queueing process of Problem 11. Show that the probability $\varphi(t, T)$ of no departures in $(t, t + T)$ satisfies the recursion relation

$$\varphi(t, T) = \int_0^t \lambda e^{-\lambda(t-\tau)} [H(\tau) + 1 - H(\tau + T)] \varphi(\tau, T) \, d\tau$$

$$+ \int_t^{t+T} \lambda e^{-\lambda \tau} [1 - H(T + t - \tau)] \varphi(0, T + t - \tau) \, d\tau + e^{-\lambda(t+T)}. \qquad (*)$$

Hint: Examine the possibilities at the instant of the first arrival.

13. Using the result of Problem 12 prove that

$$\varphi(t, T) = \exp\left[-\lambda \int_t^{t+T} H(\xi) \, d\xi \right].$$

Hint: Derive a first-order differential equation (in the variable t) for $\varphi(t, T)$ and solve.

14. (continuation). Let $\varphi_n(t, T)$ denote the probability of n departures in time $(t, t + T)$. Derive an integral equation for φ_n in terms of φ_{n-1} in the spirit of $(*)$. Then show that

$$\varphi_n(t, T) = \frac{1}{n!}\left[\lambda \int_t^{t+T} H(\xi) \, d\xi \right]^n \exp\left[-\lambda \int_t^{t+T} H(\xi) \, d\xi \right].$$

15. Problem 10 was concerned with an infinitely many-server queue with Poisson arrival pattern. Consider the dual system $G/M/\infty$ where the interarrival times are independent and identically distributed with density function $h(x)$, and the service times are independent and exponentially distributed with parameter μ. There are infinitely many servers. Determine the transition probability matrix for the embedded Markov chain whose state variable η_n is the number of busy servers at the times of successive arrivals.

Answer:

$$\Pr\{\eta_{n+1} = j \mid \eta_n = i\} = P_{ij} = \binom{i+1}{j} \int_0^\infty e^{-j\mu x}(1 - e^{-\mu x})^{i+1-j} h(x)\, dx.$$

16. In the $M/G/1$ system let $B_1, B_2, \ldots$ be a sequence of independent and identically distributed random variables whose distribution function $B(x)$ coincides with that of the busy period of the system. Suppose that the first customer of a busy period has service time X (with distribution function $H(x) = \Pr\{X \le x\}$ and that n other customers arrive during his service period. Show that

$$B(x) = \Pr\{X + B_1 + B_2 + \cdots + B_n \le x\}.$$

From this establish that the Laplace transform

$$\tilde{B}(\theta) = \int_0^\infty e^{-\theta x}\, dB(x)$$

satisfies the functional equation

$$\tilde{B}(\theta) = \psi(\theta + \lambda(1 - \tilde{B}(\theta))),$$

where

$$\psi(\theta) = \int_0^\infty e^{-\theta x}\, dH(x).$$

Use this result to find the mean duration of a busy period.

Hint: The busy period is independent of the mode of service. Suppose (as $n = 0$ is trivial) that $n > 0$ customers arrive during the initial service. With the first new arrival, begin another busy period; after completing that busy period go back to the second arrival of first service period and start another busy period with this customer; repeat n times.

Answer:

$$\text{Mean length of busy period} = \frac{\alpha}{1 - \lambda\alpha}, \qquad \alpha = \int_0^\infty x\, dH(x).$$

17. Under the same conditions as in Problem 3 with $\lambda < \mu$, consider a customer just arriving and compute the probability that exactly n other customers are served during his waiting time, given that he does not find the server free on his arrival.

Hint: Imitate the method of Problem 16 to show that the probability generating function $g(s)$ of the number of customers served during a busy period satisfies the functional equation $g(s) = \mu s[\mu + \lambda - \lambda g(s)]^{-1}$.

Answer:

$$g(s) = \sum_{r=1}^\infty \binom{\frac{1}{2}}{r}(-1)^{r+1}\left(\frac{4\lambda\mu}{(\lambda+\mu)^2}\right)^r\left(\frac{\lambda+\mu}{2\lambda}\right)s^r.$$

18. Consider a queueing process where customers are arriving regularly at times n/λ, $n = 0, 1, 2, \ldots$. Assume that the service time X_j for the jth customer is exponentially distributed with expectation $1/\mu$. Assume $\lambda > \mu$. Determine the probability that the server stays busy for an infinite length of time when there is one customer in the queue at time 0.

Hint: Show that the desired probability is

$$\Pr\{X_1 + X_2 + \cdots + X_i \geq i \text{ for all } i = 1, 2, \ldots\},$$

where X_i are independent r.v.'s with exponential distribution with parameter λ/μ (use Lemma 8.1 in the appropriate manner).

Answer: $1 - \mu/\lambda$.

19.   (preemptive priority queueing).   Consider a single-server queueing process that has two classes of customers (priority and nonpriority) with independent Poisson arrival rates with parameters λ_1 and $\lambda_2(\lambda_1 + \lambda_2 = 1)$ and with service times independent and exponentially distributed with parameters μ_1 and μ_2, respectively. Within classes there is a first-come first-served queue discipline and the service of priority customers is never interrupted. If a priority customer arrives during the service of a nonpriority customer, then the latter's service is immediately stopped in favor of the priority customer. The interrupted customer's service is resumed when there are no priority customers present. Let $p_{m,n}$ be the equilibrium probability that there are m priority and n nonpriority customers in the system. Equilibrium is achieved when $\rho_1 + \rho_2 < 1$ ($\rho_1 = \lambda_1/\mu_1, \rho_2 = \lambda_2/\mu_2$). Establish that the $p_{m,n}$ satisfy the system of equations

$$[\lambda_1 + \lambda_2 + \mu_1(1 - \delta_{m0}) + \mu_2(1 - \delta_{n0})\delta_{m0}]p_{m,n}$$

$$= \lambda_1 p_{m-1,n} + \lambda_2 p_{m,n-1} + \mu_1 p_{m+1,n} + \mu_2 \delta_{m0} p_{m,n+1} \qquad (m, n = 0, 1, 2, \ldots),$$

where δ_{ij} is the Kronecker delta, and where it is understood that any p with a negative suffix is zero. Using this equation show that the mean number of nonpriority customers is

$$\sum_{M=0}^{\infty} \sum_{n=0}^{\infty} n p_{m,n} = \frac{\rho_2}{1 - \rho_1 - \rho_2} \left[1 + \frac{\mu_2 \rho_1}{\mu_1(1 - \rho_1)} \right].$$

20.   Show for the $M/M/1$ queueing process in a stationary state that the distribution of time between successive departures has the same (exponential) distribution as the interarrival time distribution (see also Problem 10, Chapter 4).

21.   Customers arrive in a queue with a general independent interarrival time distribution. Examine the structure of the queue, specifically at regeneration points, of the following two systems: (i) there are s servers and the same exponential service time distribution for each server; (ii) there is a single server and an Erlangian service time distribution.

22.   Consider the following generalization of the GI/G/1 queueing system with interarrival distribution function $A(t)$ and service time distribution function $B(t)$, with finite means a and b, respectively: A customer who arrives to find the server idle waits a time with distribution function $V(t)$ and finite mean v before commencing service.
    If $F_n(x)$ is the distribution function of the waiting time for the nth arrival show that the limit $F(x) = \lim_{n \to \infty} F_n(x)$ exists and
    (i)   if $b - a > 0$, then $F(x) \equiv 0$;
    (ii)   if $b - a < 0$, then $F(x)$ is a distribution function.

23. We extend the idea of Problem 9 to a junction of two one-way one-lane roads A and B with traffic in A having absolute right of way; there is a stop sign in road B. As before, vehicles in A pass the junction in the manner of a Poisson process with parameter μ, and vehicles in B arrive at the junction in the manner of a Poisson process with parameter λ and queue up, waiting to enter the junction. When a road B vehicle reaches the head of the queue it waits until the first gap between A vehicles of length at least T appears in the A traffic, and it then takes time T to enter the junction. Find the probability generating function of the distribution of the number of B vehicles in the queue when the system is in a stationary state, and find the expected stationary queue size.

Hint: We have an example of the $M/G/1$ queueing system, and it is sufficient to find the "service time" distribution of road B vehicles.

Answer:

$$\pi(s) = \frac{(1 - \rho)(s - 1)K(s)}{s - K(s)}, \qquad K(s) = \tilde{B}(\lambda - \lambda s),$$

$$\tilde{B}(\theta) = \int_0^\infty e^{-\theta x}\, dH(x) = \frac{(\mu + \theta)e^{-(\mu + \theta)T}}{\theta + \mu e^{-(\mu + \theta)T}}.$$

NOTES

The literature of queueing theory is voluminous. An elegant monograph reviewing this theory and its applications is that of Cox and Smith [1].

We also direct the student to the advanced books by Takács [2] and Riordan [3].

A compendium of results on queueing theory is contained in Saaty [4]. This reference also includes an extensive bibliography.

Applications to congestion theory and telephone trunking problems can be found in Siski [5].

Some special mathematical aspects of queueing theory are developed in the monograph by Beneš [6].

An updated treatment of queueing processes of wide scope in applications is the recent book of Kleinrock [7]. The intriguing topic of network queues with its implications is amply discussed therein.

REFERENCES

1. D. R. Cox and W. L. Smith, "Queues," Methuen, London, 1961.
2. L. Takács, "Introduction to the Theory of Queues." Oxford Univ. Press, London and New York, 1962.
3. J. Riordan, "Stochastic Service Systems." Wiley, New York, 1962.
4. T. L. Saaty, "Elements of Queueing Theory with Applications." McGraw-Hill, New York, 1961.

5. E. Syski, "Congestion Theory." Wiley, New York, 1960.
6. V. E. Beneš, "General Stochastic Processes in the Theory of Queues." Addison-Wesley, Reading, Massachusetts, 1963.
7. L. Kleinrock, "Queueing Systems," Vol. I: "Theory," Vol. II: "Computer Applications." Wiley (Interscience), New York, 1976.

MISCELLANEOUS PROBLEMS

Several of the following problems, as well as some from preceding chapters, are based on research papers of the recent literature. As noted in the preface, we regretfully cannot present the many relevant references, confining ourselves to appropriate books cited at the end of each chapter.

A star-designated problem is more difficult, although accessible by the methods we have presented.

---

1. Consider standard Brownian motion in n dimensions starting at the origin. Let S be the first passage time to the surface of the unit sphere and let $X(S)$ be the position at that moment. Establish the "intuitive" proposition that S and $X(S)$ are independent random variables.

2. For Brownian motion in three dimensions show that the occupation time of the unit ball has the same probability distribution as that of the first passage time out of the unit circle for Brownian motion in one dimensions.

*3. Consider a standard Brownian motion $\{B(t)\}$ and for a time point t define

$$\beta_t = \sup\{s < t; B(s) = 0\} = \text{the last zero prior to } t;$$

$$\gamma_t = \inf\{s > t; B(s) = 0\} = \text{the first zero after } t.$$

It is well known (see Chapter 7) that $\Pr\{0 < \beta_t < t < \gamma_t < \infty\} = 1$. The interval (β_t, γ_t) is called the excursion interval straddling t and $\{|B(s)|; \beta_t < s < \gamma_t\}$ is called the excursion process. For $a \geq 0$ and $\varepsilon > 0$ define

$$S(t, a, \varepsilon) = \int_{\beta_t}^{\gamma_t} \mathbf{1}_{[a, a+\varepsilon)}(|B(s)|) \, ds,$$

which is the amount of time the excursion process spends in $[a, a + \varepsilon)$. Show that $\lim_{\varepsilon \downarrow 0} \varepsilon^{-1} S(t, a, \varepsilon) = S(t, a)$ exists, where $S(t, a)$ is called the local time of the excursion process for the level a. [Show that the moments of $\varepsilon^{-1} S(t, a, \varepsilon)$ converge as $\varepsilon \downarrow 0$]. Prove that

$$\lim_{a \downarrow 0} E[e^{-\lambda S(t, a)/a}] = (1 + 2\lambda)^{-2}.$$

4. Consider a regular diffusion $\{X(t)\}$ on $I = (l, r)$ and let $T(a)$ be the hitting time to a state a, and let $T(a, b) = T(a) \wedge T(b)$ be the hitting time to a or b. For $l < a < x < b < r$ *define the Laplace transforms*

$$\varphi_a(x) = E[e^{-\lambda T(a)} | X(0) = x], \qquad \varphi_{ab}(x) = E[e^{-\lambda T(a, b)} | X(0) = x],$$

etc. Establish the following relation between the Laplace transform $\varphi_{ab}(x)$ of the two-point hitting time and the Laplace transforms for several one-point hitting times:

$$\varphi_{ab}(x) = \frac{\varphi_a(x)[\varphi_b(a) - 1] + \varphi_b(x)[\varphi_a(b) - 1]}{\varphi_a(b)\varphi_b(a) - 1}.$$

Hint: Let $A = E[e^{-\lambda T(a)} I_a | X(0) = x]$ and $B = E[e^{-\lambda T(b)} I_b | X(0) = x]$ where $I_a = 1 - I_b$ is unity if $T_a < T_b$ and zero if $T_b < T_a$. Show that $\varphi_a(x) = A + B\varphi_a(b)$ and $\varphi_b(x) = A\varphi_b(a) + B$. Solve for A and B and then $\varphi_{ab}(x) = A + B$.

5. Consider a regular diffusion $\{X(t)\}$, let T_a be the hitting time to a state a, let $\tau_a(x) = E[T_a | X(0) = x]$, and let $\pi_a(x) = \Pr\{T_a < T_b | X(0) = x\}$. Establish the formula

$$\pi_a(x) = \frac{\tau_a(b) + \tau_b(x) - \tau_a(x)}{\tau_a(b) + \tau_b(a)}.$$

Hint:

$$\tau_a(x) = E[\min\{T_a, T_b\} | X(0) = x] + \pi_b(x)\tau_a(b)$$

and

$$\tau_b(x) = E[\min\{T_a, T_b\} | X(0) = x] + \pi_a(x)\tau_b(a).$$

Subtract and use $\pi_b(x) = 1 - \pi_a(x)$.

6. Let T be exponentially distributed with parameter $\rho > 0$ and consider *the observation process*

$$Y(t) = \begin{cases} B(t) & \text{for} \quad 0 \le t < T, \\ (t - T) + B(t) & \text{for} \quad T \le t, \end{cases}$$

where $B(t)$ is standard Brownian motion. In a quality control model, T represents the unobservable time that an undesirable disturbance enters a system causing a displacement in the infinitesimal drift from 0 to 1. It is desired to detect the disturbance as quickly as possible after it occurs. Introduce the posterior probability that the disturbance has occurred given the observation process,

$$X(t) = \Pr\{T \le t | Y(s), 0 \le s \le t\}.$$

Formally show that $\{X(t)\}$ is a diffusion process on $[0, 1]$ having infinitesimal coefficients

$$\mu(x) = (1 - x)[\rho - x(1 - x)], \qquad \sigma^2(x) = x^2(1 - x)^2.$$

Hint: Conditioned on $X(t) = x$, the observation increment $\Delta Y = Y(t + \Delta t) - Y(t)$ is normally distributed with mean Δt and variance Δt with probability x, and is normally distributed with mean zero and variance Δt with probability $1 - x$. The conditional density for ΔY is

$$f_x(\Delta y) = \frac{1}{\sqrt{2\pi(\Delta t)}} \{x e^{-(1/2)[(\Delta y) - (\Delta t)]^2/\Delta t} + (1 - x) e^{-(1/2)(\Delta y)^2/\Delta t}\}.$$

From Bayes rule, $g(x, \Delta y) = \Pr\{T < t \mid X(t) = x, \Delta Y = \Delta y\}$ is given by $g(x, \Delta y) = xf_1(\Delta y)/f_x(\Delta y)$ and the $X(t + \Delta t) = q(x, \Delta y) + [1 - q(x, \Delta y)]\rho \, \Delta t + O(\Delta t)^2$. Now evaluate $E[\Delta X]$ and $E[(\Delta X)^2]$ carrying terms up to $O(\Delta t)^2$.

7 (*continuation*). Suppose that action is taken with respect to the process whenever $X(t) \geq \xi^*$ for some critical value ξ^*. A "false alarm" occurs when action is taken where no disturbance has occurred. Give a conditional probability argument that $\Pr\{\text{false alarm}\} \equiv \xi^*$ under this control rule.

8 (*continuation*). The likelihood ratio is the statistic $Z(t) = X(t)/[1 - X(t)]$. Show that $Z(t)$ is a diffusion process for which

$$\mu_Z(z) = (1 + z)\rho + (z^2 - z)/(1 + z), \qquad \sigma_Z^2(z) = z^2.$$

9 (*continuation*). Since the situation where disturbances are rare is of interest, we consider the case $\rho \to 0$, first forming the process $W_\rho(t) = Z(t)/\rho$. Show formally that $W(t) = \lim_{\rho \to 0} W_\rho(t)$ is a diffusion process on $I = [0, \infty)$ with infinitesimal coefficients

$$\mu_W(w) = 1 - w, \qquad \sigma_W^2(w) = w^2.$$

*10. Consider the recursion formula

$$X_{n+1}^{(N)} - X_n^{(N)} = \frac{1}{N} f(X_n^{(N)}) + \frac{1}{\sqrt{N}} g[X_n^{(N)} + \lambda(X_{n+1}^{(N)} - X_n^{(N)})]\eta_n^{(N)}$$

where $X_0^{(N)} = 0$ and $\eta_i^{(N)}$ are independent and identically distributed with mean 0 and variance 1. Show informally that $X_{[Nt]}^{(N)} \to X(t)$ where $\{X(t)\}$ is a diffusion with infinitesimal parameters $\mu(x) = f(x) + \lambda g(x)g'(x)$ and $\sigma^2(x) = [g(x)]^2$.

11. Let g be a bounded and continuous function on $(0, \infty)$ for which

$$\lim_{T \to \infty} \frac{1}{T} \int_0^T g(s) \, ds = A \neq 0.$$

Let $Y(t)$ be a stationary Ornstein–Uhlenbeck process and for each $\varepsilon > 0$ form the weighted integral

$$B_\varepsilon(t) = \frac{1}{\varepsilon} \int_0^t g\left(\frac{s}{\varepsilon^2}\right) Y\left(\frac{s}{\varepsilon^2}\right) ds$$

$$= \varepsilon \int_0^{t/\varepsilon^2} g(v) Y(v) \, dv.$$

Show that $B_\varepsilon(t)$ is a zero mean Gaussian process and that $B_\varepsilon(t)$ converges to standard Brownian motion as $\varepsilon \downarrow 0$.

Hint: The $B_\varepsilon(t)$ constitute a family of Gaussian processes. So show that the covariance matrix for any finite set of time points converges.

12. Consider the diffusion on $I = [0, 1]$ with $\mu(x) = -\gamma_2(1 - x) + \gamma_1 x$ and $\sigma^2(x) = 2\beta x(1 - x)$. Determine the spectral expansion of the transition density when 0 and 1 are entrance boundaries.

Hint: Identify the eigenfunctions and eigenvalues of

$$\beta x(1 - x)\varphi'' + [\gamma_1 x - \gamma_2(1 - x)]\varphi' = -\lambda_n \varphi$$

subject to the boundary conditions

$$\frac{d}{dS}\bigg|_{x=0+} = x^{\gamma_1-1}\frac{d\varphi}{dx}\bigg|_{x=0+} = 0, \qquad \frac{d\varphi}{dS}\bigg|_{x=1-} = (1-x)^{\gamma_2-1}\frac{d\varphi}{dx}\bigg|_{x=1-} = 0$$

as the Jacobi polynomials

$$\varphi_n(x) = P_n^{(\alpha,\,\beta)}(1-2x), \qquad \alpha = \gamma_2 - 1, \quad \beta = \gamma_1 - 1$$

and eigenvalues $\lambda_n = \beta n(n + \gamma_1 + \gamma_2 - 1)$.

Answer:

$$p(t, x, y) = My^{\gamma_2-1}(1-y)^{\gamma_1-1}\sum_{n=0}^{\infty} e^{-\lambda_n t}\varphi_n(x)\varphi_n(y)\pi_n,$$

$$\pi_n = \frac{\Gamma(n+\gamma_2)\Gamma(n+\gamma_1+\gamma_2-1)}{\Gamma(n+\gamma_1)\Gamma(n+1)}(2n+\gamma_1+\gamma_2-1),$$

where M is a suitable constant.

13. Compare the Ito and Stratonovich solutions of the stochastic differential equation $dX(t) = X(t)[1 - X(t)]\,dB(t)$, where $B(t)$ is standard Brownian motion.

Answer: The Ito solution is the diffusion process on $I = (0, 1)$ with infinitesimal parameters $\mu(x) = 0$ and $\sigma^2(x) = x^2(1-x)^2$. Both boundaries are natural. The Stratonovich solution is the diffusion process with infinitesimal parameters $\mu(x) = x(1-x)(\frac{1}{2} - x)$ and $\sigma^2(x) = x^2(1-x)^2$.

14. Consider a diffusion on $[0, \infty)$, $\sigma^2(x) = \sigma$ and $\mu(x) = \mu x, \mu > 0$ where 0 is considered a regular exit boundary. This process is called by some business analysts a compounded Brownian motion, where x can be interpreted as the assets which increase through interest earnings at a rate μ modulo Brownian motion fluctuations.

   Find the probability $u(x)$ of ruin (i.e., the probability that the process reaches 0 from an initial level $x > 0$).

15. Consider the k-gene haploid selection–mutation diffusion model analogous to Example VI in Section 2 of Chapter 15. This is a diffusion on the k-dimensional simplex I,

$$I = \{\mathbf{x} = (x_1, \ldots, x_{k-1}); x_i \geq 0, x_1 + \cdots + x_{k-1} \leq 1\},$$

having diffusion coefficients

$$\mu_i(\mathbf{x}) = \gamma_i - (\gamma_1 + \cdots + \gamma_k)x_i + \sigma(1 - x_i) + 2\sigma x_i(x_1^2 + \cdots + x_k^2) - x_i$$

and

$$\sigma_{ij}(\mathbf{x}) = \begin{cases} x_i(1 - x_i), & i = j, \\ -x_i x_j, & i \neq j. \end{cases}$$

Here $\gamma_i > 0$ for $i = 1, \ldots, k$ and $x_k = 1 - (x_1 + \cdots + x_{k-1})$. Show that the density

$$\varphi(\mathbf{x}) = C\left(\prod_{i=1}^{k} x_i^{\gamma_i-1}\right)\exp\left(-\sigma\sum_{i=1}^{k} x_i^2\right)$$

satisfies the forward Kolmogorov equation stationarity condition

$$\frac{1}{2}\sum_{i,j=1}^{k}\frac{\partial^2[\sigma_{ij}(\mathbf{x})\varphi(\mathbf{x})]}{\partial x_i\,\partial x_j} - \sum_{i=1}^{k}\frac{\partial}{\partial x_i}[\mu_i(\mathbf{x})\varphi(\mathbf{x})] = 0.$$

16. Consider for standard Brownian motion $\{B(t), t \geq 0\}$ the random variable U as the occupation time of the positive axis taken until $B(t)$ first attains the level 1. Let T^* be the first passage time to the exterior of $[-1, 1]$. Show that U and T^* have the same distribution.

17. Consider Brownian motion $\{X(t), t \geq 0\}$, with negative drift parameter $c < 0$ and variance coefficient σ^2, i.e., $X(t) = \sigma B(t) + ct$, $B(t)$ standard Brownian motion. Define $M(t) = \sup\{X(s); 0 \leq s \leq t\}$. Determine the limiting distribution $\lim_{t \to \infty} \Pr\{M(t) \leq x\}$.

Answer: exponential distribution of parameter $\sigma^2/2|c|$.

18. Let $\{X(t), t \geq 0, X(0) = 0\}$ be Brownian motion with drift, i.e., with infinitesimal parameters $\mu(x, t) = ct, \sigma^2(x, t) = \sigma^2$ so that $X(t) = \sigma B(t) + ct$. Show that for fixed s

$$\max_{0 \leq t \leq s} (M(t) - X(t)) \quad \text{with} \quad M(t) = \max_{0 \leq \tau \leq t} X(\tau)$$

has the same distribution as $\max_{0 \leq t \leq s} Y(t)$, where $Y(t)$ is Brownian motion process with drift parameters $-c$, that is,

$$Y(t) = \sigma B(t) - ct$$

and exhibiting a reflecting barrier at 0.

19. Let $B(t)$ be standard Brownian motion. For fixed $s > 0$ determine the joint density of

$$M(s) = \max_{0 \leq t \leq s} B(t), \qquad \theta(s) = \{\text{the first time value } \theta \text{ where } X(\theta) = M(s)\}$$

and the endpoint value of $B(s)$.

Answer:

$$p(\theta, y, x) \, d\theta \, dy \, dx = \Pr\{\theta \in d\theta, M(s) \in dy, B(s) \in dx\}$$

$$= \frac{1}{\pi} \frac{y(y - x)}{[\theta(s - \theta)]^{3/2}} \exp\left\{-\frac{y^2}{2\theta} - \frac{(y - x)^2}{2(s - \theta)}\right\} d\theta \, dy \, dx, \quad 0 < \theta < s, \quad x < y.$$

20. Let $X(t), t \geq 0$ be Brownian motion with drift (mean coefficient μt and variance coefficient $\sigma^2 t$. Let $M(t) = \max_{0 \leq \tau \leq t} X(\tau)$ and T_a the time of first passage through level $a > 0$. Prove

$$\Pr\left\{\max_{0 \leq t \leq T_a} \frac{M(t) - X(t)}{a} \leq y \,|\, X(0) = 0\right\} = \exp\left\{-\frac{2ay}{\sigma^2} (e^{ye^{-2}} - 1)^{-1}\right\}.$$

*21. Consider two types of objects evolving according to the stochastic differential equations

$$dL = L\alpha_1 \, dt + \sqrt{a_1^2 L^2 + b_1^2 L} \, dB^{(1)}, \qquad dM = M\alpha_2 \, dt + \sqrt{a_2^2 M^2 + b_2^2 M} \, dB^{(2)},$$

where $B^{(1)}$ and $B^{(2)}$ are independent Brownian motions. Let $N(t) = L(t) + M(t)$ be the total population size, and $X(t) = L(t)/N(t)$ be the fraction of type L. By use of the vector Ito transformation formula, derive a stochastic differential equation satisfied by $X(t)$.

22. Let $\{X_n\}$ be a Markov chain with transition matrix $\mathbf{P} = \|p_{ij}\|$. Define $g(i, j) = \sum_{n=0}^{\infty} p_{ij}^{(n)} \leq \infty$. For a fixed state j, show that $Y_n = g(X_n, j)$ defines a nonnegative (possibly infinite valued) supermartingale with respect to $\{X_n\}$.

Hint: Use the identity $\mathbf{G} = \mathbf{I} + \mathbf{PG}$.

23. Let $T_1, T_2, \ldots$ be the times of events in a Poisson process of rate λ. Independently, let $X_1, X_2, \ldots$ be nonnegative independent identically distributed random variables having distribution function F. Determine the distribution of $Z = \min\{T_1 + X_1, T_2 + X_2, \ldots\}$.

Hint: Condition on the number of events of the Poisson process up to time z.

Answer: $\Pr\{Z \le z\} = 1 - \exp(-\lambda \int_0^z F(y)\,dy)$.

24. Let $X_1, X_2, \ldots$ be independent random variables with common distribution function $F(t)$. Given that $\lim_{t \to \infty} t^2[1 - F(t)] = 0$, prove that

$$\max_{1 \le i \le n} \frac{X_i}{\sqrt{n}} \to 0 \quad \text{in probability.}$$

Hint: Fix $\varepsilon > 0$. Let $\alpha_n = \Pr\{\max_{1 \le i \le n}(X_i/n) \le \varepsilon\}$. Then

$$\alpha_n = F^n(\varepsilon\sqrt{n}) = \exp[-n \log(1 - (1 - F(\varepsilon\sqrt{n})))] \to 0 \quad \text{as} \quad n \to \infty.$$

25. Consider the following simplified model of neuron firing. Electrical impulses arriving along the input fibers of a nerve cell can be one of two types, either a stimulus or an inhibitor. If one or more inhibitors arrive immediately preceding a stimulus, the cell does not respond. Otherwise it responds to the stimulus. Assume the stimuli arrive according to a Poisson process with parameter λ and the inhibitors according to a Poisson process with parameter μ, independent of each other. Calculate the Laplace transform of the response distribution if at time $t = 0$ we begin with a stimulus and thus a response.

Answer: $H(\theta) = \lambda(\lambda + \theta)/[\theta(\theta + \lambda + \mu) + \lambda(\lambda + \theta)]$.

26. Consider the following generalization to variable population sizes of the basic Wright–Fisher genetic model. Let $M_n > 0$ be a random variable denoting the population size at time n and assume that there are two types, A and a. Let X_n be the number of A-types at time n. Assume (a) the process $\{(X_n, M_n)\}$ is Markov and (b) the distribution of X_{n+1} given M_n, M_{n+1} and X_n is binomial with distribution

$$\binom{M_{n+1}}{j} \left(\frac{(1 + \sigma)X_n}{M_n + \sigma X_n}\right)^j \left(\frac{M_n - X_n}{M_n + \sigma X_n}\right)^{M_{n+1} - j}, \quad 0 \le j \le M_{n+1},$$

where $\sigma > 0$ represents a selection parameter. Show that fixation ($X_n = 0$ or $X_n = M_n$) occurs with probability 1.

Hint: Show that $\{Y_n = (X_n/M_n)\}$ is a submartingale and that $Y = \lim Y_n$ satisfies $E[Y(1 - Y)/(1 + \sigma Y)] = 0$.

27. In a population of two types, let X_n be the number of type A and Y_n the number of type a in the nth generation. Assume that $M_n = X_n + Y_n$, the total population size, grows deterministically and the X_{n+1} is binomially distributed with parameters $p_{n+1} = X_n/M_n$ and M_{n+1}, given X_n. Start with $X_0 = Y_0 = 1$. Show that fixation ($X_n = 0$ or $X_n = M_n$ for some n) occurs with probability 1 if and only if $\sum_{n=1}^{\infty} 1/M_n = \infty$.

Hint: $Z_n = X_n/M_n$ is a bounded martingale, converging to z, say. Then

$$h_n = E[Z_n(1 - Z_n)] \to h = E[Z(1 - Z)].$$

But

$$h_{n+1} = h_n\left(1 - \frac{1}{M_{n+1}}\right) = h_0 \prod_{i=1}^{n+1}\left(1 - \frac{1}{M_i}\right).$$

28. Birds arrive at an indefinitely long telegraph wire, their positions behaving like points of a line Poisson process with rate λ. Independently insects lands on the wire, their positions being those of an independent line Poisson process of rate μ. A bird can eat any insect closer to itself than to any other bird. Find the distribution of the meal size of a bird.

Answer: Probability of meal size consisting of k insects is

$$4(k + 1)\left(\frac{\mu}{2\lambda + \mu}\right)^k \frac{\lambda^2}{(2\lambda + \mu)^2}.$$

*29. Consider a continuous time birth and death process $X(t)$ with infinitesimal birth and death rates $\{\lambda_i\}$ and $\{\mu_i\}$ for $i = 0, 1, \ldots (\mu_0 = 0)$. Let $P_{ij}(t)$ be the transition probability. Consider two particles independently undergoing the process $X(t)$. the first starting at i, the second at j. Let $R_{ij}(\tau)$ be the probability density function for the time T that the particles first coincide (simultaneously occupy the same position). Show that

$$R_{ij}(\tau) = \sum_{k=0}^{\infty} (\lambda_k + \mu_{k+1})[P_{ik}(\tau)P_{jk+1}(\tau) - P_{ik+1}(\tau)P_{jk}(\tau).$$

30. A discrete-state continuous time Markov process with transition function $P_{ij}(t)$ is called symmetrizable if there exists $\alpha_i > 0$ for which $\alpha_i P_{ij}(t) = \alpha_j P_{ji}(t)$ for all i, j and $t \geq 0$. (Every birth and death process is symmetrizable with $\alpha_i = \pi_i$ of Section 5, Chapter 4). A continuous-state process with transition density $p(t, x, y)$ is symmetrizable if there exists $\alpha(x) > 0$ for which $\alpha(x)p(t, x, y) = \alpha(y)p(t, y, x)$ for all x, y and $t \geq 0$. Show that a regular diffusion process on (l, r) is symmetrizable with the choice $\alpha(x) = m(x)$, the speed density.

31. Let $\{X_n, n \geq 0\}$ be a finite time-homogeneous aperiodic Markov chain and $\{Y_n, n \geq 1\}$ a sequence of identically distributed random variables which are mutually independent and also independent of the Markov chain, with $\Pr\{Y_n = r\} = a_r$ for $r = 1, 2, \ldots$ with $\sum_{r=1}^{\infty} a_r = 1$. Let $S_0 = 0$, $S_n = Y_1 + Y_2 + \cdots + Y_n$ $(n \geq 1)$ and $Z_n = X_{S_n}$ $(n \geq 0)$.
    (a) Show that $\{Z_n, n \geq 0\}$ is a Markov chain.
    (b) Show that its transition probability matrix is given by $Q = \sum_{r=1}^{\infty} a_r P^r$, where P is the transition probability matrix of $\{X_n\}$.
    (c) Show that if P is irreducible, so is Q.
    (d) If P is irreducible and recurrent, then show that $\{Z_n\}$ has the same stationary distribution as $\{X_n\}$.

32. Let $X_1, \ldots, X_n$ and Y be random variables having finite second moments. Let Z be the minimum mean square error linear predictor of Y given $X_1, \ldots, X_n$ and for $k = 1, \ldots, n$, let $\hat{Z}_k$ be the minimum mean square error linear predictor of Z given $X_1, \ldots, X_k$. Show that $\hat{Z}_k = \hat{Y}_k$, where $\hat{Y}_k$ is the minimum mean square error linear predictor of Y given $X_1, \ldots, X_k$. (Roughly speaking, a best predictor of a best predictor is a best predictor.)

33. Suppose that $X_1, X_2, \ldots$ are independent and identically distributed random variables having the distribution function F and that N is a positive integer-valued random variable, independent of $X_1, X_2, \ldots$, and having the generating function

$$\psi(s) = \sum_{k=1}^{\infty} s^k \Pr\{N = k\}.$$

With $U = \min\{X_1, \ldots, X_N\}$, and $V = \max\{X_1, \ldots, X_N\}$, show that

$$\Pr\{V \leq v\} = \psi[F(v)],$$
$$\Pr\{U \leq u\} = 1 - \psi[1 - F(u)],$$

and

$$\Pr\{u \le U, V \le v\} = \psi[F(v) - F(u)].$$

34. A particle starting at the origin moves equally likely in any direction on the lattice of integer pairs. Letting the successive coordinate be denoted by (X_n, Y_n), compute the generating function $g(\theta) = E[\theta^{Y_{T(a)}}]$ where $T(a)$ is the first time n that $X_n = a$.

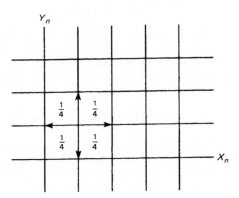

35. Suppose $X_1, \ldots, X_n$ are independent exponentially distributed random variables with parameters $\lambda_1, \ldots, \lambda_k$. Assume $\lambda_j \ne \lambda_k$ unless $j = k$. Set $T_n = X_1 + \cdots + X_n$. Establish the formula

$$\Pr\{T_n > t\} = A_{1.n}e^{-\lambda_1 t} + \cdots + A_{n.n}e^{-\lambda_n t}, \qquad t > 0,$$

where

$$A_{k.n} = \frac{\lambda_1 \cdots \lambda_{k-1}\lambda_{k+1} \cdots \lambda_n}{(\lambda_1 - \lambda_k) \cdots (\lambda_{k-1} - \lambda_k)(\lambda_{k+1} - \lambda_k) \cdots (\lambda_n - \lambda_k)}$$

$$= \prod_{\substack{1 \le j < n \\ j \ne k}} \frac{\lambda_j}{\lambda_j - \lambda_k}.$$

Hint: Work with the probability density functions. A calculation can be avoided by using a symmetry argument.

36. Let $\{X(t), t \ge 0\}$ be a pure birth process with distinct parameters $\lambda_1, \lambda_2, \ldots$ and starting from $X(0) = 1$. Verify the marginal probability distribution

$$\Pr\{X(t) = n\} = \lambda_1 \cdots \lambda_{n-1}[B_{1.n}e^{-\lambda_1 t} + \cdots + B_{n.n}e^{-\lambda_n t}],$$

where

$$B_{k.n} = 1/[(\lambda_1 - \lambda_k) \cdots (\lambda_{k-1} - \lambda_k)(\lambda_{k+1} - \lambda_k) \cdots (\lambda_n - \lambda_k)].$$

37. Let $\{X(t), t \ge 0\}$ be a modified pure birth process for which

$$\Pr\{X(t + h) = n + 1 | X(t) = n\} = \lambda_n h + o(h),$$

$$\Pr\{X(t + h) = 0 | X(t) = n\} = \theta_n h + o(h),$$

and

$$\Pr\{X(t + h) = n \,|\, X(t) = n\} = 1 - (\lambda_n + \theta_n)h + o(h).$$

Assume that $\lambda_0 = 0$ so that 0 is an absorbing state or trap, and suppose $X(0) = 1$. Establish the formula

$$\Pr\{X(t) = n\} = \sum_{k=1}^{n} C_{k,n} \exp[-(\lambda_k + \theta_k)t],$$

where

$$C_{k,n} = \frac{\lambda_k}{\lambda_n} \prod_{\substack{1 \le l \le n \\ l \ne k}} \frac{\lambda_l}{(\lambda_l + \theta_l - \lambda_k - \theta_k)},$$

the formula holding whenever $\lambda_l + \theta_l \ne \lambda_j + \theta_j$ unless $l = j$.

38. The "Peter principle" asserts that a worker will be promoted until first reaching a position in which he or she is incompetent. When this happens, they stay in that job until retirement. Examine the following single job model of the Peter principle: A person is selected at random from the population and placed in the job. If he or she is competent, he or she remains in the job for a random time having cumulative distribution function F and mean μ and is promoted. If incompetent, this person remains for a random time having cumulative distribution function G and mean $v > \mu$, and retires. Once the job is vacated, another person is selected at random and the process repeats. Assume that the infinite population contains the fraction p of competent people and $q = 1 - p$ incompetent ones. The following two questions can and may be done in any order.

(a) In the long run, what fraction of time is the position held by an incompetent person?

(b) Establish a renewal equation for $A(t) = \Pr\{$a competent person is in the job at time $t\}$ in terms of p, q, F, G and $H(t) = pF(t) + qG(t)$.

39. Suppose that X_0 is uniformly distributed on $[0, 1]$, and, given X_n, suppose that X_{n+1} is uniformly distributed on $[0, X_n]$. Show that $Z_n = 2^n X_n$ is a nonnegative martingale and find its limit $Z_\infty = \lim_{n \to \infty} Z_n$.

40. A process $\{X(t), t \ge 0\}$ at time t has value either $+t$ or $-t$. Given that $X(t) = +t$, a jump to $-t$ occurs in the infinitesimal interval dt with probability $p(t)\,dt$. Specify $p(t)$ for $t > 0$ so that $\{X(t), t \ge 0\}$ is a martingale.

41. Let $X_1, X_2, \ldots$ be independent random variables sharing the exponential distribution in which $\Pr\{X_k > t\} = e^{-t}$, $t \ge 0$. A *record value* is an observation X_k that exceeds all previous observations and its index k is called a *record mark*. Let $K(1)$, $K(2)$, $\ldots$ be the successive record marks and Z_1, Z_2, $\ldots$ be the record values. That is, beginning with $K(1) = 1$ and $Z_1 = X_1$, we have $Z_n = X_{K(n)}$ where

$$K(n) = \min\{k : X_k > X_{K(n-1)}\}.$$

(a) Determine the mean value function $m_n = E[Z_n]$.

(b) Determine the mean value function $\mu_n = E[K(n)]$.

42. A single fiber subjected to the time varying tensile load $l(t)$ fails at a random time T. We postulate the failure time distribution

$$\Pr\{T \leq t\} = 1 - \exp\left\{-\int_0^t K[l(s)]\,ds\right\},$$

which corresponds to the *failure rate* or *hazard rate* of

$$r(t) = K[l(t)].$$

That is, a single fiber not having failed prior to time t and carrying load $l(t)$ will fail during the interval $[t, t + \Delta t)$ with probability $K[l(t)]\,\Delta t + o(\Delta t)$.

The function K, called the breakdown rule, expresses how changes in the load affect the failure probability. We will be concerned with the *power law breakdown rule* in which $K(l) = l^\beta$, for some fixed positive β. Under a constant load $l(t) \equiv l$, the failure time is exponentially distributed with mean $E[T|l] = 1/K(l) = l^{-\beta}$. Under the assumed power law breakdown, a plot of mean failure time versus load is linear on log–log axes, a commonly observed phenomenon in fatigue studies.

Now place n of these fibers in parallel and subject the resulting bundle to a total load, constant in time, of nL, where L is the nominal load per fiber. Assuming that all nonfailed fibers share the total load nL equally, model the number $N(t)$, of unfailed fibers at time t, as a pure death process. Specify the death parameters μ_k. Since the fibers are in parallel, the bundle failure time T_n equals the failure time of the last fiber. Express $E[T_n]$ as a sum whose terms depend on L and β. Determine $\lim_{n\to\infty} E[T_n]$ when $\beta > 1$.

Answer: $\mu_k = k(nL/k)^\beta, k = 1, \ldots, n$.

$$E[T_n] = \sum_{k=1}^n \left(\frac{1}{\mu_k}\right) = L^{-\beta} \sum_{k=1}^n \left(\frac{k}{n}\right)^{\beta-1} \frac{1}{n} \to L^{-\beta} \int_0^1 x^{\beta-1}\,dx = \frac{1}{\beta}L^{-\beta}.$$

43. A "k out of n; F" repairable system consists of n repairable units operating in parallel which are serviced by a single repairman. System failure occurs when k units are simultaneously inoperable. Each unit fails with a constant failure rate λ so that failure times are exponentially distributed. Repair times follow an arbitrary distribution F. Determine the mean time to first system failure when $k = 2$.

Solution: Let A be the mean failure time, T_1 the time of the first failure, $T_1 + T_2$ the time of the second failure, and R the first repair time. Then, conditioning on the first fail–repair cycle, we obtain

$$A = E[T_1] + E[\min\{T_2, R\}] + A\,\Pr\{R < T_2\},$$

and

$$A = \frac{E[T_1] + E[\min\{T_2, R\}]}{\Pr\{R \geq T_2\}}$$

$$= \frac{\dfrac{1}{n\lambda} + \displaystyle\int_0^\infty e^{-(n-1)\lambda t}[1 - F(t)]\,dt}{\displaystyle\int_0^\infty [1 - F(t)](n-1)\lambda e^{-(n-1)\lambda t}\,dt}$$

$$= \frac{\dfrac{1}{n\lambda} + \bar{F}^*[(n-1)\lambda]}{(n-1)\lambda\bar{F}^*[(n-1)\lambda]},$$

where

$$\bar{F}^*(s) = \int_0^\infty e^{-st}[1 - F(t)] \, dt.$$

44. Let $\{B(t)\}$ be a standard Brownian motion. For $t \geq 0$ and $-\infty < x < \infty$, find the Markov time T^* that maximizes $E[r(x + B(T), t + T)]$ where $r(x, t) = (x - t)^+ = \max\{0, x - t\}$.

Answer: $T^* = \inf\{s \geq 0, X(s) - s \geq \frac{1}{2} + t - x\}$ and

$$v(x, t) = \sup E[r(x + B(T), t + T)] = \begin{cases} x - t & \text{for } x - t \geq \frac{1}{2}, \\ \frac{1}{2}e^{2(x-t)-1} & \text{for } x - t < \frac{1}{2}. \end{cases}$$

45. Let $\{B(t)\}$ be a standard Brownian motion. For fixed $\alpha, \lambda > 0$, $-\infty < x < \infty$, find the Markov time T that maximizes $E[e^{-\lambda T}\{x + B(T)\}^\alpha]$.

Answer: $T^* = \inf\{t \geq 0, x + B(t) \geq \alpha/\sqrt{2\lambda}\}$ and

$$v(x) = \sup E[e^{-\lambda T}\{x + B(T)\}^\alpha] = \begin{cases} x^\alpha & \text{for } x \geq \alpha/\sqrt{2\lambda}, \\ \left(\dfrac{\alpha}{\sqrt{2\lambda}}\right)^\alpha e^{\sqrt{2\lambda}x - \alpha} & \text{for } x < \alpha/\sqrt{2\lambda}. \end{cases}$$

INDEX